Tensor Computation for Data Analysis

Yipeng Liu • Jiani Liu • Zhen Long • Ce Zhu

Tensor Computation for Data Analysis

Yipeng Liu
University of Electronic Science and
Technology of China (UESTC)
Chengdu, China

Jiani Liu
University of Electronic Science and
Technology of China (UESTC)
Chengdu, China

Zhen Long
University of Electronic Science and
Technology of China (UESTC)
Chengdu, China

Ce Zhu
University of Electronic Science and
Technology of China (UESTC)
Chengdu, China

ISBN 978-3-030-74388-8 ISBN 978-3-030-74386-4 (eBook)
https://doi.org/10.1007/978-3-030-74386-4

This Springer imprint is published by the registered company Springer Nature Switzerland AG
The registered company address is: Gewerbestrasse 11, 6330 Cham, Switzerland

Preface

High dimensionality is a feature of big data. Classical data analysis methods rely on representation and computation in the form of vectors and matrices, where multi-dimensional data are unfolded into matrix for processing. However, the multi-linear structure would be lost in such vectorization or matricization, which leads to sub-optimal performance in processing. As a multidimensional array, a tensor is a generalization of vectors and matrices, and it is a natural representation for multi-dimensional data. The tensor computation based machine learning methods can avoid multi-linear data structure loss in classical matrix based counterparts. In addition, the computational complexity in tensor subspaces can be much less than that in the original form. The recent advances in applied mathematics allow us to move from classical matrix based methods to tensor based methods for many data related applications, such as signal processing, machine learning, neuroscience, communication, quantitative finance, psychometric, chemometrics, quantum physics, and quantum chemistry.

This book will first provide a basic coverage of tensor notations, preliminary operations, and main tensor decompositions and their properties. Based on them, a series of tensor analysis methods are presented as the multi-linear extensions of classical matrix based techniques. Each tensor technique is demonstrated in some practical applications.

The book contains 13 chapters and 1 appendix. It has 2 chapters in Part I for preliminaries on tensor computation.

- Chapter 1 gives an introduction of basic tensor notations, graphical representation, some special operators, and their properties.
- Chapter 2 summarizes a series of tensor decompositions, including canonical polyadic decomposition, Tucker decomposition and block term decomposition, tensor singular value decomposition, tensor networks, hybrid tensor decomposition, and scalable tensor decomposition.

Part II discusses technical aspects of tensor based data analysis methods and various applications, and it has 11 chapters.

- Chapter 3 summarizes sparse representations of tensor by a few atoms. It extends the classical dictionary learning into its tensor versions with different decomposition forms, and demonstrates the applications on image denoising and data fusion.
- Chapter 4 exploits the tensor subspace properties to estimate the missing entries of observed tensor. It extends matrix completion and gives some examples in visual data recovery, recommendation system, knowledge graph completion, and traffic flow prediction.
- Chapter 5 mainly deals with multi-modal data with shared latent information in a coupled way, thus benefits the applications in image fusion, link prediction, and visual data recovery.
- Chapter 6 extends the classical principal component analysis, which keeps a few tensor subspaces for dimensionality reduction, and shows applications in background extraction, video rain streak removal, and infrared small target detection.
- Chapter 7 extends the classical regression techniques to explore multi-directional correlations between tensors, and demonstrates its applications in weather forecasting and human pose prediction.
- Chapter 8 extends the classical statistical classification techniques by replacing the linear hyperplane with the tensor form based tensor planes for classification. Applications on digit recognition from visual images and biomedical classification from fMRI images are discussed.
- Chapter 9 extends subspace clustering by incorporating self-representation tensor, and demonstrates applications on heterogeneous information networks, multichannel ECG signal clustering, and multi-view data clustering.
- Chapter 10 summarizes different tensor decompositions used in different deep networks, and demonstrates its application in efficient deep neural networks.
- Chapter 11 summarizes the deep learning methods for tensor estimation, and demonstrates its applications in tensor rank approximation, snapshot compressive imaging, and low-rank tensor completion.
- Chapter 12 extends the Gaussian graphical model, and demonstrates its applications on environmental prediction and mice aging study.
- Chapter 13 introduces tensor sketch which succinctly approximates the original data using random projection, and demonstrates its applications in Kronecker product regression and tensorized neural network approximations.

At the end of this book, the existing software for tensor computation and its data analysis applications are summarized in the appendix.

This book has introductory parts on tensor computation, and advanced parts on machine learning techniques and practical applications. It can serve as a text book for graduates to systematically understand most of the tensor based machine learning techniques. The researchers can also refer to it to follow the state of the art on these specific techniques. As the book provides a number of examples, industrial engineers can also consult it for developing practical data processing applications.

We would like to particularly thank the current and past group members in this area. These include Lanlan Feng, Zihan Li, Zhonghao Zhang, Huyan Huang, Yingyue Bi, Hengling Zhao, Xingyu Cao, Shenghan Wang, Yingcong Lu, Bin Chen, Shan Wu, Longxi Chen, Sixing Zeng, Tengteng Liu, and Mingyi Zhou. We also thank our editors Charles Glaser and Arjun Narayanan for their advice and support. This project is supported in part by the National Natural Science Foundation of China under grant No. 61602091.

Chengdu, China Yipeng Liu

Chengdu, China Jiani Liu

Chengdu, China Zhen Long

Chengdu, China Ce Zhu

March 2021

Contents

Acronyms

ADMM	Alternating direction method of multipliers
ALS	Alternating least squares
AMP	Approximate message passing
AR	Autoregression
ASR	Automatic speech recognition
AUC	Area under curve
BCD	Block coordinate descent
BP	Basis pursuit
BTD	Block term decomposition
CMTF	Coupled matrix and tensor factorization
CNN	Convolutional neural network
CPD	Canonical polyadic decomposition
CS	Compressed sensing/count sketch
CV	Computer version
DCT	Discrete cosine transform
DNN	Deep neural network
ECoG	Electrocorticography
EEG	Electroencephalography
ERGAS	Relative dimensionless global error in synthesis
EVD	Eigenvalue decomposition
FDA	Fisher discriminant analysis
FDP	False discovery proportion
FDR	False discovery rate
FFT	Fast Fourier Transform
fMRI	Functional magnetic resonance imaging
FOCUSS	Focal under-determined system solver
GAN	Generative adversarial networks
GGM	Gaussian graphical model
GLM	Generalized linear models
GP	Gaussian process
HCL	Honey-comb lattice

HCS	Higher-order count sketch
HOOI	Higher-order orthogonal iteration
HOSVD	Higher-order singular value decomposition
HSI	Hyperspectral image
HT	Hierarchical Tucker
i.i.d.	Independently identically distributed
KLD	Kullback-Leibler Divergence
K-SVD	K-singular value decomposition
LS	Least squares
LSTM	Long short-term memory
MAE	Mean absolute error
MAP	Maximum a-posteriori probability
MERA	Multi-scale entanglement renormalization ansatz
MLD	Maximum likelihood methods
MLMTL	Multilinear multitask learning
MP	Matching pursuit
MSE	Mean squared error
MSI	Multi-spectral image
NLP	Natural language processing
NMF	Nonnegative matrix factorization
OMP	Orthogonal matching pursuit
PCA	Principal component analysis
PDF	Probability density function
PEPS	Projectedentangled-pairstates
PGM	Probabilistic graphical model
PLS	Partial least squares
PnP	Plug-and-play
RMSE	Root mean square error
RNN	Recurrent neural network
RPCA	Robust principal component analysis
RPTCA	Robust principal tensor component analysis
SAM	Spectral angle mapper
SRI	Super-resolution image
SSIM	Structural similarity
STM	Support tensor machine
SVD	Singular value decomposition
SVM	Support vector machine
TFNN	Tensor factorization neural network
TISTA	Tensor-based fast iterative shrinkage thresholding algorithm
TNN	Tensor nuclear norm/tensorized neural network
TNs	Tensor networks
TPG	Tensor projected gradient
TR	Tensor ring
TS	Tensor sketch
t-SVD	Tensor singular value decomposition

t-SVT	Tensor singular value thresholding
TT	Tensor train
TTNS	Tree tensor network state
UIQI	Universal image quality index
WT	Wavelet transform

Symbols

i, j, k	Subscripts for running indices
I, J, K	The upper bound for an index
$a, b, \cdots$	Scalar
$\mathbf{a}, \mathbf{b}, \cdots$	Vector
$\mathbf{A}, \mathbf{B}, \cdots$	Matrix
$\mathcal{A}, \mathcal{B}, \cdots$	Tensor
$\mathbf{A}_{(n)}$	Mode-n unfolding matrix of tensor $\mathcal{A}$
$\mathbf{A}_{\langle n \rangle}$	n-unfolding matrix of tensor $\mathcal{A}$
$\mathbb{R}$	Real field
$\mathbb{C}$	Complex field
$\lvert\cdot\rvert$	Absolute value or modulus
$\lVert\cdot\rVert_{\mathrm{F}}$	Frobenius norm
$\lVert\cdot\rVert_1$	ℓ_1 norm
$\lVert \cdot \rVert_2$	ℓ_2 norm
$\lVert \cdot \rVert_p$	ℓ_p norm
$\lVert \cdot \rVert_\infty$	Max norm
$\lVert \cdot \rVert_{1,\mathrm{off}}$	Off-diagonal ℓ_1 norm
$\lVert\cdot\rVert_*$	Nuclear norm
$\lfloor\cdot\rfloor$	Round down the number
$\lceil\cdot\rceil$	Round up the number
$\cdot^{\mathrm{T}}$	Transpose
$\cdot^{\mathrm{H}}$	Hermitian transpose
$\cdot^{-1}$	Inverse
$\cdot^{\dagger}$	Moore-Penrose inverse
$\mathrm{med}(\cdot)$	Median value
$\mathrm{vec}(\cdot)$	Column-wise vectorization
$\mathrm{tr}(\cdot)$	Trace
$\mathbf{I}$ or $\mathbf{I}_N$	Identity matrix (of size $N \times N$)
$\mathbf{0}_N$	Vector of zeros with length N
$\mathbf{1}_N$	Vector of ones with length N
$\mathbf{0}_{M\times N}$	Matrix of zeros of size $M \times N$

$\mathbf{1}_{M \times N}$	Matrix of ones of size $M \times N$
mod	Modular operation
$\otimes$	Kronecker product
$\odot$	Column-wise Khatri-Rao product
$\times_n$	Mode-n product
$\langle \cdot, \cdot \rangle$	Inner product
$\circ$	Outer product
$\circledast$	Hadamard or element-wise product
$*$	t-product
$\boxtimes$	Convolution
$\boxtimes_N$	Mode-N circular convolution
$\mathrm{E}(\cdot)$	Expected value
$\mathrm{Prob}(\cdot)$	Probability
$\mathrm{Cov}(\cdot)$	Covariance

Chapter 1
Tensor Computation

1.1 Notations

Tensors [4, 8, 9], as multi-dimensional generalizations of vectors and matrices, are shown to be a natural representation for big data with multiple indices, such as multi-view images and video sequence in computer vision; electroencephalogram (EEG) signal; magnetic resonance images (MRI) in neuroscience; measurements in climatology, sociology, and geography; etc.

Throughout this book, scalar, vector, matrix, and tensor are denoted by lowercase letters a, boldface lowercase letters $\mathbf{a}$, boldface capital letters $\mathbf{A}$, and calligraphy letters $\mathcal{A}$, respectively. In fact, vector and matrix can be regarded as first-order tensor and second-order tensor, respectively, as shown in Fig. 1.1. For vector $\mathbf{a} \in \mathbb{R}^I$, matrix $\mathbf{A} \in \mathbb{R}^{I \times J}$, and N-way tensor $\mathcal{A} \in \mathbb{R}^{I_1 \times I_2 \times \cdots \times I_N}$, its i-th, $\{i, j\}$-th, $\{i_1, i_2, \ldots, i_N\}$-th entries are denoted as $\mathbf{a}(i)$, $\mathbf{A}(i, j)$ and $\mathcal{A}(i_1, i_2, \ldots, i_N)$ or a_i, $a_{i,j}$ and $a_{i_1,i_2,\ldots,i_N}$, where $i = 1, \ldots, I$, $j = 1, \ldots, J$, $i_n = 1, \ldots, I_n$, $n = 1, \ldots, N$. The order N represents the number of dimensions, and each dimension $n \in \{1, \ldots, N\}$ is called as mode or way [5, 11]. For each mode n, its size is denoted as I_n.

Fixing some modes of tensor $\mathcal{A}$ can yield some subtensors like fibers or slices. For easy understanding, Fig. 1.2 provides a graphical illustration of fibers and slices for a third-order tensor $\mathcal{A} \in \mathbb{R}^{I_1 \times I_2 \times I_3}$.

Specifically, if the frontal slices for a third-order tensor $\mathcal{A} \in \mathbb{R}^{3 \times 3 \times 3}$ are as follows:

$$\mathcal{A}(:, :, 1) = \begin{bmatrix} 1 & 14 & 15 \\ 23 & 6 & 20 \\ 24 & 18 & 8 \end{bmatrix}, \mathcal{A}(:, :, 2) = \begin{bmatrix} 15 & 8 & 7 \\ 28 & 12 & 17 \\ 21 & 29 & 23 \end{bmatrix}, \mathcal{A}(:, :, 3) = \begin{bmatrix} 9 & 5 & 13 \\ 7 & 22 & 26 \\ 21 & 1 & 19 \end{bmatrix}, \tag{1.1}$$

Y. Liu et al., *Tensor Computation for Data Analysis*,
https://doi.org/10.1007/978-3-030-74386-4_1

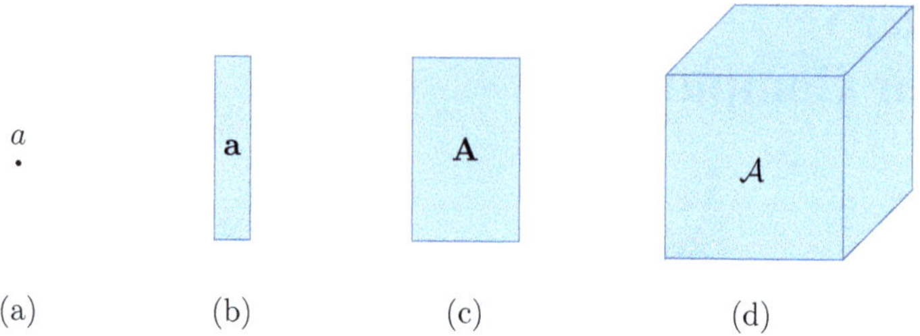

Fig. 1.1 A graphical illustration for (**a**) scalar, (**b**) vector, (**c**) matrix, and (**d**) tensor

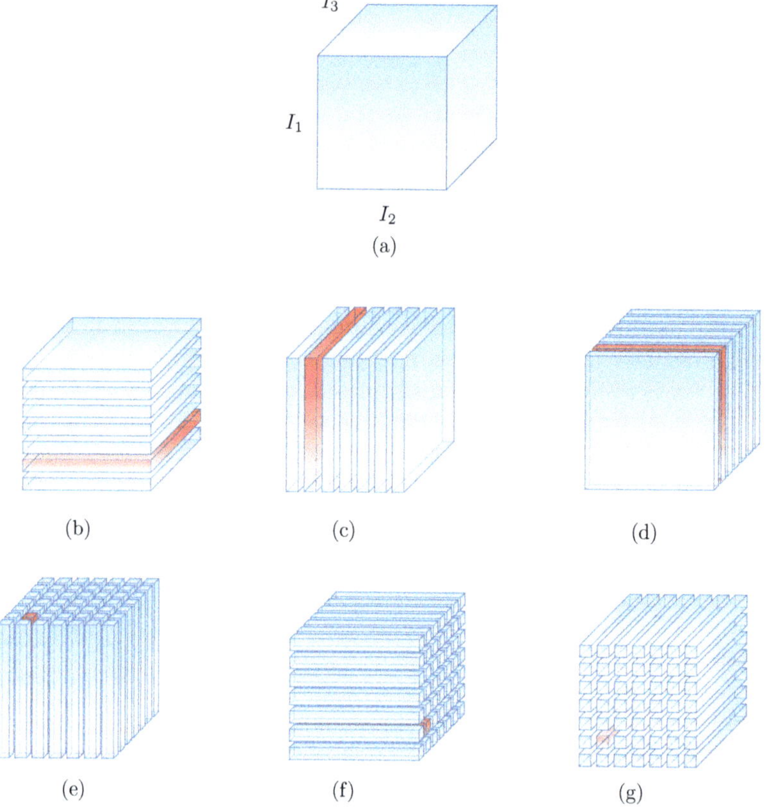

Fig. 1.2 A graphical illustration for fibers and slices of a third-order tensor $\mathcal{A}$. (**a**) third-order tensor $\mathcal{A}$. (**b**) Horizontal slice $\mathcal{A}(2, :, :)$. (**c**) Lateral slice $\mathcal{A}(:, 2, :)$. (**d**) Frontal slice $\mathcal{A}(:, :, 2)$. (**e**) Mode-1 fiber $\mathcal{A}(:, 2, 2)$. (**f**) Mode-2 fiber $\mathcal{A}(2, :, 2)$. (**g**) Mode-3 fiber $\mathcal{A}(2, 2, :)$

the corresponding horizontal slices and lateral slices are as follows:

$$\mathcal{A}(1,:,:) = \begin{bmatrix} 1 & 14 & 15 \\ 15 & 8 & 7 \\ 9 & 5 & 13 \end{bmatrix}, \mathcal{A}(2,:,:) = \begin{bmatrix} 23 & 6 & 20 \\ 28 & 12 & 17 \\ 7 & 22 & 26 \end{bmatrix}, \mathcal{A}(3,:,:) = \begin{bmatrix} 24 & 18 & 8 \\ 21 & 29 & 23 \\ 21 & 1 & 19 \end{bmatrix}, \tag{1.2}$$

$$\mathcal{A}(:,1,:) = \begin{bmatrix} 1 & 23 & 24 \\ 15 & 28 & 21 \\ 9 & 7 & 21 \end{bmatrix}, \mathcal{A}(:,2,:) = \begin{bmatrix} 14 & 6 & 18 \\ 8 & 12 & 29 \\ 5 & 22 & 1 \end{bmatrix}, \mathcal{A}(:,3,:) = \begin{bmatrix} 15 & 20 & 8 \\ 7 & 17 & 23 \\ 13 & 26 & 19 \end{bmatrix}, \tag{1.3}$$

and the mode-1 fiber $\mathcal{A}(:, 2, 2)$ is the vector $[8, 12, 29]^{\mathrm{T}}$, mode-2 fiber $\mathcal{A}(2, :, 2)$ is the vector $[28, 12, 17]^{\mathrm{T}}$, and mode-3 fiber $\mathcal{A}(2, 2, :)$ is the vector $[6, 12, 22]^{\mathrm{T}}$.

1.1.1 Special Examples

Definition 1.1 (f-Diagonal Tensor) A tensor $\mathcal{A}$ is f-diagonal if each frontal slice of $\mathcal{A}$ is a diagonal matrix, as shown in Fig. 1.3.

Definition 1.2 (Diagonal Tensor) The diagonal tensor has only nonzero elements in its super-diagonal line. It means that for diagonal tensor $\mathcal{A} \in \mathbb{R}^{I_1 \times \cdots \times I_N}$, its elements satisfy $\mathcal{A}(i_1, \ldots, i_N) \neq 0$ only if $i_1 = i_2 = \cdots = i_N$. Figure 1.4 provides an example of a third-order diagonal tensor.

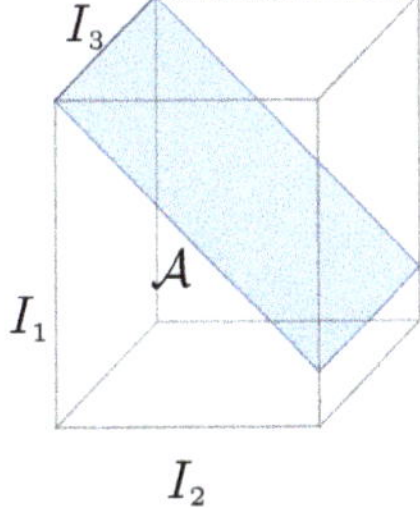

Fig. 1.3 An example for third-order f-diagonal tensor

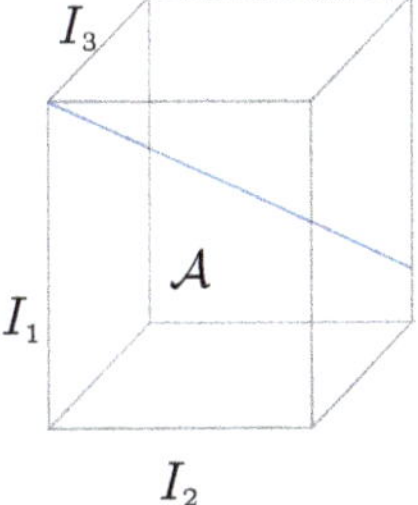

Fig. 1.4 An example for a third-order diagonal tensor

1.2 Basic Matrix Computation

Before introducing the tensor-based operations, several necessary matrix computation operators are briefly defined in this part. For matrix $\mathbf{A}$, its transpose, Hermitian transpose, inverse, and Moore-Penrose inverse are denoted as $\mathbf{A}^{\mathrm{T}}$, $\mathbf{A}^{\mathrm{H}}$, $\mathbf{A}^{-1}$, and $\mathbf{A}^{\dagger}$, respectively.

Definition 1.3 (Matrix Trace) The trace of matrix $\mathbf{A} \in \mathbb{R}^{I \times I}$ is calculated by $\operatorname{tr}(\mathbf{A}) = \sum_{i=1}^{I} \mathbf{A}(i, i)$.

Definition 1.4 (Matrix Inner Product) The inner product of matrices $\mathbf{A}$ and $\mathbf{B}$ in the same size yields

$$c = \langle \mathbf{A}, \mathbf{B} \rangle = \operatorname{vec}(\mathbf{A})^{\mathrm{T}} \operatorname{vec}(\mathbf{B}) \in \mathbb{R}, \tag{1.4}$$

where $\operatorname{vec}(\cdot)$ represents the column-wise vectorization operation.

Definition 1.5 (Frobenius Norm) The Frobenius norm of matrix $\mathbf{A}$ is denoted as $\|\mathbf{A}\|_{\mathrm{F}} = \sqrt{\langle \mathbf{A}, \mathbf{A} \rangle} \in \mathbb{R}$.

Definition 1.6 (Vector Outer Product) For vectors $\mathbf{a} \in \mathbb{R}^{I}$ and $\mathbf{b} \in \mathbb{R}^{J}$, the outer product is defined as

$$\mathbf{C} = \mathbf{a} \circ \mathbf{b} \in \mathbb{R}^{I \times J}. \tag{1.5}$$

Similarly, extending to multidimensional space, the outer product for vectors $\mathbf{a}_n \in \mathbb{R}^{I_n}$, $n = 1, \ldots, N$ will result in a multidimensional tensor as follows:

$$\mathcal{C} = \mathbf{a}_1 \circ \mathbf{a}_2 \circ \cdots \circ \mathbf{a}_N \in \mathbb{R}^{I_1 \times I_2 \times \cdots \times I_N}. \tag{1.6}$$

For example, the outer product of vectors $\mathbf{a} = [2, 6, 1]^{\mathrm{T}}$ and $\mathbf{b} = [3, 5, 8]^{\mathrm{T}}$ yields a matrix

$$\mathbf{C} = \begin{bmatrix} 6 & 10 & 16 \\ 18 & 30 & 48 \\ 3 & 5 & 8 \end{bmatrix}.$$

However, with adding the outer product of another vector $\mathbf{c} = [2, 1, 3]^{\mathrm{T}}$, we can obtain a third-order tensor as follows:

$$\mathcal{C}(:,:,1) = \begin{bmatrix} 12 & 20 & 32 \\ 36 & 60 & 96 \\ 6 & 10 & 16 \end{bmatrix}, \mathcal{C}(:,:,2) = \begin{bmatrix} 6 & 10 & 16 \\ 18 & 30 & 48 \\ 3 & 5 & 8 \end{bmatrix}, \mathcal{C}(:,:,3) = \begin{bmatrix} 18 & 30 & 48 \\ 54 & 90 & 144 \\ 9 & 15 & 24 \end{bmatrix}.$$

Definition 1.7 (Hadamard Product) For matrices $\mathbf{A}$ and $\mathbf{B}$ with the same size $I \times J$, the Hadamard product is the element-wise product as follows:

$$\begin{aligned} \mathbf{C} &= \mathbf{A} \circledast \mathbf{B} \in \mathbb{R}^{I \times J} \\ &= \begin{bmatrix} a_{1,1}b_{1,1} & a_{1,2}b_{1,2} & \dots & a_{1,J}b_{1,J} \\ a_{2,1}b_{2,1} & a_{2,2}b_{2,2} & \dots & a_{2,J}b_{2,J} \\ \vdots & \vdots & \ddots & \vdots \\ a_{I,1}b_{I,1} & a_{I,2}b_{I,2} & \dots & a_{I,J}b_{I,J} \end{bmatrix}. \end{aligned} \tag{1.7}$$

Definition 1.8 (Matrix Concatenation) $[\mathbf{A}, \mathbf{B}] \in \mathbb{R}^{I \times (J+K)}$ represents column-wise concatenation of matrices $\mathbf{A} \in \mathbb{R}^{I \times J}$ and $\mathbf{B} \in \mathbb{R}^{I \times K}$, while $[\mathbf{A}; \mathbf{B}] \in \mathbb{R}^{(I+J) \times K}$ represents row-wise concatenation of matrices $\mathbf{A} \in \mathbb{R}^{I \times K}$ and $\mathbf{B} \in \mathbb{R}^{J \times K}$.

Definition 1.9 (Kronecker Product) The Kronecker product of matrices $\mathbf{A} \in \mathbb{R}^{I \times J}$ and $\mathbf{B} \in \mathbb{R}^{K \times L}$ can be defined by

$$\begin{aligned} \mathbf{C} &= \mathbf{A} \otimes \mathbf{B} \in \mathbb{R}^{IK \times JL} \\ &= \begin{bmatrix} a_{1,1}\mathbf{B} & a_{1,2}\mathbf{B} & \dots & a_{1,J}\mathbf{B} \\ a_{2,1}\mathbf{B} & a_{2,2}\mathbf{B} & \dots & a_{2,J}\mathbf{B} \\ \vdots & \vdots & \ddots & \vdots \\ a_{I,1}\mathbf{B} & a_{I,2}\mathbf{B} & \dots & a_{I,J}\mathbf{B} \end{bmatrix}. \end{aligned} \tag{1.8}$$

For example, the Kronecker product of matrices $\mathbf{A} = \begin{bmatrix} 2 & 6 \\ 2 & 1 \end{bmatrix}$ and $\mathbf{B} = \begin{bmatrix} 1 & 5 \\ 3 & 8 \end{bmatrix}$ is

$$\mathbf{C} = \begin{bmatrix} 2\mathbf{B} & 6\mathbf{B} \\ 2\mathbf{B} & \mathbf{B} \end{bmatrix} = \begin{bmatrix} 2 & 10 & 6 & 30 \\ 6 & 16 & 18 & 48 \\ 2 & 10 & 1 & 5 \\ 6 & 16 & 3 & 8 \end{bmatrix}.$$

Theorem 1.1 *[9] For any matrices* $\mathbf{A}$, $\mathbf{B}$, $\mathbf{C}$, *and* $\mathbf{D}$, *Kronecker product has some useful properties as follows:*

$$(\mathbf{A} \otimes \mathbf{B}) \otimes \mathbf{C} = \mathbf{A} \otimes (\mathbf{B} \otimes \mathbf{C}) \tag{1.9}$$

$$\mathbf{AC} \otimes \mathbf{BD} = (\mathbf{A} \otimes \mathbf{B})(\mathbf{C} \otimes \mathbf{D}) \tag{1.10}$$

$$(\mathbf{A} \otimes \mathbf{C})^{\mathrm{T}} = \mathbf{A}^{\mathrm{T}} \otimes \mathbf{C}^{\mathrm{T}} \tag{1.11}$$

$$(\mathbf{A} \otimes \mathbf{C})^{\dagger} = \mathbf{A}^{\dagger} \otimes \mathbf{C}^{\dagger} \tag{1.12}$$

$$\mathrm{vec}(\mathbf{ABC}) = (\mathbf{C}^{\mathrm{T}} \otimes \mathbf{A})\, \mathrm{vec}(\mathbf{B}), \tag{1.13}$$

where (1.13) is commonly used to deal with the linear least squares problems in the form of $\min_{\mathbf{X}} \|\mathbf{Y} - \mathbf{A}\mathbf{X}\mathbf{B}\|_F^2$.

Definition 1.10 (Khatri-Rao Product) The Khatri-Rao product of matrices $\mathbf{A} \in \mathbb{R}^{I\times K}$ and $\mathbf{B} \in \mathbb{R}^{J\times K}$ is denoted as

$$\mathbf{A} \odot \mathbf{B} = [\mathbf{a}_1 \otimes \mathbf{b}_1, \mathbf{a}_2 \otimes \mathbf{b}_2, \ldots, \mathbf{a}_K \otimes \mathbf{b}_K] \in \mathbb{R}^{IJ\times K}, \tag{1.14}$$

where $\mathbf{a}_k$ and $\mathbf{b}_k$ are the k-th column of matrices $\mathbf{A}$ and $\mathbf{B}$.

For example, from the definition, we can first obtain the columns of matrices $\mathbf{A}$ and $\mathbf{B}$ as

$$\mathbf{a}_1 = \begin{bmatrix} 2 \\ 2 \end{bmatrix}; \mathbf{a}_2 = \begin{bmatrix} 6 \\ 1 \end{bmatrix}; \mathbf{b}_1 = \begin{bmatrix} 1 \\ 3 \end{bmatrix}; \mathbf{b}_2 = \begin{bmatrix} 5 \\ 8 \end{bmatrix}.$$

Then the Khatri-Rao product of matrices $\mathbf{A}$ and $\mathbf{B}$ can be computed by

$$\mathbf{A} \odot \mathbf{B} = [\mathbf{a}_1 \otimes \mathbf{b}_1, \mathbf{a}_2 \otimes \mathbf{b}_2] = \begin{bmatrix} 2\mathbf{b}_1 & 6\mathbf{b}_2 \\ 2\mathbf{b}_1 & \mathbf{b}_2 \end{bmatrix} = \begin{bmatrix} 2 & 30 \\ 6 & 48 \\ 2 & 5 \\ 6 & 8 \end{bmatrix}.$$

Theorem 1.2 *For any matrices* $\mathbf{A}$, $\mathbf{B}$, $\mathbf{C}$, *and* $\mathbf{D}$, *we have*

$$(\mathbf{A} \odot \mathbf{B}) \odot \mathbf{C} = \mathbf{A} \odot (\mathbf{B} \odot \mathbf{C}) \tag{1.15}$$

$$(\mathbf{A} \odot \mathbf{C})^{\mathrm{T}}(\mathbf{A} \odot \mathbf{C}) = \mathbf{A}^{\mathrm{T}}\mathbf{A} \circledast \mathbf{B}^{\mathrm{T}}\mathbf{B} \tag{1.16}$$

$$(\mathbf{A} \otimes \mathbf{B})(\mathbf{C} \odot \mathbf{D}) = \mathbf{AC} \odot \mathbf{BD} \tag{1.17}$$

and

$$\mathrm{vec}(\mathbf{ADC}) = (\mathbf{C}^{\mathrm{T}} \odot \mathbf{A})\mathbf{d} \tag{1.18}$$

holds when $\mathbf{D}$ *is a diagonal matrix with* $\mathbf{D} = \mathrm{diag}(\mathbf{d})$, *i.e.,* $\mathbf{D}(i,i) = \mathbf{d}(i)$ *for* $i = 1, \ldots, I$ *if* $\mathbf{d} \in \mathbb{R}^I$.

Definition 1.11 (Convolution) The convolution of matrices $\mathbf{A} \in \mathbb{R}^{I_1\times I_2}$ and $\mathbf{B} \in \mathbb{R}^{J_1\times J_2}$ can be defined by

$$\mathbf{C} = \mathbf{A} \boxtimes \mathbf{B} \in \mathbb{R}^{K_1\times K_2} \tag{1.19}$$

with $K_n = I_n + J_n - 1$, $n = 1, 2$. The entries of $\mathbf{C}$ satisfy

$$\mathbf{C}(k_1, k_2) = \sum_{j_1, j_2} \mathbf{B}(j_1, j_2)\mathbf{A}(k_1 - j_1 + 1, k_2 - j_2 + 1), \tag{1.20}$$

where $k_n = 1, \ldots, K_n, n = 1, 2$.

Definition 1.12 (Mode-N Circular Convolution) For vectors $\mathbf{a} \in \mathbb{R}^I$ and $\mathbf{b} \in \mathbb{R}^J$, its mode-$N$ circular convolution is denoted as

$$\mathbf{c} = \mathbf{a} \boxtimes_N \mathbf{b} \in \mathbb{R}^N, \tag{1.21}$$

whose elements satisfy

$$\mathbf{c}(n) = \sum_{\text{mod}(i+j,N)=n-1} \mathbf{a}(i)\mathbf{b}(j), \tag{1.22}$$

where $n = 1, \ldots, N$.

Definition 1.13 (ℓ_p Norm) For matrix $\mathbf{A} \in \mathbb{R}^{I \times J}$, its ℓ_p norm is denoted as

$$\|\mathbf{A}\|_p = \left(\sum_{i,j} a_{i,j}^p \right)^{1/p}, \tag{1.23}$$

which boils down to the classical ℓ_1 norm and ℓ_0 norm when $p = 1$ and $p = 0$, respectively. Compared with ℓ_1 norm, ℓ_p with $0 < p < 1$ is shown to be better approximation of ℓ_0 norm.

Definition 1.14 (Off-diagonal ℓ_1 Norm) The off-diagonal ℓ_1 norm of matrix $\mathbf{A} \in \mathbb{R}^{I \times J}$ is defined by

$$\|\mathbf{A}\|_{1,\text{off}} = \sum_{i \neq j} |a_{i,j}|. \tag{1.24}$$

Definition 1.15 (Max Norm) The max norm of matrix $\mathbf{A} \in \mathbb{R}^{I \times J}$ is defined by

$$\|\mathbf{A}\|_\infty = \max \{|a_{i,j}|, 1 \leq i \leq I, 1 \leq j \leq J\}. \tag{1.25}$$

Definition 1.16 (Nuclear Norm) The nuclear norm of matrix $\mathbf{A} \in \mathbb{R}^{I \times J}$ is defined by

$$\|\mathbf{A}\|_* = \sum_{k=1}^{\min(I,J)} \sigma_k, \tag{1.26}$$

where σ_k is the k-th singular value of matrix $\mathbf{A}$.

1.3 Tensor Graphical Representation

As it can be seen from these notations, it may be complicated to represent high-order tensors' operations, especially when the tensor orders are really high. Graph networks can provide a natural language to represent multiway tensor and its operations. Specially, tensors can be represented by nodes with different edges as modes (see examples in Fig. 1.5). As it is shown in Fig. 1.5, the number of outgoing edges represents the order of the tensor. And the value of each outgoing line represents the dimensionality along that specific mode. In this way, the tensor-based operations and existing tensor decompositions (including tensor networks) can be intuitively and visually represented by diagrams with different nodes as tensors, edges expressing the connections between different tensors, and the dimensionality of each tensor included in the graph. The edge connects two nodes implies the contraction operation which will be described below. This graph representation greatly benefits the representation, understanding, and implementation of tensor-based computation and related research. As an example, the product of matrices $\mathbf{A} \in \mathbb{R}^{I \times J}$ and $\mathbf{B} \in \mathbb{R}^{J \times K}$ can be represented by the graph, as it does in Fig. 1.6.

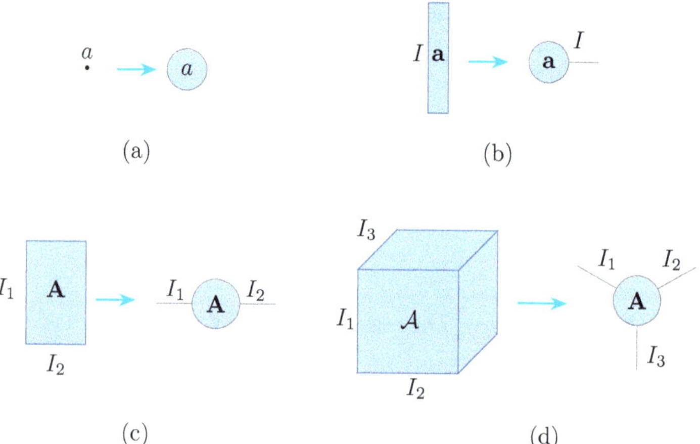

Fig. 1.5 The graphical representation of tensors. (**a**) Scalar. (**b**) Vector. (**c**) Matrix. (**d**) Tensor

Fig. 1.6 The graphical representation of matrix product

1.4 Tensor Unfoldings

Tensor unfolding operator unfolds multidimensional tensors into matrices or tensors along specific modes. We summarize several commonly used tensor unfolding operators and briefly demonstrate their differences.

1.4.1 Mode-n Unfolding

Definition 1.17 (Mode-n Unfolding) For tensor $\mathcal{A} \in \mathbb{R}^{I_1 \times I_2 \times \cdots \times I_N}$, its mode-$n$ unfolding matrix is expressed as $\mathbf{A}_{(n)}$ or $\mathbf{A}_{[n]}$. The mode-n unfolding operator arranges the n-th mode of $\mathcal{A}$ as the row while the rest modes as the column of the mode-n unfolding matrix. Mathematically, the elements of $\mathbf{A}_{(n)}$ and $\mathbf{A}_{[n]}$ satisfy

$$\mathbf{A}_{(n)}(i_n, j_1) = \mathcal{A}(i_1, \ldots, i_n, \ldots, i_N) \tag{1.27}$$

$$\mathbf{A}_{[n]}(i_n, j_2) = \mathcal{A}(i_1, \ldots, i_n, \ldots, i_N), \tag{1.28}$$

where $j_1 = \overline{i_1, \ldots, i_{n-1}, i_{n+1}, \ldots, i_N}$, $j_2 = \overline{i_{n+1}, \ldots, i_N, i_1, \ldots, i_{n-1}}$ are multi-indices. For example, the mathematical expression of multi-index $i = \overline{i_1, \ldots, i_N}$ can be either little-endian convention in reverse lexicographic order

$$\overline{i_1, \ldots, i_N} = i_1 + (i_2 - 1)I_1 + (i_3 - 1)I_1 I_2 + \cdots + (i_N - 1)I_1 \cdots I_{N-1}, \tag{1.29}$$

or big-endian in colexicographic order

$$\overline{i_1, \ldots, i_N} = i_N + (i_{N-1} - 1)I_N + \cdots + (i_2 - 1)I_3 \cdots I_N + (i_1 - 1)I_2 \cdots I_N. \tag{1.30}$$

But it should be noted that the employed order must be consistent during the whole algorithm. In this monograph, unless otherwise stated, we will use little-endian notation.

The only difference between $\mathbf{A}_{(n)}$ and $\mathbf{A}_{[n]}$ is the order of the remaining dimensions in the column. For better understanding, we take the tensor $\mathcal{A} \in \mathbb{R}^{3 \times 3 \times 3}$ used before as an example. The mode-n unfolding matrices are as follows:

$$\mathbf{A}_{(1)} = \mathbf{A}_{[1]} = \left[\begin{array}{ccc|ccc|ccc} 1 & 14 & 15 & 15 & 8 & 7 & 9 & 5 & 13 \\ 23 & 6 & 20 & 28 & 12 & 17 & 7 & 22 & 26 \\ 24 & 18 & 8 & 21 & 29 & 23 & 21 & 1 & 19 \end{array}\right] \tag{1.31}$$

$$\mathbf{A}_{(2)} = \left[\begin{array}{ccc|ccc|ccc} 1 & 23 & 24 & 15 & 28 & 21 & 9 & 7 & 21 \\ 14 & 6 & 18 & 8 & 12 & 29 & 5 & 22 & 1 \\ 15 & 20 & 8 & 7 & 17 & 23 & 13 & 26 & 19 \end{array}\right] \tag{1.32}$$

$$\mathbf{A}_{[2]} = \left[\begin{array}{ccc|ccc|ccc} 1 & 15 & 9 & 23 & 28 & 7 & 24 & 21 & 21 \\ 14 & 8 & 5 & 6 & 12 & 22 & 18 & 29 & 1 \\ 15 & 7 & 13 & 20 & 17 & 26 & 8 & 23 & 19 \end{array}\right] \quad (1.33)$$

$$\mathbf{A}_{(3)} = \mathbf{A}_{[3]} = \left[\begin{array}{ccc|ccc|ccc} 1 & 23 & 24 & 14 & 6 & 18 & 15 & 20 & 8 \\ 15 & 28 & 21 & 8 & 12 & 29 & 7 & 17 & 23 \\ 9 & 7 & 21 & 5 & 22 & 1 & 13 & 26 & 19 \end{array}\right]. \quad (1.34)$$

As shown in ((1.31), (1.34)), there is no difference between $\mathbf{A}_{(n)}$ and $\mathbf{A}_{[n]}$ when $n = 1, N$. Otherwise, the order of the remaining indices in the column of mode-n unfolding matrix does make a sense, as shown in ((1.32), (1.33)). For this unfolding operator, there is a clear disadvantage that the resulting mode-n unfolding matrix is extremely unbalanced. Simply processing the mode-n unfolding matrices to get the multiway structure information will bring a series of numerical problems.

1.4.2 Mode-n_1n_2 Unfolding

Definition 1.18 (Mode-n_1n_2 Unfolding) For a multiway tensor $\mathcal{A} \in \mathbb{R}^{I_1 \times I_2 \times \cdots \times I_N}$, its mode-$n_1n_2$ unfolding tensor is represented as $\mathcal{A}_{(n_1n_2)} \in \mathbb{R}^{I_{n_1} \times I_{n_2} \times \prod_{s \neq n_1, n_2} I_s}$, whose frontal slices are the lexicographic ordering of the mode-n_1n_2 slices of $\mathcal{A}$. Mathematically,

$$\mathcal{A}_{(n_1n_2)}(i_{n_1}, i_{n_2}, j) = \mathcal{A}(i_1, i_2, \ldots, i_N), \quad (1.35)$$

the mapping relation of $(i_1, i_2, \ldots, i_N)$-th entry of $\mathcal{A}$ and $\left(i_{n_1}, i_{n_2}, j\right)$-th entry of $\mathcal{A}_{(n_1n_2)}$ is

$$j = 1 + \sum_{\substack{s=1 \\ s \neq n_1, s \neq n_2}}^{N} (i_s - 1) J_s \text{ with } J_s = \prod_{\substack{m=1 \\ m \neq n_1, m \neq n_2}}^{s-1} I_m. \quad (1.36)$$

1.4.3 n-Unfolding

Definition 1.19 (n-Unfolding) The n-unfolding operator directly reshapes the original tensor into a matrix with the first n modes as the row while the rest modes as the column. For a Nth-order tensor $\mathcal{A} \in \mathbb{R}^{I_1 \times \cdots \times I_N}$, it can be expressed as $\mathbf{A}_{\langle n \rangle}$

whose entries meet

$$\mathbf{A}_{\langle n \rangle}(j_1, j_2) = \mathcal{A}(i_1, \ldots, i_n, \ldots, i_N), \tag{1.37}$$

where $j_1 = \overline{i_1, \ldots, i_n}$, $j_2 = \overline{i_{n+1}, \ldots, i_N}$.

For example, the n-unfolding matrices of the third-order tensor $\mathcal{A}$ used before can be expressed as

$$\mathbf{A}_{\langle 1 \rangle} = \mathbf{A}_{[1]} = \left[\begin{array}{ccc|ccc|ccc} 1 & 14 & 15 & 15 & 8 & 7 & 9 & 5 & 13 \\ 23 & 6 & 20 & 28 & 12 & 17 & 7 & 22 & 26 \\ 24 & 18 & 8 & 21 & 29 & 23 & 21 & 1 & 19 \end{array}\right] \tag{1.38}$$

$$\mathbf{A}_{\langle 2 \rangle} = \mathbf{A}_{[3]}{}^{\mathrm{T}} = \begin{bmatrix} 1 & 15 & 9 \\ 23 & 28 & 7 \\ 24 & 21 & 21 \\ 14 & 8 & 5 \\ 6 & 12 & 22 \\ 18 & 29 & 1 \\ 15 & 7 & 13 \\ 20 & 17 & 26 \\ 8 & 23 & 19 \end{bmatrix} \tag{1.39}$$

$$\mathbf{A}_{\langle 3 \rangle} = \operatorname{vec}(\mathcal{A}) = \operatorname{vec}(\mathbf{A}_{[1]}). \tag{1.40}$$

1.4.4 l-Shifting n-Unfolding

Definition 1.20 (l-Shifting) For a tensor $\mathcal{A} \in \mathbb{R}^{I_1 \times \cdots \times I_N}$, left shifting it by l yields a tensor $\overleftarrow{\mathcal{A}}^l \in \mathbb{R}^{I_l \times \cdots \times I_N \times I_1 \times \cdots \times I_{l-1}}$. It can be implemented by the command $\overleftarrow{\mathcal{A}}^l =$ permute($\mathcal{A}$, $[l, \ldots, N, 1, \ldots, l-1]$) in MATLAB.

Definition 1.21 (l-Shifting n-Unfolding) [6, 12] The l-shifting n-unfolding operator first shifts the original tensor $\mathcal{A}$ by l and unfolds the tensor along the n-th mode. Denote $\mathbf{A}_{\langle l,n \rangle}$ as the l-shifting n-unfolding matrix of tensor $\mathcal{A}$, if $\mathcal{A} \in \mathbb{R}^{I_1 \times \cdots \times I_N}$. $\mathbf{A}_{\langle l,n \rangle}$ is in the size of $I_l \cdots I_{\mathrm{mod}(l+n,N)} \times I_{\mathrm{mod}(l+n+1,N)} \cdots I_{l-1}$ with its elements defined by

$$\mathbf{A}_{\langle l,n \rangle}(j_1, j_2) = \mathcal{A}(i_1, \ldots, i_N) \tag{1.41}$$

where $j_1 = \overline{i_l, \ldots, i_{\mathrm{mod}(l+n,N)}}$, $j_2 = \overline{i_{\mathrm{mod}(l+n+1,N)}, \ldots, i_{l-1}}$, mod$(\cdot)$ represents the modulus operation.

For tensors with order lower than or equal to 3, there is no distinct difference between these tensor unfolding methods. But for high-order tensors, l-shifting n-

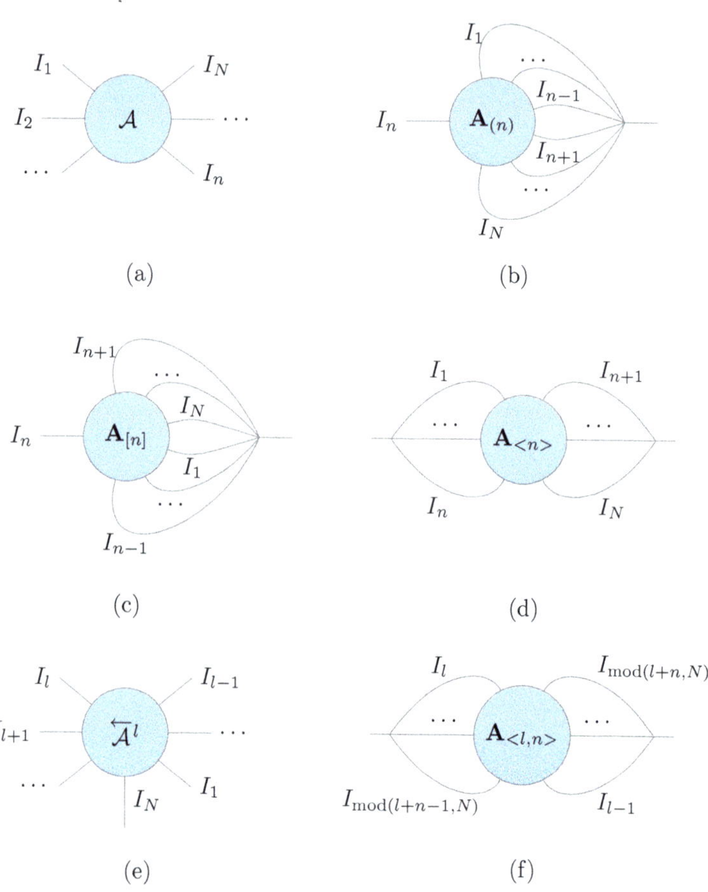

Fig. 1.7 The graphical representation of tensor unfolding methods. (**a**) Nth-order tensor $\mathcal{A}$. (**b**) Mode-n unfolding matrix $\mathbf{A}_{(n)}$. (**c**) Mode-n unfolding matrix $\mathbf{A}_{[n]}$. (**d**) n-Unfolding matrix $\mathbf{A}_{\langle n \rangle}$. (**e**) l-shifting Nth-order tensor $\overleftarrow{\mathcal{A}}^l$. (**f**) l-shifting n-unfolding matrix $\mathbf{A}_{\langle l,n \rangle}$

unfolding operator with appropriate section of l and n can generate more balanced matrices while remaining the structure information within the multidimensional data. As shown in the examples of N-th order tensor $\mathcal{A} \in \mathbb{R}^{I \times \cdots \times I}$ in Fig. 1.7, mode-n unfolding matrices are extremely unbalanced; the n-unfolding matrix is only balanced if $n = N/2$, while the l-shifting n-unfolding matrix can lead to an almost square matrix in all modes with a suitable l and n.

1.5 Tensor Products

1.5.1 Tensor Inner Product

Definition 1.22 (Tensor Inner Product) The inner product of tensor $\mathcal{A}$ and $\mathcal{B}$ with the same size is formulated as follows:

$$c = \langle \mathcal{A}, \mathcal{B} \rangle = \langle \mathrm{vec}(\mathcal{A}), \mathrm{vec}(\mathcal{B}) \rangle \in \mathbb{R}, \tag{1.42}$$

where $\mathrm{vec}(\mathcal{A}) = \mathrm{vec}(\mathbf{A}_{(1)})$.

Based on the tensor inner product definition, the Frobenius norm of tensor $\mathcal{A}$ is given by $\|\mathcal{A}\|_{\mathrm{F}} = \sqrt{\langle \mathcal{A}, \mathcal{A} \rangle}$.

1.5.2 Mode-n Product

Definition 1.23 (Mode-n Product) The mode-n product of tensor $\mathcal{A} \in \mathbb{R}^{I_1 \times \cdots \times I_N}$ and matrix $\mathbf{B} \in \mathbb{R}^{J \times I_n}$ is defined by

$$\mathcal{C} = \mathcal{A} \times_n \mathbf{B} \in \mathbb{R}^{I_1 \times \cdots \times I_{n-1} \times J \times I_{n+1} \times \cdots \times I_N}, \tag{1.43}$$

whose entries are denoted as

$$\mathcal{C}(i_1, \ldots, j, \ldots, i_N) = \sum_{i_n=1}^{I_n} \mathcal{A}(i_1, \ldots, i_n, \ldots, i_N) \mathbf{B}(j, i_n). \tag{1.44}$$

Based on the mode-n unfolding definition, the mode-n product can be expressed in matrix form as

$$\mathbf{C}_{(n)} = \mathbf{B}\mathbf{A}_{(n)}. \tag{1.45}$$

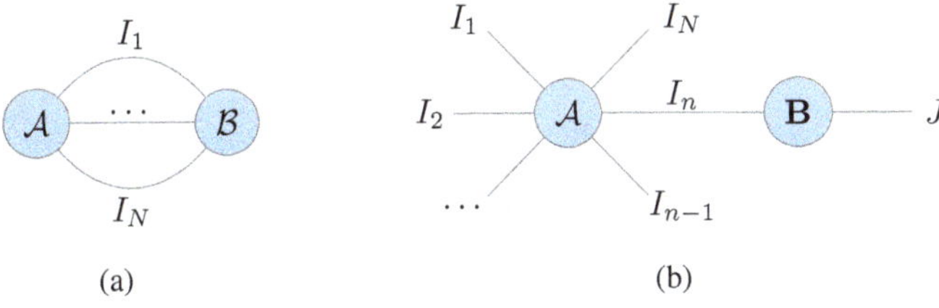

Fig. 1.8 The graphical representation for (**a**) tensor inner product and (**b**) mode-n product

Figure 1.8 provides a graphical illustration for the tensor inner product and the commonly used mode-n product in dimensionality reduction. Taking the example defined in (1.1), the mode-2 multiplication with matrix $\mathbf{B} = \begin{bmatrix} 2 & 0 & 5 \\ 4 & 7 & 1 \end{bmatrix} \in \mathbb{R}^{3\times 2}$ yields a tensor $\mathcal{C} \in \mathbb{R}^{3\times 2\times 3}$ with frontal slices as follows:

$$\mathcal{C}(:,:,1) = \begin{bmatrix} 77 & 117 \\ 146 & 154 \\ 88 & 230 \end{bmatrix} \tag{1.46}$$

$$\mathcal{C}(:,:,2) = \begin{bmatrix} 65 & 123 \\ 141 & 213 \\ 157 & 310 \end{bmatrix} \tag{1.47}$$

$$\mathcal{C}(:,:,3) = \begin{bmatrix} 83 & 84 \\ 144 & 208 \\ 137 & 110 \end{bmatrix}. \tag{1.48}$$

1.5.3 Tensor Contraction

Definition 1.24 (Tensor Contraction) For tensors $\mathcal{A} \in \mathbb{R}^{I_1\times\cdots\times I_N\times J_1\times\cdots\times J_L}$ and $\mathcal{B} \in \mathbb{R}^{J_1\times\cdots\times J_L\times K_1\times\cdots\times K_M}$, contracting their common indices $\{J_1, \ldots, J_L\}$ will result in a tensor $\mathcal{C} = \langle\mathcal{A}, \mathcal{B}\rangle_L \in \mathbb{R}^{I_1\times\cdots\times I_N\times K_1\times\cdots\times K_M}$ whose entries are calculated by

$$c_{i_1,\ldots,i_N,k_1,\ldots,k_M} = \sum_{j_1,\ldots,j_L} a_{i_1,\ldots,i_N,j_1,\ldots,j_L} b_{j_1,\ldots,j_L,k_1,\ldots,k_M}. \tag{1.49}$$

As shown in Fig. 1.9, tensor contraction is a high-dimensional extension of matrix product into tensor field. If $D = L = M = 1$, the tensor contraction operator degrades into the standard matrix product.

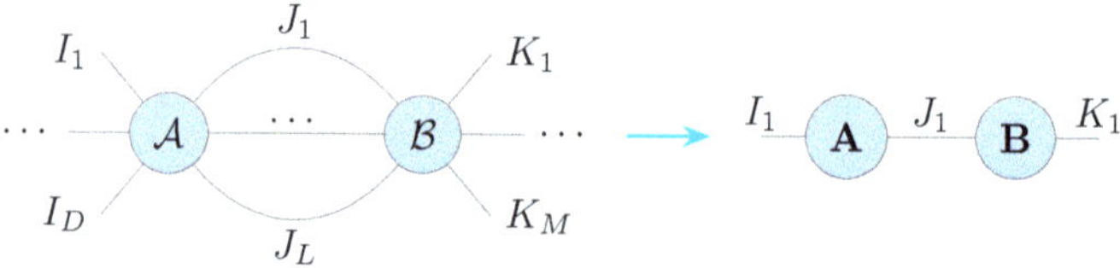

Fig. 1.9 The graphical representation for tensor contraction

1.5.4 t-Product

Definition 1.25 (t-Product) [7] The t-product for tensors $\mathcal{A} \in \mathbb{R}^{I_1 \times I_2 \times I_3}$ and $\mathcal{B} \in \mathbb{R}^{I_2 \times I_4 \times I_3}$ is defined as

$$\mathcal{C} = \mathcal{A} * \mathcal{B} \in \mathbb{R}^{I_1 \times I_4 \times I_3}, \tag{1.50}$$

whose entries satisfy

$$\mathcal{C}(i_1, i_4, :) = \sum_{i_2=1}^{I_2} \mathcal{A}(i_1, i_2, :) \boxtimes_{I_3} \mathcal{B}(i_2, i_4, :). \tag{1.51}$$

Definition 1.26 (Block Diagonal Matrix Form of Third-Order Tensor in Fourier Domain) [13] The block diagonal matrix $\overline{\mathbf{A}}$ for the tensor $\mathcal{A}$ in Fourier domain is denoted as

$$\overline{\mathbf{A}} \triangleq \begin{bmatrix} \hat{\mathcal{A}}^{(1)} & & & \\ & \hat{\mathcal{A}}^{(2)} & & \\ & & \ddots & \\ & & & \hat{\mathcal{A}}^{(I_3)} \end{bmatrix}, \tag{1.52}$$

where $\hat{\mathcal{A}}$ is the fast Fourier transform (FFT) of $\mathcal{A}$ along the third mode and $\hat{\mathcal{A}}^{(i_3)}$ is the i_3-th frontal slice of $\hat{\mathcal{A}}$.

Theorem 1.3 *For any tensor $\mathcal{A} \in \mathbb{R}^{I_1 \times I_2 \times I_3}$ and $\mathcal{B} \in \mathbb{R}^{I_2 \times I_4 \times I_3}$, their t-product can be efficiently calculated through simple matrix product in Fourier domain as follows:*

$$\mathcal{C} = \mathcal{A} * \mathcal{B} \Longleftrightarrow \overline{\mathbf{A}}\,\overline{\mathbf{B}} = \overline{\mathbf{C}}, \tag{1.53}$$

where $\overline{\mathbf{A}}$, $\overline{\mathbf{B}}$, $\overline{\mathbf{C}}$ are the block diagonal matrices of tensors $\mathcal{A}$, $\mathcal{B}$, $\mathcal{C}$ in Fourier domain.

Definition 1.27 (Identity Tensor) The identity tensor, expressed by $\mathcal{I}$, is the tensor whose first frontal slice is an identity matrix, while all other frontal tensors are all zeros. The mathematical expression is as follows:

$$\mathcal{I}(:, :, 1) = \mathbf{I};\ \mathcal{I}(:, :, 2 : L) = \mathbf{0}.$$

If $\mathcal{A} * \mathcal{B} = \mathcal{I}$ and $\mathcal{B} * \mathcal{A} = \mathcal{I}$, we can call that the inverse of tensor $\mathcal{A}$ is the tensor $\mathcal{B}$.

Definition 1.28 (Tensor Transpose) For a third-order tensor $\mathcal{A} \in \mathbb{R}^{I_1 \times I_2 \times I_3}$, the $\mathcal{A}^{\mathrm{T}} \in \mathbb{R}^{I_2 \times I_1 \times I_3}$ is obtained by transposing frontal slices and then reversing the order of transposed frontal slices from 2 to I_3.

For example, the tensor transpose of $\mathcal{A} \in \mathbb{R}^{I_1 \times I_2 \times 4}$ is

$$\mathcal{A}^{\mathrm{T}}(:,:,1) = (\mathcal{A}(:,:,1))^{\mathrm{T}};\ \mathcal{A}^{\mathrm{T}}(:,:,2) = (\mathcal{A}(:,:,4))^{\mathrm{T}};$$
$$\mathcal{A}^{\mathrm{T}}(:,:,3) = (\mathcal{A}(:,:,3))^{\mathrm{T}};\ \mathcal{A}^{\mathrm{T}}(:,:,4) = (\mathcal{A}(:,:,2))^{\mathrm{T}}. \tag{1.54}$$

Definition 1.29 (Orthogonal Tensor) Based on the definition of t-product, we say a tensor $\mathcal{Q}$ is orthogonal if it satisfies

$$\mathcal{Q}^{\mathrm{T}} * \mathcal{Q} = \mathcal{Q} * \mathcal{Q}^{\mathrm{T}} = \mathcal{I}. \tag{1.55}$$

1.5.5 3-D Convolution

Definition 1.30 (3-D Convolution) [5] Similar to the matrix convolution, the 3-D convolution of tensors $\mathcal{A} \in \mathbb{R}^{I_1 \times I_2 \times I_3}$ and $\mathcal{B} \in \mathbb{R}^{J_1 \times J_2 \times J_3}$ can be defined by

$$\mathcal{C} = \mathcal{A} \boxtimes \mathcal{B} \in \mathbb{R}^{K_1 \times K_2 \times K_3} \tag{1.56}$$

with $K_n = I_n + J_n - 1$, $n = 1, 2, 3$. The entries of $\mathcal{C}$ satisfy

$$\mathcal{C}(k_1, k_2, k_3) = \sum_{j_1, j_2, j_3} \mathcal{B}(j_1, j_2, j_3)\mathcal{A}(k_1 - j_1 + 1, k_2 - j_2 + 1, k_3 - j_3 + 1), \tag{1.57}$$

where $k_n = 1, \ldots, K_n$, $n = 1, 2, 3$. For easy understanding, a graphical representation for the 3-D convolution is provided in Fig. 1.10.

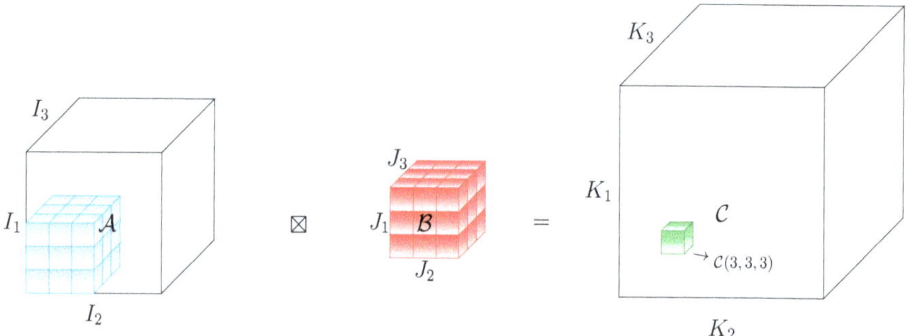

Fig. 1.10 The graphical representation of 3-D convolution

1.6 Summary

In this chapter, we have overviewed the concept of tensor representation and several popular tensor operations, including unfolding and multilinear products. It is fundamental for tensor analysis and helps the understanding of specific tensor-based methods. It should be noted that not all tensor-based operations are covered in this monologue. We only focus on the popular ones which are employed in the following chapters. There also exist some other self-defined concepts or advanced products. We refer to the interested reader for corresponding papers, such as tensor augmentation [2] or semi-tensor products [3]. In addition, some literature aim at providing fast algorithms for tensor products, such as the fast tensor contraction [1, 10].

References

1. Abdelfattah, A., Baboulin, M., Dobrev, V., Dongarra, J., Earl, C., Falcou, J., Haidar, A., Karlin, I., Kolev, T., Masliah, I., et al.: High-performance tensor contractions for GPUs. Procedia Comput. Sci. **80**, 108–118 (2016)
2. Bengua, J.A., Tuan, H.D., Phien, H.N., Do, M.N.: Concatenated image completion via tensor augmentation and completion. In: 2016 10th International Conference on Signal Processing and Communication Systems (ICSPCS), pp. 1–7. IEEE, Piscataway (2016)
3. Cheng, D., Qi, H., Xue, A.: A survey on semi-tensor product of matrices. J. Syst. Sci. Complex. **20**(2), 304–322 (2007)
4. Cichocki, A., Mandic, D., De Lathauwer, L., Zhou, G., Zhao, Q., Caiafa, C., Phan, H.A.: Tensor decompositions for signal processing applications: from two-way to multiway component analysis. IEEE Signal Process. Mag. **32**(2), 145–163 (2015)
5. Cichocki, A., Lee, N., Oseledets, I., Phan, A.H., Zhao, Q., Mandic, D.P., et al.: Tensor networks for dimensionality reduction and large-scale optimization: Part 1 low-rank tensor decompositions. Found. Trends Mach. Learn. **9**(4–5), 249–429 (2016)
6. Huang, H., Liu, Y., Liu, J., Zhu, C.: Provable tensor ring completion. Signal Process. **171**, 107486 (2020)
7. Kilmer, M.E., Braman, K., Hao, N., Hoover, R.C.: Third-order tensors as operators on matrices: a theoretical and computational framework with applications in imaging. SIAM J. Matrix Anal. Appl. **34**(1), 148–172 (2013)
8. Kolda, T.G., Bader, B.W.: Tensor decompositions and applications. SIAM Rev. **51**(3), 455–500 (2009)
9. Sidiropoulos, N.D., De Lathauwer, L., Fu, X., Huang, K., Papalexakis, E.E., Faloutsos, C.: Tensor decomposition for signal processing and machine learning. IEEE Trans. Signal Process. **65**(13), 3551–3582 (2017)
10. Solomonik, E., Demmel, J.: Fast bilinear algorithms for symmetric tensor contractions. Comput. Methods Appl. Math. **21**(1), 211–231 (2020)
11. Tucker, L.R.: Implications of factor analysis of three-way matrices for measurement of change. Probl. Meas. Change **15**, 122–137 (1963)
12. Yu, J., Zhou, G., Zhao, Q., Xie, K.: An effective tensor completion method based on multilinear tensor ring decomposition. In: 2018 Asia-Pacific Signal and Information Processing Association Annual Summit and Conference (APSIPA ASC), pp. 1344–1349. IEEE, Piscataway (2018)
13. Zhang, Z., Aeron, S.: Exact tensor completion using t-SVD. IEEE Trans. Signal Process. **65**(6), 1511–1526 (2016)

Chapter 2
Tensor Decomposition

2.1 Introduction

Tensor decomposition can break a large-size tensor into many small-size factors, which can reduce the storage and computation complexity during data processing. Tensor decompositions were originated by Hitchcock in 1927. The idea of a multi-way model is firstly proposed by Cattell in 1944 [11, 12]. These concepts received successive attention until the 1960s when Tucker published three works about tensor decomposition [71–73]. Carroll, Chang [10] and Harshman [31] proposed canonical factor decomposition (CANDECOMP) and parallel factor decomposition (PARAFAC), respectively, in 1970. These works first appeared in psychometrics literatures. In addition, some other tensor decompositions such as block term decomposition [18, 19, 21] and t-SVD [43, 44] are proposed to improve and enrich tensor decompositions. With the sensor development, many data we obtained are large scale, which drives the development of tensor networks [6, 15, 27, 29, 46, 53, 55, 56, 58, 60, 76, 81] and scalable tensor decomposition [7, 23, 35, 41, 65].

In the last two decades, tensor decompositions have attracted much interest in fields such as numerical linear algebra [9, 49, 61], signal processing [16, 50, 64, 67], data mining [2, 45, 54, 77], graph analysis [25, 37, 69], neuroscience [30, 36, 79], and computer vision [17, 26, 66].

2.2 Canonical Polyadic Decomposition

The first idea of canonical polyadic decomposition is from Hitchcock [33, 34] in 1927, which expresses a tensor as a sum of a finite number of rank-1 tensors. After that, Cattell [11, 12] proposed ideas for parallel proportional analysis and multiple axes for analysis. In 1970, the form of CANDECOMP (canonical decomposition)

Y. Liu et al., *Tensor Computation for Data Analysis*,
https://doi.org/10.1007/978-3-030-74386-4_2

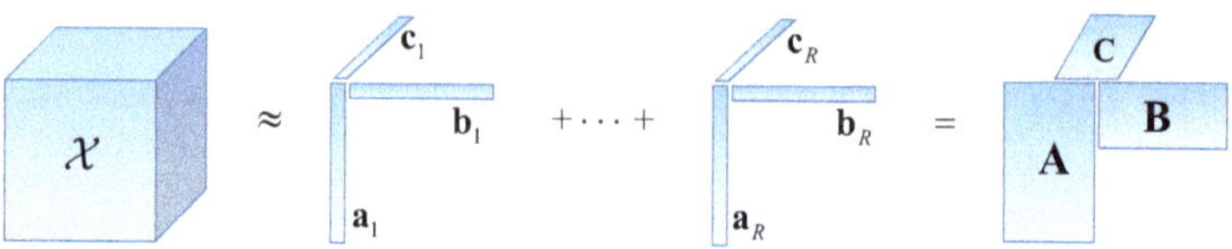

Fig. 2.1 An example for CP decomposition of a third-order tensor

[10] and PARAFAC (parallel factors) [31] is proposed by Carroll, Chang, and Harshman in the psychometrics community.

Definition 2.1 (Canonical Polyadic (CP) Decomposition) Given an Nth-order tensor $\mathcal{X} \in \mathbb{R}^{I_1 \times \cdots \times I_N}$, the CP decomposition is defined by

$$\mathcal{X} = \sum_{r=1}^{R} \mathbf{u}_r^{(1)} \circ \mathbf{u}_r^{(2)} \circ \cdots \circ \mathbf{u}_r^{(N)} = [\![\mathbf{U}^{(1)}, \mathbf{U}^{(2)}, \ldots, \mathbf{U}^{(N)}]\!], \tag{2.1}$$

or element wisely as

$$x_{i_1,i_2,\ldots,i_N} = \sum_{r=1}^{R} u_{i_1,r}^{(1)} u_{i_2,r}^{(2)} \cdots u_{i_N,r}^{(N)}, \tag{2.2}$$

where $\mathbf{U}^{(n)} = [\mathbf{u}_1^{(n)}, \mathbf{u}_2^{(n)}, \ldots \mathbf{u}_R^{(n)}] \in \mathbb{R}^{I_n \times R}, n = 1, \ldots, N$ are factor matrices.

For example, the CP decomposition of a third-order tensor is as follows:

$$\mathcal{X} = \sum_{r=1}^{R} \mathbf{a}_r \circ \mathbf{b}_r \circ \mathbf{c}_r = [\![\mathbf{A}, \mathbf{B}, \mathbf{C}]\!]. \tag{2.3}$$

In this case, the mode-n matrix versions are as follows:

$$\mathbf{X}_{(1)} = \mathbf{A}(\mathbf{C} \odot \mathbf{B})^{\mathrm{T}}, \mathbf{X}_{(2)} = \mathbf{B}(\mathbf{A} \odot \mathbf{C})^{\mathrm{T}}, \mathbf{X}_{(3)} = \mathbf{C}(\mathbf{B} \odot \mathbf{A})^{\mathrm{T}}. \tag{2.4}$$

Figure 2.1 shows a graphical representation of CP decomposition for a third-order tensor.

2.2.1 *Tensor Rank*

The rank of a tensor $\mathcal{X} \in \mathbb{R}^{I_1 \times \cdots \times I_N}$, namely, rank($\mathcal{X}$), is defined as the smallest number of rank-1 tensors in the CP decomposition as follows:

$$\mathcal{X} = \sum_{r=1}^{R} \mathcal{U}_r, \tag{2.5}$$

where $\mathcal{U}_r$ is a rank-1 tensor and can be represented as $\mathcal{U}_r = \mathbf{u}_r^{(1)} \circ \mathbf{u}_r^{(2)} \circ \cdots \circ \mathbf{u}_r^{(N)}$ and R is the smallest value for Eq. (2.5). The definition of tensor rank is analogue to the definition of matrix rank, but the properties of matrix and tensor ranks are quite different. One difference is that the rank of a real-valued tensor may actually be different over $\mathbb{R}$ and $\mathbb{C}$.

For example, let $\mathcal{X} \in \mathbb{R}^{2\times2\times2}$ be a tensor whose frontal slices are defined by

$$\mathcal{X}_{:,:,1} = \begin{bmatrix} 1 & 0 \\ 0 & 1 \end{bmatrix}, \quad \mathcal{X}_{:,:,2} = \begin{bmatrix} 0 & -1 \\ 1 & 0 \end{bmatrix}.$$

The tensor $\mathcal{X}$ can be factorized over $\mathbb{R}$ as follows:

$$\mathcal{X} = \sum_{r=1}^{3} \mathbf{a}_r \circ \mathbf{b}_r \circ \mathbf{c}_r = [\![\mathbf{A}, \mathbf{B}, \mathbf{C}]\!], \tag{2.6}$$

where $\mathbf{a}_r$, $\mathbf{b}_r$, $\mathbf{c}_r$ are the r-th column of factor matrices $\mathbf{A}$, $\mathbf{B}$, and $\mathbf{C}$, respectively.

$$\mathbf{A} = \begin{bmatrix} 1 & 0 & 1 \\ 0 & 1 & -1 \end{bmatrix}, \quad \mathbf{B} = \begin{bmatrix} 1 & 0 & 1 \\ 0 & 1 & 1 \end{bmatrix}, \quad \mathbf{C} = \begin{bmatrix} 1 & 1 & 0 \\ -1 & 1 & 1 \end{bmatrix}.$$

In this case, $\text{rank}(\mathcal{X}) = 3$ over $\mathbb{R}$. However, over $\mathbb{C}$, we can have $\text{rank}(\mathcal{X}) = 2$. $\mathcal{X}$ can be factorized as sum of two rank-1 tensors, as follows:

$$\mathcal{X} = \sum_{r=1}^{2} \mathbf{a}_r \circ \mathbf{b}_r \circ \mathbf{c}_r = [\![\mathbf{A}, \mathbf{B}, \mathbf{C}]\!], \tag{2.7}$$

where

$$\mathbf{A} = \frac{1}{\sqrt{2}} \begin{bmatrix} 1 & 1 \\ -j & j \end{bmatrix}, \quad \mathbf{B} = \frac{1}{\sqrt{2}} \begin{bmatrix} 1 & 1 \\ j & -j \end{bmatrix}, \quad \mathbf{C} = \begin{bmatrix} 1 & 1 \\ -j & j \end{bmatrix}.$$

It's difficult to determine the tensor rank. In practice, we usually use CP decomposition with tensor rank given in advance.

2.2.2 CPD Computation

Letting $\mathcal{X} \in \mathbb{R}^{I_1\times I_2\times I_3}$, the goal of CP decomposition is to find the optimal factor matrices to approximate $\mathcal{X}$ with predefined tensor rank, as follows:

$$\min_{\mathbf{A},\mathbf{B},\mathbf{C}} \|\mathcal{X} - [\![\mathbf{A}, \mathbf{B}, \mathbf{C}]\!]\|_F^2. \tag{2.8}$$

Directly solving the problem is difficult, and the alternating least squares (ALS) method is popular for it. The ALS approach fixes **B** and **C** to solve for **A**, then fixes **A** and **C** to solve for **B**, then fixes **A** and **B** to solve for **C**, and continues to repeat the entire procedure until a certain convergence criterion is satisfied. Under this framework, the problem (2.8) can be split into three subproblems, as follows:

$$
\begin{aligned}
&\min_{\mathbf{A}} \|\mathbf{X}_{(1)} - \mathbf{A}(\mathbf{C} \odot \mathbf{B})^{\mathrm{T}}\|_{\mathrm{F}}^2, \\
&\min_{\mathbf{B}} \|\mathbf{X}_{(2)} - \mathbf{B}(\mathbf{A} \odot \mathbf{C})^{\mathrm{T}}\|_{\mathrm{F}}^2, \\
&\min_{\mathbf{C}} \|\mathbf{X}_{(3)} - \mathbf{C}(\mathbf{B} \odot \mathbf{A})^{\mathrm{T}}\|_{\mathrm{F}}^2.
\end{aligned}
\tag{2.9}
$$

The solutions of problem (2.9) are as follows:

$$
\begin{aligned}
\mathbf{A} &= \mathbf{X}_{(1)}(\mathbf{C} \odot \mathbf{B})(\mathbf{C}^{\mathrm{T}}\mathbf{C} \circledast \mathbf{B}^{\mathrm{T}}\mathbf{B})^{\dagger}, \\
\mathbf{B} &= \mathbf{X}_{(2)}(\mathbf{A} \odot \mathbf{C})(\mathbf{A}^{\mathrm{T}}\mathbf{A} \circledast \mathbf{C}^{\mathrm{T}}\mathbf{C})^{\dagger}, \\
\mathbf{C} &= \mathbf{X}_{(3)}(\mathbf{B} \odot \mathbf{A})(\mathbf{B}^{\mathrm{T}}\mathbf{B} \circledast \mathbf{A}^{\mathrm{T}}\mathbf{A})^{\dagger}.
\end{aligned}
\tag{2.10}
$$

The details can be concluded in Algorithm 1.

Algorithm 1: ALS for CP decomposition of a third-order tensor

Input: $\mathcal{X} \in \mathbb{R}^{I_1 \times I_2 \times I_3}$ and CP-rank R

initialize **A**, **B**, **C**

repeat

1. $\mathbf{A} = \mathbf{X}_{(1)}(\mathbf{C} \odot \mathbf{B})(\mathbf{C}^{\mathrm{T}}\mathbf{C} \circledast \mathbf{B}^{\mathrm{T}}\mathbf{B})^{\dagger}$
2. $\mathbf{B} = \mathbf{X}_{(2)}(\mathbf{A} \odot \mathbf{C})(\mathbf{A}^{\mathrm{T}}\mathbf{A} \circledast \mathbf{C}^{\mathrm{T}}\mathbf{C})^{\dagger}$
3. $\mathbf{C} = \mathbf{X}_{(3)}(\mathbf{B} \odot \mathbf{A})(\mathbf{B}^{\mathrm{T}}\mathbf{B} \circledast \mathbf{A}^{\mathrm{T}}\mathbf{A})^{\dagger}$

until fit ceases to improve or maximum iterations exhausted

Output: **A**, **B**, **C**

The ALS method is simple to understand and easy to implement, but can take many iterations to converge. Moreover, it is not guaranteed to converge to a global minimum, only to a solution where the objective function of (2.8) decreases. To solve this issue, Li et al. [49] propose a regularized alternating least squares method for CP decomposition (CP-RALS). In this framework, the subproblems in $(k+1)$-th iteration can be written as follows:

$$
\begin{aligned}
\mathbf{A}^{k+1} &= \underset{\mathbf{A}}{\operatorname{argmin}} \|\mathbf{X}_{(1)} - \mathbf{A}(\mathbf{C}^k \odot \mathbf{B}^k)^{\mathrm{T}}\|_{\mathrm{F}}^2 + \tau\|\mathbf{A} - \mathbf{A}^k\|_{\mathrm{F}}^2, \\
\mathbf{B}^{k+1} &= \underset{\mathbf{B}}{\operatorname{argmin}} \|\mathbf{X}_{(2)} - \mathbf{B}(\mathbf{A}^{k+1} \odot \mathbf{C}^k)^{\mathrm{T}}\|_{\mathrm{F}}^2 + \tau\|\mathbf{B} - \mathbf{B}^k\|_{\mathrm{F}}^2, \\
\mathbf{C}^{k+1} &= \underset{\mathbf{C}}{\operatorname{argmin}} \|\mathbf{X}_{(3)} - \mathbf{C}(\mathbf{B}^{k+1} \odot \mathbf{A}^{k+1})^{\mathrm{T}}\|_{\mathrm{F}}^2 + \tau\|\mathbf{C} - \mathbf{C}^k\|_{\mathrm{F}}^2,
\end{aligned}
\tag{2.11}
$$

where τ is a trade-off parameter.

The solutions can be updated as:

$$\begin{aligned}
\mathbf{A}^{k+1} &= (\mathbf{X}_{(1)}(\mathbf{C}^k \odot \mathbf{B}^k) + \tau\mathbf{A}^k)((\mathbf{C}^k)^{\mathrm{T}}\mathbf{C}^k \circledast (\mathbf{B}^k)^{\mathrm{T}}\mathbf{B}^k + \tau\mathbf{I})^{-1}, \\
\mathbf{B}^{k+1} &= (\mathbf{X}_{(2)}(\mathbf{A}^{k+1} \odot \mathbf{C}^k) + \tau\mathbf{B}^k)((\mathbf{A}^{k+1})^{\mathrm{T}}\mathbf{A}^{k+1} \circledast (\mathbf{C}^k)^{\mathrm{T}}\mathbf{C}^k + \tau\mathbf{I})^{-1}, \\
\mathbf{C}^{k+1} &= (\mathbf{X}_{(3)}(\mathbf{B}^{k+1} \odot \mathbf{A}^{k+1}) + \tau\mathbf{C}^k)((\mathbf{B}^{k+1})^{\mathrm{T}}\mathbf{B}^{k+1} \circledast (\mathbf{A}^{k+1})^{\mathrm{T}}\mathbf{A}^{k+1} + \tau\mathbf{I})^{-1}.
\end{aligned} \tag{2.12}$$

It can be concluded in Algorithm 2.

Algorithm 2: RALS for CP decomposition of a third-order tensor

Input: $\mathcal{X} \in \mathbb{R}^{I_1 \times I_2 \times I_3}$ and CP-rank R
initialize $\mathbf{A}^0, \mathbf{B}^0, \mathbf{C}^0, K, \tau$
for $k = 1, \dots K$
1. $\mathbf{A}^{k+1} = (\mathbf{X}_{(1)}(\mathbf{C}^k \odot \mathbf{B}^k) + \tau\mathbf{A}^k)((\mathbf{C}^k)^{\mathrm{T}}\mathbf{C}^k \circledast (\mathbf{B}^k)^{\mathrm{T}}\mathbf{B}^k + \tau\mathbf{I})^{-1}$
2. $\mathbf{B}^{k+1} = (\mathbf{X}_{(2)}(\mathbf{A}^{k+1} \odot \mathbf{C}^k) + \tau\mathbf{B}^k)((\mathbf{A}^{k+1})^{\mathrm{T}}\mathbf{A}^{k+1} \circledast (\mathbf{C}^k)^{\mathrm{T}}\mathbf{C}^k + \tau\mathbf{I})^{-1}$
3. $\mathbf{C}^{k+1} = (\mathbf{X}_{(3)}(\mathbf{B}^{k+1} \odot \mathbf{A}^{k+1}) + \tau\mathbf{C}^k)((\mathbf{B}^{k+1})^{\mathrm{T}}\mathbf{B}^{k+1} \circledast (\mathbf{A}^{k+1})^{\mathrm{T}}\mathbf{A}^{k+1} + \tau\mathbf{I})^{-1}$

until fit ceases to improve or maximum iterations exhausted
Output: $\mathbf{A}, \mathbf{B}, \mathbf{C}$

Compared with ALS, the advantage of RALS has twofolds. One is that the solutions are more stable, and the other is for preventing mutations in results.

2.2.3 *Uniqueness*

CP decomposition has two kinds of indeterminacy: one is the permutation indeterminacy, and the other is the scaling indeterminacy.

First, the permutation indeterminacy means that the rank-one tensors can be reordered arbitrarily. For example, the rank-one tensor $\mathcal{X}_1 = \mathbf{a}_1 \circ \mathbf{b}_1 \circ \mathbf{c}_1$ and $\mathcal{X}_3 = \mathbf{a}_3 \circ \mathbf{b}_3 \circ \mathbf{c}_3$ are two components of a third-order tensor. If we exchange these two rank-one tensors, we can observe that Eq. (2.3) also keeps the same. It refers to the following formulation:

$$\mathcal{X} = [\![\mathbf{A}, \mathbf{B}, \mathbf{C}]\!] = [\![\mathbf{AP}, \mathbf{BP}, \mathbf{CP}]\!] \ \ \forall \text{ permutation matrix } \mathbf{P} \in \mathbb{R}^{R \times R}, \tag{2.13}$$

where $\mathbf{A} = [\mathbf{a}_1, \dots, \mathbf{a}_R]$, $\mathbf{B} = [\mathbf{b}_1, \dots, \mathbf{b}_R]$ and $\mathbf{C} = [\mathbf{c}_1, \dots, \mathbf{c}_R]$.

In addition, as long as $\alpha_r\beta_r\gamma_r = 1, r = 1, \dots R$, we can obtain the following equation:

$$\mathcal{X} = \sum_{r=1}^{R} (\alpha_r \mathbf{a}_r) \circ (\beta_r \mathbf{b}_r) \circ (\gamma_r \mathbf{c}_r) = \sum_{r=1}^{R} \mathbf{a}_r \circ \mathbf{b}_r \circ \mathbf{c}_r. \tag{2.14}$$

In practice, permutation indeterminacy and scaling indeterminacy are allowed. Therefore, uniqueness means that it is the only possible combination of rank-one tensors whose cumulative sum is $\mathcal{X}$. According to Kruskal rank [38, 68], CP decomposition can be unique under mild condition.

Definition 2.2 ((Kruskal Rank of a Matrix) [47]) The Kruskal rank of a matrix $\mathbf{A}$, denoted by $K_{\mathbf{A}}$, is the maximal number K such that any set of K columns of $\mathbf{A}$ is linearly independent.

Kruskal's result [38, 68] says that a sufficient condition for uniqueness for the CP decomposition in (2.3) is

$$K_{\mathbf{X}_{(1)}} + K_{\mathbf{X}_{(2)}} + K_{\mathbf{X}_{(3)}} \geq 2R + 2, \tag{2.15}$$

where R is the CP rank. Berge and Sidiropoulos [70] gave a proof that Kruskal's result is a sufficient and necessary condition for tensor when $R = 2$ and $R = 3$, but not for $R > 3$. Sidiropoulos and Bro [63] recently extended Kruskal's result to Nth-order tensors. Let $\mathcal{X} \in \mathbb{R}^{I_1 \times I_2 \times \cdots I_N}$ be an Nth-order tensor with rank R and suppose that its CP decomposition is

$$\mathcal{X} = \sum_{r=1}^{R} \mathbf{u}_r^{(1)} \circ \mathbf{u}_r^{(2)} \circ \cdots \circ \mathbf{u}_r^{(N)}, \tag{2.16}$$

then a sufficient condition for uniqueness is

$$\sum_{n=1}^{N} K_{\mathbf{X}_{(n)}} \geq 2R + (N-1), \tag{2.17}$$

where $K_{\mathbf{X}_{(n)}}$ is the Kruskal rank of $\mathbf{X}_{(n)} \in \mathbb{R}^{I_n \times I_1 \cdots I_{n-1} I_{n+1} \cdots I_N}$.

According to mode-n matricization of the third-order tensor in (2.4), Liu and Sidiropoulos [51] showed that a necessary condition for uniqueness of the CP decomposition is

$$\min(\text{rank}(\mathbf{A} \odot \mathbf{B}), \text{rank}(\mathbf{B} \odot \mathbf{C}), \text{rank}(\mathbf{C} \odot \mathbf{A})) = R. \tag{2.18}$$

More generally, they showed that for the N-way case, a necessary condition for uniqueness of the CP decomposition in (2.1) is

$$\min_{n=1\cdots N} \text{rank}(\mathbf{U}^{(1)} \odot \cdots \mathbf{U}^{(n-1)} \odot \mathbf{U}^{(n+1)} \odot \cdots \odot \mathbf{U}^{(N)}) = R, \tag{2.19}$$

according to mode-n matrices $\mathbf{X}_{(n)} = \mathbf{U}^{(n)}(\mathbf{U}^{(1)} \odot \cdots \mathbf{U}^{(n-1)} \odot \mathbf{U}^{(n+1)} \odot \cdots \odot \mathbf{U}^{(N)})^{\mathrm{T}}$. Since

$$\text{rank}(\mathbf{A} \odot \mathbf{B}) \leq \text{rank}(\mathbf{A} \otimes \mathbf{B}) \leq \text{rank}(\mathbf{A})\text{rank}(\mathbf{B}) \tag{2.20}$$

a simpler necessary condition is

$$\min_{n=1,\ldots N}\left(\prod_{m=1,m\neq n}^{N}\operatorname{rank}(\mathbf{U}^{(m)})\right)\geq R. \tag{2.21}$$

In addition, De Lathauwer proved that the CP decomposition of a third-order tensor $\mathcal{X}\in\mathbb{R}^{I_1\times I_2\times I_3}$ in (2.3) is generically unique if

$$R\leq I_3 \text{ and } R(R-1)\leq\frac{I_1(I_1-1)I_2(I_2-1)}{2}. \tag{2.22}$$

Similarly, a fourth-order tensor $\mathcal{X}\in\mathbb{R}^{I_1\times I_2\times I_3\times I_4}$ of rank R has a CP decomposition that is generically unique if

$$R\leq I_4 \text{ and } R(R-1)\leq\frac{I_1I_2I_3(3I_1I_2I_3-I_1I_2-I_3-I_1-I_2I_3-I_1I_3-I_2+3)}{4}. \tag{2.23}$$

2.3 Tucker Decomposition

Tucker decomposition is another higher-order generalization of matrix singular value decomposition (SVD). It is firstly introduced by Tucker in 1963 [71] and refined in subsequent articles by Levin [48] and Tucker [72, 73]. Tucker decomposition is also known as higher-order singular value decomposition (HOSVD). It decomposes a tensor into a core tensor multiplied by a matrix along each mode, which can be defined by

Definition 2.3 (Tucker Decomposition) For a tensor $\mathcal{X}\in\mathbb{R}^{I_1\times I_2\times\cdots\times I_N}$, the Tucker decomposition is

$$\mathcal{X}=\mathcal{G}\times_1\mathbf{U}^{(1)}\times_2\mathbf{U}^{(2)}\cdots\times_N\mathbf{U}^{(N)}=[\![\mathcal{G};\mathbf{U}^{(1)},\mathbf{U}^{(2)},\ldots,\mathbf{U}^{(N)}]\!], \tag{2.24}$$

or element wisely as

$$x_{i_1,i_2,\ldots,i_N}=\sum_{r_1}^{R_1}\sum_{r_2}^{R_2}\cdots\sum_{r_N}^{R_N}g_{r_1,r_2,\ldots,r_N}u_{i_1,r_1}^{(1)}u_{i_2,r_2}^{(2)}\cdots u_{i_N,r_N}^{(N)}, \tag{2.25}$$

where $\mathbf{U}^{(n)}\in\mathbb{R}^{I_n\times R_n}$, $n=1,\ldots,N$ are factor matrices, $R_n\leq I_n$ and $\mathbf{U}^{(n)\,\mathrm{T}}\mathbf{U}^{(n)}=\mathbf{I}_{R_n}$, and the tensor $\mathcal{G}\in\mathbb{R}^{R_1\times R_2\times\cdots R_N}$ is called core tensor. It is noticed that the core tensor $\mathcal{G}$ is a full tensor and its entries represent the level of interaction between the different components $\mathbf{U}^{(n)}$, $n=1,\ldots,N$.

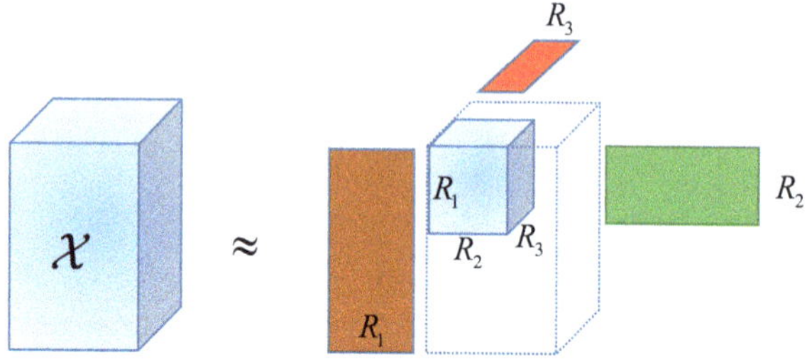

Fig. 2.2 Tucker decomposition for a third-order tensor

For example, letting $\mathcal{X} \in \mathbb{R}^{I_1 \times I_2 \times I_3}$, we can get $\mathbf{A} \in \mathbb{R}^{I_1 \times R_1}$, $\mathbf{B} \in \mathbb{R}^{I_2 \times R_2}$, $\mathbf{C} \in \mathbb{R}^{I_3 \times R_3}$ and a core tensor $\mathcal{G} \in \mathbb{R}^{R_1 \times R_2 \times R_3}$ by Tucker decomposition:

$$\mathcal{X} = \mathcal{G} \times_1 \mathbf{A} \times_2 \mathbf{B} \times_3 \mathbf{C} = \sum_{r_1=1}^{R_1} \sum_{r_2=1}^{R_2} \sum_{r_3=1}^{R_3} g_{r_1,r_2,r_3} \mathbf{a}_{r_1} \circ \mathbf{b}_{r_2} \circ \mathbf{c}_{r_3} = [\![\mathcal{G}; \mathbf{A}, \mathbf{B}, \mathbf{C}]\!]. \tag{2.26}$$

The graphical representation can be found in Fig. 2.2.

2.3.1 The n-Rank

Letting $\mathcal{X} \in \mathbb{R}^{I_1 \times I_2 \times \cdots \times I_N}$, the n-rank or multilinear rank of $\mathcal{X}$, denoted by $\text{rank}_n(\mathcal{X})$, is the column rank of $\mathcal{X}_{(n)}$. In other words, the n-rank is the size of the vector space spanned by the mode-n fibers.

For example, the n-rank of a third-order tensor $\mathcal{X} \in \mathbb{R}^{I_1 \times I_2 \times I_3}$, represented by R_n, is denoted as

$$\begin{aligned} R_1 &= \dim(\text{span}_{\mathbb{R}}\{\mathbf{x}_{:,i_2,i_3} | i_2 = 1, \ldots, I_2, i_3 = 1, \ldots, I_3\}), \\ R_2 &= \dim(\text{span}_{\mathbb{R}}\{\mathbf{x}_{i_1,:,i_3} | i_1 = 1, \ldots, I_1, i_3 = 1, \ldots, I_3\}), \\ R_3 &= \dim(\text{span}_{\mathbb{R}}\{\mathbf{x}_{i_1,i_2,:} | i_1 = 1, \ldots, I_1, i_2 = 1, \ldots, I_2\}), \end{aligned} \tag{2.27}$$

where $\text{span}_{\mathbb{R}}$ is an operator that spans the vector into a space and dim is the size of the space. $\mathbf{x}_{:,i_2,i_3}, \mathbf{x}_{i_1,:,i_3}, \mathbf{x}_{i_1,i_2,:}$ are the mode-1, mode-2, and mode-3 fibers of tensor $\mathcal{X}$, respectively. In this case, we can say that $\mathcal{X}$ is a rank-(R_1, R_2, R_3) tensor. More generally, if $R_n = \text{rank}_n(\mathcal{X}), n = 1, \ldots, N$, we can say that $\mathcal{X}$ is a rank-$(R_1, R_2, \ldots, R_N)$ tensor.

It is noticed that the row rank and the column rank are the same for the matrix. But, if $i \neq j$, the i-rank and j-rank are different for the higher-order tensor.

2.3.2 Computation and Optimization Model

In the work [20], De Lathauwer et al. showed that HOSVD is a convincing generalization of the matrix SVD and discussed ways to efficiently compute the leading left singular vectors of $\mathcal{X}_{(n)}$. When $R_n < \text{rank}_n(\mathcal{X})$ for one or more n, the decomposition is called the truncated HOSVD. The truncated HOSVD is not optimal in terms of giving the best fit as measured by the norm of the difference, but it is a good starting point for an iterative ALS algorithm.

Algorithm 3: Higher-order singular value decomposition (HOSVD)

Input: $\mathcal{X} \in \mathbb{R}^{I_1 \times I_2 \times \cdots \times I_N}$ and multilinear rank is $(R_1, R_2, \ldots, R_N)$
Output: $\mathcal{G}, \mathbf{U}^{(1)}, \mathbf{U}^{(2)}, \ldots, \mathbf{U}^{(N)}$
For $n = 1, \ldots, N$ **do**
  $\mathbf{U}^{(n)} \leftarrow R_n$ leading left singular vectors of $\mathbf{X}_{(n)}$,
End for
$\mathcal{G} \leftarrow \mathcal{X} \times_1 \mathbf{U}^{(1)\mathrm{T}} \times_2 \mathbf{U}^{(2)\mathrm{T}} \cdots \times_N \mathbf{U}^{(N)\mathrm{T}}$

De Lathauwer et al. [22] proposed more efficient techniques for calculating the factor matrices and called it the higher-order orthogonal iteration (HOOI). Assuming that $\mathcal{X} \in \mathbb{R}^{I_1 \times I_2 \times \cdots \times I_N}$, the optimization problem is

$$\begin{aligned} &\min_{\mathcal{G}, \mathbf{U}^{(1)}, \ldots, \mathbf{U}^{(N)}} \left\| \mathcal{X} - [\![\mathcal{G}; \mathbf{U}^{(1)}, \mathbf{U}^{(2)}, \ldots, \mathbf{U}^{(N)}]\!] \right\|_\mathrm{F}^2 \\ &\text{s. t. } \mathcal{G} \in \mathbb{R}^{R_1 \times R_2 \cdots \times R_N},\ \mathbf{U}^{(n)} \in \mathbb{R}^{I_n \times R_n},\ \mathbf{U}^{(n)\mathrm{T}} \mathbf{U}^{(n)} = \mathbf{I}_{R_n \times R_n}. \end{aligned} \tag{2.28}$$

In the above optimization problem, we have

$$\begin{aligned} &\left\| \mathcal{X} - [\![\mathcal{G}; \mathbf{U}^{(1)}, \mathbf{U}^{(2)}, \ldots, \mathbf{U}^{(N)}]\!] \right\|_\mathrm{F}^2 \\ &= \|\mathcal{X}\|_\mathrm{F}^2 - 2\langle \mathcal{X}, [\![\mathcal{G}; \mathbf{U}^{(1)}, \mathbf{U}^{(2)}, \ldots, \mathbf{U}^{(N)}]\!] \rangle + \left\| [\![\mathcal{G}; \mathbf{U}^{(1)}, \mathbf{U}^{(2)}, \ldots, \mathbf{U}^{(N)}]\!] \right\|_\mathrm{F}^2 \\ &= \|\mathcal{X}\|_\mathrm{F}^2 - 2\langle \mathcal{X} \times_1 \mathbf{U}^{(1)\mathrm{T}} \cdots \times_N \mathbf{U}^{(N)\mathrm{T}}, \mathcal{G} \rangle + \left\| [\![\mathcal{G}; \mathbf{U}^{(1)}, \mathbf{U}^{(2)}, \ldots, \mathbf{U}^{(N)}]\!] \right\|_\mathrm{F}^2 \\ &= \|\mathcal{X}\|_\mathrm{F}^2 - 2\langle \mathcal{G}, \mathcal{G} \rangle + \|\mathcal{G}\|_\mathrm{F}^2 \\ &= \|\mathcal{X}\|_\mathrm{F}^2 - \|\mathcal{G}\|_\mathrm{F}^2 \\ &= \|\mathcal{X}\|_\mathrm{F}^2 - \left\| \mathcal{X} \times_1 \mathbf{U}^{(1)\mathrm{T}} \cdots \times_N \mathbf{U}^{(N)\mathrm{T}} \right\|_\mathrm{F}^2. \end{aligned} \tag{2.29}$$

So the problem boils down to

$$\max_{\mathbf{U}^{(n)}} \left\| \mathcal{X} \times_1 \mathbf{U}^{(1)\mathrm{T}} \cdots \times_N \mathbf{U}^{(N)\mathrm{T}} \right\|_{\mathrm{F}}^2$$
$$\text{s. t. } \mathbf{U}^{(n)} \in \mathbb{R}^{I_n \times R_n}, \mathbf{U}^{(n)\mathrm{T}} \mathbf{U}^{(n)} = \mathbf{I}_{R_n \times R_n}, \tag{2.30}$$

which can be rewritten into a matrix form as

$$\max_{\mathbf{U}^{(n)}} \left\| \mathbf{U}^{(n)\mathrm{T}} \mathbf{W} \right\|_{\mathrm{F}}^2$$
$$\text{s. t. } \mathbf{U}^{(n)} \in \mathbb{R}^{I_n \times R_n}, \ \mathbf{U}^{(n)\mathrm{T}} \mathbf{U}^{(n)} = \mathbf{I}_{R_n \times R_n}, \tag{2.31}$$

where $\mathbf{W} = \mathbf{X}_{(n)}(\mathbf{U}^{(N)} \otimes \cdots \otimes \mathbf{U}^{(n+1)} \otimes \mathbf{U}^{(n-1)} \otimes \cdots \otimes \mathbf{U}^{(1)})$. The solution can be determined using the SVD, and $\mathbf{U}^{(n)}$ is constructed by the R_n leading left singular vectors of $\mathbf{W}$. The details of HOOI are concluded in Algorithm 4.

Algorithm 4: Higher-order orthogonal iteration (HOOI)

Input: $\mathcal{X} \in \mathbb{R}^{I_1 \times I_2 \times \cdots \times I_N}$ and multilinear rank is $(R_1, R_2, \ldots, R_N)$
Output: $\mathcal{G}, \mathbf{U}^{(1)}, \mathbf{U}^{(2)}, \ldots, \mathbf{U}^{(N)}$
initialize $\mathbf{U}^{(n)} \in \mathbb{R}^{I_n \times R_n}$ for $n = 1, \ldots, N$ using HOSVD
repeat
For $n = 1, \ldots, N$ **do**
  $\mathcal{Y} \leftarrow \mathcal{X} \times_1 \mathbf{U}^{(1)\mathrm{T}} \cdots \times_{n-1} \mathbf{U}^{(n-1)\mathrm{T}} \times_{n+1} \mathbf{U}^{(n+1)\mathrm{T}} \cdots \times_N \mathbf{U}^{(N)\mathrm{T}}$
  $\mathbf{U}^{(n)} \leftarrow R_n$ leading left singular vectors of $\mathbf{Y}_{(n)}$,
End for
until fit ceases to improve or maximum iterations exhausted
$\mathcal{G} \leftarrow \mathcal{X} \times_1 \mathbf{U}^{(1)\mathrm{T}} \times_2 \mathbf{U}^{(2)\mathrm{T}} \cdots \times_N \mathbf{U}^{(N)\mathrm{T}}$

2.3.3 *Uniqueness*

Tucker decompositions are not unique. Here we consider the Tucker decomposition of a three-way tensor. Letting $\mathbf{W} \in \mathbb{R}^{R_1 \times R_1}$, $\mathbf{U} \in \mathbb{R}^{R_2 \times R_2}$, $\mathbf{V} \in \mathbb{R}^{R_3 \times R_3}$ be nonsingular matrices, we have

$$[\![\mathcal{G}; \mathbf{A}, \mathbf{B}, \mathbf{C}]\!] = [\![\mathcal{G} \times_1 \mathbf{W} \times_2 \mathbf{U} \times_3 \mathbf{V}; \mathbf{A}\mathbf{W}^{-1}, \mathbf{B}\mathbf{U}^{-1}, \mathbf{C}\mathbf{V}^{-1}]\!]. \tag{2.32}$$

We can modify the core tensor $\mathcal{G}$ without affecting the fit as long as we apply the inverse modification to the factor matrices. This freedom opens the door for choosing transformations that simplify the core structure in some way so that most of the elements of $\mathcal{G}$ are zero, thereby eliminating interactions between corresponding components and improving uniqueness.

2.4 Block Term Decomposition

As introduced in Sects. 2.2 and 2.3, CP decomposition and Tucker decomposition are connected with two different tensor generalizations of the concept of matrix rank. The Tucker decomposition/HOSVD is linked with the n-ranks, which generalize column rank, row rank, etc. The CP decomposition linked the rank with the meaning of the minimal number of rank-1 terms that are needed in an expansion of the matrix/tensor. To unify the Tucker decomposition and CP decomposition, block term decomposition (BTD) is proposed [18, 19, 21]. It decomposes a tensor into a sum of low multilinear rank terms, which is defined as

Definition 2.4 (Block Term Decomposition) For a tensor $\mathcal{X} \in \mathbb{R}^{I_1 \times I_2 \times \cdots \times I_N}$, block term decomposition is:

$$\mathcal{X} = \sum_{r=1}^{R} \mathcal{G}_r \times_1 \mathbf{U}_r^{(1)} \times_2 \mathbf{U}_r^{(2)} \cdots \times_N \mathbf{U}_r^{(N)}, \tag{2.33}$$

or element wisely as

$$\mathcal{X}(i_1, i_2, \ldots, i_N) = \sum_{r=1}^{R} \sum_{j_1}^{J_1^r} \sum_{j_2}^{J_2^r} \cdots \sum_{j_N}^{J_N^r} \mathcal{G}_r(j_1, j_2, \ldots, j_N) \mathbf{U}_r^{(1)}(i_1, j_1) \cdots \mathbf{U}_r^{(N)}(i_N, j_N), \tag{2.34}$$

in which $\mathbf{U}_r^{(n)} \in \mathbb{R}^{I_n \times J_n^r}$ represents the n-th factor in the r-th term, and $\mathcal{G}_r \in \mathbb{R}^{J_1^r \times J_2^r \times \cdots \times J_N^r}$ is the core tensor in the r-th term.

For example, BTD of a third-order tensor $\mathcal{X} \in \mathbb{R}^{I \times J \times K}$ is a sum of rank-(L_r, M_r, N_r) terms:

$$\mathcal{X} = \sum_{r=1}^{R} \mathcal{S}_r \times_1 \mathbf{U}_r \times_2 \mathbf{V}_r \times_3 \mathbf{W}_r, \tag{2.35}$$

in which $\mathcal{S}_r \in \mathbb{R}^{L_r \times M_r \times N_r}$ are full rank (L_r, M_r, N_r) and $\mathbf{U}_r \in \mathbb{R}^{I \times L_r}$ with $(I \geq L_r)$, $\mathbf{V}_r \in \mathbb{R}^{J \times M_r}$ with $(J \geq M_r)$, and $\mathbf{W}_r \in \mathbb{R}^{K \times N_r}$ with $(K \geq N_r)$ are full column rank, $1 \leq r \leq R$. For simplicity, we assume $L_r = L$, $M_r = M$, $N_r = N$, $r = 1, \ldots, R$ hereafter.

Under this decomposition, the matrix representations of $\mathcal{X}$ can be represented as

$$\mathbf{X}_{(1)} = \mathbf{U}\,\text{blockdiag}((\mathbf{S}_1)_{(1)}, \ldots, (\mathbf{S}_R)_{(1)})(\mathbf{V} \odot_b \mathbf{W})^{\mathrm{T}}, \tag{2.36}$$

$$\mathbf{X}_{(2)} = \mathbf{V}\,\text{blockdiag}((\mathbf{S}_1)_{(2)}, \ldots, (\mathbf{S}_R)_{(2)})(\mathbf{W} \odot_b \mathbf{U})^{\mathrm{T}}, \tag{2.37}$$

$$\mathbf{X}_{(3)} = \mathbf{W}\,\text{blockdiag}((\mathbf{S}_1)_{(3)}, \ldots, (\mathbf{S}_R)_{(3)})(\mathbf{U} \odot_b \mathbf{V})^{\mathrm{T}}, \tag{2.38}$$

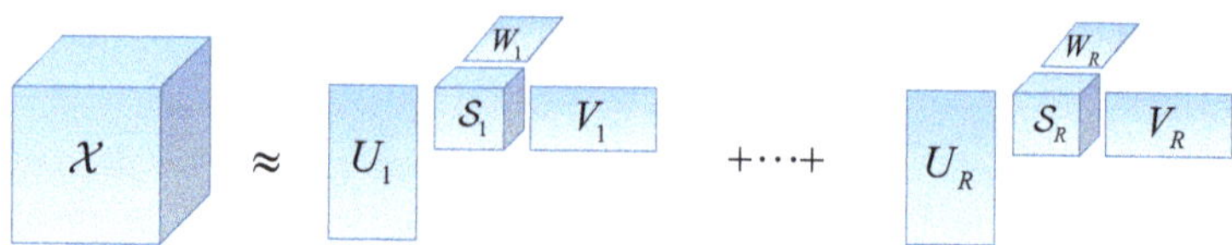

Fig. 2.3 BTD for a third-order tensor

where partitioned matrices $\mathbf{U} = [\mathbf{U}_1, \ldots, \mathbf{U}_R]$, $\mathbf{V} = [\mathbf{V}_1, \ldots, \mathbf{V}_R]$, and $\mathbf{W} = [\mathbf{W}_1, \ldots, \mathbf{W}_R]$, $(\mathbf{S}_r)_{(n)}$ is the mode-n unfolding matrix of $\mathcal{S}_r$, and here $\odot_{\mathrm{b}}$ denotes the blockwise Khatri-Rao product and it's the partition-wise Kronecker product, i.e.,

$$\mathbf{A} \odot_{\mathrm{b}} \mathbf{B} = (\mathbf{A}_1 \otimes \mathbf{B}_1, \ldots, \mathbf{A}_R \otimes \mathbf{B}_R). \tag{2.39}$$

In terms of the vector representation of $\mathcal{X}$, the decomposition can be written as

$$\mathrm{vec}(\mathcal{X}) = (\mathbf{U} \odot_{\mathrm{b}} \mathbf{V} \odot_{\mathrm{b}} \mathbf{W})[\mathrm{vec}(\mathcal{S}_1); \ldots; \mathrm{vec}(\mathcal{S}_R)]. \tag{2.40}$$

A visual representation of BTD for a third-order tensor is shown in Fig. 2.3.

In this way, there are two kinds of special cases in BTD for third-order tensor. One is decomposition in rank-$(L, L, 1)$ terms, and the other is type-2 decomposition in rank-$(L, M, \cdot)$ terms.

Definition 2.5 (**$(L, L, 1)$-decomposition**) $(L, L, 1)$-decomposition of a third-order tensor $\mathcal{X} \in \mathbb{R}^{I \times J \times K}$ is a sum of rank-$(L, L, 1)$ terms, which is denoted as

$$\mathcal{X} = \sum_{r}^{R} \mathbf{S}_r \circ \mathbf{c}_r, \tag{2.41}$$

where matrices $\mathbf{S}_r$, $r = 1, \ldots, R$ are rank-L.

In most cases, the different matrices $\mathbf{S}_r$, $r = 1, \ldots, R$ do not necessarily all have the same rank, and we assume their ranks are L_r, $r = 1, \ldots, R$, which develops $(L_r, L_r, 1)$ decomposition.

Definition 2.6 (**$(L_r, L_r, 1)$-decomposition**) $(L_r, L_r, 1)$-decomposition of a third-order tensor $\mathcal{X} \in \mathbb{R}^{I \times J \times K}$ is a sum of rank-$(L_r, L_r, 1)$ terms, which is denoted as

$$\mathcal{X} = \sum_{r=1}^{R} \mathbf{S}_r \circ \mathbf{c}_r, \tag{2.42}$$

where the matrix $\mathbf{S}_r$ is rank-L_r, $r = 1 \ldots, R$. If we factorize $\mathbf{S}_r$ as $\mathbf{A}_r \mathbf{B}_r^{\mathrm{T}}$ where the matrix $\mathbf{A}_r \in \mathbb{R}^{I \times L_r}$ and the matrix $\mathbf{B}_r \in \mathbb{R}^{J \times L_r}$, $L_r \leq I$ and $L_r \leq J$, $r = 1, \ldots, R$,

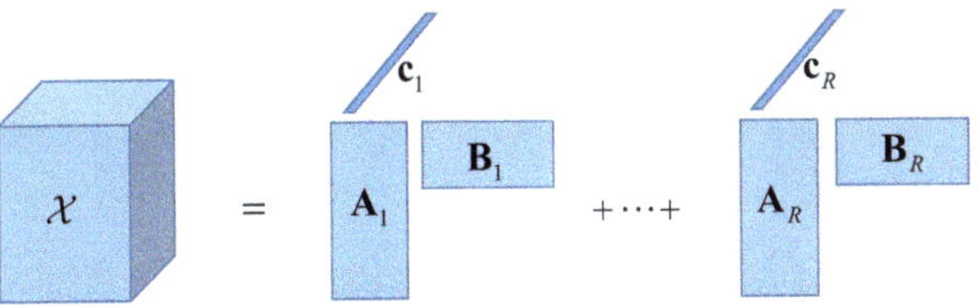

Fig. 2.4 $(L_r, L_r, 1)$-decomposition for a third-order tensor

then we can write (2.42) as

$$\mathcal{X} = \sum_{r=1}^{R} (\mathbf{A}_r \mathbf{B}_r^{\mathrm{T}}) \circ \mathbf{c}_r, \tag{2.43}$$

The standard matrix representations of $\mathcal{X}$ are

$$\mathbf{X}_{(1)} = \mathbf{A}(\mathbf{B} \odot_{\mathrm{b}} \mathbf{C})^{\mathrm{T}}, \tag{2.44}$$

$$\mathbf{X}_{(2)} = \mathbf{B}(\mathbf{C} \odot_{\mathrm{b}} \mathbf{A})^{\mathrm{T}}, \tag{2.45}$$

$$\mathbf{X}_{(3)} = \mathbf{C}[(\mathbf{A}_1 \odot \mathbf{B}_1)\mathbf{1}_{L_1}, \ldots, (\mathbf{A}_R \odot \mathbf{B}_R)\mathbf{1}_{L_R}]^{\mathrm{T}}. \tag{2.46}$$

where partitioned matrices $\mathbf{A} = [\mathbf{A}_1, \ldots, \mathbf{A}_R]$, $\mathbf{B} = [\mathbf{B}_1, \ldots, \mathbf{B}_R]$, and $\mathbf{C} = [\mathbf{c}_1, \ldots, \mathbf{c}_R]$.

A visual representation of this decomposition for the third-order case is shown in Fig. 2.4.

Definition 2.7 ($(L, M, \cdot)$-decomposition) A $(L, M, \cdot)$-decomposition of a tensor $\mathcal{X} \in \mathbb{R}^{I \times J \times K}$ is defined as

$$\mathcal{X} = \sum_{r=1}^{R} \mathcal{S}_r \times_1 \mathbf{A}_r \times_2 \mathbf{B}_r, \tag{2.47}$$

where $\mathcal{S}_r \in \mathbb{R}^{L \times M \times K}$ and $\mathbf{A}_r \in \mathbb{R}^{I \times L}$ and $\mathbf{B}_r \in \mathbb{R}^{J \times M}$, $r = 1, \ldots, R$ are of full column rank.

The standard matrix representations of $\mathcal{X}$ are

$$\mathbf{X}_{(1)} = \mathbf{A}[(\mathcal{S}_1 \times_2 \mathbf{B}_1)_{(1)}; \ldots; (\mathcal{S}_R \times_2 \mathbf{B}_R)_{(1)}], \tag{2.48}$$

$$\mathbf{X}_{(2)} = \mathbf{B}[(\mathcal{S}_1 \times_1 \mathbf{A}_1)_{(2)}; \ldots; (\mathcal{S}_R \times_1 \mathbf{A}_R)_{(2)}], \tag{2.49}$$

$$\mathbf{X}_{(3)} = [(\mathbf{S}_1)_{(3)}, \ldots, (\mathbf{S}_R)_{(3)}](\mathbf{A} \odot_{\mathrm{b}} \mathbf{B})^{\mathrm{T}}, \tag{2.50}$$

where partitioned matrices $\mathbf{A} = [\mathbf{A}_1, \ldots, \mathbf{A}_R]$, $\mathbf{B} = [\mathbf{B}_1, \ldots, \mathbf{B}_R]$.

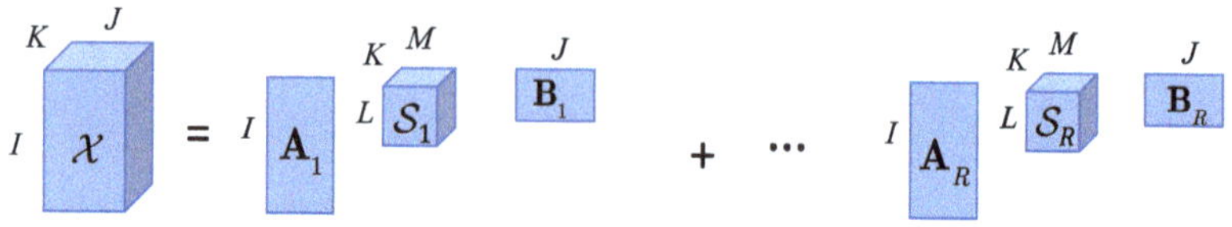

Fig. 2.5 $(L, M, \cdot)$-decomposition for a third-order tensor

A visual representation of this decomposition for the third-order case is shown in Fig. 2.5.

2.4.1 The Computation of BTD

(L, M, N)-decomposition of a third-order tensor $\mathcal{X}$ is a quadrilinear in its factors $\mathbf{U}$, $\mathbf{V}$, $\mathbf{W}$, and $\mathcal{S}$, as follows:

$$\min_{\mathbf{U},\mathbf{V},\mathbf{W},\mathcal{S}} \left\| \mathcal{X} - \sum_{r=1}^{R} \mathcal{S}_r \times_1 \mathbf{U}_r \times_2 \mathbf{V}_r \times_3 \mathbf{W}_r \right\|_{\mathrm{F}}^2. \tag{2.51}$$

Like CP decomposition under ALS framework, Lathauwer et al. [19] iteratively update each variable with other variables fixed. In this case, the (2.51) can be split into the following four subproblems:

$$\begin{aligned}
&\min_{\mathbf{U}} \|\mathbf{X}_{(1)} - \mathbf{U}\,\mathrm{blockdiag}((\mathbf{S}_1)_{(1)}, \ldots, (\mathbf{S}_R)_{(1)})(\mathbf{V} \odot_{\mathrm{b}} \mathbf{W})^{\mathrm{T}}\|_{\mathrm{F}}^2, \\
&\min_{\mathbf{V}} \|\mathbf{X}_{(2)} - \mathbf{V}\,\mathrm{blockdiag}((\mathbf{S}_1)_{(2)}, \ldots, (\mathbf{S}_R)_{(2)})(\mathbf{W} \odot_{\mathrm{b}} \mathbf{U})^{\mathrm{T}}\|_{\mathrm{F}}^2, \\
&\min_{\mathbf{W}} \|\mathbf{X}_{(3)} - \mathbf{W}\,\mathrm{blockdiag}((\mathbf{S}_1)_{(3)}, \ldots, (\mathbf{S}_R)_{(3)})(\mathbf{U} \odot_{\mathrm{b}} \mathbf{V})^{\mathrm{T}}\|_{\mathrm{F}}^2, \\
&\min_{\mathcal{S}} \|\mathrm{vec}(\mathcal{X}) - (\mathbf{U} \odot_{\mathrm{b}} \mathbf{V} \odot_{\mathrm{b}} \mathbf{W})[\mathrm{vec}(\mathcal{S}_1); \ldots; \mathrm{vec}(\mathcal{S}_R)]\|_{\mathrm{F}}^2,
\end{aligned} \tag{2.52}$$

where partitioned matrices $\mathbf{U} = [\mathbf{U}_1, \ldots, \mathbf{U}_R]$, $\mathbf{V} = [\mathbf{V}_1, \ldots, \mathbf{V}_R]$, and $\mathbf{W} = [\mathbf{W}_1, \ldots, \mathbf{W}_R]$. The details are concluded in Algorithm 5. It is noted that it is often advantageous to alternate between a few updates of $\mathbf{U}$ and $\mathcal{S}$, then alternate between a few updates of $\mathbf{V}$ and $\mathcal{S}$, and so on.

As a special case, the computation of $(L_r, L_r, 1)$-decomposition can be formulated as

$$\min_{\mathbf{A},\mathbf{B},\mathbf{C}} \left\| \mathcal{X} - \sum_{r=1}^{R} (\mathbf{A}_r \mathbf{B}_r^{\mathrm{T}}) \circ \mathbf{c}_r \right\|_{\mathrm{F}}^2, \tag{2.53}$$

Algorithm 5: ALS algorithm for decomposition in rank-(L, M, N) terms

Input: $\mathcal{X} \in \mathbb{R}^{I\times J\times K}$ and CP-rank R
Output: $\mathbf{U}, \mathbf{V}, \mathbf{W}, \mathcal{S}$
Initialize $\mathbf{U}, \mathbf{V}, \mathbf{W}, \mathcal{S}$
Iterate until convergence:
1. Update $\mathbf{U}$:
   $\tilde{\mathbf{U}} = \mathbf{X}_{(1)}((\mathbf{V} \odot \mathbf{W})^{\dagger})^{\mathrm{T}}\text{blockdiag}((\mathbf{S}_1)^{\dagger}_{(1)}, \dots, (\mathbf{S}_R)^{\dagger}_{(1)})$
   For $r = 1, \dots, R$:
   QR-factorization: $[\mathbf{Q}, \mathbf{R}] = \text{qr}(\tilde{\mathbf{U}}_r)$
   $\mathbf{U}_r \leftarrow \mathbf{Q}$
   End for
2. Update $\mathbf{V}$:
   $\tilde{\mathbf{V}} = \mathbf{X}_{(2)}((\mathbf{W} \odot \mathbf{U})^{\dagger})^{\mathrm{T}}\text{blockdiag}((\mathbf{S}_1)^{\dagger}_{(2)}, \dots, (\mathbf{S}_R)^{\dagger}_{(2)})$
   For $r = 1, \dots, R$:
   QR-factorization: $[\mathbf{Q}, \mathbf{R}] = \text{qr}(\tilde{\mathbf{V}}_r)$
   $\mathbf{V}_r \leftarrow \mathbf{Q}$
   End for
3. Update $\mathbf{W}$:
   $\tilde{\mathbf{W}} = \mathbf{X}_{(3)}((\mathbf{U} \odot \mathbf{V})^{\dagger})^{\mathrm{T}}\text{blockdiag}((\mathbf{S}_1)^{\dagger}_{(3)}, \dots, (\mathbf{S}_R)^{\dagger}_{(3)})$
   For $r = 1, \dots, R$:
   QR-factorization: $[\mathbf{Q}, \mathbf{R}] = \text{qr}(\tilde{\mathbf{W}}_r)$
   $\mathbf{W}_r \leftarrow \mathbf{Q}$
   End for
4. Update $\mathcal{S}$:
   $[\text{vec}(\mathcal{S}_1); \dots; \text{vec}(\mathcal{S}_R)] = (\mathbf{U} \odot_b \mathbf{V} \odot_b \mathbf{W})^{\dagger}\text{vec}(\mathcal{X})$

where partitioned matrices $\mathbf{A} = [\mathbf{A}_1, \dots, \mathbf{A}_R]$, $\mathbf{B} = [\mathbf{B}_1, \dots, \mathbf{B}_R]$, and $\mathbf{C} = [\mathbf{c}_1, \dots, \mathbf{c}_R]$. Similarly, this problem can be divided into the following three subproblems:

$$\min_{\mathbf{A}} \|\mathbf{X}_{(1)} - \mathbf{A}(\mathbf{B} \odot_b \mathbf{C})^{\mathrm{T}}\|_{\mathrm{F}}^2,$$
$$\min_{\mathbf{B}} \|\mathbf{X}_{(2)} - \mathbf{B}(\mathbf{C} \odot_b \mathbf{A})^{\mathrm{T}}\|_{\mathrm{F}}^2,$$
$$\min_{\mathbf{C}} \|\mathbf{X}_{(3)} - \mathbf{C}[(\mathbf{A}_1 \odot \mathbf{B}_1)\mathbf{1}_{L_1}, \dots, (\mathbf{A}_R \odot \mathbf{B}_R)\mathbf{1}_{L_R}]^{\mathrm{T}}\|_{\mathrm{F}}^2. \quad (2.54)$$

The detailed updating scheme can be concluded in Algorithm 6. It is noticed that the normalization in step 2 prevents the submatrices of $\mathbf{B}$ from becoming ill-conditioned. Besides, $(\mathbf{B} \odot_b \mathbf{C})$, $(\mathbf{C} \odot_b \mathbf{A})$, and $[(\mathbf{A}_1 \odot \mathbf{B}_1)\mathbf{1}_{L_1} \cdots (\mathbf{A}_R \odot \mathbf{B}_R)\mathbf{1}_{L_R}]$ have to be of full column rank.

Algorithm 6: ALS algorithm for decomposition in rank-$(L_r, L_r, 1)$ terms

Input: $\mathcal{X} \in \mathbb{R}^{I \times J \times K}$ and CP-rank R
Output: $\mathbf{A}, \mathbf{B}, \mathbf{C}$
initialize $\mathbf{B}, \mathbf{C}$
Iterate until convergence:
1. Update $\mathbf{A}$:
   $\mathbf{A} \leftarrow \mathbf{X}_{(1)}((\mathbf{B} \odot \mathbf{C})^{\mathrm{T}})^{\dagger}$
2. Update $\mathbf{B}$:
   $\tilde{\mathbf{B}} = \mathbf{X}_{(2)}((\mathbf{C} \odot \mathbf{A})^{\mathrm{T}})^{\dagger}$
   For $r = 1, \ldots, R$:
   QR-factorization: $[\mathbf{Q}, \mathbf{R}] = \mathrm{qr}(\tilde{\mathbf{B}}_r)$
   $\mathbf{B}_r \leftarrow \mathbf{Q}$
   End for
3. Update $\mathbf{C}$:
   $\tilde{\mathbf{C}} = \mathbf{X}_{(3)}([(\mathbf{A}_1 \odot_c \mathbf{B}_1)\mathbf{1}_{L_1}, \ldots, (\mathbf{A}_R \odot_c \mathbf{B}_R)\mathbf{1}_{L_R}]^{\mathrm{T}})^{\dagger}$
   For $r = 1, \ldots, R$:
   $\mathbf{c}_r \leftarrow \tilde{\mathbf{c}}_r / \|\tilde{\mathbf{c}}_r\|$
   End for

2.4.2 *Uniqueness*

For a third-order tensor $\mathcal{X} \in \mathbb{R}^{I \times J \times K}$, the (L, M, N)-decomposition is

$$\mathcal{X} = \sum_{r=1}^{R} \mathcal{S}_r \times_1 \mathbf{U}_r \times_2 \mathbf{V}_r \times_3 \mathbf{W}_r, \tag{2.55}$$

which has essential uniqueness if

1. $L = M$ and $I \geq LR$ and $J \geq MR$ and $N \geq 3$ and $\mathbf{W}_r$ is full column rank, $1 \leq r \leq R$;
2. $I \geq LR$ and $N > L + M - 2$ and $\min(\lfloor \frac{J}{M} \rfloor, R) + \min(\lfloor \frac{K}{N} \rfloor, R) \geq R + 2$
   or
   $J \geq MR$ and $N > L + M - 2$ and $\min(\lfloor \frac{I}{L} \rfloor, R) + \min(\lfloor \frac{K}{N} \rfloor, R) \geq R + 2$;
3. $N > L + M - 2$ and $\min(\lfloor \frac{I}{L} \rfloor, R) + \min(\lfloor \frac{K}{N} \rfloor, R) + \min(\lfloor \frac{J}{M} \rfloor, R) \geq R + 2$.

For decompositions in general $(L, L, 1)$-decomposition, we have essential uniqueness if

1. $\min(I, J) \geq LR$ and $\mathbf{C}$ does not have proportional columns;
2. $K \geq R$ and $\min(\lfloor \frac{J}{L} \rfloor, R) + \min(\lfloor \frac{J}{L} \rfloor, R) \geq R + 2$;
3. $I \geq LR$ and $\min(\lfloor \frac{J}{L} \rfloor, R) + \min(K, R) \geq R + 2$
   or
   $J \geq LR$ and $\min(\lfloor \frac{I}{L} \rfloor, R) + \min(K, R) \geq R + 2$;
4. $\lfloor \frac{IJ}{L^2} \rfloor$ and $\min(\lfloor \frac{I}{L} \rfloor, R) + \min(\lfloor \frac{J}{L} \rfloor, R) + \min(K, R) \geq R + 2$.

2.5 Tensor Singular Value Decomposition

Kilmer et al. [43, 44] firstly proposed tensor singular value decomposition (t-SVD) to build approximation to a given tensor. Different from CP decomposition and Tucker decomposition, a representation of a tensor in t-SVD framework is the tensor product of three tensors. Before introducing t-SVD, we will give several important definitions, as follows:

Definition 2.8 (t-Product) For third-order tensors $\mathcal{A} \in \mathbb{R}^{I_1 \times I_2 \times I_3}$ and $\mathcal{B} \in \mathbb{R}^{I_2 \times J \times I_3}$, the t-product $\mathcal{A} * \mathcal{B}$ is a tensor sized of $I_1 \times J \times I_3$:

$$\mathcal{A} * \mathcal{B} = \text{fold}(\text{circ}(\mathcal{A})\text{MatVec}(\mathcal{B})), \tag{2.56}$$

where

$$\text{circ}(\mathcal{A}) = \begin{bmatrix} \mathcal{A}^{(1)} & \mathcal{A}^{(I_3)} & \cdots & \mathcal{A}^{(2)} \\ \mathcal{A}^{(2)} & \mathcal{A}^{(1)} & \cdots & \mathcal{A}^{(3)} \\ \vdots & \vdots & \vdots & \vdots \\ \mathcal{A}^{(I_3)} & \mathcal{A}^{(I_3-1)} & \cdots & \mathcal{A}^{(1)} \end{bmatrix},$$

and

$$\text{MatVec}(\mathcal{B}) = \begin{bmatrix} \mathcal{B}^{(1)} \\ \mathcal{B}^{(2)} \\ \vdots \\ \mathcal{B}^{(I_3)} \end{bmatrix},$$

where $\mathcal{A}^{(i_3)}$ and $\mathcal{B}^{(i_3)}$, $i_3 = 1, \ldots, I_3$ are the frontal slices of $\mathcal{A}$ and $\mathcal{B}$, respectively.

It is well-known that block circulant matrices can be block diagonalized by the Fourier transform, which derives the following equation:

$$\overline{\mathbf{A}} = (\mathbf{F} \otimes \mathbf{I}_1)\text{circ}(\mathcal{A})(\mathbf{F}^* \otimes \mathbf{I}_2) = \begin{bmatrix} \hat{\mathcal{A}}^{(1)} & 0 & \cdots & 0 \\ 0 & \hat{\mathcal{A}}^{(2)} & 0 & \cdots \\ \vdots & \vdots & \vdots & \vdots \\ 0 & \cdots & 0 & \hat{\mathcal{A}}^{(I_3)} \end{bmatrix}, \tag{2.57}$$

where $\mathbf{F} \in \mathbb{R}^{I_3 \times I_3}$ is the discrete Fourier transform (DFT) matrix and $\mathbf{F}^*$ is the conjugate transpose of $\mathbf{F}$, $\mathbf{I}_1 \in\in \mathbb{R}^{I_1 \times I_1}$ and $\mathbf{I}_2 \in \mathbb{R}^{I_2 \times I_2}$, $\hat{\mathcal{A}}$ is the fast Fourier transform (FFT) of $\mathcal{A}$ along the third mode.

$$\text{circ}(\mathcal{A})\text{MatVec}(\mathcal{B}) = (\mathbf{F}^* \otimes \mathbf{I}_1)((\mathbf{F} \otimes \mathbf{I}_1)\text{circ}(\mathcal{A})(\mathbf{F}^* \otimes \mathbf{I}_2))(\mathbf{F} \otimes \mathbf{I}_2)\text{MatVec}(\mathcal{B}) \tag{2.58}$$

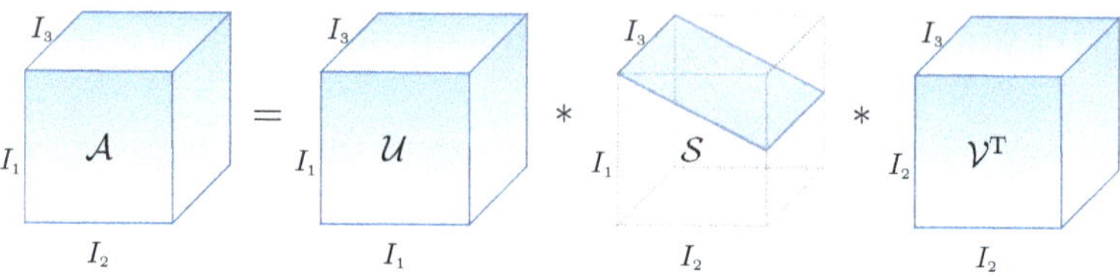

Fig. 2.6 A demonstration of t-SVD for a third-order tensor

where $(\mathbf{F} \otimes \mathbf{I}_2)\mathrm{MatVec}(\mathcal{B})$ can be computed by applying FFT along mode-3 of $\mathcal{B}$ and $\hat{\mathcal{B}}$ is denoted as its result of FFT. In this case, the t-product can be computed by multiplying each frontal slice of $\hat{\mathcal{A}}$ with each frontal slice of $\hat{\mathcal{B}}$; then take an inverse FFT along the tubes to obtain the result.

Definition 2.9 (t-SVD Decomposition) Let $\mathcal{A} \in \mathbb{R}^{I_1 \times I_2 \times I_3}$; then $\mathcal{A}$ can be factorized into

$$\mathcal{A} = \mathcal{U} * \mathcal{S} * \mathcal{V}^{\mathrm{T}},$$

where $\mathcal{U} \in \mathbb{R}^{I_1 \times I_1 \times I_3}$ and $\mathcal{V} \in \mathbb{R}^{I_2 \times I_2 \times I_3}$ are orthogonal tensors, i.e., $\mathcal{U}^{\mathrm{T}} * \mathcal{U} = \mathcal{V}^{\mathrm{T}} * \mathcal{V} = \mathcal{I}$. In Fourier domain, $\hat{\mathcal{A}}^{(i_3)} = \hat{\mathcal{U}}^{(i_3)} \hat{\mathcal{S}}^{(i_3)} \hat{\mathcal{V}}^{(i_3)}$, $i_3 = 1, \ldots, I_3$.

A graphical illustration can be seen in Fig. 2.6, and the computation of t-SVD can be summarized in Algorithm 7. Recently, Lu et al. [52] proposed a new way to efficiently compute t-SVD shown in Algorithm 8.

Algorithm 7: t-SVD for 3-way data

Input: $\mathcal{A} \in \mathbb{R}^{I_1 \times I_2 \times I_3}$
$\mathcal{D} \leftarrow$ fft($\mathcal{A}$,[],3),
for $i_3 = 1$ to I_3 **do**
  [**U**, **S**, **V**] = SVD ($\mathcal{D}(:, :, i_3)$),
  $\hat{\mathcal{U}}(:, :, i_3)$=**U**, $\hat{\mathcal{S}}(:, :, i_3)$ = **S**, $\hat{\mathcal{V}}(:, :, i_3)$ = **V**,
end for
$\mathcal{U} \leftarrow$ ifft($\hat{\mathcal{U}}$, [], 3), $\mathcal{S} \leftarrow$ ifft($\hat{\mathcal{S}}$, [], 3), $\mathcal{V} \leftarrow$ ifft($\hat{\mathcal{V}}$, [], 3),
Output: $\mathcal{U}, \mathcal{S}, \mathcal{V}$

Definition 2.10 (Tensor Tubal Rank) The tensor tubal rank, denoted by $\mathrm{rank}_{\mathrm{t}}(\mathcal{A})$, is defined as the number of nonzero singular tubes of $\mathcal{S}$, where $\mathcal{S}$ is from t-SVD $\mathcal{A} = \mathcal{U} * \mathcal{S} * \mathcal{V}^{\mathrm{T}}$, i.e.,

$$\mathrm{rank}_{\mathrm{t}}(\mathcal{A}) = \#\{i : \mathcal{S}(i, i, :) \neq 0\}. \tag{2.59}$$

where $\mathcal{A} \in \mathbb{R}^{I_1 \times I_2 \times I_3}$.

Algorithm 8: New: t-SVD for three-way tensor

Input: $\mathcal{A} \in \mathbb{R}^{I_1 \times I_2 \times I_3}$.
1 $\hat{\mathcal{A}} = \text{fft}(\mathcal{A}, [], 3)$,
2 **for** $i_3 = 1, \ldots, \lceil \frac{I_3+1}{2} \rceil$ **do**
3 | $[\hat{\mathcal{U}}^{(i_3)}, \hat{\mathcal{S}}^{(i_3)}, \hat{\mathcal{V}}^{(i_3)}] = \text{SVD}(\hat{\mathcal{A}}^{(i_3)})$;
4 **end**
5 **for** $i_3 = \lceil \frac{I_3+1}{2} \rceil + 1, \ldots, I_3$ **do**
6 | $\hat{\mathcal{U}}^{(i_3)} = \text{conj}(\hat{\mathcal{U}}^{(I_3-i_3+2)})$;
7 | $\hat{\mathcal{S}}^{(i_3)} = \hat{\mathcal{S}}^{(I_3-i_3+2)}$;
8 | $\hat{\mathcal{V}}^{(i_3)} = \text{conj}(\hat{\mathcal{V}}^{(I_3-i_3+2)})$;
9 **end**
10 $\mathcal{U} = \text{ifft}(\hat{\mathcal{U}}, [], 3)$, $\mathcal{S} = \text{ifft}(\hat{\mathcal{S}}, [], 3)$, $\mathcal{V} = \text{ifft}(\hat{\mathcal{V}}, [], 3)$.
Output: $\mathcal{U}, \mathcal{S}, \mathcal{V}$.

2.6 Tensor Networks

Owing to the increasingly affordable recording devices and large-scale data volumes, multidimensional data are becoming ubiquitous across the science and engineering disciplines. Such massive datasets may have billions of entries and be of much high order, which causes the curse of dimensionality with the order increased. This has spurred a renewed interest in the development of tensor-based algorithms that are suitable for very high-order datasets. Tensor networks (TNs) decompose higher-order tensors into sparsely interconnected low-order core tensors. These TNs provide a natural sparse and distributed representation for big data.

In this section, we mainly introduce two important networks. One is hierarchical Tucker (HT) and the other is tensor train (TT) network.

2.6.1 Hierarchical Tucker Decomposition

Hierarchical Tucker (HT) decomposition has been firstly introduced in [29] and developed by [6, 27, 46, 53, 58]. It decomposes a higher-order (order > 3) tensor into several lower-order (third-order or less) tensors by splitting the modes of a tensor in a hierarchical way, leading to a binary tree containing a subset of modes at each branch.

In order to define the HT decomposition format, we have to introduce several important definitions.

Definition 2.11 (Dimension Tree) A dimension tree $\mathbb{T}$ of order D is a binary tree with root $\mathbb{D}$, $\mathbb{D} := \{1, \ldots, D\}$, such that each node $\mathbb{C}_q \subset \mathbb{T}$, $q = 1, \ldots, Q$ has the following properties:

1. A node with only one entry is called a leaf, i.e., $\mathbb{C}_p = \{d\}$. The set of all leaves is denoted by

$$\mathbb{F}(\mathbb{T}) = \{\mathbb{C}_p \subset \mathbb{T} \mid \mathbb{C}_p \text{ is a leaf node of } \mathbb{T},\ p = 1, \ldots, P\},$$

where P is the number of leaves.

2. An interior node is the union of two disjoint successors. The set of all interior nodes can be denoted as

$$\mathbb{E}(\mathbb{T}) = \mathbb{T} \backslash \mathbb{F}(\mathbb{T}).$$

The number of interior nodes is $Q - P$.

Definition 2.12 (Matricization for Dimension Tree) For a tensor $\mathcal{X} \in \mathbb{R}^{I_1 \times \cdots \times I_D}$, a node of dimension indices $\mathbb{C}_q \subset \mathbb{D}$, and the complement $\bar{\mathbb{C}}_q := \mathbb{D} \backslash \mathbb{C}_q$, the matricization of given tensor $\mathcal{A} \in \mathbb{R}^{I_1 \times \cdots \times I_D}$ for that dimension tree is defined as

$$\mathbf{X}^{(q)} \in \mathbb{R}^{I_{\mathbb{C}_q} \times I_{\bar{\mathbb{C}}_q}},$$

where

$$I_{\mathbb{C}_q} := \prod_{c \in \mathbb{C}_q} I_c, \quad I_{\bar{\mathbb{C}}_q} := \prod_{\bar{c} \in \bar{\mathbb{C}}_q} I_{\bar{c}}.$$

Definition 2.13 (Hierarchical Rank) Letting $\mathcal{X} \in \mathbb{R}^{I_1 \times \cdots \times I_D}$, the tensor tree rank $\mathbf{k}$ is defined as

$$\mathbf{k} = \{k_q | k_q = \text{rank}(\mathbf{X}^{(q)}), \quad \forall\ \mathbb{C}_q \subset \mathbb{T}\},$$

where "rank" denotes the standard matrix rank. With the hierarchical rank $\mathbf{k}$, the set of all tensors with hierarchical rank no larger than $\mathbf{k}$ is defined as

$$\mathbb{X}_{\mathbf{k}} = \{\mathcal{X} \in \mathbb{R}^{I_1 \times \cdots \times I_D} | \text{rank}(\mathbf{X}^{(q)}) \leq k_q, \quad \forall\ \mathbb{C}_q \subset \mathbb{T}\}.$$

Lemma 2.1 (Nestedness Property) *Letting $\mathcal{X} \in \mathbb{X}_{\mathbf{k}}$, for each node $\mathbb{C}_q$ and its complement $\bar{\mathbb{C}}_q$, we can get a subspace*

$$\mathbb{U}_q := \text{span}\{\mathbf{x} \in \mathbb{R}^{I_{\mathbb{C}_q}} | \mathbf{x} \textit{ is the left singular vector of } \mathbf{X}^{(q)}\},$$

where $\mathbf{X}^{(q)} \in \mathbb{R}^{I_{\mathbb{C}_q} \times I_{\bar{\mathbb{C}}_q}}$. For each $\mathbb{C}_q \subset \mathbb{E}(\mathbb{T})$ with two successors $\mathbb{C}_{q1}$, $\mathbb{C}_{q2}$, the space $\mathbb{U}_q$ naturally decouples into

$$\mathbb{U}_q = \mathbb{U}_{q1} \otimes \mathbb{U}_{q2},$$

where $\otimes$ denotes the Kronecker product.

Based on these definitions, the hierarchical Tucker decomposition is defined as follows.

Definition 2.14 (Hierarchical Tucker Decomposition) Letting $\mathcal{X} \in \mathbb{X}_{\mathbf{k}}$, for every node $\mathbb{C}_q \subset \mathbb{T}$, we can get a representation of $\mathbf{X}^{(q)}$ as

$$\mathbf{X}^{(q)} = \mathbf{U}_{\mathbb{C}_q} \mathbf{V}_{\mathbb{C}_q}^{\mathrm{T}}, \quad \mathbf{U}_{\mathbb{C}_q} \in \mathbb{R}^{I_{\mathbb{C}_q} \times k_{\mathbb{C}_q}},$$

where $k_{\mathbb{C}_q}$ is the rank of $\mathbf{X}^{(q)}$. For $\mathbb{C}_q = \{\mathbb{C}_{q1}, \mathbb{C}_{q2}\}$, the column vectors $\mathbf{U}_{\mathbb{C}_q}(:, l)$ of $\mathbf{U}_{\mathbb{C}_q}$ fulfill the nestedness property that every vector $\mathbf{U}_{\mathbb{C}_q}(:, l)$ with $1 \leq l \leq k_{\mathbb{C}_q}$ holds

$$\mathbf{U}_{\mathbb{C}_q}(:, l) = \sum_{l_1=1}^{k_{\mathbb{C}_{q1}}} \sum_{l_2=1}^{k_{\mathbb{C}_{q2}}} \mathcal{B}_{\mathbb{C}_q}(l, l_1, l_2) \mathbf{U}_{\mathbb{C}_{q1}}(:, l_1) \otimes \mathbf{U}_{\mathbb{C}_{q2}}(:, l_2), \tag{2.60}$$

where $\mathcal{B}_{\mathbb{C}_q}$ is the coefficient of the linear combination.

For example, for a fourth-order tensor $\mathcal{X} \in \mathbb{R}^{I_1 \times I_2 \times I_3 \times I_4}$ in hierarchical Tucker format, the root node is $\{1, 2, 3, 4\}$. We can obtain the representation of root node $\mathbf{u}_{\{1,2,3,4\}} \in \mathbb{R}^{I_1 I_2 I_3 I_4 \times 1}$. Since the root node $\{1, 2, 3, 4\}$ is an interior node, it has two disjoint successors, such as $\{1, 2\}, \{3, 4\}$ or $\{1, 3\}, \{2, 4\}$ or $\{1, 4\}, \{2, 3\}$. The best choice of two disjoint successors can be referred in [28]. In this case, we choose the group $\{1, 2\}, \{3, 4\}$ for explanation. According to Eq. (2.60), the root node $\mathbf{u}_{\{1,2,3,4\}}$ can be decomposed into

$$\mathbf{u}_{\{1,2,3,4\}} = \sum_{l_1=1}^{k_{\{1,2\}}} \sum_{l_2=1}^{k_{\{3,4\}}} \mathcal{B}_{\{1,2,3,4\}}(l_1, l_2) \mathbf{U}_{\{1,2\}}(:, l_1) \otimes \mathbf{U}_{\{3,4\}}(:, l_2). \tag{2.61}$$

Similarly, the column vectors $\mathbf{U}_{\{1,2\}}(:, l)$, $1 \leq l \leq k_{\{1,2\}}$ can be decomposed into

$$\mathbf{U}_{\{1,2\}}(:, l) = \sum_{l_1=1}^{k_{\{1\}}} \sum_{l_2=1}^{k_{\{2\}}} \mathcal{B}_{\{1,2\}}(l, l_1, l_2) \mathbf{U}_{\{1\}}(:, l_1) \otimes \mathbf{U}_{\{2\}}(:, l_2), \tag{2.62}$$

where $\mathbf{U}_{\{d\}} \in \mathbb{R}^{I_d \times k_{\{d\}}}$, $d = 1, \ldots, 4$. In this way, the leaf node representations of $\mathcal{X}$ are $\mathbf{U}_{\{1\}}$,$\mathbf{U}_{\{2\}}$,$\mathbf{U}_{\{3\}}$,$\mathbf{U}_{\{4\}}$ and the representations of interior nodes are $\mathcal{B}_{\{1,2\}}$, $\mathcal{B}_{\{3,4\}}$, and $\mathcal{B}_{\{1,2,3,4\}}$. Figure 2.7 shows the graphical illustration.

As can be seen in Fig. 2.7, we need to store the transfer tensor $\mathcal{B}$ for all interior nodes and the matrix $\mathbf{U}$ for all leaves. Therefore, for all nodes $q = 1, \ldots, Q$, assuming all $k_{\mathbb{C}_q} = k$, the storage complexity is $\sum_{p=1}^{P} Ik + \sum_{q=1}^{Q-P} k^3$.

2.6.1.1 Computation of Hierarchical Tucker Decomposition

Letting $\mathcal{X} \in \mathbb{R}^{I_1 \times I_2 \times \cdots \times I_N}$, the goal of hierarchical Tucker decomposition is to find the optimal $\mathcal{B}$ and $\mathbf{U}$ in hierarchical Tucker form to approximate $\mathcal{X}$. Grasedyck [27]

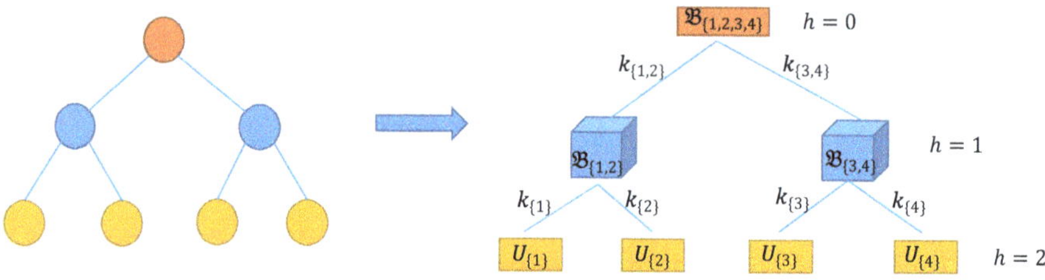

Fig. 2.7 An illustration of hierarchical Tucker decomposition for a fourth-order tensor

provided two ways to update $\mathcal{B}$ and $\mathbf{U}$. One is root-to-leaves updating way and the other is leaves-to-root way.

The first one is the root-to-leaves updating way, as summarized in Algorithm 9, where operation diag($\mathbf{S}$) means the diagonal elements of matrix $\mathbf{S}$ and Length($\cdot$) means the size of specific vector. The computational complexity of this scheme for

Algorithm 9: Root-to-leaves updating for hierarchical Tucker decomposition

Input: $\mathcal{X} \in \mathbb{R}^{I_1 \times \cdots \times I_N}$
parfor $p = 1, \ldots, P$
  $[\mathbf{U}_p, \mathbf{S}_p, \mathbf{V}_p]$=SVD($\mathbf{X}^{(p)}$)
  $k_p = \text{rank}(\mathbf{S}_p)$
end for
for $h = (H-1), \ldots, 0$ **do**
  parfor $q_h = 1, \ldots, Q_h$ **do**
    if (the q_h-th node is an interior one)
      $[\mathbf{U}_{q_h}, \mathbf{S}_{q_h}, \mathbf{V}_{q_h}]$=SVD($\mathbf{X}^{(q_h)}$)
      $k_{q_h} = \text{rank}(\mathbf{S}_{q_h})$
      $\mathcal{B}_{q_h}$=reshape($\langle \mathbf{U}_{q_h}, \mathbf{U}_{q_{h,1}} \otimes \mathbf{U}_{q_{h,2}} \rangle, k_{q_h}, k_{q_{h,1}}, k_{q_{h,2}}$)
    end if
  end for
end for
$\hat{\mathcal{X}}$ can be constructed from $\mathcal{B}_{q_h}$ and $\mathbf{U}_p$
Output: $\hat{\mathcal{X}}$ in tensor tree format

a tensor $\mathcal{X} \in \mathbb{R}^{I_1 \times I_2 \times \cdots \times I_N}$ and dimension tree $\mathbb{T}$ of depth H is $\text{O}((\prod_{n=1}^{N} I_n)^{3/2})$.

Algorithm 10 summarizes the leaves-to-root updating way. In this case, the computational complexity is $\text{O}(I^N)$.

2.6.1.2 Generalization of Hierarchical Tucker Decomposition

As a generalization of the hierarchical Tucker decomposition, the tree tensor network state (TTNS), where all nodes are of third order or higher order, are proposed in quantum chemistry [55, 60]. It is a variational method to efficiently

Algorithm 10: Leaves-to-root updating for hierarchical Tucker decomposition

Input: $\mathcal{X} \in \mathbb{R}^{I_1 \times \cdots \times I_N}$
parfor $p = 1, \ldots, P$
  $[\mathbf{U}_p, \mathbf{S}_p, \mathbf{V}_p]$=SVD($\mathbf{X}^{(p)}$)
  $k_p = \text{rank}(\mathbf{S}_p)$
end for
  $\mathcal{C}_{H-1} = \mathcal{X} \times_1 \mathbf{U}_1 \times_2 \mathbf{U}_2 \cdots \times_P \mathbf{U}_P$
for $h = (H-1), \ldots, 1$ **do**
  parfor q_h= $1, \ldots, Q_h$ **do**
    if (the q_h-th node is an interior one)
      $[\mathbf{U}_{q_h}, \mathbf{S}_{q_h}, \mathbf{V}_{q_h}]$=SVD($\mathbf{X}^{(q_h)}$)
      $k_{q_h} = \text{rank}(\mathbf{S}_{q_h})$
      $\mathcal{B}_{q_h}$=reshape($\mathbf{U}_{q_h}, k_{q_h}, k_{q_{h,1}}, k_{q_{h,2}}$)
    end if
  end for
    $\mathcal{C}_{h-1} = \mathcal{C}_h \times_1 \mathbf{U}_1 \times_2 \mathbf{U}_2 \cdots \times_{(Q_h-P_h)} \mathbf{U}_{(Q_h-P_h)}$
end for
$\hat{\mathcal{X}}$ can be constructed from $\mathcal{B}_{q_h}$ and $\mathbf{U}_p$
Output: $\hat{\mathcal{X}}$ in tensor tree format

approximate complete active space (CAS) configuration interaction (CI) wave functions in a tensor product form. The graphical illustration of TNNS is shown in Figs. 2.8 and 2.9.

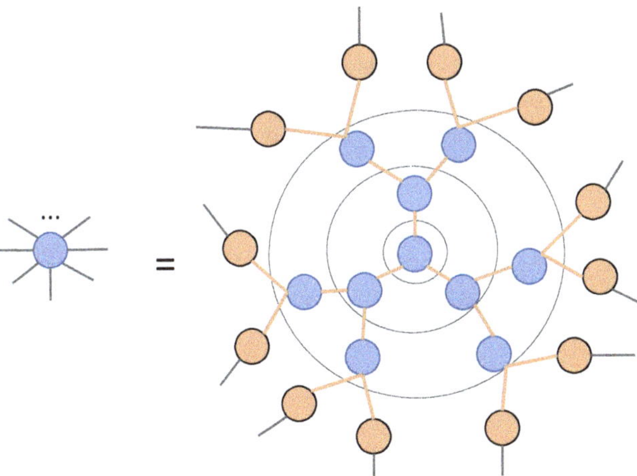

Fig. 2.8 The tree tensor network state (TTNS) with third-order cores for the representation of 12th-order tensors

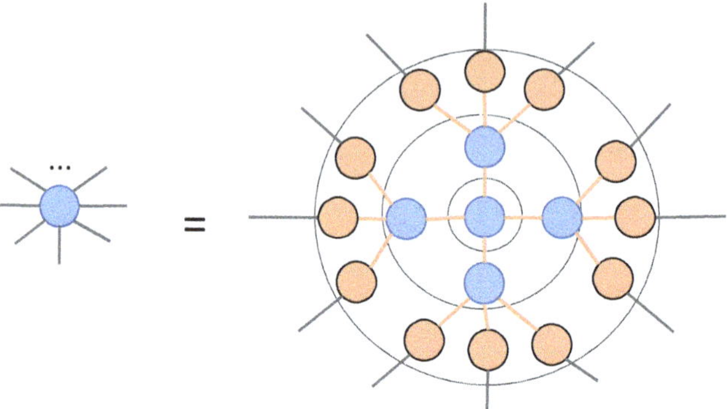

Fig. 2.9 The tree tensor network state (TTNS) with fourth-order cores for the representation of 12th-order tensors

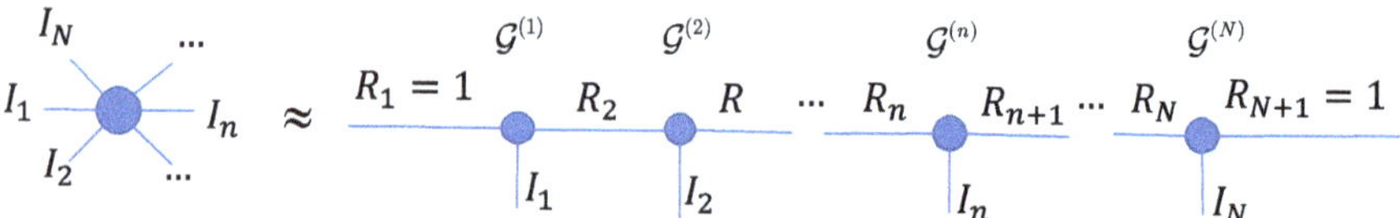

Fig. 2.10 Tensor train decomposition

2.6.2 *Tensor Train Decomposition*

As a special case of the HT decomposition, tensor train (TT) decomposition has been firstly proposed in [57] and developed by Zhang et al. [80], and Bigoni et al. [8]. TT format is known as the matrix product state (MPS) representation in the quantum physics community [62, 75]. It decomposes a higher-order tensor into several third-order tensors, which is defined as

Definition 2.15 For an Nth-order tensor $\mathcal{X} \in \mathbb{R}^{I_1 \times \cdots \times I_N}$, the TT decomposition is

$$\mathcal{X} = \sum_{r_1=1}^{R_1} \cdots \sum_{r_{N+1}=1}^{R_{N+1}} \mathcal{G}^{(1)}(r_1, :, r_2) \circ \mathcal{G}^{(2)}(r_2, :, r_3) \circ \cdots \circ \mathcal{G}^{(N)}(r_N, :, r_{N+1})$$

or element wisely as

$$\mathcal{X}(i_1, i_2 \ldots, i_N) = \mathcal{G}^{(1)}(:, i_1, :)\mathcal{G}^{(2)}(:, i_2, :) \cdots \mathcal{G}^{(N)}(:, i_N, :) \tag{2.63}$$

where $\mathcal{G}_n \in \mathbb{R}^{R_n \times I_n \times R_{n+1}}$, $n = 1, \ldots, N$ $(R_1 = R_{N+1} = 1)$ are the core factors. Figure 2.10 presents the concept of TT decomposition of an Nth-order tensor.

Using the canonical matricization of tensor in Eq. (1.37), we can obtain the definition of TT rank as follows.

Definition 2.16 (TT Rank) Letting $\mathcal{X} \in \mathbb{R}^{I_1 \times I_2 \times \cdots \times I_N}$, the TT rank of $\mathcal{X}$, denoted as $\mathrm{rank}_{\mathrm{TT}}(\mathcal{X})$, is the rank of $\mathcal{X}_{\langle n \rangle}$. For example,

$$\mathrm{rank}_{\mathrm{TT}}(\mathcal{X}) = [R_1, \ldots, R_{N+1}],$$

where $R_n = \mathrm{rank}(\mathcal{X}_{\langle n-1 \rangle})$ for $n = 2, \ldots, N$, $R_1 = R_{N+1} = 1$.

Assuming all $I_n = I$ and $R_n = R$, the storage complexity in the TT model is $\mathrm{O}((N-2)IR^2 + 2IR)$ where we only need to store the third-order core factors.

2.6.2.1 Computation of Tensor Train Decomposition

For an Nth-order tensor $\mathcal{X} \in \mathbb{R}^{I_1 \times I_2 \times \cdots \times I_N}$, the construction of TT decomposition is based on the sequential SVD of unfolding matrix $\mathbf{A}_{\langle n \rangle}$ form Eq. (1.37) with the given tensor train rank.

For example, considering the unfolding matrix $\mathbf{X}_{\langle 1 \rangle}$, if $\mathrm{rank}(\mathbf{X}_{\langle 1 \rangle}) = R_2$, we can obtain

$$\mathbf{X}_{\langle 1 \rangle} = \mathbf{U}\mathbf{V}^{\mathrm{T}} \tag{2.64}$$

with elements as

$$\mathbf{X}_{\langle 1 \rangle}(i_1; j) = \sum_{r_2=1}^{R_2} \mathbf{U}(i_1, r_2)\mathbf{V}(r_2, j) \tag{2.65}$$

where $j = \overline{i_2, \ldots, i_d}$. In this case, we let $\mathcal{G}_1 = \mathbf{U}$ with size of $\mathcal{G}_1 \in \mathbb{R}^{R_1 \times I_1 \times R_2}$, $R_1 = 1$. After that, matrix $\mathbf{V}$ can be reshaped as an (N-1)-th order tensor $\hat{\mathcal{X}} \in \mathbb{R}^{R_2 I_2 \times I_3 \times \cdots \times I_N}$. We can obtain

$$\hat{\mathbf{X}}_{\langle 1 \rangle} = \mathbf{U}\mathbf{V}^{\mathrm{T}} \tag{2.66}$$

with rank $(\hat{\mathbf{X}}_{\langle 1 \rangle}) = R_3$. The $\mathcal{G}_2 = \mathbf{U}$ with size of $\mathcal{G}_2 \in \mathbb{R}^{R_2 \times I_2 \times R_3}$. Following this sequential SVD, we can obtain the core factors in TT format. Algorithm 11 gives the details of sequential SVD for TT decomposition, where the operations Length($\cdot$) and Reshape($\cdot$) are all MATLAB functions.

In most cases, given the tensor train rank, we also can obtain core factors in TT format by ALS. For example, letting $\mathcal{X} \in \mathbb{R}^{I_1 \times I_2 \times \cdots \times I_N}$, the optimization problem is to find the best core factors in TT format to fit $\mathcal{X}$, as follows:

$$\min_{\mathcal{G}_1, \ldots, \mathcal{G}_N} \|\mathcal{X} - \sum_{r_1=1}^{R_1} \cdots \sum_{r_{N+1}=1}^{R_{N+1}} \mathcal{G}^{(1)}(r_1, :, r_2) \circ \mathcal{G}^{(2)}(r_2, :, r_3) \circ \cdots \circ \mathcal{G}^{(N)}(r_N, :, r_{N+1})\|_{\mathrm{F}}^2 \tag{2.67}$$

Algorithm 11: Sequential SVD for TT decomposition (SSVD)

Input: $\mathcal{X} \in \mathbb{R}^{I_1 \times \cdots \times I_N}$, TT ranks R_n, $n = 1, \ldots, N$
 1. $\mathbf{M}_0 = \mathbf{X}_{(1)}$, $[\mathbf{U}, \mathbf{S}, \mathbf{V}] = \text{SVD}(\mathbf{M}_0)$, $R_1 = R_{N+1}$, $\mathbf{G}^{(1)} = \mathbf{U}$, $\mathbf{M}_1 = \mathbf{S}\mathbf{V}$.
repeat
 2. $[\mathbf{U}, \mathbf{S}, \mathbf{V}] = \text{SVD}(\text{Reshape}(\mathbf{M}_{n-1}, R_n I_n, []))$.
 3. $R = \text{Length}(\text{diag}(\mathbf{S}))$.
 4. $R_{n+1} = R$, $\mathcal{G}^{(n)} = \text{Reshape}(\mathbf{U}, [R_n, I_n, R_{n+1}])$, $\mathbf{M}_n = \mathbf{S}\mathbf{V}$.
until $n = N - 1$
 5. $\mathbf{G}^{(N)} = \mathbf{M}_{N-1}$.
Output: $\mathbf{G}^{(1)}, \ldots, \mathcal{G}^{(n)}, \ldots, \mathbf{G}^{(N)}$

The detailed solutions are concluded in Algorithm 12, where the operation Permute$(\cdot)$ is a MATLAB function.

Algorithm 12: Basic ALS for TT decomposition of a tensor

Input: $\mathcal{X} \in \mathbb{R}^{I_1 \times \cdots \times I_N}$ and TT-rank threshold $R_{\max}$
initialize run SSVD for TT initialization
repeat
 for $n = 1, \ldots N$ **do**
  if $n == N$
   $\mathcal{B}_n = \mathcal{G}^{(1)}$
  else if
   $\mathcal{B}_n = \mathcal{G}^{(n+1)}$
  for $m = [n+2, \ldots N, 1, \ldots, n-1]$ **do**
   $\mathcal{B}_n = \langle \mathcal{B}_n, \mathcal{G}^{(m+1)} \rangle_1$
  end for
  $\mathcal{B}_n = \text{Reshape}(\mathcal{B}_n, [R_{n+1}, \prod_{m=1, m\neq n}^{N} I_m, R_n])$
  $\mathbf{B}_n = \text{Reshape}(\text{Permute}(\mathcal{B}_n, [3, 1, 2]), R_n R_{n+1}, [])$.
  update $\mathcal{G}^{(n)}$ by solving least square $\min_{\mathcal{G}^{(n)}} \|\left(\mathbf{G}_{(2)}^{(n)} \mathbf{B}_n\right) - \mathbf{X}_{(n)}\|_{\text{F}}^2$.
 end for
until fit ceases to improve or maximum iterations exhausted
Output: $\mathcal{G}^{(n)}$, $n = 1, \ldots, N$

2.6.2.2 Generations of Tensor Train Decomposition

The generations of tensor train decomposition in this part contain loop structure including tensor ring (TR) decomposition [81], projected entangled-pair states (PEPS) [56, 76], the honeycomb lattice (HCL) [1, 5, 32], multi-scale entanglement renormalization ansatz (MERA) [15], and so on. Among them, we will emphasize on TR decomposition due to its simpleness. For other tensor networks, we will give simple introduction.

To alleviate the ordering uncertain problem in TT decomposition, periodic boundary conditions (PBC) was first proposed, driving a new tensor decomposition named tensor ring decomposition (called tensor chain in physical field) [59], which was developed by Zhao et al. [81].

Definition 2.17 (Tensor Ring Decomposition) For an N-th order tensor $\mathcal{X} \in \mathbb{R}^{I_1 \times \cdots \times I_N}$, the TR decomposition is defined as

$$\mathcal{X} = \sum_{r_1=1}^{R_1} \cdots \sum_{r_N=1}^{R_N} \mathcal{G}^{(1)}(r_1, :, r_2) \circ \mathcal{G}^{(2)}(r_2, :, r_3) \circ \cdots \circ \mathcal{G}^{(N)}(r_N, :, r_1), \quad (2.68)$$

with elements as

$$\mathcal{X}(i_1, i_2, \ldots, i_N) = \mathrm{tr}(\mathcal{G}^{(1)}(:, i_1, :)\mathcal{G}^{(2)}(:, i_2, :) \cdots \mathcal{G}^{(N)}(:, i_N, :)), \quad (2.69)$$

where the $\mathcal{G}^{(n)} \in \mathbb{R}^{R_n \times I_n \times R_{n+1}}$, $n = 1, \ldots, N$ are the core factors and the TR ranks are defined as $[R_1, \ldots, R_N]$. We use $\mathcal{F}(\mathcal{G}^{(1)}, \ldots, \mathcal{G}^{(N)})$ to represent the tensor ring decomposition. Figure 2.11 shows a representation of tensor ring decomposition. Similar to TT decomposition, the storage complexity is NIR^2 assuming all $I_n = I$ and $R_n = R$ in the TR model.

Like the TT decomposition model, the detailed computation of tensor ring decomposition is concluded in Algorithm 14, where $\mathcal{B}_n = \overline{\otimes}_{i=1, i\neq n}^{N} \mathcal{G}^{(i)}$ is defined as the tensor contraction product of tensor $\mathcal{G}^{(i)}$, $i \neq n$, $i = 1, \ldots, N$.

Sometimes, ranks in TT form could be increased rapidly with the increasing order of a data tensor, which may be less effective for a compact representation. To alleviate this problem, PEPS are proposed by hierarchical two-dimensional TT models. The graphical illustration can be found in Fig. 2.12. In this case, the ranks are kept considerably smaller at a cost of employing fourth- or even fifth-order core

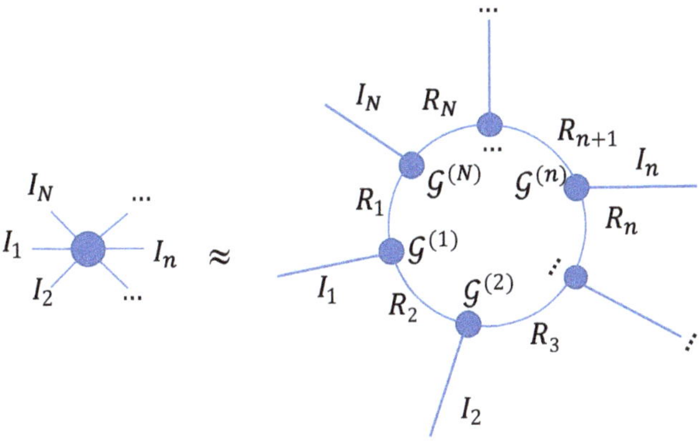

Fig. 2.11 Tensor ring decomposition

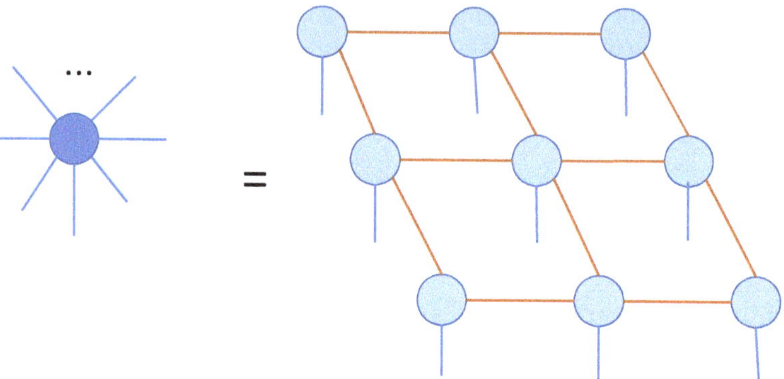

Fig. 2.12 Projected entangled-pair states (PEPS) for a ninth-order tensor

---

Algorithm 13: Sequential SVD for TR initialization

Input: $\mathcal{X} \in \mathbb{R}^{I_1 \times \cdots \times I_N}$, TR-rank threshold $R_{\max}$
initialize $\mathcal{G}^{(n)}$, $n = 1, \ldots, N$
1. $\mathbf{M}_0 = \mathbf{X}_{(1)}$, $[\mathbf{U}, \mathbf{S}, \mathbf{V}] = \text{SVD}(\mathbf{M}_0)$, $R_1 = 1$, $R_2 = \min\{R, R_{\max}\}$,
$\mathcal{G}^{(1)}(1 : R_1, :, 1 : R_2) = \mathbf{U}(:, 1 : R_2)$, $\mathbf{M}_1 = \mathbf{SV}$.
repeat
2. $[\mathbf{U}, \mathbf{S}, \mathbf{V}] = \text{SVD}(\text{Reshape}(\mathbf{M}_{n-1}, R_n I_n, []))$.
3. $R_{n+1} = \min\{R, R_{\max}\}$, $\mathcal{G}^{(n)} = \text{Reshape}(\mathbf{U}, [R_n, I_n, R_{n+1}])$.
4. $\mathbf{M}_n = \mathbf{SV}$
until $n = N - 1$
5. $\mathcal{G}^{(N)}(1 : R_N, :, 1 : R_1) = \mathbf{M}_{N-1}$.
Output: $\mathcal{G}^{(n)}$, $n = 1, \ldots, N$

---

tensors. However, for very high-order tensors, the ranks may increase rapidly with an increase in the desired accuracy of approximation. For further control of the ranks, alternative tensor networks can be employed including the HCL which uses third-order cores and can be found in Fig. 2.13 and the MERA which consists of both third- and fourth-order core tensors as shown in Fig. 2.14.

2.7 Hybrid Decomposition

2.7.1 Hierarchical Low-Rank Tensor Ring Decomposition

In [3], Ahad et al. proposed a hierarchical low-rank tensor ring decomposition. For the first layer, the traditional tensor ring decomposition model is used to factorize a tensor into many third-order subtensors. For the second layer, each third-order tensor is further decomposed by the t-SVD [42]. Figure 2.15 shows the details of hierarchical tensor ring decomposition.

Algorithm 14: Basic ALS for TR decomposition of a tensor

Input: $\mathcal{X} \in \mathbb{R}^{I_1 \times \cdots \times I_N}$ and TR-rank threshold $R_{\max}$
initialize run SSVD for TR initialization in Algorithm 13
repeat
  for $n = 1, \ldots N$ **do**
    if $n == N$
      $\mathcal{B}_n = \mathcal{G}^{(1)}$
    else if
      $\mathcal{B}_n = \mathcal{G}^{(n+1)}$
    for $m = [n+2, \ldots N, 1, \ldots, n-1]$ **do**
      $\mathcal{B}_n = \langle \mathcal{B}_n, \mathcal{G}^{(m+1)} \rangle_1$
    end for
    $\mathcal{B}_n = \text{Reshape}(\mathcal{B}_n, [R_{n+1}, \prod_{m=1, m \neq n}^{N} I_m, R_n])$
    $\mathbf{B}_n = \text{Reshape}\left(\text{Permute}\left(\mathcal{B}_n, [3, 1, 2]\right), R_n R_{n+1}, []\right)$.
    update $\mathcal{G}^{(n)}$ by solving least square $\min_{\mathcal{G}^{(n)}} \|\left(\mathbf{G}_{(2)}^{(n)} \mathbf{B}_n\right) - \mathbf{X}_{(n)}\|_{\mathrm{F}}^2$.
  end for
until fit ceases to improve or maximum iterations exhausted
Output: $\mathcal{G}^{(n)}, \; n = 1, \ldots, N$

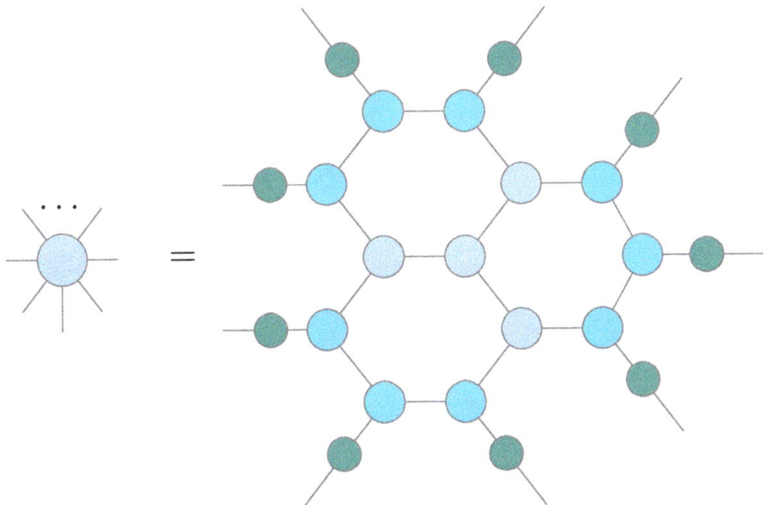

Fig. 2.13 Honeycomb lattice (HCL) for a ninth-order tensor

The low-rank approximation based on the proposed hierarchical tensor-ring decomposition of $\mathcal{X}$ can be formulated as follows:

$$\min_{\mathcal{G}^{(n)}, n=1,\ldots,N} \frac{1}{2} \|\mathcal{X} - \mathcal{F}(\mathcal{G}^{(1)}, \ldots, \mathcal{G}^{(N)})\|_{\mathrm{F}}^2 + \sum_{n=1}^{N} \text{rank}_{\text{tubal}}(\mathcal{G}^{(n)}). \tag{2.70}$$

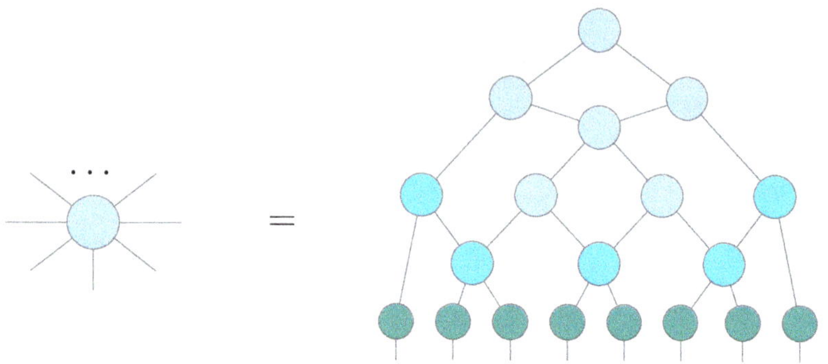

Fig. 2.14 Multi-scale entanglement renormalization ansatz (MERA) for an eighth-order tensor

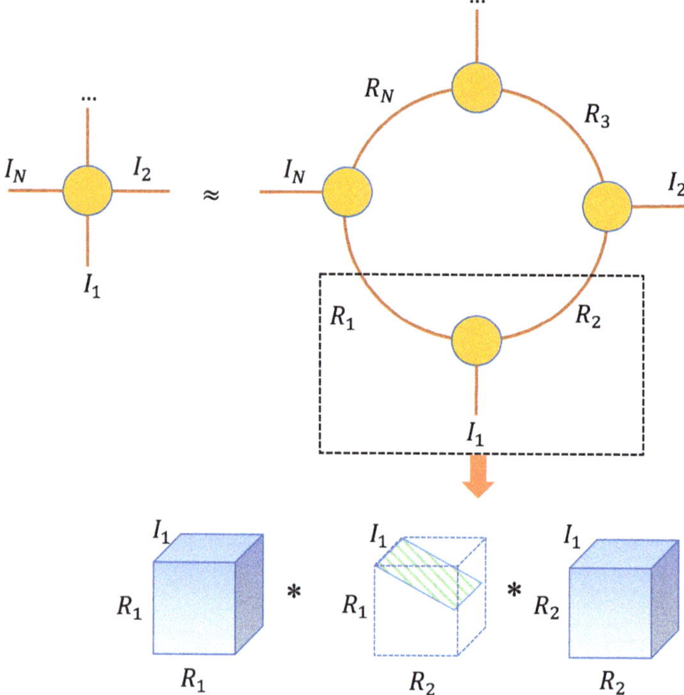

Fig. 2.15 Graphical representation of a hierarchical tensor ring (TR) model

This optimization problem can be solved by alternating direction method of multipliers (ADMM).

2.8 Scalable Tensor Decomposition

Tensor networks are efficient to deal with high-order data but fail to handle the large data where it contains billions of entries in each mode. Therefore, developing scalable tensor algorithms to handle these big data is interesting and necessary.

In this part, we will introduce some scalable tensor decomposition methods. According to the characteristics of the data (sparse or low-level), we can divide them into two categories. One is to exploit the sparsity of tensors, and the other is a subdivision of the large-scale tensor into smaller ones based on low-rank assumption.

2.8.1 *Scalable Sparse Tensor Decomposition*

When dealing with large-scale tensor, the intermediate data explosion problem often occurs. For example, using CP-ALS algorithm to deal with large-scale data $\mathcal{X} \in \mathbb{R}^{I\times J\times K}$, the computational complexity mainly comes from updating factors, as follows:

$$\mathbf{A} = \mathbf{X}_{(1)}(\mathbf{C} \odot \mathbf{B})(\mathbf{C}^{\mathrm{T}}\mathbf{C} \circledast \mathbf{B}^{\mathrm{T}}\mathbf{B})^{\dagger} \tag{2.71}$$

$$\mathbf{B} = \mathbf{X}_{(2)}(\mathbf{A} \odot \mathbf{C})(\mathbf{A}^{\mathrm{T}}\mathbf{A} \circledast \mathbf{C}^{\mathrm{T}}\mathbf{C})^{\dagger} \tag{2.72}$$

$$\mathbf{C} = \mathbf{X}_{(3)}(\mathbf{B} \odot \mathbf{A})(\mathbf{B}^{\mathrm{T}}\mathbf{B} \circledast \mathbf{A}^{\mathrm{T}}\mathbf{A})^{\dagger}. \tag{2.73}$$

To simplify it, we mainly focus on Eq. (2.71) to illustrate the intermediate data explosion problem in detail. To compute Eq. (2.71), we will first calculate $\mathbf{C} \odot \mathbf{B}$ and $\mathbf{C}^{\mathrm{T}}\mathbf{C} \circledast \mathbf{B}^{\mathrm{T}}\mathbf{B}$. For large-scale data, the matrix $\mathbf{C} \odot \mathbf{B} \in \mathbb{R}^{JK\times R}$ is very large and dense and cannot be stored in multiple disks, causing intermediate data explosion.

GigaTensor [41] is the first work to avoid intermediate data explosion problem when we handle the large-scale sparse tensor. The idea of this work is to decouple the $\mathbf{C}\odot\mathbf{B}$ in the Khatri-Rao product and perform algebraic operations involving $\mathbf{X}_{(1)}$ and $\mathbf{C}$, $\mathbf{X}_{(1)}$, and $\mathbf{B}$ and then combine the results. In this work, the authors give the proof that computing $\mathbf{X}_{(1)}(\mathbf{C} \odot \mathbf{B})$ is equivalent to computing $(\mathbf{F}_1 \circledast \mathbf{F}_2)\mathbf{1}_{JK}$, where $\mathbf{F}_1 = \mathbf{X}_{(1)} \circledast (\mathbf{1}_I \circ (\mathbf{C}(:,r)^{\mathrm{T}} \otimes \mathbf{1}_J^{\mathrm{T}}))$, $\mathbf{F}_2 = \mathrm{bin}(\mathbf{X}_{(1)}) \circledast (\mathbf{1}_I \circ (\mathbf{1}_K^{\mathrm{T}} \otimes \mathbf{B}(:,r)^{\mathrm{T}}))$, and $\mathbf{1}_{JK}$ is an all-one vector of size JK, bin() is an operator that converts any nonzero value into 1. The detailed fast computation is concluded in Algorithm 15.

Using this process, the flops of computing $\mathbf{X}_{(1)}(\mathbf{C}\odot\mathbf{B})$ are reduced from $JKR+2mR$ to $5mR$, and the intermediate data size is reduced from $JKR+m$ to $\max(J+m, K+m)$, where m is the number of nonzeros in $\mathbf{X}_{(1)}$.

Following this work, HaTen2 [35] unifies Tucker and CP decompositions into a general framework. Beutel et al. [7] propose FlexiFaCT, a flexible tensor decomposition method based on distributed stochastic gradient descent method. FlexiFaCT supports various types of decompositions such as matrix decomposition,

Algorithm 15: Calculating $\mathbf{X}_{(1)}(\mathbf{C} \odot \mathbf{B})$

Input: $\mathbf{X}_{(1)} \in \mathbb{R}^{I \times JK}$, $\mathbf{C} \in \mathbb{R}^{K \times R}$, $\mathbf{B} \in \mathbb{R}^{J \times R}$
initialize $\mathbf{M}_1 = \mathbf{0}$;
for $r = 1, \ldots, R$ **do**
  1. $\mathbf{F}_1 = \mathbf{X}_{(1)} \circledast (\mathbf{1}_I \circ (\mathbf{C}(:, r)^{\mathrm{T}} \otimes \mathbf{1}_J^{\mathrm{T}}))$;
  2. $\mathbf{F}_2 = \text{bin}(\mathbf{X}_{(1)}) \circledast (\mathbf{1}_I \circ (\mathbf{1}_K^{\mathrm{T}} \otimes \mathbf{B}(:, r)^{\mathrm{T}}))$;
  3. $\mathbf{F}_3 = \mathbf{F}_1 \circledast \mathbf{F}_2$;
  4. $\mathbf{M}_1(:, r) = \mathbf{F}_3 \mathbf{1}_{JK}$;
end for
Output: $\mathbf{M}_1 = \mathbf{X}_{(1)}(\mathbf{C} \odot \mathbf{B})$

CP decomposition, and coupled matrix-tensor factorization. The main idea in those papers is to avoid intermediate product explosion when computing sequential tensor-matrix (mode) products. However, these methods operate on the whole tensor in each iterative step, which is still prohibitive when the tensor is very large.

2.8.2 *Strategy on Dense Tensor Decomposition*

To handle the data which is large-scale and not sparse, we need to divide large-scale data into small-scale one for processing. Based on the low-rank assumption on big data, the existing algorithms for dealing with large-scale low-rank tensor are mainly divided into two groups. One is based on the parallel distributed techniques, and the other is based on the projection technique.

2.8.2.1 Scalable Distributed Tensor Decomposition

In this group, the idea to handle large-scale data is to develop a "divide and conquer" strategy. It consists of three steps: the one breaks the large-scale tensor into small-scale tensors, finds factors of the small-scale tensors, and combines the factors of the small-scale tensors to recover the factors of the original large-scale tensor. PARCUBE [65] is the first approach to use the "divide and conquer" strategy to process the large-scale tensor. The architecture of PARCUBE has three parts. Firstly multiple small-scale tensors are parallel subsampled from the large-scale tensor, and each small-scale tensor is independently factorized by CP decomposition secondly. Finally, the factors of small-scale tensors are joined via a master linear equation.

For example, given a large-scale data $\mathcal{X} \in \mathbb{R}^{I \times J \times K}$, multiple small-scale tensors are parallel subsampled from the large-scale tensor firstly. The subsampled process can be found in Fig. 2.16, where the sample matrices $\mathbf{U}_n \in \mathbb{R}^{I \times I_n}$,$\mathbf{V}_n \in \mathbb{R}^{J \times J_n}$,$\mathbf{W}_n \in \mathbb{R}^{K \times K_n}$ are randomly drawn from an absolutely continuous distribution and the elements of these sample matrices are independent and identically distributed Gaussian random variables with zero mean and unit variance.

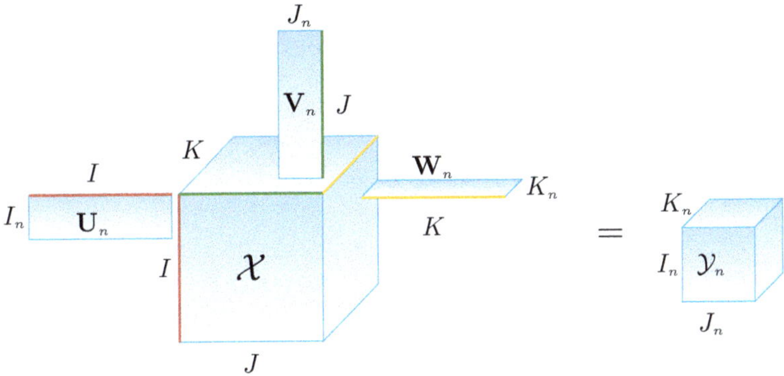

Fig. 2.16 The random sample process

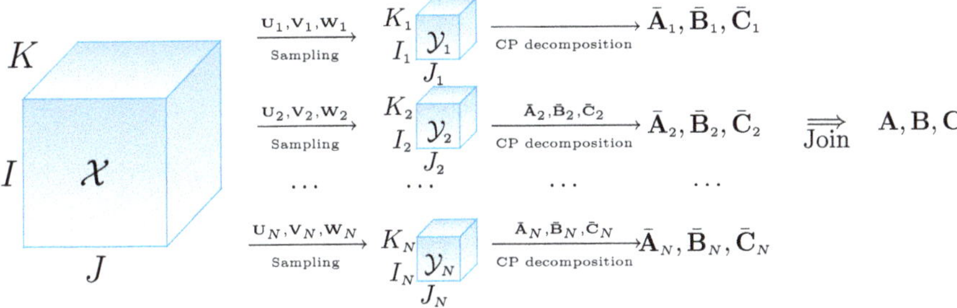

Fig. 2.17 A schematic illustration of the PARCUBE architecture

After subsampling N times, we obtain N small-scale tensors and factorize these small-scale tensors on different machines in parallel by CP decomposition. Next, we need to recover the factor matrices of the original tensor from the factors obtained from the parallel decomposition step. The whole process can be found in Fig. 2.17 for illustration.

Following the idea of PARCUBE, some similar works are proposed to improve the accuracy of decomposition or to complete the identifiability theory [40, 78].

2.8.2.2 Randomized Tensor Decomposition

Different from the scalable distributed tensor decomposition, the idea of the randomized tensor decomposition methods is firstly to apply random projections to obtain compressed tensor, and the obtained compressed tensor can be further factorized by different tensor decompositions. Finally, the tensor decomposition of original tensor can be obtained by projecting factor matrices of the compressed tensor back.

For example, in [23], the authors use randomized projection on CP decomposition for large-scale data as follows. Assuming $\mathcal{X} \in \mathbb{R}^{I_1 \times \cdots \times I_N}$ is an N-th order tensor with large $I_n,\ n = 1, \ldots, N$, the first step of randomized CP decomposition is to obtain a small tensor $\mathcal{Y} \in \mathbb{R}^{J \times \cdots \times J}$ which can preserve almost all the multidimensional information of the original tensor. The key step to obtain the compressed tensor is to seek a natural basis in the form of a set of orthonormal matrices $\{\mathbf{U}_n \in \mathbb{R}^{I_n \times J}\}_{n=1}^{N}$, so that

$$\mathcal{X} \approx \mathcal{X} \times_1 \mathbf{U}_1 \mathbf{U}_1^{\mathrm{T}} \times_2 \cdots \times_N \mathbf{U}_N \mathbf{U}_N^{\mathrm{T}}. \tag{2.74}$$

Given a fixed target rank J, these basis matrices $\{\mathbf{U}_n \in \mathbb{R}^{I_n \times J}\}_{n=1}^{N}$ can be efficiently obtained using a randomized algorithm. First, a random sample matrix $\mathbf{W} \in \mathbb{R}^{\prod_{d \neq n} I_d \times J}$ should be constructed, where each column vector is drawn from a Gaussian distribution. Then, the random sample matrix is used to sample the column space of $\mathbf{X}_{(n)} \in \mathbb{R}^{I_n \times \prod_{d \neq n} I_d}$ as follows:

$$\mathbf{Z} = \mathbf{X}_{(n)} \mathbf{W}, \tag{2.75}$$

where $\mathbf{Z} \in \mathbb{R}^{I_n \times J}$ is the sketch. The sketch can represent the approximate basis for the row space of $\mathcal{X}_{(n)}$. According to the probability theory [4, 39], if each column vector of sample matrix $\mathbf{W} \in \mathbb{R}^{\prod_{d \neq n} I_d \times J}$ is linearly independent with high probability, the random projection $\mathbf{Z}$ will efficiently sample the row space of $\mathcal{X}_{(n)}$. Finally, the orthonormal basis can be obtained by

$$\mathbf{U}_n = \mathrm{QR}(\mathbf{Z}), \tag{2.76}$$

where QR means QR decomposition.

After getting the n-mode orthogonal basis, we can project the $\mathcal{X}$ to the low dimensional subspace as follows:

$$\mathcal{Y} = \mathcal{X} \times_n \mathbf{U}_n^{\mathrm{T}}. \tag{2.77}$$

After N iterations, we can obtain the compressed tensor $\mathcal{Y}$ and a set of orthonormal matrices $\{\mathbf{U}_n \in \mathbb{R}^{I_n \times J}\}_{n=1}^{N}$. The detailed solutions are concluded in Algorithm 16.

The next step performs CP decomposition on $\mathcal{Y}$, and obtains compressed factor matrices $\bar{\mathbf{A}}_n \in \mathbb{R}^{J \times R}, n = 1 \ldots, N$. The factor matrices of original tensor can be recovered by

$$\mathbf{A}_n \approx \mathbf{U}_n \bar{\mathbf{A}}_n, n = 1, \ldots, N, \tag{2.78}$$

where $\mathbf{U}_n \in \mathbb{R}^{I_n \times J}, n = 1, \ldots, N$ denote the orthonormal basis matrices. In addition, the oversampling and power methods can be used to improve the solution accuracy of randomized algorithms.

Algorithm 16: Randomized tensor compression algorithm

Input: $\mathcal{X} \in \mathbb{R}^{I_1 \times \cdots \times I_N}$ and a desired target rank J
initialize compressed tensor $\mathcal{Y} = \mathcal{X}$
For $n = 1, \ldots, N$
1. generate random sample matrix $\mathbf{W} \in \mathbb{R}^{\prod_{d \neq n} I_d \times J}$
2. $\mathbf{Z} = \mathbf{X}_{(n)} \mathbf{W}$
3. obtain n-mode orthogonal basis $\mathbf{U}_n = \text{QR}(\mathbf{Z})$
4. update compressed tensor $\mathcal{Y} = \mathcal{Y} \times_n \mathbf{U}_n^{\text{T}}$.

End for
Output: compressed tensor $\mathcal{Y} \in \mathbb{R}^{J \times \cdots \times J}$, orthonormal matrices $\{\mathbf{U}_n \in \mathbb{R}^{I_n \times J}\}_{n=1}^{N}$.

In the same way, random Tucker decompositions for large-scale data are proposed, such as random projection HOSVD algorithm [13] and random projection orthogonal iteration [82]. Similarly, other random tensor decompositions using random projection can refer to [24].

2.9 Summary and Future Work

Tensor decompositions are very powerful tools, which are ubiquitous in image processing, machine learning, and computer vision. In addition, different decompositions have varied applications. For example, due to the fact that CP decomposition is unique under mild condition, it is suitable for extracting interpretable latent factors. Tucker has a good ability of compressing, which is frequently used in compressing data. In addition, with tensor order increases, some tensor networks, including tensor train decomposition, tensor tree decomposition, and tensor ring decomposition are utilized to alleviate the curse of dimensionality and reduce the storage memory.

However, there still exist some challenges in tensor decomposition. We summarize them in the following parts.

- Which tensor decomposition is the best one to exploit spatial or temporal structure of data? For example, in work [74], the authors use nonnegative CP decomposition to represent the hyperspectral images, and authors [14] apply Tucker decomposition to capture spatial-temporal information of traffic speed data. Is there a generic way to incorporate such modifications in a tensor model and enable it to handle these data effectively? Furthermore, is there an optimal tensor network structure which can adaptively represent data?
- In real world, there exist many heterogeneous information networks such as social networks and knowledge graph. These networks can be well represented by graphs where nodes reveal different types of data and edges present the relationship between them. Compared with graph, is there a way to represent

these networks by tensor? Connecting these networks with tensor decomposition is an interesting direction in the future.

- With the development of sensors, we can capture information about the same object by different ways. How to use tensor decomposition to analyze the multimodal data is a promising way in data mining and machine learning.
- With data size increases, there is an urgent need to develop effective methods to handle large-scale data. From a software perspective, the development of distributed tensor calculations, online decomposition, and high-performance tensor calculation methods is a promising direction to alleviate this problem. From a hardware point of view, designing some new quantum circuits to accelerate the tensor operator computation is a hopeful way.

References

1. Ablowitz, M.J., Nixon, S.D., Zhu, Y.: Conical diffraction in honeycomb lattices. Phys. Rev. A **79**(5), 053830–053830 (2009)
2. Acar, E., Dunlavy, D.M., Kolda, T.G.: Link prediction on evolving data using matrix and tensor factorizations. In: 2009 IEEE International Conference on Data Mining Workshops, pp. 262–269. IEEE, New York (2009)
3. Ahad, A., Long, Z., Zhu, C., Liu, Y.: Hierarchical tensor ring completion (2020). arXiv e-prints, pp. arXiv–2004
4. Ahfock, D.C., Astle, W.J., Richardson, S.: Statistical properties of sketching algorithms. Biometrika (2020). https://doi.org/10.1093/biomet/asaa062
5. Bahat-Treidel, O., Peleg, O., Segev, M.: Symmetry breaking in honeycomb photonic lattices. Opt. Lett. **33**(19), 2251–2253 (2008)
6. Ballani, J., Grasedyck, L., Kluge, M.: Black box approximation of tensors in hierarchical Tucker format. Linear Algebra Appl. **438**(2), 639–657 (2013)
7. Beutel, A., Talukdar, P.P., Kumar, A., Faloutsos, C., Papalexakis, E.E., Xing, E.P.: Flexifact: scalable flexible factorization of coupled tensors on hadoop. In: Proceedings of the 2014 SIAM International Conference on Data Mining, pp. 109–117. SIAM, Philadelphia (2014)
8. Bigoni, D., Engsig-Karup, A.P., Marzouk, Y.M.: Spectral tensor-train decomposition. SIAM J. Sci. Comput. **38**(4), A2405–A2439 (2016)
9. Brachat, J., Comon, P., Mourrain, B., Tsigaridas, E.: Symmetric tensor decomposition. Linear Algebra Appl. **433**(11–12), 1851–1872 (2010)
10. Carroll, J.D., Chang, J.J.: Analysis of individual differences in multidimensional scaling via an n-way generalization of "Eckart-Young" decomposition. Psychometrika **35**(3), 283–319 (1970)
11. Cattell, R.B.: "parallel proportional profiles" and other principles for determining the choice of factors by rotation. Psychometrika **9**(4), 267–283 (1944)
12. Cattell, R.B.: The three basic factor-analytic research designs—their interrelations and derivatives. Psychol. Bull. **49**(5), 499–520 (1952)
13. Che, M., Wei, Y.: Randomized algorithms for the approximations of Tucker and the tensor train decompositions. Adv. Comput. Math. **45**(1), 395–428 (2019)
14. Chen, X., He, Z., Wang, J.: Spatial-temporal traffic speed patterns discovery and incomplete data recovery via SVD-combined tensor decomposition. Transp. Res. Part C Emerg. Technol. **86**, 59–77 (2018)
15. Cincio, L., Dziarmaga, J., Rams, M.: Multiscale entanglement renormalization ansatz in two dimensions: quantum ising model. Phys. Rev. Lett. **100**(24), 240603–240603 (2008)

16. Cong, F., Lin, Q.H., Kuang, L.D., Gong, X.F., Astikainen, P., Ristaniemi, T.: Tensor decomposition of EEG signals: a brief review. J. Neurosci. Methods **248**, 59–69 (2015)
17. Cyganek, B., Gruszczyński, S.: Hybrid computer vision system for drivers' eye recognition and fatigue monitoring. Neurocomputing **126**, 78–94 (2014)
18. De Lathauwer, L.: Decompositions of a higher-order tensor in block terms—part I: lemmas for partitioned matrices. SIAM J. Matrix Anal. Appl. **30**(3), 1022–1032 (2008)
19. De Lathauwer, L.: Decompositions of a higher-order tensor in block terms—part II: definitions and uniqueness. SIAM J. Matrix Anal. Appl. **30**(3), 1033–1066 (2008)
20. De Lathauwer, L., De Moor, B., Vandewalle, J.: A multilinear singular value decomposition. SIAM J. Matrix Anal. Appl. **21**(4), 1253–1278 (2000)
21. De Lathauwer, L., Nion, D.: Decompositions of a higher-order tensor in block terms—part III: alternating least squares algorithms. SIAM J. Matrix Anal. Appl. **30**(3), 1067–1083 (2008)
22. De Lathauwer, L., Vandewalle, J.: Dimensionality reduction in higher-order signal processing and rank-(r1, r2,..., rn) reduction in multilinear algebra. Linear Algebra Appl. **391**, 31–55 (2004)
23. Erichson, N.B., Manohar, K., Brunton, S.L., Kutz, J.N.: Randomized CP tensor decomposition. Mach. Learn. Sci. Technol. **1**(2), 025012 (2020)
24. Fonał, K., Zdunek, R.: Distributed and randomized tensor train decomposition for feature extraction. In: 2019 International Joint Conference on Neural Networks (IJCNN), pp. 1–8. IEEE, New York (2019)
25. Franz, T., Schultz, A., Sizov, S., Staab, S.: Triplerank: ranking semantic web data by tensor decomposition. In: International Semantic Web Conference, pp. 213–228. Springer, Berlin (2009)
26. Govindu, V.M.: A tensor decomposition for geometric grouping and segmentation. In: 2005 IEEE Computer Society Conference on Computer Vision and Pattern Recognition (CVPR'05), vol. 1, pp. 1150–1157. IEEE, New York (2005)
27. Grasedyck, L.: Hierarchical singular value decomposition of tensors. SIAM J. Matrix Anal. Appl. **31**(4), 2029–2054 (2010)
28. Grelier, E., Nouy, A., Chevreuil, M.: Learning with tree-based tensor formats (2018). Preprint, arXiv:1811.04455
29. Hackbusch, W., Kühn, S.: A new scheme for the tensor representation. J. Fourier Anal. Appl. **15**(5), 706–722 (2009)
30. Hardoon, D.R., Shawe-Taylor, J.: Decomposing the tensor kernel support vector machine for neuroscience data with structured labels. Mach. Learn. **79**(1–2), 29–46 (2010)
31. Harshman, R.: Foundations of the PARAFAC procedure: models and conditions for an "explanatory" multimodal factor analysis. In: UCLA Working Papers in Phonetics, vol. 16, pp. 1–84 (1970)
32. Herbut, I.: Interactions and phase transitions on graphene's honeycomb lattice. Phys. Rev. Lett. **97**(14), 146401–146401 (2006)
33. Hitchcock, F.L.: The expression of a tensor or a polyadic as a sum of products. J. Math. Phys. **6**(1–4), 164–189 (1927)
34. Hitchcock, F.L.: Multiple invariants and generalized rank of a p-way matrix or tensor. J. Math. Phys. **7**(1–4), 39–79 (1928)
35. Jeon, I., Papalexakis, E.E., Kang, U., Faloutsos, C.: Haten2: billion-scale tensor decompositions. In: 2015 IEEE 31st International Conference on Data Engineering, pp. 1047–1058. IEEE, New York (2015)
36. Ji, H., Li, J., Lu, R., Gu, R., Cao, L., Gong, X.: EEG classification for hybrid brain-computer interface using a tensor based multiclass multimodal analysis scheme. Comput. Intell. Neurosci. **2016**, 1732836–1732836 (2016)
37. Jiang, B., Ding, C., Tang, J., Luo, B.: Image representation and learning with graph-Laplacian tucker tensor decomposition. IEEE Trans. Cybern. **49**(4), 1417–1426 (2018)
38. Jiang, T., Sidiropoulos, N.D.: Kruskal's permutation lemma and the identification of CANDECOMP/PARAFAC and bilinear models with constant modulus constraints. IEEE Trans. Signal Process. **52**(9), 2625–2636 (2004)

39. Johnson, W.B., Lindenstrauss, J.: Extensions of Lipschitz mappings into a Hilbert space. Contemp. Math. **26**(189–206), 1 (1984)
40. Kanatsoulis, C.I., Sidiropoulos, N.D.: Large-scale canonical polyadic decomposition via regular tensor sampling. In: 2019 27th European Signal Processing Conference (EUSIPCO), pp. 1–5. IEEE, New York (2019)
41. Kang, U., Papalexakis, E., Harpale, A., Faloutsos, C.: Gigatensor: scaling tensor analysis up by 100 times-algorithms and discoveries. In: Proceedings of the 18th ACM SIGKDD International Conference on Knowledge Discovery and Data Mining, pp. 316–324 (2012)
42. Kilmer, M.E., Braman, K., Hao, N., Hoover, R.C.: Third-order tensors as operators on matrices: a theoretical and computational framework with applications in imaging. SIAM J. Matrix Anal. Appl. **34**(1), 148–172 (2013)
43. Kilmer, M.E., Martin, C.D.: Factorization strategies for third-order tensors. Linear Algebra Appl. **435**(3), 641–658 (2011)
44. Kilmer, M.E., Martin, C.D., Perrone, L.: A third-order generalization of the matrix SVD as a product of third-order tensors. Tufts University, Department of Computer Science, Tech. Rep. TR-2008-4 (2008)
45. Kolda, T.G., Sun, J.: Scalable tensor decompositions for multi-aspect data mining. In: 2008 Eighth IEEE International Conference on Data Mining, pp. 363–372. IEEE, New York (2008)
46. Kressner, D., Tobler, C.: htucker—a matlab toolbox for tensors in hierarchical Tucker format. Mathicse, EPF Lausanne (2012)
47. Kruskal, J.B.: Three-way arrays: rank and uniqueness of trilinear decompositions, with application to arithmetic complexity and statistics. Linear Algebra Appl. **18**(2), 95–138 (1977)
48. Levin, J.: Three-mode factor analysis. Psychol. Bull. **64**(6), 442–452 (1965)
49. Li, N., Kindermann, S., Navasca, C.: Some convergence results on the regularized alternating least-squares method for tensor decomposition. Linear Algebra Appl. **438**(2), 796–812 (2013)
50. Lim, L.H., Comon, P.: Multiarray signal processing: tensor decomposition meets compressed sensing. C. R. Mec. **338**(6), 311–320 (2010)
51. Liu, X., Sidiropoulos, N.D.: Cramér-Rao lower bounds for low-rank decomposition of multidimensional arrays. IEEE Trans. Signal Process. **49**(9), 2074–2086 (2001)
52. Lu, C., Feng, J., Chen, Y., Liu, W., Lin, Z., Yan, S.: Tensor robust principal component analysis with a new tensor nuclear norm. IEEE Trans. Pattern Anal. Mach. Intell. **42**(4), 925–938 (2019)
53. Lubich, C., Rohwedder, T., Schneider, R., Vandereycken, B.: Dynamical approximation by hierarchical Tucker and tensor-train tensors. SIAM J. Matrix Anal. Appl. **34**(2), 470–494 (2013)
54. Mørup, M.: Applications of tensor (multiway array) factorizations and decompositions in data mining. Wiley Interdiscip. Rev. Data Min. Knowl. Disc. **1**(1), 24–40 (2011)
55. Murg, V., Verstraete, F., Legeza, Ö., Noack, R.: Simulating strongly correlated quantum systems with tree tensor networks. Phys. Rev. B **82**(20), 1–21 (2010)
56. Orús, R.: A practical introduction to tensor networks: matrix product states and projected entangled pair states. Ann. Phys. **349**, 117–158 (2014)
57. Oseledets, I.V.: Tensor-train decomposition. SIAM J. Sci. Comput. **33**(5), 2295–2317 (2011)
58. Perros, I., Chen, R., Vuduc, R., Sun, J.: Sparse hierarchical tucker factorization and its application to healthcare. In: 2015 IEEE International Conference on Data Mining, pp. 943–948. IEEE, New York (2015)
59. Pirvu, B., Verstraete, F., Vidal, G.: Exploiting translational invariance in matrix product state simulations of spin chains with periodic boundary conditions. Phys. Rev. B **83**(12), 125104 (2011)
60. Pižorn, I., Verstraete, F., Konik, R.M.: Tree tensor networks and entanglement spectra. Phys. Rev. B **88**(19), 195102 (2013)
61. Qi, L., Sun, W., Wang, Y.: Numerical multilinear algebra and its applications. Front. Math. China **2**(4), 501–526 (2007)
62. Schollwöck, U.: The density-matrix renormalization group in the age of matrix product states. Ann. Phys. **326**(1), 96–192 (2011)

63. Sidiropoulos, N.D., Bro, R.: On the uniqueness of multilinear decomposition of n-way arrays. J. Chemometrics J. Chemometrics Soc. **14**(3), 229–239 (2000)
64. Sidiropoulos, N.D., De Lathauwer, L., Fu, X., Huang, K., Papalexakis, E.E., Faloutsos, C.: Tensor decomposition for signal processing and machine learning. IEEE Trans. Signal Process. **65**(13), 3551–3582 (2017)
65. Sidiropoulos, N.D., Papalexakis, E.E., Faloutsos, C.: Parallel randomly compressed cubes: a scalable distributed architecture for big tensor decomposition. IEEE Signal Process. Mag. **31**(5), 57–70 (2014)
66. Sobral, A., Javed, S., Jung, S.K., Bouwmans, T., Zahzah, E.h.: Online stochastic tensor decomposition for background subtraction in multispectral video sequences. In: 2015 IEEE International Conference on Computer Vision Workshop (ICCVW), pp. 946–953. IEEE, New York (2015)
67. Sørensen, M., De Lathauwer, L.: Blind signal separation via tensor decomposition with vandermonde factor: canonical polyadic decomposition. IEEE Trans. Signal Process. **61**(22), 5507–5519 (2013)
68. Stegeman, A., Sidiropoulos, N.D.: On Kruskal's uniqueness condition for the Candecomp/Parafac decomposition. Linear Algebra Appl. **420**(2–3), 540–552 (2007)
69. Sun, J., Tao, D., Faloutsos, C.: Beyond streams and graphs: dynamic tensor analysis. In: Proceedings of the 12th ACM SIGKDD International Conference on Knowledge Discovery and Data Mining, pp. 374–383 (2006)
70. Ten Berge, J.M., Sidiropoulos, N.D.: On uniqueness in CANDECOMP/PARAFAC. Psychometrika **67**(3), 399–409 (2002)
71. Tucker, L.R.: Implications of factor analysis of three-way matrices for measurement of change. Probl. Meas. Change **15**, 122–137 (1963)
72. Tucker, L.R.: Some mathematical notes on three-mode factor analysis. Psychometrika **31**(3), 279–311 (1966)
73. Tucker, L.R., et al.: The extension of factor analysis to three-dimensional matrices. Contrib. Math. Psychol. **110119** (1964)
74. Veganzones, M.A., Cohen, J.E., Farias, R.C., Chanussot, J., Comon, P.: Nonnegative tensor CP decomposition of hyperspectral data. IEEE Trans. Geosci. Remote Sens. **54**(5), 2577–2588 (2015)
75. Verstraete, F., Cirac, J.I.: Matrix product states represent ground states faithfully. Phys. Rev. B **73**(9), 094423 (2006)
76. Verstraete, F., Murg, V., Cirac, J.I.: Matrix product states, projected entangled pair states, and variational renormalization group methods for quantum spin systems. Adv. Phys. **57**(2), 143–224 (2008)
77. Wang, Y., Tung, H.Y., Smola, A.J., Anandkumar, A.: Fast and guaranteed tensor decomposition via sketching. In: Advances in Neural Information Processing Systems, vol. 28, pp. 991–999 (2015)
78. Yang, B., Zamzam, A., Sidiropoulos, N.D.: Parasketch: parallel tensor factorization via sketching. In: Proceedings of the 2018 SIAM International Conference on Data Mining, pp. 396–404. SIAM, Philadelphia (2018)
79. Yang, G., Jones, T.L., Barrick, T.R., Howe, F.A.: Discrimination between glioblastoma multiforme and solitary metastasis using morphological features derived from the p: q tensor decomposition of diffusion tensor imaging. NMR Biomed. **27**(9), 1103–1111 (2014)
80. Zhang, Z., Yang, X., Oseledets, I.V., Karniadakis, G.E., Daniel, L.: Enabling high-dimensional hierarchical uncertainty quantification by ANOVA and tensor-train decomposition. IEEE Trans. Comput. Aided Des. Integr. Circuits Syst. **34**(1), 63–76 (2014)
81. Zhao, Q., Zhou, G., Xie, S., Zhang, L., Cichocki, A.: Tensor ring decomposition (2016). arXiv e-prints, pp. arXiv–1606
82. Zhou, G., Cichocki, A., Xie, S.: Decomposition of big tensors with low multilinear rank (2014). arXiv e-prints, pp. arXiv–1412

Chapter 3
Tensor Dictionary Learning

3.1 Matrix Dictionary Learning

Dictionary learning, also called sparse coding, is a representation learning method [52]. It learns a set of vector basis from training data, and each signal can be represented by a linear combination of the vectors in the learned set. The matrix stacking from learned vectors is called dictionary, and the vector is called atom. To allow more flexibility, the vectors can be non-orthogonal, and this kind of overcomplete dictionaries may achieve better representation performance [1].

Compared with fixed dictionaries like discrete Fourier transform matrix, the learned dictionary can give more sparser representation and better fit the specific data by learning, which provides an enhanced performance in many applications.

Dictionary learning has been successfully used in signal reconstruction [22], enhancement [56], classification [37, 41], and unsupervised clustering [44, 49] and benefits data processing applications, such as compressive sensing [66], image classification [60], audio processing [70], data compression [30], texture synthesis [42], biomedical data analysis [48], etc.

3.1.1 Sparse and Cosparse Representation

Sparse models are generally classified into two categories: synthetic sparse models and analytic sparse models. In synthetic sparse model, a signal $\mathbf{x}$ is represented as a linear combination of certain atoms of an overcomplete dictionary. The analytic sparse model characterizes the signal $\mathbf{x}$ by multiplying it with an analytic overcomplete dictionary, leading to a sparse outcome. Analytic sparse model is also called cosparse model. Figures 3.1 and 3.2 are schematic diagrams of two models with overcomplete dictionaries.

Y. Liu et al., *Tensor Computation for Data Analysis*,
https://doi.org/10.1007/978-3-030-74386-4_3

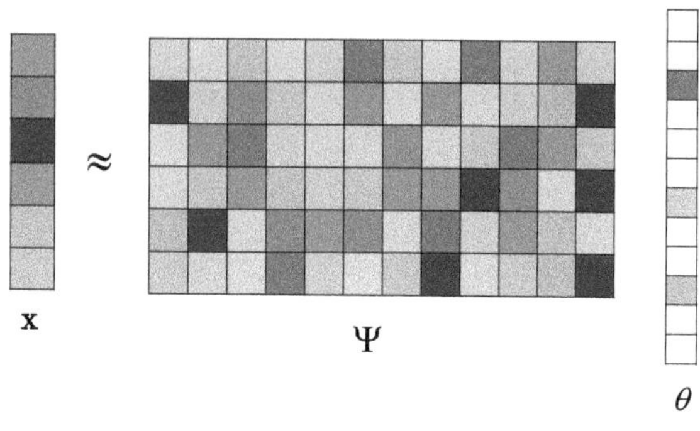

Fig. 3.1 The synthetic sparse model

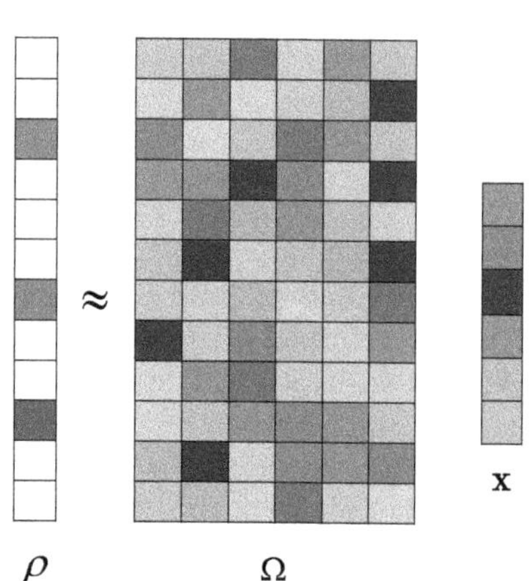

Fig. 3.2 The analytic sparse model

The sparse representation can be formulated as follows [33]:

$$\mathbf{x} = \boldsymbol{\Psi}\boldsymbol{\theta} + \mathbf{e}_\mathrm{s} = \sum_{n=1}^{N} \theta_n \boldsymbol{\psi}_n + \mathbf{e}_\mathrm{s}, \tag{3.1}$$

where $\boldsymbol{\Psi} \in \mathbb{R}^{I \times N}$ is the dictionary, $\boldsymbol{\psi}_n$, $n = 1, \ldots, N$ is the atom, $\boldsymbol{\theta} \in \mathbb{R}^N$ is the sparse coefficient, and $\mathbf{e}_\mathrm{s}$ is the approximation error. In an overcomplete $\boldsymbol{\Psi}$, the number of columns is larger than the number of rows, which implies that the sparsest solution $\boldsymbol{\theta}$ may not be unique.

According to the sparse representation model, a signal $\mathbf{x}$ can be reconstructed by selecting the sparsest one from all the candidates which satisfy the data fitting model as follows:

$$\min_{\boldsymbol{\theta}} \|\boldsymbol{\theta}\|_0, \quad \text{s. t. } \|\mathbf{x} - \boldsymbol{\Psi}\boldsymbol{\theta}\|_2 \leq \alpha, \tag{3.2}$$

where $\|\boldsymbol{\theta}\|_0$ is the pseudo-ℓ_0 norm which counts the nonzero entries and α bounds the approximation error in sparse representation. Under mild conditions, this optimization problem can be reformulated into an equivalent form:

$$\min_{\boldsymbol{\theta}} \|\mathbf{x} - \boldsymbol{\Psi}\boldsymbol{\theta}\|_2^2, \quad \text{s. t.} \ \ \|\boldsymbol{\theta}\|_0 \leq \beta, \tag{3.3}$$

where β bounds the number of nonzero entries.

This problem for ℓ_0 optimization is NP-hard. Fortunately, there exist some polynomial time algorithms for suboptimal solutions, such as matching pursuit (MP) [31], orthogonal matching pursuit (OMP) [35], basis pursuit (BP) [9], focal underdetermined system solver (FOCUSS) [21], etc.

In fact, the performance of the ℓ_0 optimization problem in terms of the approximation error and the coefficient sparsity vector depends not only on the algorithms for sparse solver but also on the dictionary $\boldsymbol{\Psi}$.

Different from the synthetic sparse model which has been investigated for a long time, research on analysis sparse model is still in its infancy. The analysis model use redundant analysis operator $\boldsymbol{\Omega} : \mathbb{R}^I \to \mathbb{R}^N, \ I \leq N$, where we assume the vector $\Omega\mathbf{x}$ is sparse. In this case, the analysis model puts emphasis on the number of zeros of the vector $\boldsymbol{\Omega}\mathbf{x}$, which is called as cosparsity.

Definition 3.1 The cosparsity ℓ of a signal $\mathbf{x} \in \mathbb{R}^I$ with $\boldsymbol{\Omega} \in \mathbb{R}^{N \times I}$ is defined by

$$\ell := N - \|\boldsymbol{\Omega}\mathbf{x}\|_0.$$

Generally, we assume that $\boldsymbol{\Omega}$ is of full column rank. In a word, $\mathbf{x}$ is cosparse or $\mathbf{x}$ has a cosparse representation when the cosparsity ℓ is large. Clearly, $\|\boldsymbol{\Omega}\mathbf{x}\|_0 \geq N - I$, and the largest ℓ approaches I. In this case, the maximization of ℓ with respect to $\boldsymbol{\Omega}$ can be transformed into the following optimization problem:

$$\min_{\boldsymbol{\Omega}} \|\boldsymbol{\Omega}\mathbf{x}\|_0. \tag{3.4}$$

By introducing the auxiliary variables, this model for cosparse recovery with a fixed dictionary $\boldsymbol{\Omega}$ can be formulated as follows:

$$\min_{\boldsymbol{\rho}} \|\boldsymbol{\rho}\|_0, \quad \text{s. t.} \ \ \|\boldsymbol{\rho} - \boldsymbol{\Omega}\mathbf{x}\|_2 \leq \epsilon, \tag{3.5}$$

where ϵ is the error tolerance related to the noise power and $\boldsymbol{\rho} = \boldsymbol{\Omega}\mathbf{x} + \mathbf{e}_{\mathrm{a}}$, where $\mathbf{e}_{\mathrm{a}}$ is the approximation error term.

All in all, in synthesis sparse model, the signal subspace consists of the columns $\boldsymbol{\psi}_n, \ n \in \mathbb{T}$, where the set $\mathbb{T}$ contains the index of nonzero coefficients. Correspondingly, in analysis sparse model, the analysis subspace consists of the rows $\boldsymbol{\omega}_n$ with $\langle \omega_n, x \rangle = 0$. In addition, when both the synthesis dictionary $\boldsymbol{\Psi}$ and the analysis $\boldsymbol{\Omega}$ are overcomplete, they are likely to be very different. However, if they have complete orthogonal basis, these two kinds of representations can be equivalent with $\boldsymbol{\Omega} = \boldsymbol{\Psi}^{-1}$.

3.1.2 *Dictionary Learning Methods*

Dictionary is the basis for sparse and cosparse representation. There are generally two groups of methods for dictionary design. One is fixed dictionary, which is characterized by mathematical functions, such as discrete cosine transform (DCT) [2], wavelet transform (WT) [5], contour transform (CT) [38], shearlet [28], grouplet [61], and parametric dictionary [59]. Although these fixed dictionaries have simple structures and low computational complexity, the basic atoms are fixed. The atom morphology is not rich enough to match some complicated data structures. Therefore, these fixed dictionaries may be nonoptimal representations for some data processing applications.

The other one learns overcomplete dictionaries from training data. Compared with the fixed dictionary based on analytical design, the data-driven dictionary is versatile, simple, and efficient. It can better match the data structure with a sparser/cosparser representation. The classical dictionary learning methods include maximum likelihood methods (MLD) [34], method of optimal directions (MOD) [18], maximum a posteriori probability (MAP) approach [25], generalized PCA (GPCA) [55], K-SVD [1], analysis K-SVD [47], etc.

3.1.2.1 The MOD Method

Proposed in 1999, MOD is one of the most classical dictionary learning methods [18]. It formulates the dictionary learning into a bilinear optimization model with a sparse constraint as follows:

$$\min_{\boldsymbol{\Psi},\boldsymbol{\Theta}} \|\mathbf{X} - \boldsymbol{\Psi}\boldsymbol{\Theta}\|_{\mathrm{F}}^2, \quad \text{s. t. } \|\boldsymbol{\theta}_t\|_0 \le \beta, \tag{3.6}$$

where $\mathbf{X} = [\mathbf{x}_1, \mathbf{x}_2, \ldots, \mathbf{x}_T]$ and $\boldsymbol{\Theta} = [\boldsymbol{\theta}_1, \boldsymbol{\theta}_2, \ldots, \boldsymbol{\theta}_T]$ contain the training samples and corresponding sparse coefficients, respectively.

The optimization problem (3.6) is solved by alternating optimization method in MOD. It alternatively solves sparse coding problem and dictionary fitting problem for updating. Fixing dictionary $\boldsymbol{\Psi}$, a number of efficient sparse coding algorithms can be used to update sparse coefficient $\boldsymbol{\Theta}$, e.g., OMP. When updating dictionary, it uses the overall representation mean square error:

$$||\mathbf{E}||_{\mathrm{F}}^2 = ||[\mathbf{e}_1, \mathbf{e}_2, \ldots, \mathbf{e}_T]||_{\mathrm{F}}^2 = ||\mathbf{X} - \boldsymbol{\Psi}\boldsymbol{\Theta}||_{\mathrm{F}}^2. \tag{3.7}$$

A close form updating rule with the minimum mean square error can be obtained:

$$\boldsymbol{\Psi} := \mathbf{X}\boldsymbol{\Theta}^{\mathrm{T}} \left(\boldsymbol{\Theta}\boldsymbol{\Theta}^{\mathrm{T}}\right)^{-1}. \tag{3.8}$$

It can be seen that the multiplication of large matrix and inverse operation in formula (3.8) may make the MOD algorithm very computationally expensive and suffer from high storage capacity.

3.1.2.2 K-Singular Value Decomposition (K-SVD)

To reduce computational complexity, K-singular value decomposition (K-SVD) is proposed [1], and (3.6) is also the optimization model.

The main contribution of K-SVD is to update one dictionary atom at a time due to the difficulty in optimizing the whole dictionary. It is realized by formulating the fitting error with the k-th atom removed, which can be formulated as follows:

$$\mathbf{E} = \mathbf{X} - \sum_{n=1}^{N} \boldsymbol{\psi}_n \boldsymbol{\theta}_n = \left(\mathbf{X} - \sum_{n \neq k} \boldsymbol{\psi}_n \boldsymbol{\theta}_n \right) - \boldsymbol{\psi}_k \boldsymbol{\theta}_k = \mathbf{E}_k - \boldsymbol{\psi}_k \boldsymbol{\theta}_k, \tag{3.9}$$

where $\boldsymbol{\theta}_n$ is a n-th row vector and $\boldsymbol{\psi}_n$ is a n-th column vector. Thus $\mathbf{X}$ can be seen as the sum of N rank-1 matrices. In K-SVD algorithm, we define $\mathbf{W} \in \mathbb{R}^{T \times |\mathbb{T}|}$, with ones at $(t, \mathbb{T}(t))$ and zeros elsewhere, where the set of index $\mathbb{T} = \{t \big| 1 \leq t \leq T,\ \theta_k(t) \neq 0\}$. By multiplying $\mathbf{W}$, the model (3.9) can be rewritten as

$$\|\mathbf{E}_k \mathbf{W} - \boldsymbol{\psi}_k \boldsymbol{\theta}_k \mathbf{W}\|_{\mathrm{F}}^2 = \|\tilde{\mathbf{E}}_k - \boldsymbol{\psi}_k \tilde{\boldsymbol{\theta}}_k\|_{\mathrm{F}}^2. \tag{3.10}$$

Then we can perform SVD on $\tilde{\mathbf{E}}_k$ as $\tilde{\mathbf{E}}_k = \mathbf{U}\boldsymbol{\Sigma}\mathbf{V}^{\mathrm{T}}$. Then the k-th dictionary atom is updated by $\boldsymbol{\psi}_k = \mathbf{u}_1$, and the coefficient is updated by $\tilde{\boldsymbol{\theta}}_k = \boldsymbol{\Sigma}(1, 1)\mathbf{v}_1$, where $\mathbf{u}_1$ and $\mathbf{v}_1$ are left and right singular vector with respect to the largest singular value, respectively.

Without matrix inverse calculation, the computational complexity of K-SVD algorithm is much lower than that of MOD algorithm. The coefficient matrix is updated jointly with dictionary atoms in the dictionary update step, which improves the convergence speed of the algorithm. K-SVD algorithm is one of the most widely used dictionary learning algorithms in practice. Inspired by K-SVD, a series of related methods such as discriminative K-SVD [64] and analysis K-SVD [47] have been developed for possible performance improvement.

3.1.2.3 Analysis K-Singular Value Decomposition (Analysis K-SVD)

Motivated by the analytic sparse model and K-SVD algorithm, analysis K-SVD algorithm is proposed to learn cosparse dictionary [47]. Similarly, the analysis dictionary learning can be divided into two tasks: analytic sparse coding and analytic dictionary learning. Given the analytic dictionary $\boldsymbol{\Omega}$ and noisy observation $\mathbf{Y} \in \mathbb{R}^{I \times T}$ of $\mathbf{X}$, the optimization model for analysis dictionary learning can be formulated as follows:

$$\begin{aligned} \min_{\boldsymbol{\Omega}, \mathbf{X}, \{\mathbb{I}_t\}_{t=1}^{T}} \quad & \|\mathbf{X} - \mathbf{Y}\|_{\mathrm{F}}^2 \\ \text{s. t.} \quad & \boldsymbol{\Omega}_{\mathbb{I}_t} \mathbf{x}_t = 0, \ \ \forall\, 1 \le t \le T, \\ & \mathrm{Rank}(\boldsymbol{\Omega}_{\mathbb{I}_t}) = I - r, \ \ \forall\, 1 \le t \le T, \\ & \|\boldsymbol{\omega}_k\|_2 = 1, \ \ \forall\, 1 \le k \le N, \end{aligned} \tag{3.11}$$

where $\mathbf{x}_t$ and $\mathbf{y}_t$ is t-th column of $\mathbf{X}$ and $\mathbf{Y}$, respectively. $\boldsymbol{\Omega}_{\mathbb{I}_t} \in \mathbb{R}^{|\mathbb{I}_t| \times I}$ is the sub-matrix of $\boldsymbol{\Omega}$, $\mathbb{I}_t$ includes indices of the rows orthogonal to $\mathbf{x}_t$, $\boldsymbol{\omega}_k$ is the k-th row of $\boldsymbol{\Omega}$, and r is the dimension of the subspace that signal $\mathbf{x}_t$ belongs to.

Fixing the dictionary $\boldsymbol{\Omega}$, the optimization problem (3.11) of $\mathbf{X}$ can be solved with respect to each column individually. Thus it is formulated as follows:

$$\begin{aligned} \min_{\mathbf{x}_t, \mathbb{I}_t} \ & \|\mathbf{x}_t - \mathbf{y}_t\|_2^2 \\ \text{s. t.} \ & \boldsymbol{\Omega}_{\mathbb{I}_t} \mathbf{x}_t = 0, \\ & \mathrm{Rank}(\boldsymbol{\Omega}_{\mathbb{I}_t}) = I - r, \end{aligned} \tag{3.12}$$

which can be solved by a number of cosparse recovery algorithms [47].

The other task is to update analysis dictionary. In [47], it is formulated as follows:

$$\begin{aligned} \min_{\boldsymbol{\omega}_k, \mathbf{X}_{\mathbb{T}}} \ & \|\mathbf{X}_{\mathbb{T}} - \mathbf{Y}_{\mathbb{T}}\|_{\mathrm{F}}^2 \\ \text{s. t.} \ & \boldsymbol{\Omega}_{\mathbb{I}_t} \mathbf{x}_t = 0, \ \ \forall\, t \in \mathbb{T} \\ & \mathrm{Rank}(\boldsymbol{\Omega}_{\mathbb{I}_t}) = I - r, \ \ \forall\, t \in \mathbb{T} \\ & \|\boldsymbol{\omega}_k\|_2 = 1, \end{aligned} \tag{3.13}$$

where $\mathbf{X}_{\mathbb{T}} \in \mathbb{R}^{I \times |\mathbb{T}|}$ and $\mathbf{Y}_{\mathbb{T}} \in \mathbb{R}^{I \times |\mathbb{T}|}$ are the sub-matrix of $\mathbf{X}$ and $\mathbf{Y}$, respectively. $\mathbb{T}$ includes indices of the columns orthogonal to $\boldsymbol{\omega}_k$. Then the problem (3.13) can be usually simplified as the following problem:

$$\begin{aligned} \min_{\boldsymbol{\omega}_t} & \|\boldsymbol{\omega}_t \mathbf{Y}_{\mathbb{T}}\|_2^2 \\ \text{s. t.} \ & \|\boldsymbol{\omega}_t\|_2 = 1, \end{aligned} \tag{3.14}$$

where $\mathbf{Y}_{\mathbb{T}}$ is the sub-matrix of $\mathbf{Y}$ and $\mathbb{T}$ includes indices of the columns orthogonal to $\boldsymbol{\omega}_t$. In fact, the analytic dictionary updating problem can be regarded as a regularized least squares problem. The solution is the eigenvector that corresponds to the smallest eigenvalue of matrix $\mathbf{Y}_{\mathbb{T}}$.

3.2 Tensor Dictionary Learning Based on Different Decompositions

Recent research has demonstrated the advantages of maintaining internal information of original data in higher-order data processing. When the signal is multidimensional, the above dictionary learning methods usually transform the high-dimensional data into the vector space by vectorizing it and perform dictionary learning using traditional matrix computation. However, this vectorization can lead to performance degeneration because it breaks the original multidimensional structure of the signal. To overcome this problem, tensor dictionary learning methods can handle input data in form of tensor and learn dictionaries based on tensor decompositions or tensor products, which can preserve the original structure and correlation along different dimensions. Besides, with the dimensionality curse, the number of entries exponentially grows, which will cause huge storage requirement and computation load. By using tensor decomposition, tensor dictionary learning model can use less parameters and computational requirement while maintaining the capability of exploring the multidimensional information of the input tensor.

There are several tensor dictionary learning algorithms based on different decompositions, such as Tucker Decomposition [13, 19, 22, 36, 39, 42, 46, 51, 69], CP decomposition [17], and t-SVD [48, 65]. In this section, we will demonstrate the fundamentals and algorithms of these tensor dictionary learning methods according to different data representation forms.

3.2.1 *Tucker Decomposition-Based Approaches*

In Tucker decomposition, a tensor can be decomposed into a core tensor with product on each mode with a factor matrix. In standard Tucker decomposition, the factor matrices are assumed to be orthonormal. In fact, the optimization model for tensor dictionary learning based on Tucker decomposition can be obtained easily by replacing some structural constraints in the decomposition model.

3.2.1.1 Classical Tucker-Based Tensor Dictionary Learning

Specifically, given a third-order tensor $\mathcal{Y}$ as an example, i.e., $\mathcal{Y} \in \mathbb{R}^{I_1 \times I_2 \times I_3}$, the optimization model for tensor dictionary learning in Tucker decomposition form can be formulated as follows [69]:

$$\begin{aligned} \min_{\mathcal{X}, \mathbf{D}_1, \mathbf{D}_2, \mathbf{D}_3} \quad & \|\mathcal{Y} - \mathcal{X} \times_1 \mathbf{D}_1 \times_2 \mathbf{D}_2 \times_3 \mathbf{D}_3\|_{\mathrm{F}}^2 \\ \text{s. t.} \quad & \mathrm{g}(\mathcal{X}) \leq K, \end{aligned} \tag{3.15}$$

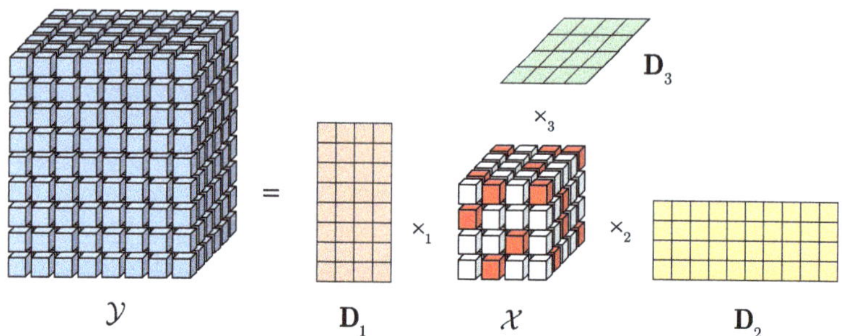

Fig. 3.3 Tensor dictionary learning in Tucker decomposition form

where matrices $\mathbf{D}_d \in \mathbb{R}^{I_d \times M_d}(d = 1, 2, 3)$ are taken as dictionaries, the core tensor $\mathcal{X} \in \mathbb{R}^{M_1 \times M_2 \times M_3}$ is the sparse coefficient, and $\mathrm{g}(\cdot)$ represents the sparsity regularization term like ℓ_0 or ℓ_1 operator. The illustration of the Tucker dictionary learning model is given in Fig. 3.3.

Then the classical Tucker-based tensor dictionary learning optimization model can be formulated as follows [13, 51]:

$$\min_{\mathcal{X},\mathbf{D}_1,\mathbf{D}_2,\mathbf{D}_3} \|\mathcal{Y} - \mathcal{X} \times_1 \mathbf{D}_1 \times_2 \mathbf{D}_2 \times_3 \mathbf{D}_3\|_{\mathrm{F}}^2, \quad \text{s. t. } \|\mathcal{X}\|_0 \leq K, \tag{3.16}$$

where K represents the maximum number of nonzero entries in the sparse coefficient $\mathcal{X}$. We can alternately optimize $\mathcal{X}$ and $\{\mathbf{D}_d,\ d = 1, 2, 3\}$ while fixing the other variables. Specifically, we have two main subproblems as follows:

1. *Sparse tensor coding*: With all dictionaries fixed, we need to solve:

$$\min_{\mathcal{X}} \|\mathcal{Y} - \mathcal{X} \times_1 \mathbf{D}_1 \times_2 \mathbf{D}_2 \times_3 \mathbf{D}_3\|_{\mathrm{F}}^2 + \lambda \|\mathcal{X}\|_0, \tag{3.17}$$

   where $\lambda > 0$. Seeking the nonzero elements of $\mathcal{X}$ sequentially, the approximation of signal can be rewritten in the following vector form [8]:

$$\mathbf{y} = (\mathbf{P}_3 \odot \mathbf{P}_2 \odot \mathbf{P}_1)\mathbf{a}, \tag{3.18}$$

   where the symbol $\odot$ is Khatri-Rao product. Matrices $\mathbf{P}_d \in \mathbb{R}^{I_d \times k}(d = 1, 2, 3)$ contain the selected atoms of mode-d dictionary in the previous k iterations, i.e., $\mathbf{P}_d(:, k) = \mathbf{D}_d(:, m_d^k)$, where m_d^k implies the selected index of mode-d dictionary in the k iteration and $\mathbf{a}$ is current nonzero elements of multidimensional data $\mathcal{X}$, i.e., $a_k = x_{m_1^k, m_2^k, m_3^k}$. So, we can obtain the update of $\mathbf{a}$ at k-th iteration from

$$\mathbf{a} = (\mathbf{P}_3 \odot \mathbf{P}_2 \odot \mathbf{P}_1)^{\dagger} \mathbf{y}, \tag{3.19}$$

where the symbol $\dagger$ represents Moore-Penrose pseudo-inverse. The proposed Kronecker-OMP is summarized in Algorithm 17. The initialized index set is empty set, that is, $\mathbb{M} = \emptyset$. Otherwise, with the block sparsity assumption, an N-way block orthogonal matching pursuit (N-BOMP) approach is proposed to solve multiway block sparsity constraint problem in [8].

Algorithm 17: Kronecker-OMP

Input: Dictionaries $\mathbf{D}_d \in \mathbb{R}^{I_d \times M_d}$ $(d = 1, 2, 3)$, signal $\mathcal{Y} \in \mathbb{R}^{I_1 \times I_2 \times I_3}$, sparsity level K, tolerance of representation ϵ.
Output: $\mathbf{a}$ and support set $\mathbb{M}$.
Initialize $\mathbb{M}_d = \emptyset (d = 1, 2, 3)$, $\mathbf{P}_d = \mathbf{0} \in \mathbb{R}^{I_d \times K} (d = 1, 2, 3)$, $\mathcal{R} = \mathcal{Y}$, $k = 1$.
while $k \leq K$ *and* $\|\mathcal{R}\|_F > \epsilon$ **do**
  Find the new index:
  $[m_1^k, m_2^k, m_3^k] = \text{argmax}_{\{m_1, m_2, m_3\}} |\mathcal{R} \times_1 \mathbf{D}_1^T(:, m_1) \times_2 \mathbf{D}_2^T(:, m_2) \times_3 \mathbf{D}_3^T(:, m_3)|$.
  Add into support set: $\mathbb{M}_d = \{\mathbb{M}_d, m_d^k\} (d = 1, 2, 3)$; $\mathbf{P}_d(:, k) = \mathbf{D}_d(:, m_d^k)$.
  Compute new solution: $\mathbf{a} = \text{argmin}_{\mathbf{a}} \|(\mathbf{P}_3 \odot \mathbf{P}_2 \odot \mathbf{P}_1)\mathbf{a} - \text{vec}(\mathcal{Y})\|_{\text{F}}^2$.
  Compute the residual: $\text{vec}(\mathcal{R}) = \text{vec}(\mathcal{Y}) - \mathbf{P}_3 \odot \mathbf{P}_2 \odot \mathbf{P}_1 \mathbf{a}$.
end

---

2. *Dictionary update*: When other variables are fixed, the dictionary $\mathbf{D}_d$ can be solved by an alternating least squares method. For example, to update $\mathbf{D}_1$, the optimization model (3.16) can be formulated as follows:

$$\min_{\mathbf{D}_1} \left\| \mathbf{Y}_{(1)} - \mathbf{D}_1 \mathbf{X}_{(1)} (\mathbf{D}_3 \otimes \mathbf{D}_2)^{\mathrm{T}} \right\|_{\mathrm{F}}^2. \tag{3.20}$$

For convenience, we set $\mathbf{C}_{(1)} = \mathbf{X}_{(1)} (\mathbf{D}_3 \otimes \mathbf{D}_2)^{\mathrm{T}}$. So we can update the dictionary $\mathbf{D}_1$ as

$$\mathbf{D}_1 = \left(\mathbf{Y}_{(1)} \mathbf{C}_{(1)}^{\mathrm{T}}\right) \left(\mathbf{C}_{(1)} \mathbf{C}_{(1)}^{\mathrm{T}}\right)^{\dagger}, \tag{3.21}$$

where $\dagger$ represents the Moore-Penrose pseudo-inverse of the matrix. $\mathbf{D}_2$ and $\mathbf{D}_3$ can be calculated in a similar way. Finally, this progress can be concluded in Algorithm 18.

Algorithm 18: Tensor-MOD dictionary learning

Input: Signal $\mathcal{Y} \in \mathbb{R}^{I_1 \times I_2 \times I_3}$, sparsity level K, tolerance of representation ϵ.
Output: Dictionaries $\mathbf{D}_d \in \mathbb{R}^{I_d \times M_d} (d = 1, 2, 3)$.
Initialize Dictionaries $\mathbf{D}_d \in \mathbb{R}^{I_d \times M_d} (d = 1, 2, 3)$ randomly;.
while *a stopping criterion is not met* **do**
  for $d = 1, \ldots, 3$:
    Update $\mathbf{D}_d$ via Eq. (3.21).
  end for
  Computing the corresponding sparse coefficient $\mathcal{X}$ via Algorithm 17.
  Updating the global error e.
end

3.2.1.2 Tucker-Based Tensor Dictionary Learning with Normalized Constraints

To avoid the pathological case in which the entries of sparse coefficients approach zero while the entries of dictionaries approach infinity, each matrix dictionary is commonly assumed to be normalized. The corresponding convex optimization model is formulated as follows [39]:

$$\begin{aligned} \min_{\mathcal{X}, \mathbf{D}_1, \mathbf{D}_2, \mathbf{D}_3} \quad & \frac{1}{2} \|\mathcal{Y} - \mathcal{X} \times_1 \mathbf{D}_1 \times_2 \mathbf{D}_2 \times_3 \mathbf{D}_3\|_{\mathrm{F}}^2 \\ \text{s. t.} \quad & \mathrm{g}(\mathcal{X}) \leq K; \\ & \|\mathbf{D}_d(:, m_d)\|_2^2 = 1, \ d = 1, 2, 3; \ m_d = 1, 2, \ldots, M_d. \end{aligned} \tag{3.22}$$

Solving such a problem, each variable can be optimized alternatively while fixing the others. In this way, problem (3.22) can be transformed into two subproblems: sparse tensor coding and tensor dictionary updating.

1. *Sparse tensor coding*: When all dictionary matrices in (3.22) are fixed, the optimization model for sparse tensor coding is as follows:

$$\min_{\mathcal{X}} \frac{1}{2} \|\mathcal{Y} - \mathcal{X} \times_1 \mathbf{D}_1 \times_2 \mathbf{D}_2 \times_3 \mathbf{D}_3\|_{\mathrm{F}}^2 + \lambda \, \mathrm{g}(\mathcal{X}), \tag{3.23}$$

which can be rewritten into

$$\min_{\mathcal{X}} \mathrm{f}(\mathcal{X}) + \lambda \, \mathrm{g}(\mathcal{X}), \tag{3.24}$$

where $\mathrm{f}(\mathcal{X}) = \frac{1}{2} \|\mathcal{Y} - \mathcal{X} \times_1 \mathbf{D}_1 \times_2 \mathbf{D}_2 \times_3 \mathbf{D}_3\|_{\mathrm{F}}^2$ is Lipschitz continuous and convex.

A tensor-based fast iterative shrinkage thresholding algorithm (TISTA) can be developed for solving (3.24). Similar to TISTA [6], at the k-th iteration, we can update the sparse tensor coefficient by solving the following problem:

$$\begin{aligned}\min_{\mathcal{X}} & \,\mathrm{f}(\mathcal{X}_{k-1}) + \langle \nabla \mathrm{f}(\mathcal{X}_{k-1}), \mathcal{X} - \mathcal{X}_{k-1} \rangle \\ & + \frac{L_k}{2} \|\mathcal{X} - \mathcal{X}_{k-1}\|_{\mathrm{F}}^2 + \lambda\, \mathrm{g}(\mathcal{X}),\end{aligned} \tag{3.25}$$

where $L_k = \eta^k \prod_{d=1}^{3} \left\|\mathbf{D}_d^{\mathrm{T}}\mathbf{D}_d\right\|_{\mathrm{F}} > 0$ with $\eta^k \geq 1$ is a Lipschitz constant and $\nabla \mathrm{f}(\cdot)$ is a gradient operator defined by

$$-\nabla \mathrm{f}(\mathcal{X}) = \mathcal{Y} \times_1 \mathbf{D}_1^{\mathrm{T}} \times_2 \mathbf{D}_2^{\mathrm{T}} \times_3 \mathbf{D}_3^{\mathrm{T}} - \mathcal{X} \times_1 \mathbf{D}_1^{\mathrm{T}}\mathbf{D}_1 \times_2 \mathbf{D}_2^{\mathrm{T}}\mathbf{D}_2 \times_3 \mathbf{D}_3^{\mathrm{T}}\mathbf{D}_3. \tag{3.26}$$

Then the problem (3.25) is equivalent to

$$\min_{\mathcal{X}} \frac{1}{2} \left\| \mathcal{X} - \left(\mathcal{X}_{k-1} - \frac{1}{L_k} \nabla \mathrm{f}(\mathcal{X}_{k-1}) \right) \right\|_{\mathrm{F}}^2 + \frac{\lambda}{L_k} \mathrm{g}(\mathcal{X}). \tag{3.27}$$

In fact, we can get different solutions with respect to different $\mathrm{g}(\mathcal{X})$. In case of $\mathrm{g}(\mathcal{X}) = \|\mathcal{X}\|_1$, the solution of (3.27) is

$$\mathcal{X}_k = \mathrm{S}_{\lambda/L_k} \left(\mathcal{X}_{k-1} - \frac{1}{L_k} \nabla \mathrm{f}(\mathcal{X}_{k-1}) \right),$$

where $\mathrm{S}_\tau(\cdot) = \mathrm{sign}(\cdot)\max(|\cdot| - \tau, 0)$ is the soft-thresholding operator. And if $\mathrm{g}(\mathcal{X}) = \|\mathcal{X}\|_0$, the solution of (3.27) is

$$\mathcal{X}_k = \mathrm{H}_{\lambda/L_k} \left(\mathcal{X}_{k-1} - \frac{1}{L_k} \nabla \mathrm{f}(\mathcal{X}_{k-1}) \right),$$

where $\mathrm{H}_\tau(\cdot) = \max(\cdot - \tau, 0)$ is the hard-thresholding operator.

In this chapter, we denote $\mathrm{G}_{\lambda/L_k}\left(\mathcal{X}_{k-1} - \frac{1}{L_k}\nabla \mathrm{f}(\mathcal{X}_{k-1})\right)$ as the solution of (3.27). Algorithm 19 presents the details of the TISTA algorithm. In order to speed up the convergence, the momentum method is considered as

$$\mathcal{C}_{k+1} = \mathcal{X}_k + \frac{t_k - 1}{t_{k+1}} (\mathcal{X}_k - \mathcal{X}_{k-1}),$$

where t_k varies with the iteration for further acceleration.

2. *Dictionary update*: For updating the dictionary $\mathbf{D}_d$ with fixed $\mathcal{X}$, the optimization model can be formulated as follows:

$$\begin{aligned}\min_{\mathbf{D}_d} & \left\| \mathbf{Y}_{(d)} - \mathbf{D}_d \mathbf{A}_{(d)} \right\|_{\mathrm{F}}^2 \\ \text{s. t.} & \;\; \|\mathbf{D}_d(:, m_d)\|_2^2 = 1,\, 1 \leq m_d \leq M_d,\end{aligned} \tag{3.28}$$

Algorithm 19: Tensor-based iterative shrinkage thresholding (TISTA)

Input: Dictionaries $\{\mathbf{D}_d\}$, $d = 1, 2, 3$, training Set $\mathcal{Y}$, number of iteration K.
Output: Sparse coefficient $\mathcal{X}$
Initialize: $\mathcal{C}_1 = \mathcal{X}_0 \in \mathbb{R}^{M_1 \times M_2 \times M_3}$, $t_1 = 1$
For $k = 1, \ldots, K$:
  Set $L^k = \eta^k \prod_{d=1}^{3} \|\mathbf{D}_d^{\mathrm{T}}\mathbf{D}_d\|_2$.
  Compute $\nabla \mathrm{f}(\mathcal{C}^k)$.
  Compute $\mathcal{X}_k$ via $\mathrm{G}_{\lambda/L_k}\left(\mathcal{C}_k - \frac{1}{L_k}\nabla \mathrm{f}(\mathcal{C}_k)\right)$.
  $t_{k+1} = \frac{1+\sqrt{1+4t_k^2}}{2}$
  $\mathcal{C}_{k+1} = \mathcal{X}_k + \frac{t_k-1}{t_{k+1}}(\mathcal{X}_k - \mathcal{X}_{k-1})$
End for

where

$$\mathcal{A} = \mathcal{X} \times_1 \mathbf{D}_1 \cdots \times_{d-1} \mathbf{D}_{d-1} \times_{d+1} \mathbf{D}_{d+1} \cdots \times_D \mathbf{D}_D \in \mathbb{R}^{I_1 \times \cdots \times I_{d-1} \times M_d \times I_{d+1} \times \cdots \times I_D},$$

so that $\mathcal{Y} \approx \mathcal{A} \times_d \mathbf{D}_d$. Unfolding $\mathcal{A}$ in d-mode, we can obtain $\mathbf{Y}_{(d)} \approx \mathbf{D}_d \mathbf{A}_{(d)}$.

Obviously, (3.28) is a quadratically constrained quadratic programming (QCQP) problem, which can be solved via the Lagrange dual method. The Lagrangian function about the problem (3.28) is formulated as

$$\begin{aligned} \mathrm{L}(\mathbf{D}_d, \boldsymbol{\delta}) = {} & \mathrm{tr}((\mathbf{Y}_{(d)} - \mathbf{D}_d \mathbf{A}_{(d)})^{\mathrm{T}} (\mathbf{Y}_{(d)} - \mathbf{D}_d \mathbf{A}_{(d)})) \\ & + \sum_{m_d=1}^{M_d} \delta_{m_d} \left(\sum_{i_d=1}^{I_d} \mathbf{D}_d(i_d, m_d)^2 - 1 \right), \end{aligned}$$

where $\boldsymbol{\delta} \in \mathbb{R}^{M_d}$ and $\delta_{m_d} > 0$ is dual parameter. Therefore, the Lagrange dual function is $D(\boldsymbol{\delta}) = \min_{\mathbf{D}_d} \mathrm{L}(\mathbf{D}_d, \boldsymbol{\delta})$. The optimal solution of $D(\boldsymbol{\delta})$ can be obtained by Newton's method or conjugate gradient. Then the optimal dictionary $\mathbf{D}_d^{\mathrm{T}} = (\mathbf{A}_{(d)} \mathbf{A}_{(d)}^{\mathrm{T}} + \Delta)^{-1} (\mathbf{Y}_{(d)} \mathbf{A}_{(d)}^{\mathrm{T}})^{\mathrm{T}}$ can be obtained by maximizing $D(\boldsymbol{\delta})$, where $\Delta = \mathrm{diag}(\boldsymbol{\delta})$.

We summarize the algorithm for tensor-based dictionary learning in Algorithm 20. As we can see, the tensor-based dictionary learning method reduces both the computational and memory costs in dealing with real-world multidimensional signals.

Algorithm 20: Tensor-based dictionary learning

Input: Training Set $\mathcal{Y}$, number of iteration K
Output: Learned dictionaries $\{\mathbf{D}_d\}$, $d = 1, 2, 3$
Initialize: Set the dictionary $\{\mathbf{D}_d\}$, $d = 1, 2, 3$.
For $k = 1, \ldots, K$:
  Sparse coding Step:
    Compute $\mathcal{X}$ via TISTA.
  Dictionary Update Step:
    $\mathcal{A} = \mathcal{X} \times_1 \mathbf{D}_1 \times_2 \mathbf{D}_2 \cdots \mathbf{D}_{d-1} \times_{d+1} \mathbf{D}_{d+1} \cdots \times_N \mathbf{D}_N$
    Get $\mathbf{A}_{(d)}$ and Update $\mathbf{D}_d$ via Eq. (3.28).
End for

3.2.1.3 Tucker Based on Tensor Dictionary Learning with Orthogonal Constraints

When the dictionaries are overcomplete, the corresponding sparse coefficient can be very sparse, and the sparse representation of original data can be improved. However, the large size of dictionary leads to high computational complexity, which makes the overcomplete tensor dictionary learning and its applications time-consuming in some applications. One effective way for processing time reduction is to design structured dictionary. Orthogonality is one of the most frequently used.

Based on the Tucker form, the optimization model for orthogonal tensor dictionary learning can be formulated as [22, 42]

$$\begin{aligned} \min_{\mathcal{X},\mathbf{D}_1,\mathbf{D}_2,\mathbf{D}_3} \quad & \frac{1}{2}\|\mathcal{Y} - \mathcal{X} \times_1 \mathbf{D}_1 \times_2 \mathbf{D}_2 \times_3 \mathbf{D}_3\|_{\mathrm{F}}^2 \\ \text{s. t.} \quad & g(\mathcal{X}) \le K, \\ & \mathbf{D}_d^{\mathrm{H}}\mathbf{D}_d = \mathbf{I}_d, \quad d = 1, 2, 3, \end{aligned} \tag{3.29}$$

which can be divided into sparse tensor coding and dictionary updating in similar way, and sparse tensor coding algorithms like OMP can be used with fixed dictionaries. When updating dictionary $\mathbf{D}_1$, an equivalent optimization model for it can be formulated as follows:

$$\begin{aligned} \min_{\mathbf{D}_1} \quad & \frac{1}{2}\|\mathbf{Y}_{(1)} - \mathbf{D}_1\mathbf{X}_{(1)} (\mathbf{D}_2 \otimes \mathbf{D}_3)^{\mathrm{H}} \|_{\mathrm{F}}^2 \\ \text{s. t.} \quad & \mathbf{D}_1^{\mathrm{H}}\mathbf{D}_1 = \mathbf{I}, \end{aligned} \tag{3.30}$$

which has a closed-form solution $\mathbf{D}_1 = \mathbf{U}\mathbf{V}^{\mathrm{H}}$, where $\mathbf{U}$ and $\mathbf{V}$ can be obtained by an SVD operation as follows:

$$\mathbf{Y}_{(1)} (\mathbf{D}_2 \otimes \mathbf{D}_3) \mathbf{X}_{(1)}{}^{\mathrm{H}} = \mathbf{U}\boldsymbol{\Sigma}\mathbf{V}^{\mathrm{H}}.$$

The rest dictionaries can be updated by solving similar optimization models.

3.2.2 CP Decomposition-Based Approaches

As the other classical tensor decompositions, canonical polyadic decomposition (CPD) motivates tensor dictionary learning too [3, 11, 17, 63, 67], which finds applications in spectral unmixing, medical image reconstruction, etc.

Before we illustrate the CP-based tensor dictionary learning models, we first introduce a definition of the mode-n product of a tensor and a vector.

Definition 3.2 The mode-n ($n = 1, 2, \ldots, N$) product of tensor $\mathcal{A} \in \mathbb{R}^{I_1 \times \cdots I_N}$ and vector $\mathbf{v} \in \mathbb{R}^{I_n}$ is defined by

$$\mathcal{B} = \mathcal{A} \times_n \mathbf{v} \in \mathbb{R}^{I_1 \times \cdots \times I_{n-1} \times I_{n+1} \times \cdots \times I_N},$$

whose entries are denoted as

$$\mathcal{B}(i_1, \ldots, i_{n-1}, i_{n+1}, \ldots, i_N) = \sum_{i_n=1}^{I_n} \mathcal{A}(i_1, i_2, \ldots, i_N)\mathbf{v}(i_n).$$

As an extension of K-SVD method, K-CPD is a new algorithm for tensor sparse coding but with tensor formulation [17, 63, 67]. Specifically, given a set of third-order training tensors $\mathcal{Y}_n \in \mathbb{R}^{I_1 \times I_2 \times I_3}$, $n = 1, 2, \ldots, N$, the optimization model for K-CPD tensor dictionary learning can be formulated as

$$\begin{aligned} &\min_{\mathcal{D}, \mathbf{x}_n} \sum_{n=1}^{N} \| \mathcal{Y}_n - \mathcal{D} \times_4 \mathbf{x}_n \|_F^2 \\ &\text{s. t. } \| \mathbf{x}_n \|_0 \leq K, \end{aligned} \tag{3.31}$$

where $\mathcal{D} \in \mathbb{R}^{I_1 \times I_2 \times I_3 \times I_0}$ is a tensor dictionary, I_0 is the number of atoms, $\mathbf{x}_n \in \mathbb{R}^{I_0}$ is the sparse coefficient vector, and K represents the sparsity level. Each $\mathcal{D}^{(i_0)} = \mathcal{D}(:, :, :, i_0) \in \mathbb{R}^{I_1 \times I_2 \times I_3}$ is the i_0-th atom, which is a rank-one tensor. The i_0-th atom can be rewritten as $\mathcal{D}^{(i_0)} = \mathbf{d}_{i_0}^{(1)} \circ \mathbf{d}_{i_0}^{(2)} \circ \mathbf{d}_{i_0}^{(3)}$, where $\mathbf{d}_{i_0}^{(1)} \in \mathbb{R}^{I_1}$, $\mathbf{d}_{i_0}^{(2)} \in \mathbb{R}^{I_2}$, and $\mathbf{d}_{i_0}^{(3)} \in \mathbb{R}^{I_3}$ are normalized vectors.

Likewise, the optimization model (3.31) can be solved in an alternating way. With fixed dictionary, the sparse tensor coding model can be formulated as follows:

$$\begin{aligned} &\min_{\mathbf{x}_n} \| \mathcal{Y}_n - \mathcal{D} \times_4 \mathbf{x}_n \|_F^2 \\ &\text{s. t. } \| \mathbf{x}_n \|_0 \leq K, \quad \forall n = 1, 2, \ldots, N. \end{aligned} \tag{3.32}$$

Following the idea of OMP [16], the multilinear orthogonal matching pursuit (MOMP) algorithm can be developed for solving (3.32). Algorithm 21 presents the details of the MOMP algorithm.

Algorithm 21: Multilinear orthogonal matching pursuit (MOMP)

Input: Tensor dictionary $\mathcal{D}$, signal $\mathcal{Y}_n \in \mathbb{R}^{I_1 \times I_2 \times I_3}$, sparsity level K,
tolerance of representation error ε.
Output: Sparse coefficient $\mathbf{x}_n$.
Initialize: $\mathcal{R}_1 = \mathcal{Y}_n$, $k = 1$, index set $\mathbb{I} = \emptyset$, $\mathbf{x}_n = \mathbf{0} \in \mathbb{R}^{I_0}$.
while $k \leq K$ & $\|\mathcal{R}_k\|_F > \varepsilon$ **do**
  Project error to each atom: $\mathcal{P}_{i_0} = \mathcal{R}_k \times_1 \mathbf{d}_{i_0}^{(1)} \times_2 \mathbf{d}_{i_0}^{(2)} \times_3 \mathbf{d}_{i_0}^{(3)}$.
  Update the dictionary by adding the atom corresponding to the maximal $\mathcal{P}_{i_0}$:
    $i = \arg\max_{i_0}(\mathcal{P}_{i_0})$, $\mathbb{I} = \{\mathbb{I}, i\}$;
    $\tilde{\mathcal{D}}_k = \mathbf{d}_i^{(1)} \circ \mathbf{d}_i^{(2)} \circ \mathbf{d}_i^{(3)}$;
    $\tilde{\mathcal{D}} = [\tilde{\mathcal{D}}_1, \ldots, \tilde{\mathcal{D}}_k]$.
  Update the sparse coefficient: $\mathbf{a}_k = \arg\min_{\mathbf{a}} \|\mathcal{Y}_n - \tilde{\mathcal{D}} \times_4 \mathbf{a}\|_{\mathrm{F}}^2$.
  Compute the residual: $\mathcal{R}_{k+1} = \mathcal{Y}_n - \tilde{\mathcal{D}} \times_4 \mathbf{a}_k$.
  Updating the counter: $k = k + 1$.
end
Return $\mathbf{x}_n(\mathbb{I}) = \mathbf{a}_k$.

To update the i_0-th atom $\mathcal{D}^{(i_0)}$, we need to fix other atoms. Analogously, those third-order tensor $\mathcal{Y}_n$ can be integrated into a fourth-order tensor $\mathcal{Y} \in \mathbb{R}^{I_1 \times I_2 \times I_3 \times N}$ with $\mathcal{Y}(:,:,:,n) = \mathcal{Y}_n$, $n = 1, \ldots, N$. The convex optimization model (3.31) can be formulated as follows:

$$\min_{\mathcal{D}} \|\mathcal{Y} - \mathcal{D} \times_4 \mathbf{X}\|_{\mathrm{F}}^2, \tag{3.33}$$

where $\mathbf{X} \in \mathbb{R}^{N \times I_0}$ represents the sparse representation coefficient matrix of $\mathcal{Y}$ and $\mathbf{x}_n$ is the n-th row of matrix $\mathbf{X}$.

Inspired from K-SVD algorithm, the optimization model can be formulated as follows:

$$\min_{\mathcal{D}^{(i_0)}} \|\mathcal{E}_{i_0} - \mathcal{D}^{(i_0)} \times_4 \mathbf{x}_{i_0}\|_{\mathrm{F}}^2, \tag{3.34}$$

where $\mathcal{E}_{i_0} = \mathcal{Y} - \sum_{j \neq i_0} \mathcal{D}^{(j)} \times_4 \mathbf{x}_j$ represents the approximation error for $\mathcal{Y}$ when the i_0-th atom $\mathcal{D}^{(i_0)}$ is removed and $\mathbf{x}_{i_0}$ is the i_0-th column of $\mathbf{X}$.

Simplifying the problem, we can just focus on the columns corresponding to the index of nonzero elements in $\mathbf{x}_{i_0}$ and denote them as $\mathcal{E}_{i_0}^{\mathrm{nz}}$, $\mathcal{Y}^{\mathrm{nz}}$, $\mathbf{X}^{\mathrm{nz}}$ and $\mathbf{x}_{i_0}^{\mathrm{nz}}$. In this way, the above formulation can be regarded as a general optimization problem of CP decomposition as follows:

$$\min_{\lambda_{i_0}, \{\mathbf{d}_{i_0}^{(k)}\}_{k=1}^4} \|\mathcal{E}_t^{\mathrm{nz}} - \lambda_{i_0} \mathbf{d}_{i_0}^{(1)} \circ \mathbf{d}_{i_0}^{(2)} \circ \mathbf{d}_{i_0}^{(3)} \circ \mathbf{d}_{i_0}^{(4)}\|_{\mathrm{F}}^2. \tag{3.35}$$

with

$$\begin{aligned} \mathcal{D}^{(i_0)} &= \mathbf{d}_{i_0}^{(1)} \circ \mathbf{d}_{i_0}^{(2)} \circ \mathbf{d}_{i_0}^{(3)}, \\ \mathbf{x}_{i_0}^{\text{nz}} &= \lambda_{i_0} \mathbf{d}_{i_0}^{(4)}. \end{aligned} \tag{3.36}$$

The detailed updating procedures are concluded in Algorithm 22.

Algorithm 22: K-CPD

Input: Training tensor $\mathcal{Y} \in \mathbb{R}^{I_1 \times I_2 \times I_3 \times N}$, sparsity level K, tolerance of representation error ε, and iteration number L.
Output: Tensor dictionary $\mathcal{D}$
Initialize: Tensor dictionary $\mathcal{D}$ and $l = 1$.
Obtain the sparse representation matrix $\mathbf{X}$ using MOMP.
For $i_0 = 1, \ldots, I_0$:
Find $\mathcal{Y}_n$, whose sparse representation coefficients on $\mathcal{D}^{(i_0)}$ are nonzero, and grouping them as a fourth-order tensor $\mathcal{Y}^{\text{nz}}$;
Calculate the error tensor $\mathcal{E}_{i_0}^{\text{nz}}$ of $\mathcal{Y}^{\text{nz}}$ without $\mathcal{D}^{(i_0)}$:
$\mathcal{E}_{i_0}^{\text{nz}} = \mathcal{Y}^{\text{nz}} - \mathcal{D} \times_4 \mathbf{X}^{\text{nz}} + \mathcal{D}^{(i_0)} \times_4 \mathbf{x}_{i_0}^{\text{nz}}$
Decompose the error tensor using Eq. (3.35);
Update the i_0-th atom and the corresponding coefficients using Eq. (3.36);
End for
$l = l + 1$
Repeat the above steps until the convergence criteria are met or $l > L$.
Return $\mathcal{D}$.

In [3], the tensor dictionary learning problem can also be solved easily and efficiently by alternating direction method of multipliers (ADMM). The augmented Lagrangian function of the problem (3.31) is formulated as follows:

$$\mathrm{L}(\mathbf{X}, \mathbf{G}, \mathcal{D}, \mathbf{Q}) = \frac{1}{2} \|\mathcal{Y} - \mathcal{D} \times_4 \mathbf{X}\|_{\mathrm{F}}^2 + \lambda \sum_{n=1}^{N} \|\mathbf{G}(n,:)\|_0 + \langle \mathbf{Q}, \mathbf{G} - \mathbf{X} \rangle + \frac{\eta}{2} \|\mathbf{G} - \mathbf{X}\|_{\mathrm{F}}^2,$$

where $\eta > 0$ is the augmented Lagrangian parameter and $\mathbf{Q}$ is the Lagrangian multiplier. Then each variable can be optimized alternatively while fixing the others.

3.2.3 T-Linear-Based Approaches

In sparse matrix representation, given an overcomplete dictionary $\mathbf{D}$, a signal $\mathbf{y}$ can be represented as a linear combination of columns of $\mathbf{D}$, i.e., $\mathbf{y} = \mathbf{D}\mathbf{x}$. Similarly, a given tensor data $\mathcal{Y}$ can also be represented as the t-product of two tensors as follows:

$$\mathcal{Y} = \mathcal{D} * \mathcal{X}, \tag{3.37}$$

where $\mathcal{Y} \in \mathbb{R}^{I_1 \times N \times I_2}$, $\mathcal{D} \in \mathbb{R}^{I_1 \times M \times I_2}$ is the tensor dictionary, $\mathcal{X} \in \mathbb{R}^{M \times N \times I_2}$ is the sparse coefficient.

Corresponding to different sparsity constraints, there are different tensor sparse coding schemes based on t-product. The optimization model for t-product-based tensor dictionary learning can be formulated as follows:

$$\begin{aligned} &\min_{\mathcal{X},\mathcal{D}} \|\mathcal{Y} - \mathcal{D} * \mathcal{X}\|_{\mathrm{F}}^2 \\ &\text{s. t. } \mathrm{g}(\mathcal{X}) \leq K, \end{aligned} \tag{3.38}$$

where $\mathrm{g}(\cdot)$ represents the sparsity regularization term like ℓ_1 defined as

$$\|\mathcal{X}\|_1 = \sum_{m,n,i_2} |\mathcal{X}(m, n, i_2)|,$$

or $\ell_{1,1,2}$ defined by

$$\|\mathcal{X}\|_{1,1,2} = \sum_{m,n} \|\mathcal{X}(m, n, :)\|_{\mathrm{F}}.$$

Figure 3.4 illustrates the t-product-based sparse tensor representation with different sparsity measurements, in which red blocks represent nonzero elements and $\overrightarrow{\mathcal{Y}}_n$ is the n-th lateral slice for $\mathcal{Y}$.

For example, in [65], Zhang et al. use $\|\cdot\|_{1,1,2}$ norm to represent tubal sparsity of a tensor, and the corresponding optimization model is given by

$$\min_{\mathcal{D},\mathcal{X}} \quad \frac{1}{2}\|\mathcal{Y} - \mathcal{D} * \mathcal{X}\|_{\mathrm{F}}^2 + \lambda\|\mathcal{X}\|_{1,1,2}, \tag{3.39}$$

where $\lambda > 0$ is penalty parameter. Motivated by the classical K-SVD algorithm, the K-TSVD algorithm is presented in Algorithm 23 to solve the t-product-based tensor dictionary learning problem (3.39).

Especially, the optimization problem with respect to $\mathcal{X}$ can be reformulated in Fourier domain as

$$\min_{\overline{\mathbf{X}}} \quad \frac{1}{2}\|\overline{\mathbf{Y}} - \overline{\mathbf{D}}\overline{\mathbf{X}}\|_{\mathrm{F}}^2 + \lambda\sqrt{I_2}\|\hat{\mathcal{X}}\|_{1,1,2}, \tag{3.40}$$

where $\overline{\mathbf{X}}$, $\overline{\mathbf{Y}}$, $\overline{\mathbf{D}}$ represent block diagonal matrix form of $\mathcal{X}$, $\mathcal{Y}$, $\mathcal{D}$ in Fourier domain respectively, $\hat{\mathcal{X}}$ is the fast Fourier transform (FFT) of $\mathcal{X}$ along the third mode and

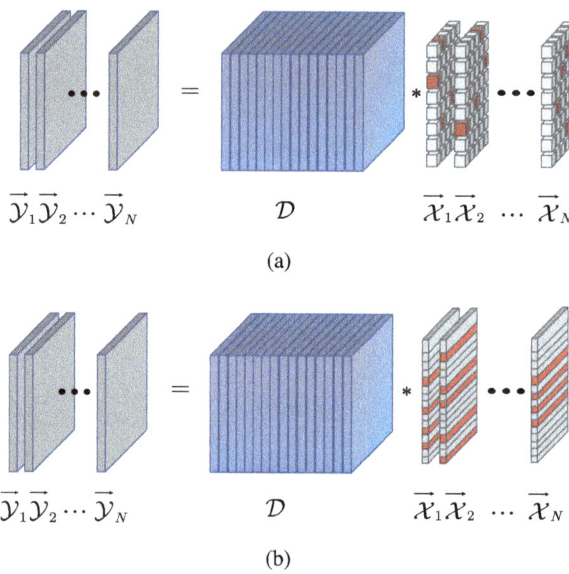

Fig. 3.4 Sparse tensor representation based on t-product (**a**) with ℓ_1 constraint or (**b**) with tubal sparsity constrained by $\ell_{1,1,2}$ norm

$\|\mathcal{X}\|_{\mathrm{F}} = \|\hat{\mathcal{X}}\|_{\mathrm{F}}/\sqrt{I_2}$. The problem (3.40) can be solved easily and efficiently by ADMM. We can divide the problem (3.40) into three subproblems as follows:

$$\begin{aligned}
\overline{\mathbf{X}}_{k+1} &= \arg\min_{\overline{\mathbf{X}}} \left\|\overline{\mathbf{Y}} - \overline{\mathbf{D}}\overline{\mathbf{X}}\right\|_{\mathrm{F}}^2 + \operatorname{tr}\left(\overline{\mathbf{L}}_k^{\mathrm{T}}\overline{\mathbf{X}}\right) + \frac{\rho}{2}\left\|\overline{\mathbf{X}} - \overline{\mathbf{Z}}_k\right\|_{\mathrm{F}}^2, \\
\mathcal{Z}_{k+1} &= \arg\min_{\mathcal{Z}} \|\mathcal{Z}\|_{1,1,2} + \frac{\rho}{2\lambda}\left\|\mathcal{X}_{k+1} + \frac{1}{\rho}\mathcal{L}_k - \mathcal{Z}\right\|_{\mathrm{F}}^2, \\
\mathcal{L}_{k+1} &= \mathcal{L}_k + \rho\left(\mathcal{X}_{k+1} - \mathcal{Z}_{k+1}\right),
\end{aligned} \tag{3.41}$$

where $\rho > 0$ is penalty parameter and $\mathcal{L}$ is Lagrangian multiplier.

Besides, based on the standard t-product-based tensor dictionary learning (3.38), some constraints can be added for improved performance. For instance, the optimization model based on normalized dictionaries can be formulated as follows: [23, 24]:

$$\begin{aligned}
\min_{\mathcal{D},\mathcal{X}} \quad & \frac{1}{2}\|\mathcal{Y} - \mathcal{D} * \mathcal{X}\|_{\mathrm{F}}^2 + \lambda\, \mathrm{g}(\mathcal{X}) \\
\text{s. t.} \quad & \|\mathcal{D}(:, m, :)\|_{\mathrm{F}}^2 \le 1, \quad m = 1, \ldots, M,
\end{aligned} \tag{3.42}$$

where $\mathrm{g}(\mathcal{X})$ is $||\mathcal{X}||_1$ in [23], or $||\mathcal{X}||_{1,1,2}$ in [24]. With given tensor dictionary, this kind of sparse tensor coding problem can be solved by a tensor-based fast iterative

Algorithm 23: K-TSVD

Input: Observed tensor data $\mathcal{Y} \in \mathbb{R}^{I_1 \times N \times I_2}$, $\lambda > 0$
Output: Trained tensor dictionary $\mathcal{D}$
Initialize: Dictionary $\mathcal{D}_0 \in \mathbb{R}^{I_1 \times M \times I_2}$
Iterate until convergence:
Update $\mathcal{X}$ by ADMM:
$\mathcal{X} = \arg\min_{\mathcal{X}} \|\mathcal{Y} - \mathcal{D} * \mathcal{X}\|_{\mathrm{F}}^2 + \lambda \|\mathcal{X}\|_{1,1,2}$
For $m = 1, \ldots, M$:
1. Let $w_m = \{n | \mathcal{X}(m, n, :) \neq 0\}$ be the set of indices where data $\mathcal{Y}$'s sparse representation coefficients on $\mathcal{D}(:, m, :)$ are nonzero.
2. Calculate the error tensor $\mathcal{E}_m$ of $\mathcal{Y}$ without $\mathcal{D}(:, m, :)$:
$\mathcal{E}_m = \mathcal{Y} - \sum_{j \neq m} \mathcal{D}(:, j, :) * \mathcal{X}(j, :, :)$
3. Restrict $\mathcal{E}_m$ by choosing only the tensor columns corresponding to w_m and obtain $\mathcal{R}_m$:
$\mathcal{R}_m = \mathcal{E}_m(:, w_m, :)$
4. Compute the t-SVD of $\mathcal{R}_m$:
$\mathcal{R}_m = \mathcal{U} * \mathcal{S} * \mathcal{V}^{\mathrm{T}}$
5. Update the m-th atom and corresponding coefficients:
$\mathcal{D}(:, m, :) = \mathcal{U}(:, 1, :)$
$\mathcal{X}(m, w_m, :) = \mathcal{S}(1, 1, :) * \mathcal{V}(:, 1, :)^{\mathrm{T}}$
End for

shrinkage thresholding algorithm (TFISTA) [24]. And the dictionary can be updated by ADMM in the frequency domain when the sparse coefficient is fixed.

3.3 Analysis Tensor Dictionary Learning

The last section lists a series of algorithms based on different tensor decompositions forms, and most of them can be regarded as the ones based on the synthetic sparse tensor models. Correspondingly, there exist some analytic tensor dictionary learning methods. For example, the analysis tensor dictionary learning in the form of Tucker decomposition can be found in [10, 40].

3.3.1 Analysis Tensor Dictionary Learning Based on Mode-n Product

3.3.1.1 Noiseless Analysis Tensor Dictionary Learning

In a similar fashion, we extract the multilinear features of $\mathcal{Y}$ by multiplying a series of analytic dictionaries $\mathbf{\Omega}_d \in \mathbb{R}^{P_d \times I_d}$, $d = 1, 2, \ldots, D$ along d-th mode to obtain a sparse tensor $\mathcal{X} \in \mathbb{R}^{P_1 \times P_2 \cdots \times P_D}$. The optimization model of analytic tensor dictionary learning in Tucker form can be formulated as follows:

$$\begin{aligned} \min_{\mathcal{X},\{\boldsymbol{\Omega}_d\}_{d=1}^D} & \|\mathcal{X} - \mathcal{Y} \times_1 \boldsymbol{\Omega}_1 \times_2 \boldsymbol{\Omega}_2 \cdots \times_D \boldsymbol{\Omega}_D\|_{\mathrm{F}}^2 \\ \text{s. t.} & \|\mathcal{X}\|_0 \le P - K, \end{aligned} \tag{3.43}$$

where $\mathcal{X}$ is K-cosparse, which has only K nonzero entries, $P = \prod_d P_d$. $\boldsymbol{\Omega}_d$ is the d-th dictionary, $P_d \ge I_d, d = 1, 2, \ldots, D$.

With the normalization regularization term, the target problem in [10] is as follows:

$$\begin{aligned} \min_{\mathcal{X},\{\Omega_d\}_{d=1}^D} & \|\mathcal{X} - \mathcal{Y} \times_1 \boldsymbol{\Omega}_1 \times_2 \boldsymbol{\Omega}_2 \cdots \times_D \boldsymbol{\Omega}_D\|_{\mathrm{F}}^2 \\ \text{s. t.} & \|\mathcal{X}\|_0 = P - K, \\ & \|\boldsymbol{\Omega}_d(p_d, :)\|_2 = 1, \quad p_d = 1, 2, \ldots, P_d; \ d = 1, 2, \ldots, D, \end{aligned} \tag{3.44}$$

which can be solved efficiently by ADMM. In analysis sparse coding, $\mathcal{X}$ can be solved by using a hard thresholding operator on $\mathcal{X}$. And in dictionary update stage, the subproblem can be transformed into matrix form

$$\begin{aligned} & \min_{\Omega_d} \|\mathbf{X}_{(d)} - \boldsymbol{\Omega}_d \mathbf{Y}_{(d)} (\boldsymbol{\Omega}_D \otimes \cdots \otimes \boldsymbol{\Omega}_{d+1} \otimes \boldsymbol{\Omega}_{d-1} \otimes \cdots \otimes \boldsymbol{\Omega}_1)^{\mathrm{T}}\|_{\mathrm{F}}^2 \\ & \text{s. t.} \ \|\boldsymbol{\Omega}_d(p_d, :)\|_2 = 1, \quad p_d = 1, 2, \ldots, P_d, \end{aligned} \tag{3.45}$$

which can be solved by simultaneous codeword optimization [12].

3.3.1.2 Analysis Tensor Dictionary Learning with Noisy Observations

In this subsection, we consider noisy observation $\mathcal{Q}$ of $\mathcal{Y}$ as

$$\mathcal{Q} = \mathcal{Y} + \mathcal{E}, \tag{3.46}$$

where $\mathcal{E}$ is the white-Gaussian additive noise. In this case, the optimization model for updating analytic dictionaries $\{\boldsymbol{\Omega}_d\}_{d=1}^D$ and recovering the clean data $\mathcal{Y}$ is as follows [40]:

$$\begin{aligned} \min_{\{\Omega_d\}_{d=1}^D,\mathcal{Y}} & \|\mathcal{Q} - \mathcal{Y}\|_{\mathrm{F}}^2 \\ \text{s. t.} & \|\mathcal{Y} \times_1 \boldsymbol{\Omega}_1 \times_2 \boldsymbol{\Omega}_2 \cdots \times_D \boldsymbol{\Omega}_D\|_0 = P - K, \\ & \|\boldsymbol{\Omega}_d(p_d, :)\|_2 = 1, \quad p_d = 1, 2, \ldots, P_d; \ d = 1, 2, \ldots, D. \end{aligned} \tag{3.47}$$

The problem (3.47) can be solved by a two-phase block coordinate relaxation algorithm.

1. *Subproblem with respect to* $\mathcal{Y}$: With fixed analytic dictionaries $\{\mathbf{\Omega}_d\}_{d=1}^{D}$, we let $\mathbf{\Omega} = \mathbf{\Omega}_D \otimes \mathbf{\Omega}_{D-1} \otimes \cdots \otimes \mathbf{\Omega}_2 \otimes \mathbf{\Omega}_1 \in \mathbb{R}^{P \times I}$ with $P = \prod_d P_d$ and $I = \prod_d I_d$, and the subproblem of (3.47) according to $\mathcal{Y}$, which is also called cosparse tensor coding problem, can be formulated as

$$\begin{aligned} \min_{\text{vec}(\mathcal{Y})} & \; \| \text{vec}(\mathcal{Q}) - \text{vec}(\mathcal{Y}) \|_{\text{F}}^2 \\ \text{s. t.} & \; \|\mathbf{\Omega} \, \text{vec}(\mathcal{Y})\|_0 \leq P - K, \; \|w_p\|_2 = 1, \; p = 1, \ldots, P, \end{aligned} \tag{3.48}$$

where w is the p-th row of $\mathbf{\Omega}$. Thus the problem (3.48) is a 1D sparse coding problem; we can solve it using backward greedy (BG) [47], greedy analysis pursuit (GAP) [32], and so on.
2. *Subproblem with respect to* $\mathbf{\Omega}_d$: The optimization model for updating $\mathbf{\Omega}_d$ is non-convex, and an alternative optimization algorithm is employed. We define

$$\mathbf{U}_d = \mathbf{Y}_{(d)} \left(\mathbf{\Omega}_D \otimes \cdots \otimes \mathbf{\Omega}_{d+1} \otimes \mathbf{\Omega}_{d-1} \otimes \cdots \otimes \mathbf{\Omega}_1 \right)^{\text{T}},$$

$$\mathbf{W}_d = \mathbf{Q}_{(d)} \left(\mathbf{\Omega}_D \otimes \cdots \otimes \mathbf{\Omega}_{d+1} \otimes \mathbf{\Omega}_{d-1} \otimes \cdots \otimes \mathbf{\Omega}_1 \right)^{\text{T}},$$

the optimization problem with respect to $\mathbf{\Omega}_d$ can be rewritten as

$$\begin{aligned} \min_{\mathbf{\Omega}_d} & \; \|\mathbf{\Omega}_d \mathbf{U}_d - \mathbf{\Omega}_d \mathbf{W}_d\|_{\text{F}}^2 \\ \text{s. t.} & \; \|\mathbf{\Omega}_d \mathbf{U}_d\|_0 = P - K, \\ & \; \|\mathbf{\Omega}_d(p_d, :)\|_2 = 1, \quad p_d = 1, 2, \ldots, P_d. \end{aligned} \tag{3.49}$$

We set $\mathbf{W}_{d,\mathbb{T}}$ as the sub-matrix of $\mathbf{W}_d$, and $\mathbb{T}$ includes indices of the columns in $\mathbf{U}_d$ that are orthogonal to $\mathbf{\Omega}_d(p_d, :)$. Then the optimization model can be formulated as follows:

$$\begin{aligned} \min_{\mathbf{\Omega}_d(p_d,:)} & \; \|\mathbf{\Omega}_d(p_d, :)\mathbf{W}_{d,\mathbb{T}}\|_2^2 \\ \text{s. t.} & \; \|\mathbf{\Omega}_d(p_d, :)\|_2 = 1. \end{aligned} \tag{3.50}$$

The solution can refer to Sect. 3.1.2.3, and the detailed updating procedures are concluded in Algorithm 24.

Algorithm 24: Multidimensional analytic dictionary learning algorithm

Input: Training set $\mathcal{Y}$, number of iteration T
Output: Learned dictionaries $\{\boldsymbol{\Omega}_d\}_{d=1}^D$
Initialize: $\{\boldsymbol{\Omega}_d\}_{d=1}^D$.
For $t = 1, \ldots, T$:
  Cosparse coding Step:
    $\boldsymbol{\Omega} = \boldsymbol{\Omega}_D \otimes \boldsymbol{\Omega}_{D-1} \otimes \cdots \boldsymbol{\Omega}_2 \otimes \boldsymbol{\Omega}_1$
    Compute $\mathcal{X}$ via Eq. (3.48).
  Dictionary update Step:
    For each row $\boldsymbol{\Omega}_d(p_d, :)$ in $\boldsymbol{\Omega}_d$, update it by Eq. (3.50).
End for

3.3.2 Tensor Convolutional Analysis Dictionary Learning Model

In addition to the Tucker form, other convolutional analysis models for dictionary learning have also been developed recently [58]. The corresponding optimization can be written as

$$\min_{\{\mathcal{O}_d\}_{d=1}^D, \mathcal{Y}} \sum_{d=1}^{D} \|\mathcal{O}_d \boxtimes \mathcal{Y}\|_0, \tag{3.51}$$

where $\mathcal{Y} \in \mathbb{R}^{I_1 \times I_2 \cdots \times I_N}$ is the input tensor, $\boxtimes$ denotes the convolution operator as defined in Chap. 1, and $\mathcal{O}_d \in \mathbb{R}^{I_1 \times I_2 \cdots \times I_N}$ $(d = 1, \ldots, D)$ are a series of analysis dictionary atoms desired to estimate.

In order to avoid trivial solutions, the N-th order tensor atom $\mathcal{O}_d$ can be decomposed as the convolution of two N-th order tensors with smaller size. For convenience, here uses a double index to represent the D dictionary atoms as $\{\mathcal{O}_{1,1}, \mathcal{O}_{1,2}, \ldots, \mathcal{O}_{I,J}\}$ where $IJ = D$. Then, the orthogonality constrained convolutional factorization of i, j-th atom $\mathcal{O}_{i,j}$ can be represented as

$$\mathcal{O}_{i,j} = \mathcal{U}_i \boxtimes \mathcal{V}_j,$$

where the factors $\mathcal{U}_i$ and $\mathcal{V}_j$ satisfy

$$\begin{aligned} \mathbb{U} &= \left\{\mathcal{U}_i | \left\langle \mathcal{U}_{i_1}, \mathcal{U}_{i_2} \right\rangle = \delta_{i_1, i_2}, \forall i_i, i_2 = 1, \ldots, I\right\}_{i=1}^{I} \\ \mathbb{V} &= \left\{\mathcal{V}_j | \left\langle \mathcal{V}_{j_1}, \mathcal{V}_{j_2}, \right\rangle = \delta_{j_1, j_2}, \forall j_1, j_2 = 1, \ldots, J\right\}_{j=1}^{J}, \end{aligned} \tag{3.52}$$

where $\mathbb{U}, \mathbb{V}$ are collections of two N-th order orthogonal factor dictionaries,

$$\delta_{i,j} = \begin{cases} 1, & i = j \\ 0, & i \neq j \end{cases}$$

is the Kronecker delta.

In this case, if $\mathcal{U}_i \in \mathbb{R}^{P\times\cdots\times P}$, $\mathcal{V}_j \in \mathbb{R}^{L\times\cdots\times L}$, the resulting $\mathcal{O}_{i,j}$ will be sized of $(P + L - 1) \times (P + L - 1) \cdots \times (P + L - 1)$ with the same order. By the way, it is necessary to note that the orthogonality of factor dictionaries $\mathbb{U}, \mathbb{V}$ is not to say that the dictionary $\mathcal{O}$ is orthogonal.

Then the optimization model (3.51) can be reformulated as follows:

$$\begin{aligned} \min_{\mathcal{Y},\mathcal{O}} \quad & \sum_{i,j=1}^{I,J} \|\mathcal{O}_{i,j} \boxtimes \mathcal{Y}\|_0 \\ \text{s. t.} \quad & \mathcal{O}_{i,j} = \mathcal{U}_i \boxtimes \mathcal{V}_j, \forall i, j \\ & \mathcal{U}_i \in \mathbb{U}, \mathcal{V}_j \in \mathbb{V}, \end{aligned} \tag{3.53}$$

which can be solved easily and efficiently by ADMM.

3.4 Online Tensor Dictionary Learning

The previously discussed tensor dictionary learning methods are offline. Each iteration needs to be calculated with all training samples, which requires high storage space and computational complexity. Especially when the samples are large-scale, the learning can be very slow. Online dictionary learning is one of the main ways for real-time processing when the data are acquired dynamically.

There are some online tensor dictionary learning algorithms for fast dictionary learning [26, 43, 50, 53, 54, 68]. The data tensor for online tensor dictionary is assumed to be $\mathcal{Y}_\tau \in \mathbb{R}^{I_1\times I_2\cdots\times I_D} (\tau = 1, \ldots, t)$, which is an observed sequence of t N-order tensors. And $\mathcal{Y} = \{\mathcal{Y}_1, \mathcal{Y}_2, \ldots, \mathcal{Y}_t\}$ accumulates a number of previously acquired training signals. Therefore, we can build up an optimization model of tensor dictionary learning in the Tucker form [53] as follows:

$$\min_{\mathcal{X},\{\mathbf{D}_d\}_{d=1}^D} \frac{1}{2} \|\mathcal{Y} - \mathcal{X} \times_1 \mathbf{D}_1 \times_2 \mathbf{D}_2 \cdots \times_D \mathbf{D}_D\|_F^2 + g_1(\mathcal{X}) + g_2(\mathbf{D}_1, \ldots, \mathbf{D}_D), \tag{3.54}$$

where g_1, g_2 are general penalty functions.

For each newly added tensor $\mathcal{Y}_t$, two processes are executed to retrain the dictionaries: sparse coding and dictionary updating.

1. *Sparse tensor coding of* $\mathcal{X}_t$: Online sparse coding is different from offline sparse coding. When new training data is arrived, only the sparse coefficient with respect to the newly added sample needs to be obtained. Therefore, with all dictionary matrices fixed and $\mathcal{X}_\tau (\tau = 1, \ldots, t-1)$ unchanged, the sparse representation $\mathcal{X}_t$ for newly added tensor $\mathcal{Y}_t$ can be computed. The optimization model for sparse tensor coding is as follows:

$$\mathcal{X}_t = \arg\min_{\mathcal{X}} \frac{1}{2} \left\| \mathcal{Y}_t - \mathcal{X} \times_1 \mathbf{D}_1^{t-1} \times_2 \mathbf{D}_2^{t-1} \cdots \times_D \mathbf{D}_D^{t-1} \right\|_{\mathrm{F}}^2 + \mathrm{g}_1(\mathcal{X}), \tag{3.55}$$

where $\left\{\mathbf{D}_d^{t-1} \in \mathbb{R}^{I_d \times M_d} (d = 1, \cdots, D)\right\}$ represent the dictionaries updated at last iteration.

We set $\mathrm{f}(\mathcal{X}) = \frac{1}{2} \left\| \mathcal{Y}_t - \mathcal{X} \times_1 \mathbf{D}_1^{t-1} \times_2 \mathbf{D}_2^{t-1} \cdots \times_D \mathbf{D}_D^{t-1} \right\|_{\mathrm{F}}^2$, which is differentiable. Therefore, (3.55) can be tackled by a proximal minimization approach [7], and the detailed updating procedures are concluded in Algorithm 25, where $\mathrm{prox}_{\mathrm{g}_1}$ is a proximal operator and

$$\begin{aligned} \frac{\partial \mathrm{f}(\mathcal{X})}{\partial \mathcal{X}} &= -\mathcal{Y}_t \times_1 (\mathbf{D}_1^{t-1})^{\mathrm{T}} \times_2 (\mathbf{D}_2^{t-1})^{\mathrm{T}} \cdots \times_D (\mathbf{D}_D^{t-1})^{\mathrm{T}} \\ &\quad + \mathcal{X} \times_1 (\mathbf{D}_1^{t-1})^{\mathrm{T}} \mathbf{D}_1^{t-1} \times_2 (\mathbf{D}_2^{t-1})^{\mathrm{T}} \mathbf{D}_2^{t-1} \cdots \times_D (\mathbf{D}_D^{t-1})^{\mathrm{T}} \mathbf{D}_D^{t-1}. \end{aligned}$$

Algorithm 25: Update the sparse coefficient

Input: New training tensor $\mathcal{Y}_t$, dictionaries $\left\{\mathbf{D}_d^{t-1}\right\}$, $d = 1, \ldots, D$, the gradient descent step η, initial sparse coefficient $\mathcal{X}^0$
Output: Sparse coefficient $\mathcal{X}_t$ corresponding to $\mathcal{Y}_t$
Initialize: $k = 0$, $\mathcal{X}^k = \mathcal{X}^0$
while *a stopping criterion is not met* **do**
  $\mathcal{X}^{k+1} = \mathrm{prox}_{\mathrm{g}_1}\left(\mathcal{X}^k - \eta \frac{\partial \mathrm{f}(\mathcal{X}^k)}{\partial \mathcal{X}}\right)$
  $k = k + 1$
end
Return $\mathcal{X} = \mathcal{X}^k$.

---

2. *Tensor dictionary update*: For updating the dictionary $\mathbf{D}_d$ with fixed $\mathcal{X}$, the optimization model can be formulated as follows:

$$\mathbf{D}_d^t = \arg\min_{\mathbf{D}_d} \mathrm{L}_{d,t}(\mathbf{D}_d), \tag{3.56}$$

with

$$\begin{aligned}\mathrm{L}_{d,t}(\mathbf{D}_d) =& \frac{1}{t}\sum_{\tau=1}^{t}\frac{1}{2}\left\|\mathcal{Y}_\tau - \mathcal{X}_\tau \times_1 \mathbf{D}_1^t \cdots \times_{d-1} \mathbf{D}_{d-1}^t \times_d \mathbf{D}_d \times_{d+1} \mathbf{D}_{d+1}^{t-1}\right. \\ &\left.\cdots \times_D \mathbf{D}_D^{t-1}\right\|_\mathrm{F}^2 \\ &+ g_2\left(\mathbf{D}_1^t, \ldots, \mathbf{D}_{d-1}^t, \mathbf{D}_d, \mathbf{D}_{d+1}^{t-1} \ldots, \mathbf{D}_D^{t-1}\right).\end{aligned} \tag{3.57}$$

With respect to $\mathbf{D}_d$, it can be updated by block coordinate descent algorithm. For convenience, the gradient of $\mathrm{L}_{d,t}\left(\mathbf{D}_d\right)$ can be given by

$$\frac{\partial \mathrm{L}_{d,t}\left(\mathbf{D}_d\right)}{\partial \mathbf{D}_d} = -\frac{\mathbf{U}_d^t}{t} + \frac{\mathbf{D}_d\mathbf{V}_d^t}{t} + \frac{\partial g_2(\mathbf{D}_1^t, \ldots, \mathbf{D}_{d-1}^t, \mathbf{D}_d, \mathbf{D}_{d+1}^{t-1} \ldots, \mathbf{D}_D^{t-1})}{\partial \mathbf{D_d}}, \tag{3.58}$$

with

$$\mathcal{M}_\tau = \mathcal{X}_\tau \times_1 \mathbf{D}_1^t \cdots \times_{d-1} \mathbf{D}_{d-1}^t \times_d \mathbf{I} \times_{d+1} \mathbf{D}_{d+1}^{t-1} \cdots \times_D \mathbf{D}_D^{t-1}, \tag{3.59}$$

$$\mathcal{N}_\tau = \mathcal{Y}_\tau \times_1 (\mathbf{D}_1^t)^\mathrm{T} \cdots \times_{d-1} (\mathbf{D}_{d-1}^t)^\mathrm{T} \times_d \mathbf{I} \times_{d+1} (\mathbf{D}_{d+1}^{t-1})^\mathrm{T} \cdots \times_D (\mathbf{D}_D^{t-1})^\mathrm{T}, \tag{3.60}$$

$$\mathbf{U}_d^t = \sum_{\tau=1}^{t} \mathbf{N}_{\tau(d)}\mathbf{X}_{\tau(d)}^\mathrm{T}, \tag{3.61}$$

$$\mathbf{V}_d^t = \sum_{\tau=1}^{t} \mathbf{M}_{\tau(d)}\mathbf{M}_{\tau(d)}^\mathrm{T}, \tag{3.62}$$

where $\mathbf{N}_{\tau(d)}, \mathbf{M}_{\tau(d)}, \mathbf{X}_{\tau(d)}$ represent the mode-d unfolding of $\mathcal{N}_\tau, \mathcal{M}_\tau, \mathcal{X}_\tau$, respectively. For convenience, $\mathbf{U}_d^t$ and $\mathbf{V}_d^t$ are obtained sequentially by

$$\begin{aligned}\mathbf{U}_d^t &= \mathbf{U}_d^{t-1} + \mathbf{N}_{t(d)}\mathbf{X}_{t(d)}^\mathrm{T}, \\ \mathbf{V}_d^t &= \mathbf{V}_d^{t-1} + \mathbf{M}_{t(d)}\mathbf{M}_{t(d)}^\mathrm{T}.\end{aligned} \tag{3.63}$$

Then the solution of the problem (3.56) is presented in Algorithm 26. Finally, the whole process of OTL method is summarized in Algorithm 27.

Besides the Tucker-based model, some other CP-based models and algorithms are also explored for online dictionary learning, such as online nonnegative CPD [50] and TensorNOODL [43].

Algorithm 26: Update dictionary atom $\mathbf{D}_d$

Input: New training tensor $\mathcal{Y}_t$, dictionaries $\left\{\mathbf{D}_1^t, \ldots, \mathbf{D}_{d-1}^t\right\}$, and $\left\{\mathbf{D}_{d+1}^{t-1}, \ldots, \mathbf{D}_D^{t-1}\right\}$, initial dictionary $\mathbf{D}_d^0$, the gradient descent step η
Output: Dictionary atom $\mathbf{D}_d$
Initialize: k=0, $\mathbf{D}_d^k = \mathbf{D}_d^0$
while *a stopping criterion is not met* **do**
  $\mathbf{D}_d^{k+1} = \left(\mathbf{D}_d^k - \eta \frac{\partial \mathrm{L}_{d,t}(\mathbf{D}_d^k)}{\partial \mathbf{D}_d}\right)$
  $k = k + 1$
end
Return $\mathbf{D}_d = \mathbf{D}_d^k$.

Algorithm 27: Online multimodal dictionary learning

Input: Training set $\left\{\mathcal{Y}_\tau \in \mathbb{R}^{I_1 \times \cdots \times I_D}\right\}$ $(\tau = 1, \ldots, t, \ldots)$
Output: Dictionary $\{\mathbf{D}_d\}$, $d = 1, \ldots, D$
Initialize: Initial dictionary matrices $\left\{\mathbf{D}_d^0\right\}$, $d = 1, \ldots, D$ and initial core tensor $\mathcal{X}_0$
while *a stopping criterion is not met* **do**
  Cosparse coding Step:
    Update the coefficient $\mathcal{X}_t$ by Algorithm 25.
  Dictionary update Step:
    For $d = 1, \ldots, D$:
      Update the statistics $\mathbf{U}_d^t$ and $\mathbf{V}_d^t$ by Eq. (3.63)
      Update dictionary atom $\mathbf{D}_d$ by Algorithm 26.
    End for
end

3.5 Applications

Tensor dictionary leaning has been widely used in data processing applications, such as dimensionality reduction, classification, medical image reconstruction, and image denoising and fusing. Here we briefly demonstrate its effectiveness on image denoising and image fusion by numerical experiments.

3.5.1 Image Denoising

In this section, we focus on the application of tensor dictionary learning on image denoising problem. Image denoising is a classical inverse problem in image processing and computer vision, which aims to extract a clean image from the observed data. In fact, real image is always affected by noise corruption. Due to its capability in restoring original image and recovering the details, image denoising has been successfully applied to remote sensing [15, 20, 27, 36] and medical image [22, 48, 51].

In this group of experiments, the used CAVE dataset [62] includes 32 scenes that are separated into 5 sections. Each hyperspectral image has spatial resolution with 512×512 and 31 spectral bands including full spectral resolution reflectance from 400 to 700 nm in 10 nm steps. This dataset is about various real-world materials and objects. As the additive white Gaussian noise (AWGN) can come from many natural sources, we perturb each image with a Gaussian noise of standard deviation $\sigma = 0.1$. Several methods are selected for comparison. All parameters involved in the above algorithms are chosen as described in the reference papers.

Comparison methods: Matrix-based dictionary learning method K-SVD [1] is chosen as a baseline. For tensor-based methods, we choose LRTA [45], PARAFAC [29], TDL [36], KBRreg [57], and LTDL [20], where TDL and LTDL are tensor dictionary learning methods. Here is a brief introduction to these five tensor-based approaches.

- LRTA [45]: a low-rank tensor approximation method based on Tucker decomposition.
- PARAFAC [29]: a low-rank tensor approximation method based on parallel factor analysis (PARAFAC).
- TDL [36]: a tensor dictionary learning model by combining the nonlocal similarity in space with global correlation in spectrum.
- KBR-denoising [57]: a tensor sparse coding model based on Kronecker-basis representation for tensor sparsity measure.
- LTDL [20]: a tensor dictionary learning model which considers the nearly low-rank approximation in a group of similar blocks and the global sparsity.

The denoising performance of different methods is compared in Table 3.1. And the denoised images by different methods are shown in Fig. 3.5.

Obviously, all tensor-based methods perform better than K-SVD[1], which illustrates that tensor can preserve the original structure of data. TDL [36], KBR [57], and LTDL [20] are based on sparsity and low-rank model, and it can be observed that the three methods outperform other methods, which are only based on sparsity or low-rank model, in details of the images, and four different performance indicators.

Table 3.1 The denoising performance comparison on CAVE dataset

Method	PSNR	SSIM	ERGAS	SAM
Noisy image	20.00	0.12	607.83	1.07
K-SVD [1]	30.28	0.56	181.44	0.70
LRTA [45]	36.24	0.85	90.65	0.40
PARAFAC [29]	32.76	0.78	135.33	0.51
TDL [36]	39.41	0.94	62.91	0.26
KBR [57]	40.12	0.91	59.60	0.55
LTDL [20]	41.57	0.97	48.98	0.15

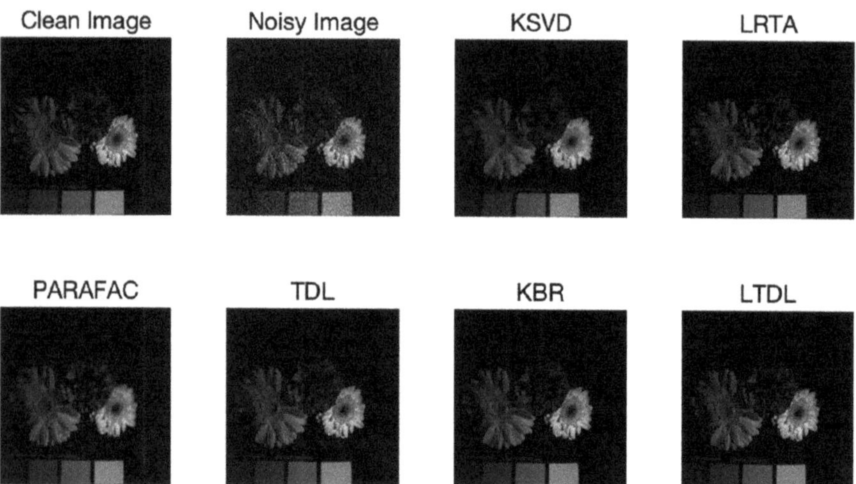

Fig. 3.5 Hyperspectral image denoising by different dictionary learning methods

3.5.2 Fusing Hyperspectral and Multispectral Images

As shown in many existing works, hyperspectral image (HSI) is commonly low rank. This can be simply verified by fast degeneration of tensor singular values from higher-order singular value decomposition of an HSI. Mathematically, a high-resolution (HR) HSI can be estimated by a low-rank tensor model as

$$\mathcal{X} = \mathcal{C} \times_1 \mathbf{D}_1 \times_2 \mathbf{D}_2 \times_3 \mathbf{D}_3, \tag{3.64}$$

where $\mathcal{X}$ is an HR-HSI data and the matrices $\mathbf{D}_d \in \mathbb{R}^{M_d \times I_d}$, $(d = 1, 2, 3)$ denote the dictionaries of the spacial width mode, height mode, and spectral mode, respectively.

In practical situation, the acquired HSI can be of low resolution (LR), and the obtained HSI sample $\mathcal{Y}$ can be regarded as the spatially downsampled version of $\mathcal{X}$ as follows:

$$\mathcal{Y} = \mathcal{X} \times_1 \mathbf{\Phi}_1 \times_2 \mathbf{\Phi}_2, \tag{3.65}$$

where $\mathbf{\Phi}_1$ and $\mathbf{\Phi}_2$ are the downsampling matrices on the spacial width mode and height mode, respectively.

If we only take a few spectral data from the HSI, the other practically acquired high-resolution multispectral image (HR-MSI) $\mathcal{Z}$ can be regarded as the spectrally downsampled version of $\mathcal{X}$ as follows:

$$\mathcal{Z} = \mathcal{X} \times_3 \mathbf{\Phi}_3 \tag{3.66}$$

where $\mathbf{\Phi}_3$ is the downsampling matrix on the spectral mode.

Substituting (3.64) into (3.65) and (3.66), the LR-HSI $\mathcal{Y}$ and HR-MSI $\mathcal{X}$ can be represented as :

$$\begin{aligned} \mathcal{Y} &= \mathcal{C} \times_1 (\mathbf{\Phi}_1 \mathbf{D}_1) \times_2 (\mathbf{\Phi}_2 \mathbf{D}_2) \times_3 \mathbf{D}_3 = \mathcal{C} \times_1 \mathbf{D}_1^* \times_2 \mathbf{D}_2^* \times_1 \mathbf{D}_3, \\ \mathcal{Z} &= \mathcal{C} \times_1 \mathbf{D}_1 \times_2 \mathbf{D}_2 \times_3 (\mathbf{\Phi}_3 \mathbf{D}_3) = \mathcal{C} \times_1 \mathbf{D}_1 \times_2 \mathbf{D}_2 \times_3 \mathbf{D}_3^*, \end{aligned} \tag{3.67}$$

where $\mathbf{D}_1^*$, $\mathbf{D}_2^*$, $\mathbf{D}_3^*$ are the downsampled width dictionary, downsampled height dictionary, and downsampled spectral dictionary, respectively.

Thus the fusion problem constrained by least-squares optimization problem can be formulated as follows:

$$\begin{aligned} \min_{\mathbf{D}_1, \mathbf{D}_2, \mathbf{D}_3, \mathcal{C}} & \|\mathcal{Y} - \mathcal{C} \times \mathbf{D}_1^* \times_2 \mathbf{D}_2^* \times_3 \mathbf{D}_3\|_{\mathrm{F}}^2 + \|\mathcal{Z} - \mathcal{C} \times {}_1\mathbf{D}_1 \times_2 \mathbf{D}_2 \times_3 \mathbf{D}_3^*\|_{\mathrm{F}}^2 \\ \text{s. t.} \quad & \mathrm{g}(\mathcal{C}) \leq K. \end{aligned} \tag{3.68}$$

Popular solutions for this problem include the matrix dictionary learning method which is a generalization of simultaneous orthogonal matching pursuit (GSOMP) [4] and tensor dictionary learning methods like non-local sparse tensor factorization (NLSTF) [15] and coupled sparse tensor factorization (CSTF) [27]. In this section, we conduct a simple experiment to verify the effectiveness of these dictionary learning methods for fusing hyperspectral and multispectral images.

In this group of experiments, we use the University of Pavia image [14] as the original HR-HSI image. It contains the urban area of the University of Pavia, Italy, and is acquired by the reflective optics system imaging spectrometer (ROSIS) optical sensor. The image has the size of $610 \times 340 \times 115$. And the Pavia University of size $256 \times 256 \times 93$ image is used as the reference image.

To generate the LR-HSI of size $32 \times 32 \times 93$, the HR-HSI is downsampled by averaging the 8×8 disjoint spatial blocks. The HR-MSI $\mathcal{Z}$ can be generated by downsampling $\mathcal{X}$ with downsampling spectral matrix $\mathbf{\Phi}_3$ in spectral model, which means selecting several bands in HR-HSI $\mathcal{X}$.

As for the quality of reconstructed HSI, six quantitative metrics are used for evaluation, which are the root mean square error (RMSE), the relative dimensionless global error in synthesis (ERGAS), the spectral angle mapper (SAM), the universal image quality index (UIQI), the structural similarity (SSIM), and the computational time. The performances are reported in Table 3.2. We can see that NLSTF [15] and CSTF [27] outperform GSOMP [4] in terms of reconstruction accuracy and computational time, which further illustrates the superiority of learned dictionaries compared with fixed basis.

Table 3.2 The fusion performance comparison on Pavia University dataset

Method	RMSE	ERGAS	SAM	UIQI	SSIM	Time(s)
GSOMP [4]	2.914	0.876	2.601	0.982	0.960	398
NLSTF [15]	1.995	0.578	1.970	0.993	0.993	38
CSTF [27]	1.732	0.490	1.780	0.995	0.9869	61
Best values	0	0	0	1	1	0

3.6 Summary

In this section, we briefly introduce the matrix dictionary learning technique. Based on it, we review the tensor dictionary learning as its extension. Generally we divide it into two kinds of subproblems: sparse tensor coding and tensor dictionary updating. From different aspects, we discuss the sparse tensor dictionary learning and cosparse dictionary learning, tensor dictionary learning based on different tensor decompositions forms, and offline and online tensor dictionary learning. The corresponding motivations, optimization models, and typical algorithms are summarized too. As the basic multilinear data-driven model for data processing, it can be widely used in signal processing, biomedical engineering, finance, computer vision, remote sensing, etc. We demonstrate its effectiveness in image denoising and data fusion in numerical experiments.

In this chapter, we mainly focus on the classical tensor decomposition forms to extend the dictionary learning. In fact, some of most recently developed tensor networks may motivate new tensor dictionary learning methods with better performance. Besides the standard ℓ_0 norm for sparse constraint, some other structural sparsity can be enforced in some applications. In addition, some other structures for the dictionaries may be used for structural dictionaries, which may make a good balance between the learning dictionary and mathematically designed dictionary. Inspired by the recently proposed deep matrix factorization, a deep tensor dictionary learning may be developed.

References

1. Aharon, M., Elad, M., Bruckstein, A.: K-SVD: an algorithm for designing overcomplete dictionaries for sparse representation. IEEE Trans. Signal Process. **54**(11), 4311–4322 (2006)
2. Ahmed, N., Natarajan, T., Rao, K.R.: Discrete cosine transform. IEEE Trans. Comput. **100**(1), 90–93 (1974)
3. Aidini, A., Tsagkatakis, G., Tsakalides, P.: Tensor dictionary learning with representation quantization for remote sensing observation compression. In: 2020 Data Compression Conference (DCC), pp. 283–292. IEEE, New York (2020)
4. Akhtar, N., Shafait, F., Mian, A.: Sparse spatio-spectral representation for hyperspectral image super-resolution. In: European Conference on Computer Vision, pp. 63–78. Springer, Berlin (2014)

5. Antonini, M., Barlaud, M., Mathieu, P., Daubechies, I.: Image coding using wavelet transform. IEEE Trans. Image Process. **1**(2), 205–220 (1992)
6. Beck, A., Teboulle, M.: A fast iterative shrinkage-thresholding algorithm for linear inverse problems. SIAM J. Imag. Sci. **2**(1), 183–202 (2009)
7. Bolte, J., Sabach, S., Teboulle, M.: Proximal alternating linearized minimization for nonconvex and nonsmooth problems. Math. Program. **146**(1–2), 459–494 (2014)
8. Caiafa, C.F., Cichocki, A.: Computing sparse representations of multidimensional signals using kronecker bases. Neural Comput. **25**(1), 186–220 (2013)
9. Chen, S.S., Donoho, D.L., Saunders, M.A.: Atomic decomposition by basis pursuit. SIAM Rev. **43**(1), 129–159 (2001)
10. Chen, G., Zhou, Q., Li, G., Zhang, X.P., Qu, C.: Tensor based analysis dictionary learning for color video denoising. In: 2019 IEEE International Conference on Signal, Information and Data Processing (ICSIDP), pp. 1–4. IEEE, New York (2019)
11. Cohen, J.E., Gillis, N.: Dictionary-based tensor canonical polyadic decomposition. IEEE Trans. Signal Process. **66**(7), 1876–1889 (2017)
12. Dai, W., Xu, T., Wang, W.: Simultaneous codeword optimization (SimCO) for dictionary update and learning. IEEE Trans. Signal Process. **60**(12), 6340–6353 (2012)
13. Dantas, C.F., Cohen, J.E., Gribonval, R.: Learning tensor-structured dictionaries with application to hyperspectral image denoising. In: 2019 27th European Signal Processing Conference (EUSIPCO), pp. 1–5. IEEE, New York (2019)
14. Dell'Acqua, F., Gamba, P., Ferrari, A., Palmason, J.A., Benediktsson, J.A., Árnason, K.: Exploiting spectral and spatial information in hyperspectral urban data with high resolution. IEEE Geosci. Remote Sens. Lett. **1**(4), 322–326 (2004)
15. Dian, R., Fang, L., Li, S.: Hyperspectral image super-resolution via non-local sparse tensor factorization. In: Proceedings of the IEEE Conference on Computer Vision and Pattern Recognition, pp. 5344–5353 (2017)
16. Donoho, D.L., Elad, M., Temlyakov, V.N.: Stable recovery of sparse overcomplete representations in the presence of noise. IEEE Trans. Inf. Theory **52**(1), 6–18 (2005)
17. Duan, G., Wang, H., Liu, Z., Deng, J., Chen, Y.W.: K-CPD: Learning of overcomplete dictionaries for tensor sparse coding. In: Proceedings of the 21st International Conference on Pattern Recognition (ICPR2012), pp. 493–496. IEEE, New York (2012)
18. Engan, K., Aase, S.O., Husoy, J.H.: Method of optimal directions for frame design. In: 1999 IEEE International Conference on Acoustics, Speech, and Signal Processing. Proceedings. ICASSP99 (Cat. No. 99CH36258), vol. 5, pp. 2443–2446. IEEE, New York (1999)
19. Fu, Y., Gao, J., Sun, Y., Hong, X.: Joint multiple dictionary learning for tensor sparse coding. In: 2014 International Joint Conference on Neural Networks (IJCNN), pp. 2957–2964. IEEE, New York (2014)
20. Gong, X., Chen, W., Chen, J.: A low-rank tensor dictionary learning method for hyperspectral image denoising. IEEE Trans. Signal Process. **68**, 1168–1180 (2020)
21. Gorodnitsky, I.F., Rao, B.D.: Sparse signal reconstruction from limited data using focuss: a re-weighted minimum norm algorithm. IEEE Trans. Signal Process. **45**(3), 600–616 (1997)
22. Huang, J., Zhou, G., Yu, G.: Orthogonal tensor dictionary learning for accelerated dynamic MRI. Med. Biol. Eng. Comput. **57**(9), 1933–1946 (2019)
23. Jiang, F., Liu, X.Y., Lu, H., Shen, R.: Efficient two-dimensional sparse coding using tensor-linear combination (2017). Preprint, arXiv:1703.09690
24. Jiang, F., Liu, X.Y., Lu, H., Shen, R.: Efficient multi-dimensional tensor sparse coding using t-linear combination. In: Proceedings of the AAAI Conference on Artificial Intelligence, vol. 32 (2018)
25. Kreutz-Delgado, K., Murray, J.F., Rao, B.D., Engan, K., Lee, T.W., Sejnowski, T.J.: Dictionary learning algorithms for sparse representation. Neural Comput. **15**(2), 349–396 (2003)
26. Li, P., Feng, J., Jin, X., Zhang, L., Xu, X., Yan, S.: Online robust low-rank tensor learning. In: Proceedings of the 26th International Joint Conference on Artificial Intelligence, pp. 2180–2186 (2017)

27. Li, S., Dian, R., Fang, L., Bioucas-Dias, J.M.: Fusing hyperspectral and multispectral images via coupled sparse tensor factorization. IEEE Trans. Image Process. **27**(8), 4118–4130 (2018)
28. Lim, W.Q.: The discrete shearlet transform: a new directional transform and compactly supported shearlet frames. IEEE Trans. Image Process. **19**(5), 1166–1180 (2010)
29. Liu, X., Bourennane, S., Fossati, C.: Denoising of hyperspectral images using the PARAFAC model and statistical performance analysis. IEEE Trans. Geosci. Remote Sens. **50**(10), 3717–3724 (2012)
30. Liu, E., Payani, A., Fekri, F.: Seismic data compression using online double-sparse dictionary learning schemes. Foreword by Directors, p. 6 (2017)
31. Mallat, S.G., Zhang, Z.: Matching pursuits with timefrequency dictionaries. IEEE Trans. Signal Process. **41**(12), 3397–3415 (1993)
32. Nam, S., Davies, M.E., Elad, M., Gribonval, R.: The cosparse analysis model and algorithms. Appl. Comput. Harmon. Anal. **34**(1), 30–56 (2013)
33. Olshausen, B.A., Field, D.J.: Emergence of simple-cell receptive field properties by learning a sparse code for natural images. Nature **381**(6583), 607–609 (1996)
34. Olshausen, B.A., Field, D.J.: Sparse coding with an overcomplete basis set: a strategy employed by v1? Vis. Res. **37**(23), 3311–3325 (1997)
35. Pati, Y.C., Rezaiifar, R., Krishnaprasad, P.S.: Orthogonal matching pursuit: recursive function approximation with applications to wavelet decomposition. In: Proceedings of 27th Asilomar Conference on Signals, Systems and Computers, pp. 40–44. IEEE, New York (1993)
36. Peng, Y., Meng, D., Xu, Z., Gao, C., Yang, Y., Zhang, B.: Decomposable nonlocal tensor dictionary learning for multispectral image denoising. In: Proceedings of the IEEE Conference on Computer Vision and Pattern Recognition, pp. 2949–2956 (2014)
37. Peng, Y., Li, L., Liu, S., Wang, X., Li, J.: Weighted constraint based dictionary learning for image classification. Pattern Recogn. Lett. **130**, 99–106 (2020)
38. Petersen, L., Sprunger, P., Hofmann, P., Lægsgaard, E., Briner, B., Doering, M., Rust, H.P., Bradshaw, A., Besenbacher, F., Plummer, E.: Direct imaging of the two-dimensional fermi contour: Fourier-transform STM. Phys. Rev. B **57**(12), R6858 (1998)
39. Qi, N., Shi, Y., Sun, X., Yin, B.: TenSR: Multi-dimensional tensor sparse representation. In: Proceedings of the IEEE Conference on Computer Vision and Pattern Recognition, pp. 5916–5925 (2016)
40. Qi, N., Shi, Y., Sun, X., Wang, J., Yin, B., Gao, J.: Multi-dimensional sparse models. IEEE Trans. Pattern Anal. Mach. Intell. **40**(1), 163–178 (2017)
41. Qiu, Q., Patel, V.M., Chellappa, R.: Information-theoretic dictionary learning for image classification. IEEE Trans. Pattern Anal. Mach. Intell. **36**(11), 2173–2184 (2014)
42. Quan, Y., Huang, Y., Ji, H.: Dynamic texture recognition via orthogonal tensor dictionary learning. In: Proceedings of the IEEE International Conference on Computer Vision, pp. 73–81 (2015)
43. Rambhatla, S., Li, X., Haupt, J.: Provable online CP/PARAFAC decomposition of a structured tensor via dictionary learning (2020). arXiv e-prints **33**, arXiv–2006
44. Ramirez, I., Sprechmann, P., Sapiro, G.: Classification and clustering via dictionary learning with structured incoherence and shared features. In: 2010 IEEE Computer Society Conference on Computer Vision and Pattern Recognition, pp. 3501–3508. IEEE, New York (2010)
45. Renard, N., Bourennane, S., Blanc-Talon, J.: Denoising and dimensionality reduction using multilinear tools for hyperspectral images. IEEE Geosci. Remote Sens. Lett. **5**(2), 138–142 (2008)
46. Roemer, F., Del Galdo, G., Haardt, M.: Tensor-based algorithms for learning multidimensional separable dictionaries. In: 2014 IEEE International Conference on Acoustics, Speech and Signal Processing (ICASSP), pp. 3963–3967. IEEE, New York (2014)
47. Rubinstein, R., Peleg, T., Elad, M.: Analysis K-SVD: A dictionary-learning algorithm for the analysis sparse model. IEEE Trans. Signal Process. **61**(3), 661–677 (2012)
48. Soltani, S., Kilmer, M.E., Hansen, P.C.: A tensor-based dictionary learning approach to tomographic image reconstruction. BIT Numer. Math. **56**(4), 1425–1454 (2016)

49. Sprechmann, P., Sapiro, G.: Dictionary learning and sparse coding for unsupervised clustering. In: 2010 IEEE International Conference on Acoustics, Speech and Signal Processing, pp. 2042–2045. IEEE, New York (2010)
50. Strohmeier, C., Lyu, H., Needell, D.: Online nonnegative tensor factorization and cp-dictionary learning for Markovian data (2020). arXiv e-prints pp. arXiv–2009
51. Tan, S., Zhang, Y., Wang, G., Mou, X., Cao, G., Wu, Z., Yu, H.: Tensor-based dictionary learning for dynamic tomographic reconstruction. Phys. Med. Biol. **60**(7), 2803 (2015)
52. Tosic, I., Frossard, P.: Dictionary learning. IEEE Signal Process. Mag. **28**(2), 27–38 (2011)
53. Traoré, A., Berar, M., Rakotomamonjy, A.: Online multimodal dictionary learning. Neurocomputing **368**, 163–179 (2019)
54. Variddhisai, T., Mandic, D.: Online multilinear dictionary learning (2017). arXiv e-prints pp. arXiv–1703
55. Vidal, R., Ma, Y., Sastry, S.: Generalized principal component analysis (GPCA). IEEE Trans. Pattern Anal. Mach. Intell. **27**(12), 1945–1959 (2005)
56. Wang, S., Zhang, L., Liang, Y., Pan, Q.: Semi-coupled dictionary learning with applications to image super-resolution and photo-sketch synthesis. In: 2012 IEEE Conference on Computer Vision and Pattern Recognition, pp. 2216–2223. IEEE, New York (2012)
57. Xie, Q., Zhao, Q., Meng, D., Xu, Z.: Kronecker-basis-representation based tensor sparsity and its applications to tensor recovery. IEEE Trans. Pattern Anal. Mach. Intell. **40**(8), 1888–1902 (2017)
58. Xu, R., Xu, Y., Quan, Y.: Factorized tensor dictionary learning for visual tensor data completion. IEEE Trans. Multimedia **23**, 1225–1238 (2020)
59. Yaghoobi, M., Daudet, L., Davies, M.E.: Parametric dictionary design for sparse coding. IEEE Trans. Signal Process. **57**(12), 4800–4810 (2009)
60. Yang, M., Zhang, L., Feng, X., Zhang, D.: Sparse representation based fisher discrimination dictionary learning for image classification. Int. J. Comput. Vis. **109**(3), 209–232 (2014)
61. Yao, B., Fei-Fei, L.: Grouplet: a structured image representation for recognizing human and object interactions. In: 2010 IEEE Computer Society Conference on Computer Vision and Pattern Recognition, pp. 9–16. IEEE, New York (2010)
62. Yasuma, F., Mitsunaga, T., Iso, D., Nayar, S.K.: Generalized assorted pixel camera: postcapture control of resolution, dynamic range, and spectrum. IEEE Trans. Image Process. **19**(9), 2241–2253 (2010)
63. Zhai, L., Zhang, Y., Lv, H., Fu, S., Yu, H.: Multiscale tensor dictionary learning approach for multispectral image denoising. IEEE Access **6**, 51898–51910 (2018)
64. Zhang, Q., Li, B.: Discriminative K-SVD for dictionary learning in face recognition. In: 2010 IEEE Computer Society Conference on Computer Vision and Pattern Recognition, pp. 2691–2698. IEEE, New York (2010)
65. Zhang, Z., Aeron, S.: Denoising and completion of 3d data via multidimensional dictionary learning (2015). Preprint, arXiv:1512.09227
66. Zhang, L., Wei, W., Zhang, Y., Shen, C., Van Den Hengel, A., Shi, Q.: Dictionary learning for promoting structured sparsity in hyperspectral compressive sensing. IEEE Trans. Geosci. Remote Sens. **54**(12), 7223–7235 (2016)
67. Zhang, Y., Mou, X., Wang, G., Yu, H.: Tensor-based dictionary learning for spectral CT reconstruction. IEEE Trans. Med. Imaging **36**(1), 142–154 (2016)
68. Zhao, R., Wang, Q.: Learning separable dictionaries for sparse tensor representation: an online approach. IEEE Trans. Circuits Syst. Express Briefs **66**(3), 502–506 (2018)
69. Zubair, S., Wang, W.: Tensor dictionary learning with sparse tucker decomposition. In: 2013 18th International Conference on Digital Signal Processing (DSP), pp. 1–6. IEEE, New York (2013)
70. Zubair, S., Yan, F., Wang, W.: Dictionary learning based sparse coefficients for audio classification with max and average pooling. Digital Signal Process. **23**(3), 960–970 (2013)

Chapter 4
Low-Rank Tensor Recovery

4.1 Introduction

Many real-world data are inevitably corrupted or missing during acquisition or transmission. For example, the famous Netflix rating matrix $\mathbf{X} \in \mathbb{R}^{I \times J}$ is a missing data in recommendation system, where each observed entry $\mathbf{X}(i, j)$ represents the rating of movie j ranked by customer i if customer i has watched movie j. Otherwise, it will be missing. The missing entries may imply how much the specific custom likes the specific movie. Therefore, how to accurately recover this missing matrix is essential for recommending movies.

The missing component analysis, also named matrix completion, can fill in the missing entries based on certain assumptions and has attracted increasing attentions in recommendation system [22, 23] and image completion [15, 18]. However, with the emergence of higher-order data, matrix completion cannot exploit the higher-order structure well due to the matricization or vectorization operator. As a natural generation of matrix completion, tensor completion is proposed and popularly used to tackle high-order data missing problems.

In this chapter, we firstly introduce the basic knowledge for matrix completion and mainly focus on providing unified and comprehensive frameworks for tensor completion including sampling, algorithms, and applications. Finally, some representative methods are employed to show the effectiveness of tensor completion in fields of image completion, recommendation system, knowledge graph completion, and traffic flow prediction.

Y. Liu et al., *Tensor Computation for Data Analysis*,
https://doi.org/10.1007/978-3-030-74386-4_4

4.2 Matrix Completion

Before we introduce tensor completion, we can take matrix completion as a simple and special case. There are two critical questions for matrix completion: one is that how to model the relationship between the known and unknown entries and the other one is how the observed samples affect the completion performance.

One natural consideration for the first question is to make it as similar as its neighbors or employ global features like low rankness. Benefiting from the low-rank assumption, matrix completion can divide into two groups: one is rank minimization model and the other is low-rank tensor factorization model.

An intuitive answer for the second question is that the larger number of samples is, the better completion can be achieved. To systematically and quantitatively answer this question, we will discuss the sample complexity for different optimization models in the following subsections.

4.2.1 Rank Minimization Model

In the low-rank matrix completion, one needs to find a matrix $\mathbf{X}$ of smallest rank, which matches the observed matrix $\mathbf{M} \in \mathbb{R}^{I_1 \times I_2}$ at all indices involved in the observation set $\mathbb{O}$. The mathematical formulation of this problem is

$$\begin{aligned} &\min_{\mathbf{X}} \ \operatorname{rank}(\mathbf{X}) \\ &\text{s. t. } \ \mathbf{X}(i_1, i_2) = \mathbf{M}(i_1, i_2), \ \forall (i_1, i_2) \in \mathbb{O}. \end{aligned} \tag{4.1}$$

Unfortunately, optimization of Eq. (4.1) is NP-hard, so some tricks are utilized to replace the rank function for efficient solution, such as the regularization penalty taking the form of a nuclear norm $R(\mathbf{X}) = \|\mathbf{X}\|_*$, which is the convex surrogate for rank function. The mathematical formulation is rewritten as

$$\begin{aligned} &\min_{\mathbf{X}} \ \|\mathbf{X}\|_* \\ &\text{s. t. } \ \mathbf{X}(i_1, i_2) = \mathbf{M}(i_1, i_2), \ \forall (i_1, i_2) \in \mathbb{O}. \end{aligned} \tag{4.2}$$

where the minimization of the nuclear norm can be achieved using the singular value thresholding algorithm, resulting in computational complexity of $\mathrm{O}\left(I^3\right)$ where $I = \max(I_1, I_2)$.

In [11], Candès and Tao proved that for a matrix $\mathbf{X}$ of rank R, required samples to exactly recover $\mathbf{M}$ are on the order of $C\mu^2 I R \log^6 I$, with probability at least $1 - \frac{1}{I^3}$, where C is a constant and μ is a parameter that satisfies the strong incoherence property.

4.2.2 *Low-Rank Matrix Completion Model*

Since the low-rank matrix completion can be also regarded as matrix factorization with missing components, another equivalent formulation is proposed as follows:

$$\begin{aligned} &\min_{\mathbf{U},\mathbf{V}} \frac{1}{2} \left\| \mathbf{U}\mathbf{V}^{\mathrm{T}} - \mathbf{M} \right\|_{\mathrm{F}}^{2} \\ &\text{s. t. } \mathbf{U}(i_1,:)\mathbf{V}(i_2,:)^{\mathrm{T}} = \mathbf{M}(i_1,i_2),\ \forall (i_1,i_2) \in \mathbb{O}, \end{aligned} \tag{4.3}$$

where $\mathbf{U} \in \mathbb{R}^{I_1 \times R}$ and $\mathbf{V} \in \mathbb{R}^{I_2 \times R}$. For this group of approaches, the rank R needs to be given in advance and thus requires a lot of time for parameter tuning.

In [25] Keshavan, Montanari, and Oh showed that by their algorithm the square error is bound as $\epsilon = \left\| \hat{\mathbf{M}} - \mathbf{M} \right\|_{\mathrm{F}}^{2} \leq C \max\{\mathbf{M}\} \sqrt{I_1 I_2} R / |\mathbb{O}|$ with probability $1 - 1/I^3$, $I = \max\{I_1, I_2\}$. Using the incoherence condition, in [21] the authors showed that their algorithm can recover an incoherent matrix $\mathbf{M}$ in $\mathrm{O}(\log(1/\epsilon))$ steps with $|\mathbb{O}| = \mathrm{O}\left((\sigma_1/\sigma_R)^6 R^7 \log(I R \|\mathbf{M}\|_{\mathrm{F}}/\epsilon)\right)$ random entries by the perturbed power method, where σ_1 and σ_R are the largest and the smallest singular values, respectively.

4.3 Tensor Completion

The success of matrix completion promotes the emergence of tensor completion accordingly, which recovers tensor from its partial observations using various tensor decomposition forms [30]. Some earliest references to tensor completion are [4, 28] in which the authors use the CP and Tucker decomposition to construct the optimization model. They show the potential of tensor completion to recover the correlated data. To deal with the difficulty of choosing CP rank, the authors in [54] propose the Bayesian CP Factorization (FBCP) for image recovery by using Bayesian inference to automatically determine the CP rank. With the development of tensor decomposition forms, the quantum-inspired decomposition methods such as tensor train (TT) and tensor ring (TR) show powerful expression ability and hence gain achievement [7, 19, 20, 31, 46] in visual data completion compared with classical decompositions.

To create a more intuitive understanding, we give a simple example for tensor completion in Fig. 4.1. The top 3 subfigures give 11 observations from a $3 \times 3 \times 3$ tensor with CP rank 1 (equivalently, TR rank 1). The blank patches represent the unknown entries. The task for tensor completion is to fill these blanks from the given measurements with the low-rank assumption on the tensor. Using the formulation of CP decomposition, it is not hard to list 11 equations, and the answer follows from solving these equations, which is shown in the bottom 3 subfigures of Fig. 4.1. Note

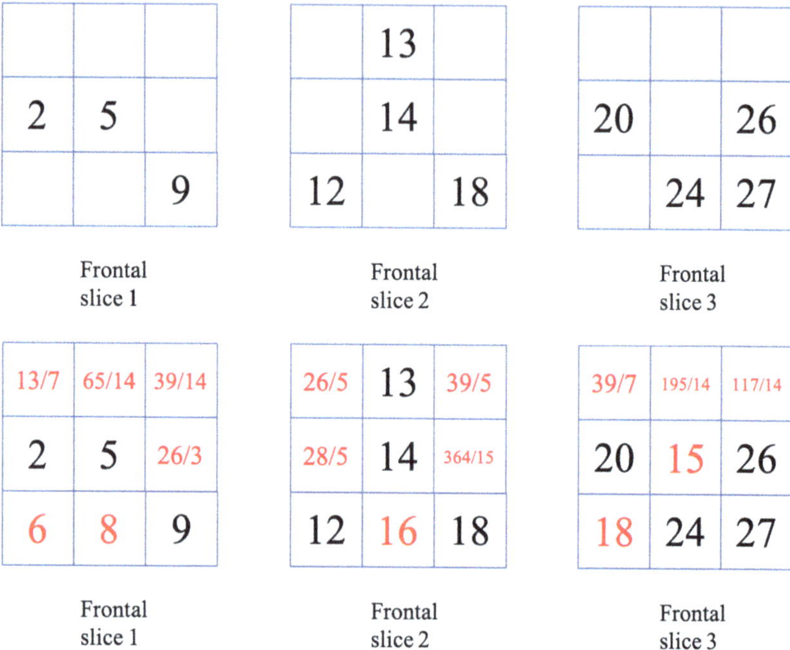

Fig. 4.1 A simple tensor completion case. Given a third-order tensor shown in the form of frontal slice with 11 observed values. Using the only one assumption that the tensor admits the rank-1 CP decomposition, the unknown entries are derived by solving 11 equations

that if we only focus on one of these frontal slices, we are actually doing the matrix completion.

Since tensors are generalization of matrices, the low-rank framework for tensor completion is similar with that of matrix completion, namely, rank minimization model and low-rank tensor factorization model.

4.3.1 Rank Minimization Model

The original rank minimization model for tensor completion of $\mathcal{T} \in \mathbb{R}^{I_1 \times \cdots \times I_D}$ can be formulated as a problem that finds a tensor $\mathcal{X}$ of the same size with minimal rank such that the projection of $\mathcal{X}$ on $\mathbb{O}$ matches the observations

$$\begin{aligned} &\min_{\mathcal{X}} \ \operatorname{rank}(\mathcal{X}) \\ &\text{s. t. } \mathrm{P}_{\mathbb{O}}(\mathcal{X}) = \mathrm{P}_{\mathbb{O}}(\mathcal{T}), \end{aligned} \tag{4.4}$$

where $\mathrm{P}_{\mathbb{O}}$ is a random sampling operator as

$$\mathrm{P}_{\mathbb{O}}(\mathcal{X}) = \begin{cases} \mathcal{X}(i_1, \ldots, i_D), & (i_1, \ldots, i_D) \in \mathbb{O} \\ 0, & \text{otherwise.} \end{cases}$$

Motivated by the fact that the matrix nuclear norm is the tightest convex relaxation of matrix rank function, convex optimization models for tensor completion mainly inherit and exploit this property [17, 19, 49]. They extend matrix nuclear norm to tensor ones in various ways, and the corresponding convex optimization model can be formulated as follows:

$$\begin{aligned} &\min_{\mathcal{X}} \|\mathcal{X}\|_* \\ &\text{s. t.} \quad \mathrm{P}_{\mathbb{O}}(\mathcal{X}) = \mathrm{P}_{\mathbb{O}}(\mathcal{T}), \end{aligned} \tag{4.5}$$

where the tensor nuclear norm varies according to different tensor decompositions. For example, the nuclear norm from Tucker decomposition is $\sum_{d=1}^{D} w_d \left\|\mathbf{X}_{(d)}\right\|_*$ [28], the nuclear norm of tensor train (TT) decomposition is formalized as $\sum_{d=1}^{D-1} w_d \left\|\mathbf{X}_{\langle d \rangle}\right\|_*$ [7], and the tensor ring (TR) nuclear norm is defined as $\sum_{d=1}^{L} w_d \left\|\mathbf{X}_{\langle d,L \rangle}\right\|_*$ [19], where $\mathbf{X}_{(d)}, \mathbf{X}_{\langle d \rangle}, \mathbf{X}_{\langle d,L \rangle}$ are the mode-d unfolding, d-unfolding, and d-shifting L-unfolding matrices of $\mathcal{X}$, respectively, with respect to different decompositions. Parameter w_d here denotes the weights of each unfolding matrix.

As another popular tensor network, the nuclear norm of HT is similar to that of TR, which can be realized by the shifting unfolding. Since HT corresponds to a tree graph in tensor network, the line that cuts off the virtual bond between any parent and son can be regarded as the rank of the matrix composed of contracting the factors on both sides of the line. A simple illustration for HT decomposition is shown in Fig. 4.2.

Table 4.1 summarizes the convex relaxations for different tensor ranks. The main difference is the unfolding matrices involved except for CP and t-SVD. Since

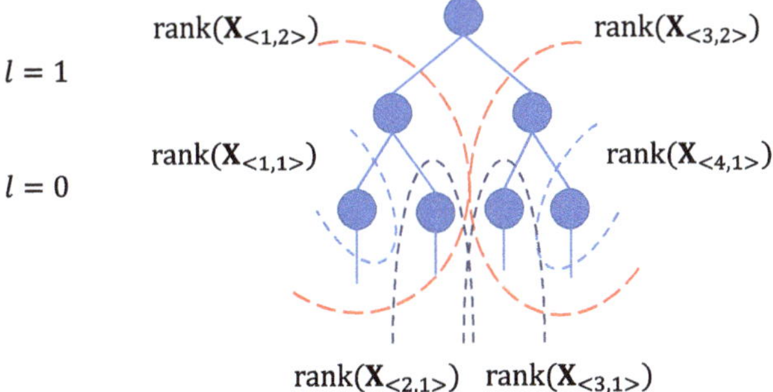

Fig. 4.2 An illustration of hierarchical Tucker decomposition for a 4D tensor

Table 4.1 Convex relaxations of tensor ranks in different tensor decompositions

Decomposition	CP	Tucker	HT	t-SVD	TT	TR
Convex relaxation	$\sum_{r=1}^{R} \lvert w_r \rvert$	$\sum_{d=1}^{D} w_d \left\Vert \mathbf{X}_{(d)} \right\Vert_*$	$\sum_{l=0}^{\log_2 D-1} \sum_{i\in[1:2^l:D]} w_{li} \left\Vert \mathbf{X}_{\langle i,2^l \rangle} \right\Vert_*$	$\sum_{i_3=1}^{I_3} \left\Vert \tilde{\mathcal{X}}(:,:,i_3) \right\Vert_*$	$\sum_{d=1}^{D-1} w_d \left\Vert \mathbf{X}_{\langle d \rangle} \right\Vert_*$	$\sum_{d=1}^{L} w_d \left\Vert \mathbf{X}_{\langle d,L \rangle} \right\Vert_*$

CP decomposition decomposes the $\mathcal{X}$ into several rank-1 tensors with w_r, $r = 1, \ldots, R$ representing the weights of each rank-1 component, its nuclear norm is defined as $\sum_{r=1}^{R} |w_r|$. The nuclear norm of $\mathcal{X}$ under t-SVD is expressed as the sum of nuclear norm of its each frontal slice in the frequency domain, namely, $\sum_{i_3=1}^{I_3} \left\| \tilde{\mathcal{X}}(:, :, i_3) \right\|_*$, which can be calculated by Fourier transformation. The term $\tilde{\mathcal{X}}$ is the fast Fourier transformation of $\mathcal{X}$ along the third mode.

A popular algorithm to solve (4.5) is the alternating direction method of multiplier (ADMM). By introducing tensors $\mathcal{M}_1, \ldots, \mathcal{M}_D$ and dual variables $\mathcal{Y}_1, \ldots, \mathcal{Y}_D$ for the augmented Lagrangian form, an equivalent optimization model to (4.5) can be formulated as follows:

$$
\begin{aligned}
&\min_{\mathcal{X}, \mathcal{M}_d, \mathcal{Y}_d} \sum_{d=1}^{D} w_d \left\| \mathcal{M}_{d,[d]} \right\|_* + \langle \mathcal{M}_d - \mathcal{X}, \mathcal{Y}_d \rangle + \frac{\rho}{2} \|\mathcal{M}_d - \mathcal{X}\|_{\mathrm{F}}^2 \\
&\text{s. t. } \mathrm{P}_{\mathbb{O}}(\mathcal{X}) = \mathrm{P}_{\mathbb{O}}(\mathcal{T})
\end{aligned} \tag{4.6}
$$

where the subscript $[d]$ denotes the d-th unfolding scheme for $\mathcal{M}_d$ according to the selected tensor decomposition and ρ is a penalty coefficient which parameter can be fixed or changed with iterations from a pre-defined value.

The procedure of ADMM is to split the original problem into pieces of small problems, each of which has a variable that can be fast solved. In this sense, (4.6) can be broken into $2D + 1$ problems with respect to variables $\mathcal{M}_d,\ d = 1, \ldots, D$, $\mathcal{X}$, and $\mathcal{Y}_d,\ d = 1, \ldots, D$.

$$
\begin{cases}
\mathcal{M}_d^* = \arg\min_{\mathcal{M}_d} w_d \left\| \mathcal{M}_{d,[d]} \right\|_* + \langle \mathcal{M}_d - \mathcal{X}, \mathcal{Y}_d \rangle + \frac{\rho}{2} \|\mathcal{M}_d - \mathcal{X}\|_{\mathrm{F}}^2,\ d = 1, \ldots, D \\
\mathcal{X}^* = \arg\min_{\mathcal{X}} \sum_{d=1}^{D} \langle \mathcal{M}_d - \mathcal{X}, \mathcal{Y}_d \rangle + \frac{\rho}{2} \|\mathcal{M}_d - \mathcal{X}\|_{\mathrm{F}}^2 \\
\qquad \text{s. t. } \mathrm{P}_{\mathbb{O}}(\mathcal{X}) = \mathrm{P}_{\mathbb{O}}(\mathcal{T}) \\
\mathcal{Y}_d = \mathcal{Y}_d + \rho (\mathcal{M}_d - \mathcal{X}),\ d = 1, \ldots, D,
\end{cases} \tag{4.7}
$$

where the problems regarding variables $\mathcal{M}_d,\ d = 1, \ldots, D$ can be solved, respectively, and in a parallel way. By setting some stop criteria such as maximal number of iteration and tolerance for certain absolute or relative change of variables or objective function, the algorithm can be converged to a global optimum. We summarize the details about ADMM for convex tensor completion in Algorithm 28, where soft is the soft thresholding operator.

The sampling analysis for recovery methods based on CP decomposition, Tucker decomposition, t-SVD, and TR can be found in [1, 19, 36, 52], respectively. The main idea is to prove that by adding an arbitrary perturbation the objective function is strictly greater than the value corresponding to the optimal solution. In

Algorithm 28: ADMM for tensor completion

Input: Observed tensor $\mathcal{T}$, support set $\mathbb{O}$, maximal iteration number K
Output: $\mathcal{X}$
set $\mathcal{X}^0 = \mathcal{T}, \mathcal{M}_d^0 = \mathcal{T}, \mathcal{Y}_d^0$ as zero tensor
for $k = 1$ **to** K
  $\mathcal{M}_d^k = \text{fold}_{[d]}\left[\text{soft}_{\frac{w_d}{\rho}}\left(\mathcal{X}_{[d]}^{k-1} - \frac{1}{\rho}\mathcal{Y}_{d[d]}^{k-1}\right)\right]$
  $\mathcal{X}^k = \mathcal{T}_{\mathbb{O}} + \left(\frac{1}{D}\sum_{d=1}^{D}\mathcal{M}_d^k + \frac{1}{\rho}\mathcal{Y}_d^{k-1}\right)_{\mathbb{O}^\perp}$
  $\mathcal{Y}_d^k = \mathcal{Y}_d^{k-1} + \rho\left(\mathcal{M}_d^k - \mathcal{X}^k\right)$
  if converged
    break
  end
end
Return $\mathcal{X}^K$

Table 4.2 Comparison of sample complexity of five popular algorithms stemmed from different tensor decompositions

Decomposition	Sampling scheme	Model	Algorithm	Complexity
TR	Uniform	Sum of nuclear norm minimization	TRBU [19]	$\mathrm{O}\left(I^{\lceil D/2 \rceil}R^2\log^7\left(I^{\lceil D/2 \rceil}\right)\right)$
TT	Uniform	First-order polynomials [2]	Newton's method [3]	$\mathrm{O}\left(D\log\left(IR^2\right)\right)$
t-SVD	Uniform	Tensor nuclear norm minimization	LRTC-TNN [52]	$\mathrm{O}\left(I^{D-1}R\log\left(I^{D-1}\right)\right)$
Tucker	Gaussian	Sum of nuclear norm minimization	SNN [36]	$\mathrm{O}\left(R^{\lfloor D/2 \rfloor}I^{\lceil D/2 \rceil}\right)$
CP	Uniform	First-order polynomials [1]	Newton's method [3]	$\mathrm{O}\left(I^2\log(DIR)\right)$

addition, reference [1] uses the algebraic geometry way, which is different from the framework used in [19, 36, 52]. Table 4.2 provides a list for recent references with comparison in sample complexity. For simplicity, $I_d = I$, $R_d = R$ for $d = 1, \ldots, D$.

4.3.2 *Low-Rank Tensor Factorization Model*

Under specific tensor decomposition form, the nonconvex optimization model for tensor completion can be formulated as a generalized data fitting problem with predefined tensor ranks, where the desired variables are latent factors of the tensor required to recover.

Assuming the tensor $\mathcal{X}$ admits $\mathcal{X} = \mathrm{f}(\mathcal{U}_1, \ldots, \mathcal{U}_D)$, where function f represents the contraction function that merges tensor factors $\mathcal{U}_1, \ldots, \mathcal{U}_D$ to a tensor, then the optimization model can be formulated as follows:

$$\min_{\mathcal{U}_1,\ldots,\mathcal{U}_D} \|\mathrm{P}_{\mathbb{O}}(\mathrm{f}(\mathcal{U}_1, \ldots, \mathcal{U}_D)) - \mathrm{P}_{\mathbb{O}}(\mathcal{T})\|_2^2 \tag{4.8}$$

where $\mathcal{X} = \mathrm{f}(\mathcal{U}_1, \ldots, \mathcal{U}_D)$ varies with different tensor decompositions such as Tucker form in [16], TT form in [45], and TR form in [46].

As shown in (4.8), it is a nonlinear fitting problem. A number of algorithms are proposed to tackle this tensor factorization problem with missing entries, e.g., block coordinate descent (BCD) algorithm [47], stochastic gradient descent algorithm [26, 38], trust region algorithm [12], and Levenberg-Marquardt algorithm [35]. Here we take the block coordinate descent algorithm for explanation. This algorithm takes the alternating minimization way, which alternately optimizes the factor $\mathcal{U}_d$ while keeping the rest factors fixed. Each subproblem can be formulated as follows:

$$\min_{\mathcal{U}_d} \|\mathrm{P}_{\mathbb{O}}(\mathrm{f}(\mathcal{U}_1, \ldots, \mathcal{U}_D)) - \mathrm{P}_{\mathbb{O}}(\mathcal{T})\|_2^2 . \tag{4.9}$$

This is recognized as a quadratic form with respect to each block variable and hence is easy to solve via the BCD.

In order to make readers better understand, we demonstrate the BCD algorithm in the case of TR decomposition. Note that the objective function regarding each TR factor is convex while the other factors are treated as fixed variables; we can specify (4.9) as

$$\min_{\mathbf{U}_d} \left\| \mathbf{W}_{\langle d,1\rangle} \circledast (\mathbf{U}_d \mathbf{B}_d) - \mathbf{W}_{\langle d,1\rangle} \circledast \mathbf{T}_{\langle d,1\rangle} \right\|_2^2 \tag{4.10}$$

by contracting all the factors except the d-th one to $\mathbf{B}_d \in \mathbb{R}^{\prod_{n\neq d}^{D} I_n \times R_d R_{d+1}}$ and converting the projection operator to a binary tensor $\mathcal{W}$ with "1" indicating the observed position and "0" for the opposite. To alleviate the computational burden, we can further split the problem (4.10) into I_d subproblems:

$$\min_{\mathbf{u}} \|\mathbf{w} \circledast (\mathbf{u}\mathbf{B}_d) - \mathbf{w} \circledast \mathbf{t}\|_2^2 , \tag{4.11}$$

where $\mathbf{u} = \mathbf{U}_d(i_d, :)$, $\mathbf{w} = \mathbf{W}_{\langle d,1\rangle}(i_d, :)$, and $\mathbf{t} = \mathbf{T}_{\langle d,1\rangle}(i_d, :)$. Extracting the columns of $\mathbf{B}_d$ that correspond to "1"s in $\mathbf{w}$, now the problem (4.11) can be constructed as a standard linear least square problem. In one iteration of ALS, we solve $\sum_{d=1}^{D} I_d$ linear squares. The stop criteria can be set as a maximal number of iterations or certain tolerance of relative/absolute change of variables, gradients, or objective function. The details about BCD method for tensor completion are presented in Algorithm 29.

Algorithm 29: BCD algorithm for tensor completion

Input: Observed $\mathcal{T}$, support set $\mathbb{O}$, maximal iteration number K
Output: $\mathcal{X}$
set $\mathcal{X}^{(0)} = \mathcal{T}$
for $k = 1$ **to** K
  $\mathcal{U}_d^k = \arg\min_{\mathcal{U}_d} \|\mathrm{P}_{\mathbb{O}}(\mathrm{f}(\mathcal{U}_1, \ldots, \mathcal{U}_D)) - \mathrm{P}_{\mathbb{O}}(\mathcal{T})\|_2^2$
  if converged
    break
  end
end
Return $\mathcal{X}^K$

The computational complexity of eight algorithms based on different decompositions is summarized in Table 4.3, including tensor ring nuclear norm minimization for tensor completion (TRNNM) [50], low-rank tensor ring completion via alternating least square (TR-ALS) [46], simple low-rank tensor completion via tensor train (SiLRTC-TT) [7], high accuracy low-rank tensor completion algorithm (HaLRTC) [28], low-rank tensor completion via tensor nuclear norm minimization (LRTC-TNN) [32], Bayesian CP Factorization (FBCP) for image recovery [54], smooth low-rank tensor tree completion (STTC) [29], and provable tensor ring completion using balanced unfolding [19]. These algorithms are based on various tensor decompositions, including CP decomposition, Tucker decomposition, t-SVD, HT, TT, and TR. In Table 4.3, D is the number of ways of the tensor data, I is the dimensional size, M is the number of samples, and R ($[R, \ldots, R]$) is the tensor rank.

4.4 Applications

Tensor completion is applicable in many fields that involve incomplete multiway data with correlated structures. For instance, natural visual data, data in recommendation system, knowledge graph, and traffic flow data often approximately meet this requirement. In this part, we will demonstrate the effectiveness of tensor completion in these applications with numerical experiments.

4.4.1 Visual Data Recovery

With the development of cameras, visual data are widely and easily acquired and used in many situations. The visual data, including RGB images and videos and light-field images, can be approximated by low-rank tensor. In fact, some other images also share the similar low-rank structure and can be processed by the tensor completion techniques for denoising, restoration, enhancement, and compressive

Table 4.3 Comparison of computational complexity of eight algorithms in one iteration

Algorithm	TRBU	TRNNM	TR-ALS	SiLRTC-TT	LRTC-TNN	FBCP	HaLRTC	STTC
Complexity	$O(DI^{3D/2})$	$O(DI^{3D/2})$	$O(DMR^4)$	$O(DI^{3D/2})$	$O(I^{D+1})$	$O(DMR^2)$	$O(DI^{3D-3})$	$O(DI^{D+1})$

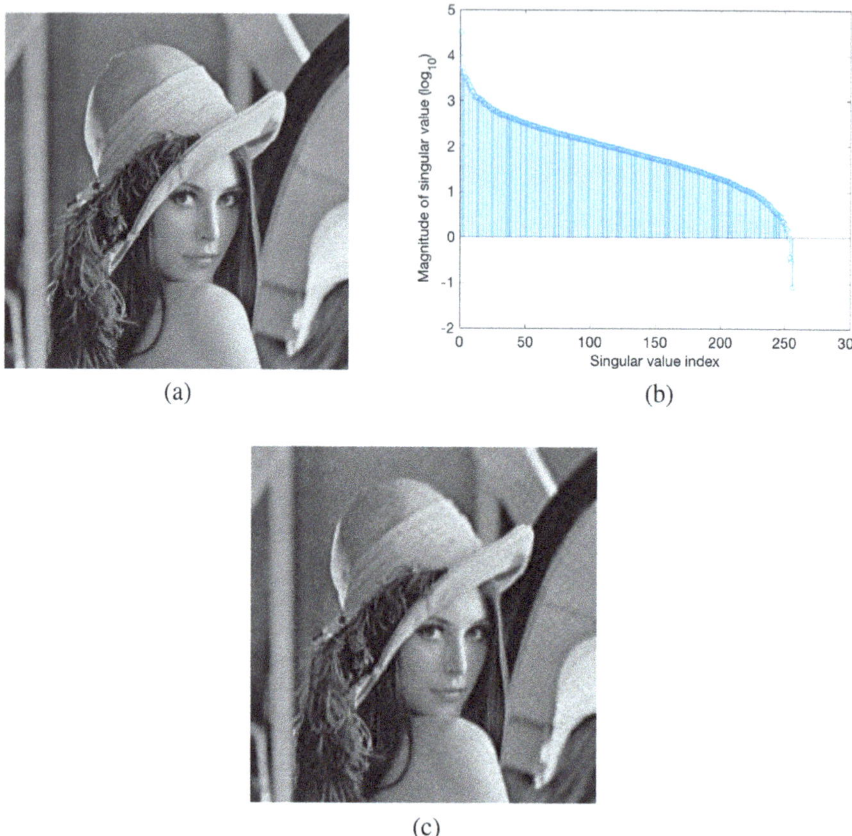

Fig. 4.3 The panel (**a**) is the gray copy of *lena* with size of 256×256. The panel (**b**) is the singular values of *lena*. The panel (**c**) is the reconstruction of (**a**) after a threshold with first 50 singular values. (**a**) Original *lena*. (**b**) The singular values of *lena*. (**c**) Truncated *lena*

sensing. They are multispectral images, hyperspectral images, magnetic resonance images, etc.

To more intuitively illustrate the low-rank structure of these visual data, we choose the *lena* gray image in size of 256×256 as an example and compute its singular value decomposition. For convenient illustration we only consider the gray image here. Figure 4.3 shows the image *lena* and its singular values given by the singular value decomposition of this image in subfigures Fig. 4.3a and b, respectively. We can see that the largest singular value is really greater than others. The larger singular values constitute the main component of the image and contain most of the information of the image. Figure 4.3c shows the reconstructed image from the first 50 singular values.

Extending to higher-order tensor data, the low-rank assumption also holds. To show the effectiveness of tensor completion in color image recovery, we employ

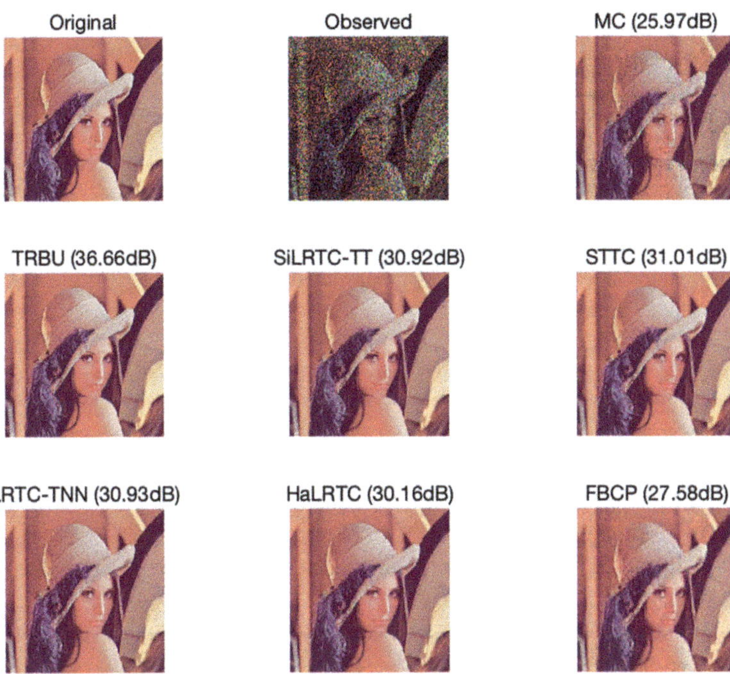

Fig. 4.4 The completion result for *lena* derived by seven methods. The tensor completion methods show different PSNRs under the same sampling scheme, which implies various expression powers of tensor decompositions

several algorithms based on different tensor decompositions for comparison with a baseline of matrix completion [10], i.e., FBCP [54], HaLRTC [28], STTC [29], LRTC-TNN [52], SiLRTC-TT [7], and TRBU [19]. The tensor data used is the RGB image *Lena* in size of $256 \times 256 \times 3$, as shown in Fig. 4.4 labeled by "Original." We uniformly pick up 50% pixels at random for each channel to result in the subsampling measurements ("Observed" in Fig. 4.4). To conduct matrix completion, we unfold the image into a matrix of size $256 \times (256 * 3)$, and for other six comparison algorithms, we use the original tensor form.

The recovery results of the missing data by the seven algorithms mentioned before are presented in Fig. 4.4, respectively, where "MC" stands for the matrix completion method. Besides the visual results, we also utilize a common metric to measure the completion performance, namely, peak signal-to-noise ratio (PSNR), which is defined as $10 \log_{10}\left(255^2/\mathrm{MSE}\right)$, and MSE is the mean squared error between the recovery and the reference.

From Fig. 4.4, we can observe that matrix completion yields the lowest PSNR. This implies the tensor-based methods can capture more low-rank structure information. The second conclusion is that different tensor-based methods gain different PSNRs, which suggests tensor decompositions of different forms possess different representative capability. This conjecture is experimentally verified in [7, 19, 46] with a theoretical corroboration in [56].

4.4.2 Recommendation System

Tensor completion can be used for recommendation system by exploiting the latent multidimensional correlations. For example, in movie rating system, due to the large number of movies and users, not everyone can see every movie, so the data usually obtained is only a very small part. Users rank a movie highly if it contains fundamental features they like, as shown in Fig. 4.5 which gives an illustration of the association between the latent factor film genre and the user-item data. It can be seen that different users will have different preferences but there may be some similarities between them. By using the user's own preferences and the correlation between different movies, unknown data can be predicted.

In this subsection, the user-centered collaborative location and activity filtering (*UCLAF*) data [55] is used. It comprises 164 users' global positioning system (GPS)

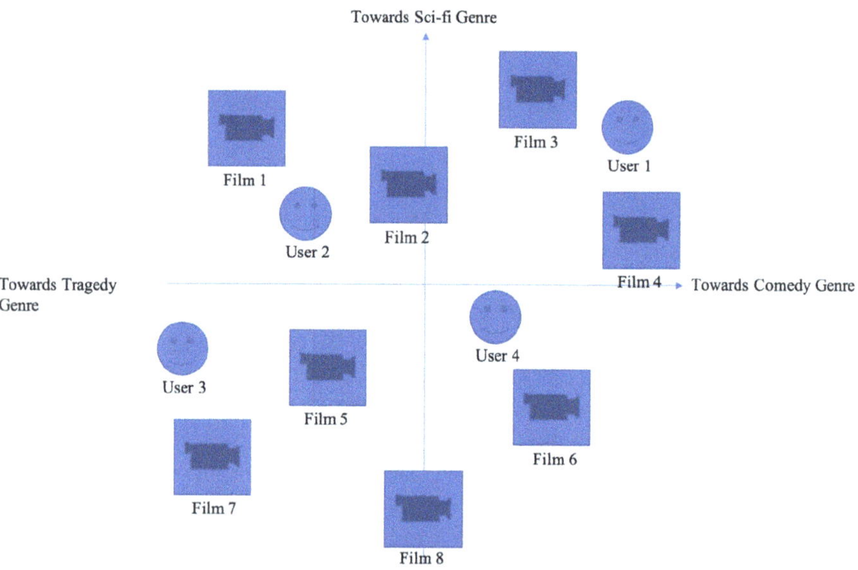

Fig. 4.5 A simple user-item demonstration. The square blocks represent films with different combinations of four basic genres, and the circles represent users. Each user tends to have distinct preference, and they are more likely to choose the films that have more common characteristics, respectively. Mathematically, the Euclidean distance (sometimes the cosine distance is favorable) is used to measure the similarity between the films and the users, which is a key idea for user and item prediction in recommendation system

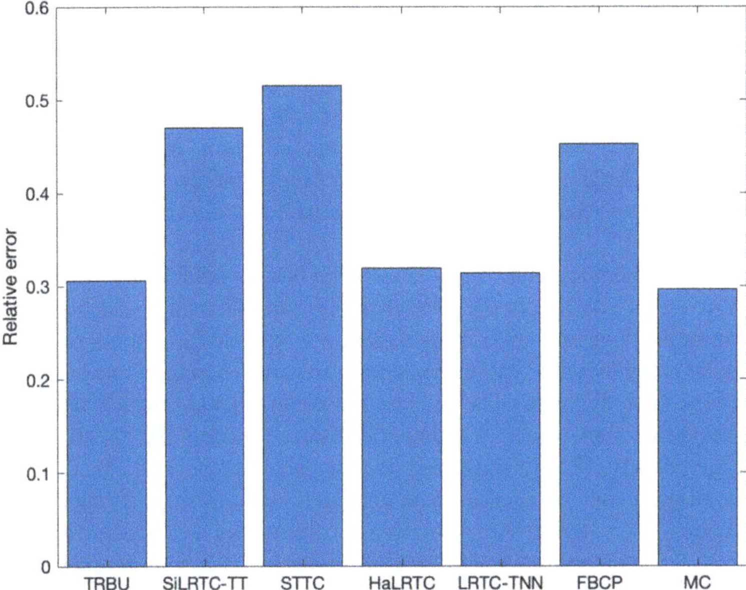

Fig. 4.6 The completion results for *UCLAF* derived by seven methods, where the label "MC" on the horizontal axis means matrix completion and it is compared with tensor-based methods as a baseline

trajectories based on their partial 168 locations and 5 activity annotations. Since people of different circles and personalities tend to be active in specific places, this *UCLAF* dataset satisfies low-rank property for similar reasons. For efficiency, we construct a new tensor of size $12 \times 4 \times 5$ by selecting the slices in three directions such that each new slice has at least 20 nonzeros entries. Then we randomly choose 50% samples from the obtained tensor for experiments.

The reconstruction error is measured by the ratio of the ℓ_2 norm of the deviation of the estimate and the ground truth to the ℓ_2 norm of the ground truth, i.e., $\left\|\hat{\mathcal{X}} - \mathcal{X}\right\| / \|\mathcal{X}\|$. Figure 4.6 shows the completion results by six tensor completion algorithms along with the matrix completion method as a baseline. A fact shown in Fig. 4.6 is that the matrix completion even performs better than some tensor-based methods. Besides, the TRBU (a TR decomposition-based algorithm) gives a similar relative error as the matrix completion method, which can be probably explained by that TRBU can be viewed as a combination of several cyclical matrix completions which in fact contains the matrix used in matrix completion.

4.4.3 Knowledge Graph Completion

Knowledge graph can be viewed as a knowledge base with graph structure. It characterizes real-world knowledge by constructing relations between entities. It is an appealing tool when dealing with recommendation systems [53] and question answering [33]. As shown in Fig. 4.7, the nodes represent real-world entities, and the connections denote the relations between them. It is a directed graph with labeled edges.

Nowadays, there are many open knowledge bases available, including WordNet [34], YAGO [14], Wikidata [42], Freebase [8], and Google's search engine [39]. Apart from the huge amount of facts stored in it, knowledge graph also has a lot of missing facts due to limitations in acquisition process. Therefore, lots of methods are put forward to predict new facts based on those existing ones, which can be roughly categorized into three types: translational distance models [9, 27, 40, 44], neural network models [6, 13, 43], and tensor decomposition models [5, 24, 37, 41, 48, 51]. Among all these methods, tensor decomposition methods show its superiority over other models due to the strong expression capability of tensor scheme [5, 41, 51].

The facts in a binary-relation knowledge graph are usually represented by triplets as (*head entity, relation, tail entity*), e.g., (*Alice, a friend of, Jane*). Under the framework of tensor decomposition models, *head entity, relation*, and *tail entity* correspond to the three modes of a third-order tensor, respectively, and each triplet corresponds to an entry in tensor, as shown in Fig. 4.8. The value of (i, j, k)-th entry in the tensor indicates whether the i-th *head entity* is linked to the k-th *tail entity* by the j-th *relation*. Typically, the entries with value "1" represent the known facts in knowledge graph, while "0" entries represent the false or missing facts.

Thus, the task of predicting missing facts in knowledge graph converts into determining the true value of "0" entries in tensor, that is, to infer which of the zero entries are true facts but originally missing. In order to achieve this, a variety

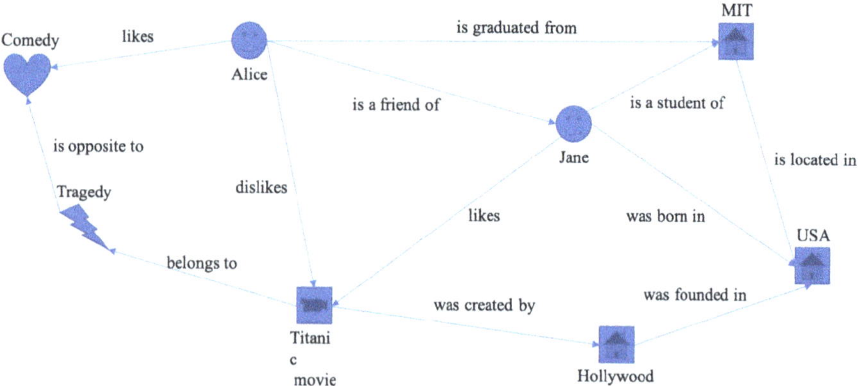

Fig. 4.7 A simple knowledge graph demonstration

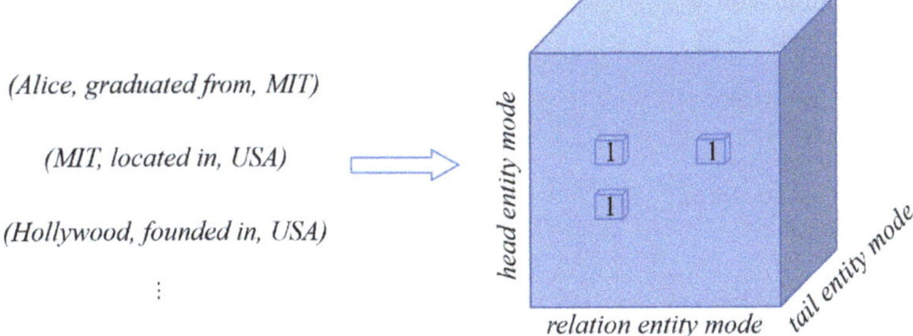

Fig. 4.8 Knowledge graph and tensor. Head entity, relation entity, and tail entity correspond to the three modes separately. Each element in tensor indicates the state of a triplet, that is, "1" stands for true fact and "0" stands for false or missing fact

Table 4.4 Scoring function utilized in different models. In scoring functions, $\mathbf{e}_h$ and $\mathbf{e}_t$ represent the head and tail entity embedding vectors, respectively, $\mathbf{r}$ represents relation embedding vector, and $\mathbf{W}_r$ denotes the relation embedding matrix

Decomposition	Model	Scoring function
CP	DistMult [48]	$\langle \mathbf{e}_h, \mathbf{r}, \mathbf{e}_t \rangle$
CP	ComplEx [41]	$\mathrm{Re}(\langle \mathbf{e}_h, \mathbf{r}, \bar{\mathbf{e}}_t \rangle)$
Tucker	TuckER [5]	$\mathcal{W} \times_1 \mathbf{e}_h \times_2 \mathbf{r} \times_3 \mathbf{e}_t$
Tensor train	TensorT [51]	$\langle \mathbf{e}_h, \mathbf{W}_r, \mathbf{e}_t \rangle$

of scoring functions are derived according to different decomposition frameworks, as shown in Table 4.4. It is usually implemented with a sigmoid function $\sigma(\cdot)$; the resulting value indicates the possibility that the triplet in the knowledge graph is a true fact. To alleviate the computational manipulation, knowledge graph is embedded into a lower-dimensional continuous vector space while preserving its inherent structure. The resulting vectors are referred to as embeddings, e.g., $\mathbf{e}_h$, $\mathbf{r}$, and $\mathbf{e}_t$ in Table 4.4.

Take TuckER [5] as an example; it employs Tucker decomposition to construct a linear scoring function and achieves great performance. Its scoring function is defined as $\phi(e_h, r, e_t) = \mathcal{W} \times_1 \mathbf{e}_h \times_2 \mathbf{r} \times_3 \mathbf{e}_t$, where $\mathbf{e}_h, \mathbf{e}_t \in \mathbb{R}^{d_e}$ represent the entity embeddings, $\mathbf{r} \in \mathbb{R}^{d_r}$ represents the relation embedding, and the elements in $\mathcal{W} \in \mathbb{R}^{d_e \times d_r \times d_e}$ can be read as the interactions between them, as shown in Fig. 4.9.

To conduct the experiment, we choose FB15k as our dataset, which is a standard dataset employed in related literature. It is the subset of Freebase with 14951 entities and 1345 relations, containing 592,213 real-world facts such as *Leonardo DiCaprio is the actor of Titanic*. The dataset are divided into three parts for different usages: training data, validation data, and test data. Based on the known facts (training data) in knowledge graph, we can embed it into a continuous vector space to get the initial embeddings. Taking advantage of the Tucker-form scoring function $\phi(\cdot)$ and sigmoid function $\sigma(\cdot)$, we assign a possibility for each triplet. Then a Binary Cross Entropy (BEC) loss is optimized to maximize the plausibility of current embeddings. By iteratively doing these steps, we can get our trained model. We

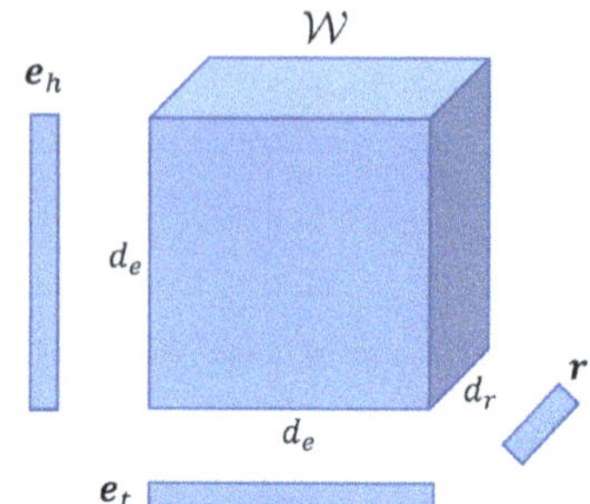

Fig. 4.9 An illustration of TuckER framework. Each multiplication of $\mathcal{W},\ \mathbf{e}_h,\ \mathbf{r},\ \mathbf{e}_t$ generates a score for a triplet. The possibility that the triplet $\{\mathbf{e}_h, \mathbf{r}, \mathbf{e}_t\}$ in the knowledge graph is a true fact is expressed by $\sigma(\mathcal{W} \times_1 \mathbf{e}_h \times_2 \mathbf{r} \times_3 \mathbf{e}_t)$

Table 4.5 Results of knowledge graph completion. TuckER and ComplEx are both tensor decomposition models, while ConvE is a translational model. The results of TuckER and ComplEx are run on our computer, and ConvE is drawn directly from the original paper since its code does not have the option of running the FB15k dataset

Model	MRR	Hits@1	Hits@3	Hits@10
TuckER [5]	**0.789**	**0.729**	**0.831**	**0.892**
ComplEx [41]	0.683	0.591	0.750	0.834
ConvE [13]	0.657	0.558	0.723	0.831

then test the learned TuckER model on test data and evaluate it by mean reciprocal rank (MRR) and hits@k, $k \in \{1,\ 3,\ 10\}$; the results are demonstrated in Table 4.5, where the best are marked in bold. MRR is the average of inverse rank of the true triplet over all other possible triplets. Hits@k indicates the proportion of the true entities ranked within the top k possible triplets. As we can see, TuckER can not only achieve better results than its counterpart ComplEx but also outperform translational model ConvE. This result mainly benefits from the co-sharing structure of its scoring function, which takes advantage of Tucker decomposition to encode underlying information in the low-rank core tensor $\mathcal{W}$.

4.4.4 Traffic Flow Prediction

Traffic flow prediction has gained more and more attention with the rapid development and deployment of intelligent transportation systems (ITS) and vehicular cyber-physical systems (VCPS). It can be decomposed into a temporal and a spatial process; thus the traffic flow data are naturally in the form of tensors. The traffic flow prediction problem can be stated as follows. Let $X\ (t, i)$ denote the observed traffic flow quantity at t-th time interval of the i-th observation location; we can get a sequence $\mathbf{X}$ of observed traffic flow data, $i = 1, 2, \ldots, I,\ t = 1, 2, \ldots, T$. Then the problem is to predict the traffic flow at time interval $(t + \Delta t)$ for some prediction horizon Δt based on the historical traffic information. Considering the seasonality of $\mathbf{X}$, we fold $\mathbf{X}$ along the time dimension in days, which results in a tensor in size of $location \times timespot \times seasonality$.

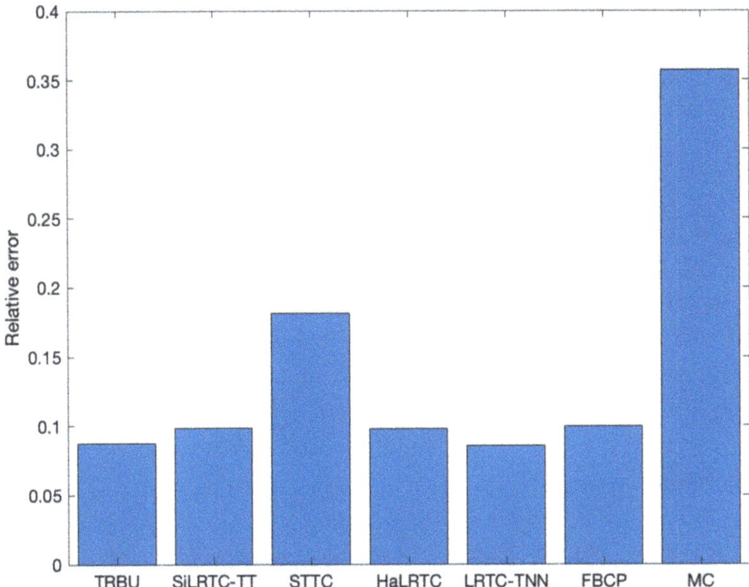

Fig. 4.10 The completion result for *traffic* data by seven methods

Specifically, we use the *traffic* dataset downloaded from the website.[1] It can be reshaped into a tensor of size $209 \times 60 \times 144$. We randomly sample 20% of the entries as observations and take six tensor-based algorithms to compare their performance with a baseline method of matrix completion. The experimental results are presented in Fig. 4.10. Since this *traffic* dataset shows low rankness in three dimensions rather than two dimensions, all tensor-based methods achieve lower relative errors than matrix completion method. Among all tensor-based methods, the TRBU and LRTC-TNN give better performance, which demonstrates our speculation in Sect. 4.4.1 accordingly that decomposition of TR form has more powerful representation ability.

4.5 Summary

This chapter presents a brief summary of current research progress on tensor completion, starting from the concept of matrix completion, presenting two main optimization frameworks, including low-rank minimization and tensor factorization based, and discussing the difference between popular solutions. Finally, some applications are given with practical examples. Since tensor is a natural representation of

[1]https://zenodo.org/record/1205229#.X5FUXi-1Gu4.

high-dimensional structured data, the expansion of tensor completion from matrix completion benefits a variety of applications. Compared with the point-to-point linear relation used by matrix completion, the advantage of tensor completion is that it reveals the missing components by exploiting multilinear relationships when dealing with the multidimensional data.

However, since tensor completion obeys the assumption that the information of a tensor is not centrally distributed (mathematically, the incoherence of tensor factors), it is difficult to apply in the scenario that the missing components have rich details. Besides, tensor completion can only learn multilinear mapping as a non-supervised learning method. It brings a limitation to manipulate missing components of nonlinear structured data, such as natural language, 3D point cloud, graph, and audio.

References

1. Ashraphijuo, M., Wang, X.: Fundamental conditions for low-CP-rank tensor completion. J. Mach. Learn. Res. **18**(1), 2116–2145 (2017)
2. Ashraphijuo, M., Wang, X.: Characterization of sampling patterns for low-tt-rank tensor retrieval. Ann. Math. Artif. Intell. **88**(8), 859–886 (2020)
3. Ashraphijuo, M., Wang, X., Zhang, J.: Low-rank data completion with very low sampling rate using Newton's method. IEEE Trans. Signal Process. **67**(7), 1849–1859 (2019)
4. Asif, M.T., Mitrovic, N., Garg, L., Dauwels, J., Jaillet, P.: Low-dimensional models for missing data imputation in road networks. In: 2013 IEEE International Conference on Acoustics, Speech and Signal Processing, pp. 3527–3531. IEEE, New York (2013)
5. Balazevic, I., Allen, C., Hospedales, T.: TuckER: Tensor factorization for knowledge graph completion. In: Proceedings of the 2019 Conference on Empirical Methods in Natural Language Processing and the 9th International Joint Conference on Natural Language Processing (EMNLP-IJCNLP), pp. 5188–5197 (2019)
6. Balažević, I., Allen, C., Hospedales, T.M.: Hypernetwork knowledge graph embeddings. In: International Conference on Artificial Neural Networks, pp. 553–565. Springer, Berlin (2019)
7. Bengua, J.A., Phien, H.N., Tuan, H.D., Do, M.N.: Efficient tensor completion for color image and video recovery: low-rank tensor train. IEEE Trans. Image Process. **26**(5), 2466–2479 (2017)
8. Bollacker, K., Evans, C., Paritosh, P., Sturge, T., Taylor, J.: Freebase: a collaboratively created graph database for structuring human knowledge. In: Proceedings of the 2008 ACM SIGMOD International Conference on Management of Data, pp. 1247–1250 (2008)
9. Bordes, A., Usunier, N., Garcia-Duran, A., Weston, J., Yakhnenko, O.: Translating embeddings for modeling multi-relational data. In: Neural Information Processing Systems (NIPS), pp. 1–9 (2013)
10. Candès, E.J., Recht, B.: Exact matrix completion via convex optimization. Found. Comput. Math. **9**(6), 717 (2009)
11. Candès, E.J., Tao, T.: The power of convex relaxation: near-optimal matrix completion. IEEE Trans. Inf. Theory **56**(5), 2053–2080 (2010)
12. Conn, A.R., Gould, N.I., Toint, P.L.: Trust Region Methods. SIAM, Philadelphia (2000)
13. Dettmers, T., Minervini, P., Stenetorp, P., Riedel, S.: Convolutional 2d knowledge graph embeddings. In: Proceedings of the AAAI Conference on Artificial Intelligence, vol. 32 (2018)

14. Fabian, M., Gjergji, K., Gerhard, W., et al.: Yago: A core of semantic knowledge unifying wordnet and wikipedia. In: 16th International World Wide Web Conference, WWW, pp. 697–706 (2007)
15. Fan, J., Cheng, J.: Matrix completion by deep matrix factorization. Neural Netw. **98**, 34–41 (2018)
16. Filipović, M., Jukić, A.: Tucker factorization with missing data with application to low-rank tensor completion. Multidim. Syst. Sign. Process. **26**(3), 677–692 (2015)
17. Gandy, S., Recht, B., Yamada, I.: Tensor completion and low-n-rank tensor recovery via convex optimization. Inverse Prob. **27**(2), 025010 (2011)
18. Hu, Y., Zhang, D., Ye, J., Li, X., He, X.: Fast and accurate matrix completion via truncated nuclear norm regularization. IEEE Trans. Pattern Anal. Mach. Intell. **35**(9), 2117–2130 (2012)
19. Huang, H., Liu, Y., Liu, J., Zhu, C.: Provable tensor ring completion. Signal Process. **171**, 107486 (2020)
20. Huang, H., Liu, Y., Long, Z., Zhu, C.: Robust low-rank tensor ring completion. IEEE Trans. Comput. Imag. **6**, 1117–1126 (2020)
21. Jain, P., Netrapalli, P., Sanghavi, S.: Low-rank matrix completion using alternating minimization. In: Proceedings of the Forty-fifth Annual ACM Symposium on Theory of Computing, pp. 665–674 (2013)
22. Jannach, D., Resnick, P., Tuzhilin, A., Zanker, M.: Recommender systems—beyond matrix completion. Commun. ACM **59**(11), 94–102 (2016)
23. Kang, Z., Peng, C., Cheng, Q.: Top-n recommender system via matrix completion. In: Proceedings of the AAAI Conference on Artificial Intelligence, vol. 30 (2016)
24. Kazemi, S.M., Poole, D.: SimplE embedding for link prediction in knowledge graphs. In: Advances in Neural Information Processing Systems, vol. 31 (2018)
25. Keshavan, R.H., Montanari, A., Oh, S.: Matrix completion from a few entries. IEEE Trans. Inf. Theory **56**(6), 2980–2998 (2010)
26. Kiefer, J., Wolfowitz, J., et al.: Stochastic estimation of the maximum of a regression function. Ann. Math. Stat. **23**(3), 462–466 (1952)
27. Lin, Y., Liu, Z., Sun, M., Liu, Y., Zhu, X.: Learning entity and relation embeddings for knowledge graph completion. In: Proceedings of the AAAI Conference on Artificial Intelligence, vol. 29 (2015)
28. Liu, J., Musialski, P., Wonka, P., Ye, J.: Tensor completion for estimating missing values in visual data. IEEE Trans. Pattern Anal. Mach. Intell. **35**(1), 208–220 (2012)
29. Liu, Y., Long, Z., Zhu, C.: Image completion using low tensor tree rank and total variation minimization. IEEE Trans. Multimedia **21**(2), 338–350 (2019)
30. Long, Z., Liu, Y., Chen, L., Zhu, C.: Low rank tensor completion for multiway visual data. Signal Process. **155**, 301–316 (2019)
31. Long Z., Zhu C., Liu, J., Liu, Y.: Bayesian low rank tensor ring for image recovery. IEEE Trans. Image Process. **30**, 3568–3580 (2021)
32. Lu, C.: A Library of ADMM for Sparse and Low-rank Optimization. National University of Singapore (2016). https://github.com/canyilu/LibADMM
33. Lukovnikov, D., Fischer, A., Lehmann, J., Auer, S.: Neural network-based question answering over knowledge graphs on word and character level. In: Proceedings of the 26th International Conference on World Wide Web, pp. 1211–1220 (2017)
34. Miller, G.A.: WordNet: a lexical database for english. Commun. ACM **38**(11), 39–41 (1995)
35. Moré, J.J.: The Levenberg-Marquardt algorithm: implementation and theory. In: Numerical Analysis, pp. 105–116. Springer, Berlin (1978)
36. Mu, C., Huang, B., Wright, J., Goldfarb, D.: Square deal: lower bounds and improved relaxations for tensor recovery. In: International Conference on Machine Learning, pp. 73–81 (2014)
37. Nickel, M., Tresp, V., Kriegel, H.P.: A three-way model for collective learning on multi-relational data. In: Proceedings of the 28th International Conference on International Conference on Machine Learning, pp. 809–816 (2011)

38. Robbins, H., Monro, S.: A stochastic approximation method. Ann. Math. Stat. **22**(3), 400–407 (1951)
39. Singhal, A.: Introducing the knowledge graph: things, not strings. Official Google Blog **5** (2012)
40. Sun, Z., Deng, Z.H., Nie, J.Y., Tang, J.: RotatE: knowledge graph embedding by relational rotation in complex space. In: International Conference on Learning Representations (2018)
41. Trouillon, T., Dance, C.R., Gaussier, É., Welbl, J., Riedel, S., Bouchard, G.: Knowledge graph completion via complex tensor factorization. J. Mach. Learn. Res. **18**(1), 4735–4772 (2017)
42. Vrandečić, D., Krötzsch, M.: Wikidata: a free collaborative knowledgebase. Commun. ACM **57**(10), 78–85 (2014)
43. Vu, T., Nguyen, T.D., Nguyen, D.Q., Phung, D., et al.: A capsule network-based embedding model for knowledge graph completion and search personalization. In: Proceedings of the 2019 Conference of the North American Chapter of the Association for Computational Linguistics: Human Language Technologies, Volume 1 (Long and Short Papers), pp. 2180–2189 (2019)
44. Wang, Z., Zhang, J., Feng, J., Chen, Z.: Knowledge graph embedding by translating on hyperplanes. In: Proceedings of the AAAI Conference on Artificial Intelligence, vol. 28 (2014)
45. Wang, W., Aggarwal, V., Aeron, S.: Tensor completion by alternating minimization under the tensor train (TT) model (2016). Preprint, arXiv:1609.05587
46. Wang, W., Aggarwal, V., Aeron, S.: Efficient low rank tensor ring completion. In: Proceedings of the IEEE International Conference on Computer Vision, pp. 5697–5705 (2017)
47. Xu, Y., Yin, W.: A block coordinate descent method for regularized multiconvex optimization with applications to nonnegative tensor factorization and completion. SIAM J. Imag. Sci. **6**(3), 1758–1789 (2013)
48. Yang, B., Yih, W.T., He, X., Gao, J., Deng, L.: Embedding entities and relations for learning and inference in knowledge bases. In: 3rd International Conference on Learning Representations, ICLR 2015, San Diego, CA, USA, 7–9 May 2015, Conference Track Proceedings (2015)
49. Yang, Y., Feng, Y., Suykens, J.A.: A rank-one tensor updating algorithm for tensor completion. IEEE Signal Process Lett. **22**(10), 1633–1637 (2015)
50. Yu, J., Li, C., Zhao, Q., Zhao, G.: Tensor-ring nuclear norm minimization and application for visual: Data completion. In: ICASSP 2019-2019 IEEE International Conference on Acoustics, Speech and Signal Processing (ICASSP), pp. 3142–3146. IEEE, New York (2019)
51. Zeb, A., Haq, A.U., Zhang, D., Chen, J., Gong, Z.: KGEL: a novel end-to-end embedding learning framework for knowledge graph completion. Expert Syst. Appl. **167**, 114164 (2020)
52. Zhang, Z., Aeron, S.: Exact tensor completion using t-SVD. IEEE Trans. Signal Process. **65**(6), 1511–1526 (2017)
53. Zhang, F., Yuan, N.J., Lian, D., Xie, X., Ma, W.Y.: Collaborative knowledge base embedding for recommender systems. In: Proceedings of the 22nd ACM SIGKDD International Conference on Knowledge Discovery and Data Mining, pp. 353–362 (2016)
54. Zhao, Q., Zhang, L., Cichocki, A.: Bayesian CP factorization of incomplete tensors with automatic rank determination. IEEE Trans. Pattern Anal. Mach. Intell. **37**(9), 1751–1763 (2015)
55. Zheng, V., Cao, B., Zheng, Y., Xie, X., Yang, Q.: Collaborative filtering meets mobile recommendation: a user-centered approach. In: Proceedings of the AAAI Conference on Artificial Intelligence, vol. 24 (2010)
56. Zniyed, Y., Boyer, R., de Almeida, A.L., Favier, G.: High-order tensor estimation via trains of coupled third-order CP and Tucker decompositions. Linear Algebra Appl. **588**, 304–337 (2020)

Chapter 5
Coupled Tensor for Data Analysis

5.1 Introduction

Multimodal signals widely exist in data acquisition with the help of different sensors. For example, in medical diagnoses, we can obtain several types of data from a patient at the same time, including electroencephalogram (EEG), electrocardiogram (ECG) monitoring data, and functional magnetic resonance imaging (fMRI) scans. Such data share some common latent components but also keep certain independent features of their own. Therefore, it may be advantageous to analyze such data in a coupled way instead of processing independently.

Joint analysis of data from multiple sources can be modeled with the help of coupled matrix/tensor decomposition, where the obtained data are represented by matrix/tensor and the common modes are linked in a coupled way. Coupled matrix/tensor component analysis has attracted much attention in data mining [2, 6] and signal processing [4, 12, 14].

For instance, coupled matrix/tensor component analysis can be used for missing data recovery when the obtained data have low-rank structure. As shown in Fig. 5.1, link prediction aims to give some recommendations according to the observed data. The cube represents the relationship of locations-tourist-activities and can be modeled as a tensor. If we only observe this data, the cold start problem would occur [9], since the collected information is not yet sufficient. If we can obtain the relationship of tourist-tourist, features-locations, activities-activities, and tourist-locations, which can be regarded as the auxiliary information and modeled as matrices, the cold start problem may be avoided. This is because the tensor and the matrices share some latent information and when the tensor is incomplete, the shared information in matrices will benefit the tensor recovery.

Y. Liu et al., *Tensor Computation for Data Analysis*,
https://doi.org/10.1007/978-3-030-74386-4_5

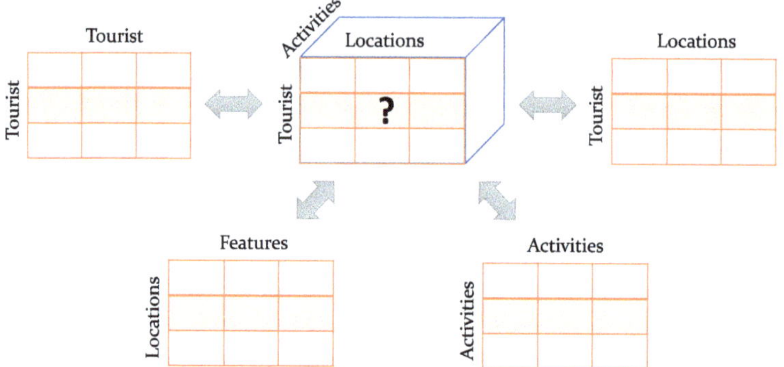

Fig. 5.1 UCLAF dataset where the tensor is incomplete and coupled with each matrix by sharing some latent information

Fig. 5.2 An example of coupled matrix and tensor factorization

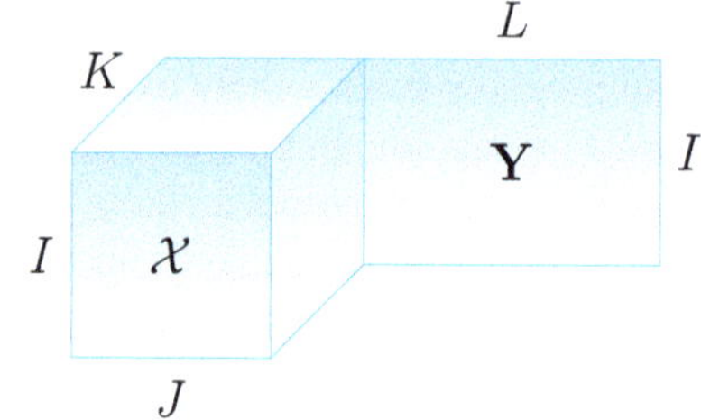

5.2 Coupled Tensor Component Analysis Methods

5.2.1 *Coupled Matrix and Tensor Factorization Model*

In multimodal systems, we can obtain different kinds of data. Taking restaurant recommendation system as an example, we can obtain a third-order rating tensor, whose entries represent the rating on different meals from different restaurants by different customers, and a matrix which represents the customers' social network, as it shows in Fig. 5.2. Coupled analysis of matrices and tensors is a powerful tool to efficiently analyze such data, which commonly appears in many applications, such as community detection, collaborative filtering, and chemometrics.

Given a third-order tensor data $\mathcal{X} \in \mathbb{R}^{I \times J \times K}$, and related matrix data $\mathbf{Y} \in \mathbb{R}^{I \times L}$, they are assumed to be coupled in the first mode. The corresponding coupled matrix and tensor factorization (CMTF) can be formulated as follows:

$$\min_{\mathbf{A},\mathbf{B},\mathbf{C},\mathbf{D}} \|\mathcal{X} - [\![\mathbf{A}, \mathbf{B}, \mathbf{C}]\!]\|_F^2 + \|\mathbf{Y} - \mathbf{A}\mathbf{D}^{\mathrm{T}}\|_F^2, \tag{5.1}$$

where $[\![\mathbf{A}, \mathbf{B}, \mathbf{C}]\!]$ represents CP decomposition with the factor matrices $\mathbf{A}$, $\mathbf{B}$ and $\mathbf{C}$, $\mathbf{A}$ and $\mathbf{D}$ are the factors of $\mathbf{Y}$. Similar to CP-ALS, problem (5.1) can be easily solved by ALS framework which updates one variable with others fixed, as concluded in Algorithm 30.

Algorithm 30: ALS for CMTF

Input: $\mathcal{X} \in \mathbb{R}^{I\times J\times K}$, $\mathbf{Y} \in \mathbb{R}^{I\times L}$ and CP-rank R
Initialize A, B, C, D
While not converged **do**
1. solve for **A** (with **B**, **C**, **D** fixed)

$$\min_{\mathbf{A}} \|[\mathbf{X}_{(1)}, \mathbf{Y}] - \mathbf{A}[(\mathbf{C}\otimes\mathbf{B})^{\mathrm{T}}, \mathbf{D}^{\mathrm{T}}]\|_{\mathrm{F}}^2$$

2. solve for **B** (with **A**, **C**, **D** fixed)

$$\min_{\mathbf{B}} \|\mathbf{X}_{(2)} - \mathbf{B}(\mathbf{C}\otimes\mathbf{A})^{\mathrm{T}}\|_{\mathrm{F}}^2$$

3. solve for **C** (with **A**, **B**, **D** fixed)

$$\min_{\mathbf{C}} \|\mathbf{X}_{(3)} - \mathbf{C}(\mathbf{B}\otimes\mathbf{A})^{\mathrm{T}}\|_{\mathrm{F}}^2$$

4. solve for **D** (with **A**, **B**, **C** fixed)

$$\min_{\mathbf{D}} \|\mathbf{Y} - \mathbf{A}\mathbf{D}^{\mathrm{T}}\|_{\mathrm{F}}^2$$

Until fit ceases to improve or maximum iterations exhausted
Output: **A**, **B**, **C**

However, this algorithm can stop at local optimal point, and overfitting may easily occur when the tensor rank is set too large. To deal with these problems, CMTF-OPT [1] is proposed to simultaneously solve all factor matrices by a gradient-based optimization.

It is noticed that CMTF models have been extended by other tensor factorizations such as Tucker [3] and BTD [15]. In this chapter, we mainly introduce CMTF model using CP for modeling higher-order tensors, and the squared Frobenius norm is employed as the loss function.

In practice, the factors can have many properties according to the physical properties of the data. For instance, nonnegativity is an important property of latent factors since many real-world tensors have nonnegative values and the hidden components have a physical meaning only when they are nonnegative. Besides, sparsity and orthogonality constraints on latent factors can identify the shared/unshared factors in coupled data. In this case, Acar et al. [1] proposed a flexible framework for CMTF as follows:

$$\begin{aligned}
\min_{\mathbf{A},\mathbf{B},\mathbf{C},\mathbf{D},\Sigma,\boldsymbol{\lambda}} \quad & \|\mathcal{X} - [\![\boldsymbol{\lambda}; \mathbf{A}, \mathbf{B}, \mathbf{C}]\!]\|_{\mathrm{F}}^2 + \|\mathbf{Y} - \mathbf{A}\Sigma\mathbf{D}^{\mathrm{T}}\|_{\mathrm{F}}^2 \\
\text{s. t.} \quad & \|\mathbf{a}_r\|_2 = \|\mathbf{b}_r\|_2 = \|\mathbf{c}_r\|_2 = \|\mathbf{d}_r\|_2 = 1, \\
& |\mathbf{a}_r^{\mathrm{T}}\mathbf{a}_s| \leq \alpha, |\mathbf{b}_r^{\mathrm{T}}\mathbf{b}_s| \leq \alpha, |\mathbf{c}_r^{\mathrm{T}}\mathbf{c}_s| \leq \alpha, |\mathbf{d}_r^{\mathrm{T}}\mathbf{d}_s| \leq \alpha, \\
& \sum_{r=1}^{R} \lambda_r \leq \beta, \sum_{r=1}^{R} \sigma_r \leq \beta, \\
& \sigma_r, \lambda_r \geq 0 \quad \forall r, s \in \{1, \ldots, R\}, r \neq s,
\end{aligned} \tag{5.2}$$

where $\mathbf{A} = [\mathbf{a}_1, \ldots, \mathbf{a}_R]$, $\mathbf{B} = [\mathbf{b}_1, \ldots, \mathbf{b}_R]$,$\mathbf{C} = [\mathbf{c}_1, \ldots, \mathbf{c}_R]$, $\mathbf{D} = [\mathbf{d}_1, \ldots, \mathbf{d}_R]$, $\|\mathbf{a}_r\|_2 = \|\mathbf{b}_r\|_2 = \|\mathbf{c}_r\|_2 = \|\mathbf{d}_r\|_2 = 1, r = 1, \ldots R$ means the columns of factor matrices have a unit norm, $|\mathbf{a}_r^{\mathrm{T}}\mathbf{a}_s| \leq \alpha$, $|\mathbf{b}_r^{\mathrm{T}}\mathbf{b}_s| \leq \alpha$, $|\mathbf{c}_r^{\mathrm{T}}\mathbf{c}_s| \leq \alpha$, $|\mathbf{d}_r^{\mathrm{T}}\mathbf{d}_s| \leq \alpha$ prevent two factors from being similar and enforce the extra factors to get zero weights, and $\boldsymbol{\sigma}$ and $\boldsymbol{\lambda}$ are the weights of rank-1 components in the matrix and the third-order tensor, $\Sigma = \mathrm{diag}(\boldsymbol{\sigma})$.

5.2.2 *Coupled Tensor Factorization Model*

Besides the above CMTF model with one tensor and one matrix, both the coupled data can be in the form of tensor. For example, in the traffic information system, we can collect different kinds of traffic data. We can obtain a third-order traffic flow data denoted as $\mathcal{X} \in \mathbb{R}^{I \times J \times K}$, where I represents the number of road segments, J represents the number of days, K represents the number of time slots, and its entries represent traffic flow. In addition, the road environment data in the same road traffic network can be collected as the side information $\mathcal{Y} \in \mathbb{R}^{I \times L \times M}$, where L represents the number of lanes and M represents the number of variables about the weather, e.g., raining, snowing, sunny, and so on. Each element denotes the frequency of accidents. Such a coupled tensor model can be found in Fig. 5.3.

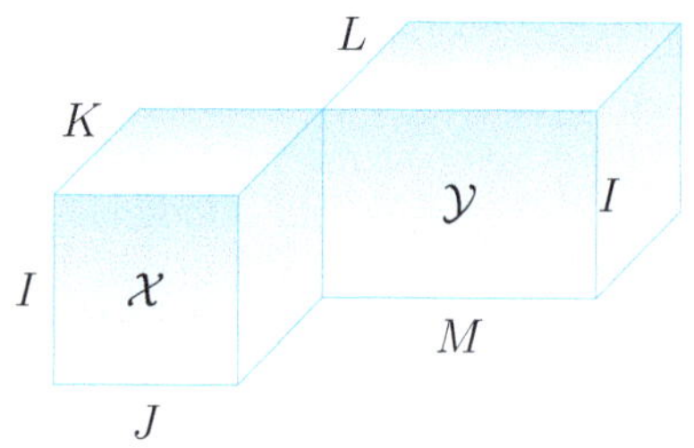

Fig. 5.3 An example of coupled tensor factorization in the first mode

As shown in Fig. 5.3, tensors $\mathcal{X}$ and $\mathcal{Y}$ are coupled in the first mode. Therefore, the coupled tensor factorization (CTF) of $\mathcal{X}$ and $\mathcal{Y}$ can be defined as

$$\min_{\mathbf{A},\mathbf{B},\mathbf{C},\mathbf{D},\mathbf{E}} \|\mathcal{X} - [\![\mathbf{A},\mathbf{B},\mathbf{C}]\!]\|_{\mathrm{F}}^2 + \|\mathcal{Y} - [\![\mathbf{A},\mathbf{D},\mathbf{E}]\!]\|_{\mathrm{F}}^2, \tag{5.3}$$

where $[\![\mathbf{A},\mathbf{B},\mathbf{C}]\!]$ represents CP decomposition for $\mathcal{X}$ with the factor matrices $\mathbf{A}$, $\mathbf{B}$ and $\mathbf{C}$ and $[\![\mathbf{A},\mathbf{D},\mathbf{E}]\!]$ represents CP decomposition for $\mathcal{Y}$. Following CMTF-OPT [1] for coupled matrix and tensor factorization, the optimization model in (5.3) can be solved in a similar way. The loss function of this group is as follows:

$$f(\mathbf{A},\mathbf{B},\mathbf{C},\mathbf{D},\mathbf{E}) = \frac{1}{2}\|\mathcal{X} - [\![\mathbf{A},\mathbf{B},\mathbf{C}]\!]\|_{\mathrm{F}}^2 + \frac{1}{2}\|\mathcal{Y} - [\![\mathbf{A},\mathbf{D},\mathbf{E}]\!]\|_{\mathrm{F}}^2, \tag{5.4}$$

and we further define $\hat{\mathcal{X}} = [\![\mathbf{A},\mathbf{B},\mathbf{C}]\!]$ and $\hat{\mathcal{Y}} = [\![\mathbf{A},\mathbf{D},\mathbf{E}]\!]$ as the estimates of $\mathcal{X}$ and $\mathcal{Y}$, respectively.

First, combining all factor matrices, we can obtain the variable $\mathbf{z}$:

$$\mathbf{z} = \left[\mathbf{a}_1^{\mathrm{T}};\cdots;\mathbf{a}_R^{\mathrm{T}};\mathbf{b}_1^{\mathrm{T}};\cdots;\mathbf{b}_R^{\mathrm{T}};\mathbf{c}_1^{\mathrm{T}};\cdots;\mathbf{c}_R^{\mathrm{T}};\mathbf{d}_1^{\mathrm{T}};\cdots;\mathbf{d}_R^{\mathrm{T}};\mathbf{e}_1^{\mathrm{T}};\cdots;\mathbf{e}_R^{\mathrm{T}}\right]. \tag{5.5}$$

Then we can calculate:

$$\begin{aligned}
\frac{\partial f}{\partial \mathbf{A}} &= \left(\hat{\mathbf{X}}_{(1)} - \mathbf{X}_{(1)}\right)(\mathbf{C}\odot\mathbf{B}) + (\hat{\mathbf{Y}}_{(1)} - \mathbf{Y}_{(1)})(\mathbf{E}\odot\mathbf{D})\\
\frac{\partial f}{\partial \mathbf{B}} &= \left(\hat{\mathbf{X}}_{(2)} - \mathbf{X}_{(2)}\right)(\mathbf{C}\odot\mathbf{A})\\
\frac{\partial f}{\partial \mathbf{C}} &= \left(\hat{\mathbf{X}}_{(3)} - \mathbf{X}_{(3)}\right)(\mathbf{B}\odot\mathbf{A})\\
\frac{\partial f}{\partial \mathbf{D}} &= \left(\hat{\mathbf{Y}}_{(2)} - \mathbf{Y}_{(2)}\right)(\mathbf{E}\odot\mathbf{A})\\
\frac{\partial f}{\partial \mathbf{E}} &= \left(\hat{\mathbf{Y}}_{(3)} - \mathbf{Y}_{(3)}\right)(\mathbf{D}\odot\mathbf{A}),
\end{aligned} \tag{5.6}$$

where $\mathbf{X}_{(n)}$ is the mode-n matricization of tensor $\mathcal{X}$. Finally the vectorization combination of the gradients of all factor matrices can be written as

$$\mathbf{g} = \left[\mathrm{vec}\left(\frac{\partial f}{\partial \mathbf{A}}\right),\cdots,\mathrm{vec}\left(\frac{\partial f}{\partial \mathbf{E}}\right)\right]^{\mathrm{T}}. \tag{5.7}$$

The nonlinear conjugate gradient (NCG) with Hestenes-Stiefel updated and the More-Thuente line search are used to optimize the factor matrices. The details of CTF-OPT algorithm are given in Algorithm 31.

Another kind of CTF is coupled along all modes by linear relationship. For example, due to the hardware limitations, it is hard to collect super-resolution

Algorithm 31: Joint optimization based coupled tensor factorizations

Input: Two 3rd-order tensors $\mathcal{X} \in \mathbb{R}^{I \times J \times K}$ and $\mathcal{Y} \in \mathbb{R}^{I \times L \times M}$, number of components R
initialize $\mathbf{A}, \mathbf{B}, \mathbf{C}, \mathbf{D}, \mathbf{E}$
1. Calculate $\mathbf{z}$ via equation (5.5)
2. Calculate f via equation (5.4)
3. Calculate $\mathbf{g}$ via equation (5.7)
4. **while** not converged **do**
   Update $\mathbf{z}$, f, $\mathbf{g}$ by using NCG and line search.
until fit ceases to improve or maximum iterations exhausted
Output: $\mathbf{A}, \mathbf{B}, \mathbf{C}, \mathbf{D}, \mathbf{E}$

image (SRI) which admits both high-spatial and high-spectral resolutions. Instead, we can obtain hyperspectral image (HSI) which has low-spatial and high-spectral resolutions. It can be denoted as $\mathcal{X} \in \mathbb{R}^{I_H \times J_H \times K}$, where I_H and J_H denote the spatial dimensions and K denotes the number of spectral bands. On the other hand, the obtained multispectral image (MSI), which can be represented as $\mathcal{Y} \in \mathbb{R}^{I \times J \times K_M}$, owns high-spatial and low-spectral resolutions. Both $\mathcal{X}$ and $\mathcal{Y}$ are third-order tensor data and coupled in all modes. The graphical illustration can be found in Fig. 5.4.

The mathematical model can be formulated as

$$\min_{\mathbf{A},\mathbf{B},\mathbf{C}} \|\mathcal{X} - [\![\mathbf{P}_1\mathbf{A}, \mathbf{P}_2\mathbf{B}, \mathbf{C}]\!]\|_F^2 + \|\mathcal{Y} - [\![\mathbf{A}, \mathbf{B}, \mathbf{P}_M\mathbf{C}]\!]\|_F^2, \tag{5.8}$$

where $\mathbf{A} \in \mathbb{R}^{I \times R}$, $\mathbf{B} \in \mathbb{R}^{J \times R}$, $\mathbf{C} \in \mathbb{R}^{K \times R}$, $\mathbf{P}_1 \in \mathbb{R}^{I_H \times I}$, and $\mathbf{P}_2 \in \mathbb{R}^{J_H \times J}$ are the spatial degradation operator and $\mathbf{P}_M \in \mathbb{R}^{K_M \times K}$ is the spectral degradation operator.

For two tensors coupled along all mode by linear relationship, ALS is an efficient way to solve this problem. According to Eq. (5.8), the loss function can be rewritten as

$$f(\mathbf{A}, \mathbf{B}, \mathbf{C}) = \frac{1}{2}\|\mathcal{X} - [\![\mathbf{P}_1\mathbf{A}, \mathbf{P}_2\mathbf{B}, \mathbf{C}]\!]\|_F^2 + \frac{1}{2}\|\mathcal{Y} - [\![\mathbf{A}, \mathbf{B}, \mathbf{P}_M\mathbf{C}]\!]\|_F^2. \tag{5.9}$$

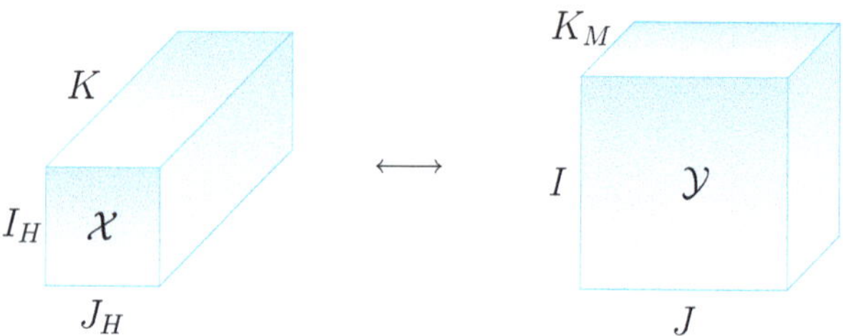

Fig. 5.4 An example of coupled tensor factorization

Under ALS framework, this problem can be divided into three subproblems as follows:

$$f(\mathbf{A}) = \frac{1}{2}\|\mathcal{X}_{(1)} - \mathbf{P}_1\mathbf{A}(\mathbf{C} \odot \mathbf{P}_2\mathbf{B})^{\mathrm{T}}\|_{\mathrm{F}}^2 + \frac{1}{2}\|\mathcal{Y}_{(1)} - \mathbf{A}(\mathbf{P}_M\mathbf{C} \odot \mathbf{B})^{\mathrm{T}}\|_{\mathrm{F}}^2, \tag{5.10}$$

$$f(\mathbf{B}) = \frac{1}{2}\|\mathcal{X}_{(2)} - \mathbf{P}_2\mathbf{B}(\mathbf{C} \odot \mathbf{P}_1\mathbf{A})^{\mathrm{T}}\|_{\mathrm{F}}^2 + \frac{1}{2}\|\mathcal{Y}_{(2)} - \mathbf{B}(\mathbf{P}_M\mathbf{C} \odot \mathbf{A})^{\mathrm{T}}\|_{\mathrm{F}}^2, \tag{5.11}$$

$$f(\mathbf{C}) = \frac{1}{2}\|\mathcal{X}_{(3)} - \mathbf{C}(\mathbf{P}_2\mathbf{B} \odot \mathbf{P}_1\mathbf{A})^{\mathrm{T}}\|_{\mathrm{F}}^2 + \frac{1}{2}\|\mathcal{Y}_{(3)} - \mathbf{P}_M\mathbf{C}(\mathbf{B} \odot \mathbf{A})^{\mathrm{T}}\|_{\mathrm{F}}^2. \tag{5.12}$$

The details can be concluded in Algorithm 32.

Algorithm 32: Coupled tensor factorizations

Input: Two 3rd-order tensors $\mathcal{X} \in \mathbb{R}^{I_H \times J_H \times K}$ and $\mathcal{Y} \in \mathbb{R}^{I \times J \times K_M}$, number of components R
initialize $\mathbf{A}, \mathbf{B}, \mathbf{C}$
while not converged **do**
1. Update $\mathbf{A}$ via equation (5.10)
2. Update $\mathbf{B}$ via equation (5.11)
3. Update $\mathbf{C}$ via equation (5.12)

until fit ceases to improve or maximum iterations exhausted
Output: $\mathbf{A}, \mathbf{B}, \mathbf{C}$

5.2.3 Generalized Coupled Tensor Factorization Model

To construct a general and practical framework for computation of coupled tensor factorizations, Yilmaz et. al. [19] proposed a generalized coupled tensor factorization model by extending the well-established theory of generalized linear models. For example, we consider multiple tensors $\mathcal{X}_i \in \mathbb{R}^{J_1 \times \cdots \times J_{M_i}}$, $i = 1, \ldots, I$, where I is the number of coupled tensors. Factorizing these tensors simultaneously based on CP decomposition will result in N latent factors $\mathbf{U}^{(n)}$, $n = 1, \ldots, N$, which are also called latent component in generalized linear models. In this way, the generalized coupled tensor factorizations for $\mathcal{X}_i, i = 1, \ldots, I$ can be defined by

$$\mathcal{X}_i(j_1, \ldots, j_{M_i}) = \sum_{r=1}^{R}\prod_{n=1}^{N}(\mathbf{U}^{(n)}(j_n, r))^{p_{i,n}}, i = 1, \ldots I, \tag{5.13}$$

where $\mathbf{P} \in \mathbb{R}^{I \times N}$ is a coupling matrix and can be denoted as

$$p_{i,n} = \begin{cases} 1 & \mathcal{X}_i \text{ and } \mathbf{U}^{(n)} \text{ are connected,} \\ 0 & \text{otherwise ,} \end{cases}$$

where $p_{i,n}$ is i, n-th element of $\mathbf{P}$.

To further illustrate this model, we take $\mathcal{X}_1 \in \mathbb{R}^{J_1 \times J_2 \times J_3}$, $\mathbf{X}_2 \in \mathbb{R}^{J_2 \times J_4}$, and $\mathbf{X}_3 \in \mathbb{R}^{J_2 \times J_5}$ as an example:

$$\mathcal{X}_1(j_1, j_2, j_3) = \sum_{r=1}^{R} \mathbf{U}^{(1)}(j_1, r)\mathbf{U}^{(2)}(j_2, r)\mathbf{U}^{(3)}(j_3, r), \tag{5.14}$$

$$\mathbf{X}_2(j_2, j_4) = \sum_{r=1}^{R} \mathbf{U}^{(2)}(j_2, r)\mathbf{U}^{(4)}(j_4, r), \tag{5.15}$$

$$\mathbf{X}_3(j_2, j_5) = \sum_{r=1}^{R} \mathbf{U}^{(2)}(j_2, r)\mathbf{U}^{(5)}(j_5, r), \tag{5.16}$$

where the number of tensors is 3 and the number of factor matrices is 5. It can be rewritten in detail as follows:

$$\mathcal{X}_1(j_1, j_2, j_3) = \sum_{r=1}^{R} \mathbf{U}^{(1)}(j_1, r)^1\mathbf{U}^{(2)}(j_2, r)^1\mathbf{U}^{(3)}(j_3, r)^1\mathbf{U}^{(4)}(j_4, r)^0\mathbf{U}^{(5)}(j_5, r)^0,$$

$$\mathcal{X}_2(j_2, j_4) = \sum_{r=1}^{R} \mathbf{U}^{(1)}(j_1, r)^0\mathbf{U}^{(2)}(j_2, r)^1\mathbf{U}^{(3)}(j_3, r)^0\mathbf{U}^{(4)}(j_4, r)^1\mathbf{U}^{(5)}(j_5, r)^0,$$

$$\mathcal{X}_3(j_2, j_5) = \sum_{r=1}^{R} \mathbf{U}^{(1)}(j_1, r)^0\mathbf{U}^{(2)}(j_2, r)^1\mathbf{U}^{(3)}(j_3, r)^0\mathbf{U}^{(4)}(j_4, r)^0\mathbf{U}^{(5)}(j_5, r)^1,$$

with the coupling matrix $\mathbf{P} \in \mathbb{R}^{3 \times 5}$ as

$$\mathbf{P} = \begin{bmatrix} 1\,1\,1\,0\,0 \\ 0\,1\,0\,1\,0 \\ 0\,1\,0\,0\,1 \end{bmatrix}.$$

In addition, the graphical illustration can be found in Fig. 5.5.

The aim of generalized coupled tensor factorization is to find the latent components following linear models. Accordingly, each element $\mathcal{X}_i(j_1, \ldots, j_{M_i})$ in coupled tensors obeys independent Gaussian distribution. The log likelihood can

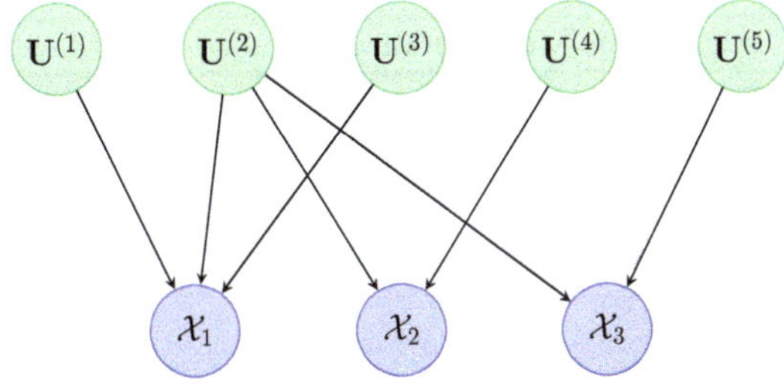

Fig. 5.5 An example of generalized coupled tensor factorization

be rewritten as

$$\mathrm{L}(\mathcal{X}_i(j_1,\ldots,j_{M_i})) = \sum_{i}^{I}\sum_{j_1,\ldots,j_{M_i}}^{J_1,\ldots,J_{M_i}} \frac{(\mathcal{X}_i(j_1,\ldots,j_{M_i}) - \hat{\mathcal{X}}_i(j_1,\ldots,j_{M_i}))^2}{2\tau^2}, \tag{5.17}$$

where $\hat{\mathcal{X}}_i(j_1,\ldots,j_{M_i}) = \sum_{r=1}^{R}\prod_{n=1}^{N}(\mathbf{U}^{(n)}(j_n,r))^{p_{i,n}}$. For the coupled factorization, the derivative of the log likelihood is

$$\frac{\partial\,\mathrm{L}}{\partial\mathbf{U}^{(n)}(j_n,r)} = -\sum_{i=1}^{I}\sum_{j_1,\ldots,j_{M_i}}^{J_1,\ldots,J_{M_i}} p_{i,n}(\mathcal{X}_i(j_1,\ldots,j_{M_i}) - \hat{\mathcal{X}}_i(j_1,\ldots,j_{M_i}))\tau^{-2} \times \frac{\partial\hat{\mathcal{X}}_i(j_1,\ldots,j_{M_i})}{\partial\mathbf{U}^{(n)}(j_n,r)},$$

and the second-order derivative of the log likelihood is

$$\frac{\partial^2\,\mathrm{L}}{\partial\mathbf{U}^{(n)}(j_n,r)^2} = \sum_{i=1}^{I}\sum_{j_1,\ldots,j_{M_i}}^{J_1,\ldots,J_{M_i}} p_{i,n}\tau^{-2}\left(\frac{\partial\hat{\mathcal{X}}_i(j_1,\ldots,j_{M_i})}{\partial\mathbf{U}^{(n)}(j_n,r)}\right)^{\mathrm{T}}\frac{\partial\hat{\mathcal{X}}_i(j_1,\ldots,j_{M_i})}{\partial\mathbf{U}^{(n)}(j_n,r)}.$$

The latent variable can be updated with others fixed, as follows:

$$\mathbf{U}^{(n)} := \mathbf{U}^{(n)} + \left(\sum_{i=1}^{I} p_{i,n}\left(\frac{\partial\hat{\mathcal{X}}_i}{\partial\mathbf{U}^{(n)}}\right)^{\mathrm{T}}\frac{\partial\hat{\mathcal{X}}_i}{\partial\mathbf{U}^{(n)}}\right)^{-1}\left(\sum_{i=1}^{I} p_{i,n}(\mathcal{X}_i - \hat{\mathcal{X}}_i)\frac{\partial\hat{\mathcal{X}}_i}{\partial\mathbf{U}^{(n)}}\right). \tag{5.18}$$

The details are concluded in Algorithm 33.

Algorithm 33: Generalized coupled tensor factorization

Input: multiple tensors $\mathcal{X}_i \in \mathbb{R}^{J_1 \times \cdots \times J_{M_i}}$, $i = 1, \ldots, I$, I is the number of tensors
initialize $\mathbf{U}^{(n)}$, $n = 1, \ldots, N$, N is the number of factor matrices
while not converged **do**
  For $n = 1$ to N
    Update $\mathbf{U}^{(n)}$ via equation (5.18)
  End For
until fit ceases to improve or maximum iterations exhausted
Output: $\mathbf{U}^{(n)}$, $n = 1, \ldots, N$

5.3 Applications

5.3.1 HSI-MSI Fusion

Hyperspectral imaging can acquire images across hundreds of different wavelengths. With high-spectral resolution, it can be used to material identification by spectroscopic analysis, which leads to a number of applications, such as target detection and spectral unmixing. However, since the fact that the camera needs to acquire photons from many spectral bands, the spatial resolution of hyperspectral image (HSI) is generally limited. Therefore, there has to be a tradeoff between spatial and spectral resolutions. In comparison, multispectral image (MSI) can have high-spatial resolution and a low spectral resolution. Therefore, it is attractive to fuse a hyperspectral image with a multispectral image to obtain a high-spatial and high-spectral super-resolution image (SRI). This procedure called HSI-MSI fusion can be graphically illustrated in Fig. 5.6.

Coupled tensor factorization in Eq. (5.8) can solve HSI-MSI fusion problem here. For example, in [8], Tucker decomposition was applied to recover SRI from HSI and MSI. Besides, coupled canonical polyadic decomposition (CPD) model [7] was proposed to tackle this problem. In [5], tensor train decomposition model was addressed by learning the correlations among the spatial, spectral, and nonlocal

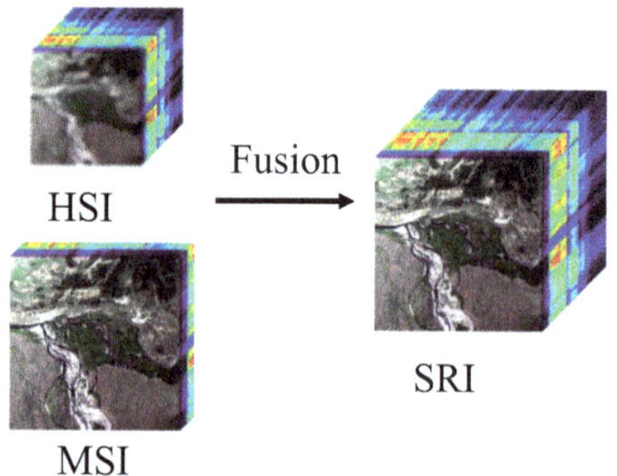

Fig. 5.6 HSI-MSI fusion procedure

modes of the nonlocal similar HSI cubes. In addition, spatial-spectral-graph-regularized low-rank tensor decomposition was proposed to solve this problem [21].

Two different datasets are used to test algorithms in this part. The first one is the Pavia University dataset[1] acquired by the ROSIS-3 optical airborne sensor in 2003. Each scene consists of 610 × 340 pixels with a ground sample distance (GSD) of 1.3 meters. Twelve noisy bands have been removed, so that a total of 103 bands covering the spectral range from 430 to 838 nm are used in experiments. In addition, we select 608 × 336 pixels from each image in experiments. Finally, the size of Pavia University dataset as SRI is 608 × 336 × 103. The second one is the Washington DC Mall dataset[2] provided with the permission of Spectral Information Technology Application Center of Virginia. The size of each scene is 1208 × 307. The dataset includes 210 bands in the spectral range from 400 to 2400 nm. We use the number of bands to 191 by removing bands where the atmosphere is opaque. And 612 × 306 pixels are selected from each image, forming the SRI with the size of 612 × 306 × 191.

The generation models of HSI and MSI can be formulated as follows:

$$\mathcal{X} = \mathcal{Z} \times_1 \mathbf{P}_1 \times_2 \mathbf{P}_2 + \mathcal{N}_H, \tag{5.19}$$

$$\mathcal{Y} = \mathcal{Z} \times_3 \mathbf{P}_M + \mathcal{N}_M, \tag{5.20}$$

where $\mathcal{N}_H$ and $\mathcal{N}_M$ are additive white Gaussian noise. We set spatial downsampling rate $P = 4$ and $\mathbf{P}_M \in \mathbb{R}^{4\times 103}$ for Pavia University data and $P = 6$ and $\mathbf{P}_M \in \mathbb{R}^{8\times 128}$ for Washington DC Mall data.

Eight state-of-the-art algorithms are used for comparison, namely, FUSE [16], CNMF [20] and HySure [11], SCOTT [10], STERRO [7] and CSTF-FUSE [8], and CNMVF [22]. The parameters of these methods are tuned to the best performance.

Figures 5.7 and 5.8 demonstrate the reconstructed images from various methods including FUSE, SCOTT, STEERO, CSTF-FUSE, HySure, CNMF, and CNMVF for Pavia University data and Washington DC Mall data, respectively. Besides, we add Gaussian noise in HSI and MSI with SNR = 15 dB. For Pavia University data, we select its 60th band to show while we choose 40th band from Washington DC Mall data. From Figs. 5.7 and 5.8, we can observe CMVTF performs best among all the state-of-the-art methods in terms of image resolution.

5.3.2 Link Prediction in Heterogeneous Data

The goal of link prediction is to predict the existence of missing connections between entries of interest. For example, a tourist arriving in Chengdu might wonder

[1]https://rslab.ut.ac.ir/data/.

[2]https://rslab.ut.ac.ir/data/.

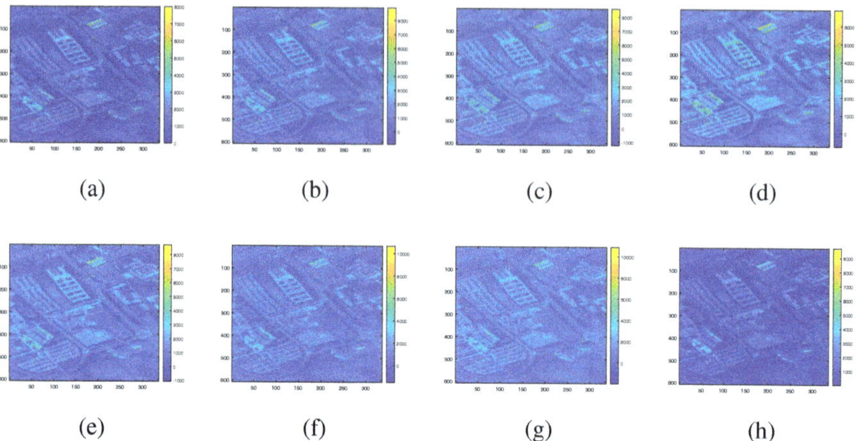

Fig. 5.7 The reconstruction of Pavia University data when SNR = 15 dB. (a) SRI. (b) CNMVF. (c) STEREO. (d) SCOTT. (e) CSTF. (f) FUSE. (g) HySure. (h) CNMF

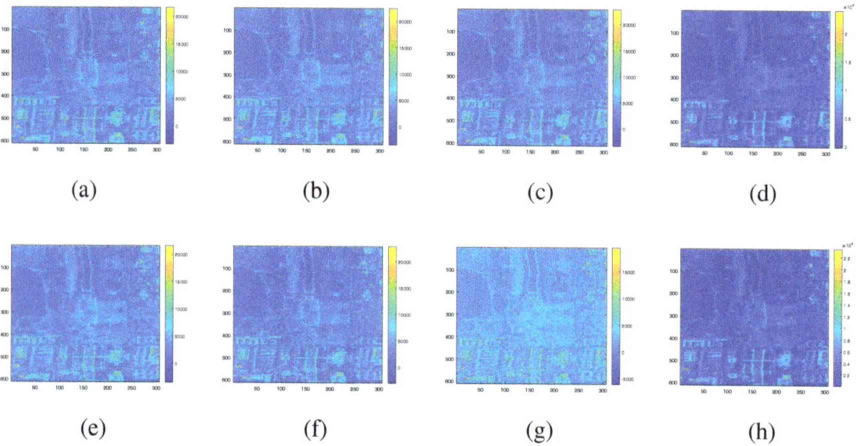

Fig. 5.8 The reconstruction of Washington DC University data when SNR = 15 dB. (a) SRI. (b) CNMVF. (c) STEREO. (d) SCOTT. (e) CSTF. (f) FUSE. (g) HySure. (h) CNMF

where to visit and what to do. The tourist, the location, and the activities can be considered to be linked. The task of recommending other scenic spots or activities to tourist can be cast as a missing link prediction problem. However, the results are likely to be poor if the prediction is done in isolation on a single view of data. Therefore, we should give the recommendation by combining multi-view of data.

In this part, we use the GPS data [24], where the relations between tourist, location, and activity are used to construct a third-order tensor $\mathcal{X}_1 \in \mathbb{R}^{164\times 168\times 5}$. To construct the dataset, GPS points are clustered into 168 meaningful locations, and the user comments attached to the GPS data are manually parsed into activity

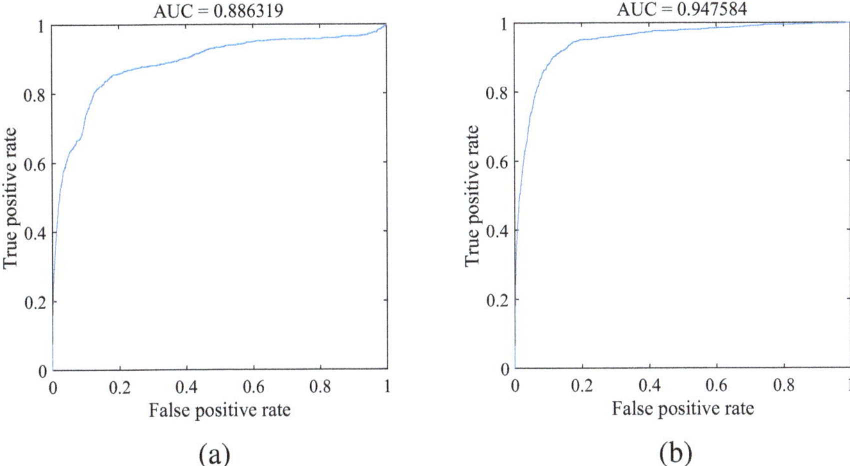

Fig. 5.9 The comparison of area under curve (AUC) using one dataset (left) and using multiple dataset (right) with 80% missing ratio. (**a**) The performance of AUC using one dataset. (**b**) The performance of AUC using multiple dataset

annotations for the 168 locations. Consequently, the data consists of 164 users, 168 locations, and 5 different types of activities, i.e., "Food and Drink," "Shopping," "Movies and Shows," "Sports and Exercise," and "Tourism and Amusement." The element $\mathcal{X}_1(i, j, k)$ represents the frequency of tourist i visiting location j and doing activity k there. The collected data also include additional side information, such as the tourist-location preferences from the GPS trajectory data and the location features from the points of interest database, represented as the matrix $\mathbf{X}_2 \in \mathbb{R}^{164\times 168}$ and $\mathbf{X}_3 \in \mathbb{R}^{168\times 14}$, respectively. Besides, the similarities on tourist-tourist $\mathbf{X}_4$ and activities-activities $\mathbf{X}_5$ can further enhance the performance of the recommendation system.

Figure 5.9 shows the area under curve (AUC) using one dataset (left) and using multiple datasets (right) with 80% missing ratio. We can see that the prediction accuracy of using multiple datasets is higher than that of using one dataset, which further illustrates the side information of different views on the same object can improve the information usage.

In addition, we have shown the prediction results of three state-of-the-art algorithms using GPS data in Fig. 5.10, where the missing ratio ranges from 10% to 70% with step 10%. These algorithms include SDF [13], CCP [17], and CTucker [18], which all consider the coupled mode. The first one uses Tucker decomposition under coupled tensor factorization framework to solve GPS data. The rest algorithms are on coupled rank minimization framework with CP and Tucker decomposition, respectively. From Fig. 5.10, we could observe the smaller the missing ratio is, the better the recovery performance is. Among these three methods, SDF performs best.

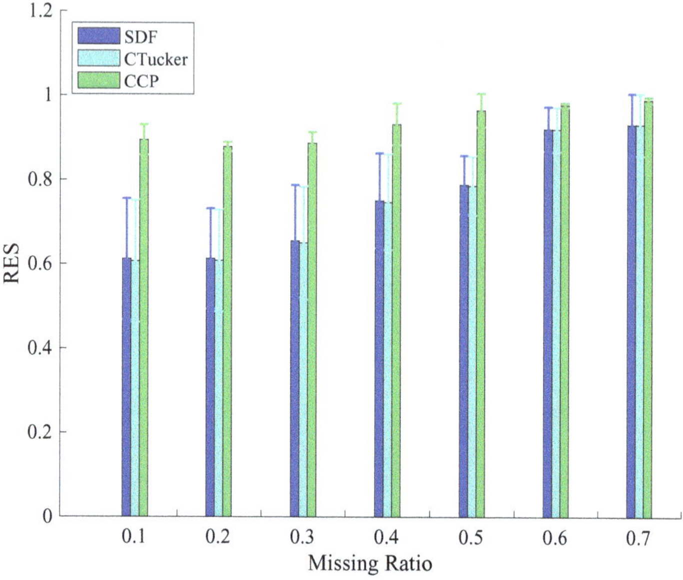

Fig. 5.10 The comparison performance of link prediction

5.3.3 *Visual Data Recovery*

In traditional visual data recovery tasks, they only consider the correlations of images and recover them by the global/local prior. However, with the development of sensors, we can obtain more information about the images. Effectively using this information can improve the recovery performance. In this part, we use the CASIA-SURF dataset [23] to illustrate that the side information can enhance the recovery results. The CASIA-SURF dataset consists of 1000 subjects and 21,000 video clips with 3 modalities (RGB, depth, infrared (IR)), each frame containing a person with the size of 256×256. For convenient computation, we choose 30 subjects with 2 modalities (RGB, IR) as shown in Fig. 5.11. Then we create a tensor $\mathcal{X} \in \mathbb{R}^{256\times 256\times 30}$ by changing each RGB image into gray image and a side information tensor $\mathcal{Y} \in \mathbb{R}^{256\times 256\times 30}$ by considering the IR information.

Figure 5.12 shows the results of three existing algorithms including SDF [13], CCP [17], and CTucker [18] on CASIA-SURF dataset. From Fig. 5.12, we could observe that CTucker outperforms CCP and SDF for image recovery. It is noted that the recovery performance of CCP keeps the same for different missing ratios, which may be caused by the CP decomposition.

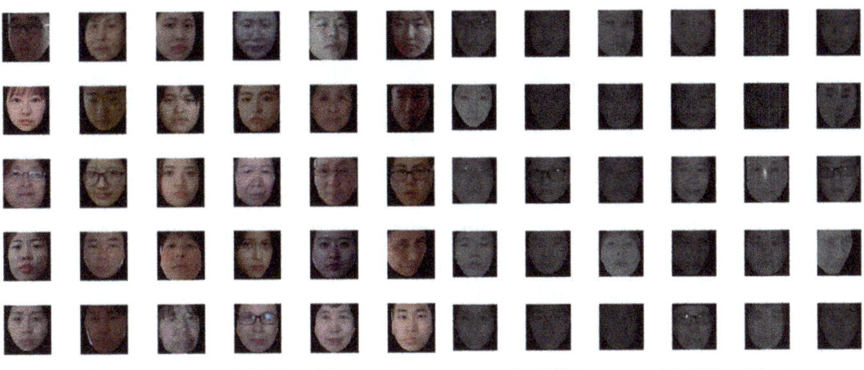

(a) RGB image with 30 subjects (b) IR image with 30 subjects

Fig. 5.11 The illustration of CASIA-SURF dataset. (**a**) RGB image with 30 subjects. (**b**) IR image with 30 subjects

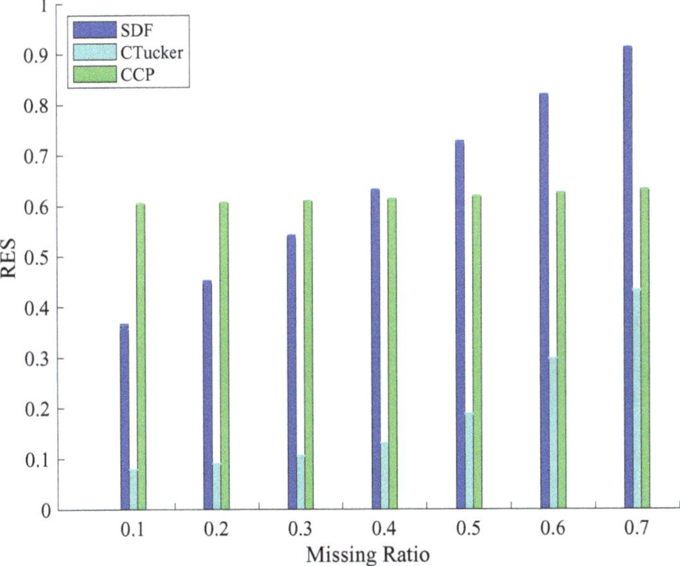

Fig. 5.12 The performance comparison on visual data recovery

5.4 Summary

As a tool to explore the data with shared latent information, coupled tensor component analysis plays an important role in signal processing and data mining. In this chapter, we only focus on two tensor component analysis models. One is coupled tensor completion with observed missing tensor. The other one is the coupled tensor fusion with a linear coupled way. In experiments, we show that with the assistance of side information, the performance of coupled tensor component analysis on link prediction and visual data recovery is better than traditional tensor component analysis.

Even though the coupled tensor decomposition has developed for many years, the works are still a few on multimodality data and lack of theory foundations. Regarding these, there are three main research directions:

- How to achieve the identifiability of coupled tensor decomposition on some mild conditions?
- How to efficiently explore the structure of multimodality data by coupled tensor decomposition?
- How to tackle these data with shared information and independent structure in a noisy condition?

References

1. Acar, E., Nilsson, M., Saunders, M.: A flexible modeling framework for coupled matrix and tensor factorizations. In: 2014 22nd European Signal Processing Conference (EUSIPCO), pp. 111–115. IEEE, Piscataway (2014)
2. Almutairi, F.M., Sidiropoulos, N.D., Karypis, G.: Context-aware recommendation-based learning analytics using tensor and coupled matrix factorization. IEEE J. Selec. Topics Signal Process. **11**(5), 729–741 (2017)
3. Bahargam, S., Papalexakis, E.E.: Constrained coupled matrix-tensor factorization and its application in pattern and topic detection. In: 2018 IEEE/ACM International Conference on Advances in Social Networks Analysis and Mining (ASONAM), pp. 91–94. IEEE, Piscataway (2018)
4. Chatzichristos, C., Davies, M., Escudero, J., Kofidis, E., Theodoridis, S.: Fusion of EEG and fMRI via soft coupled tensor decompositions. In: 2018 26th European Signal Processing Conference (EUSIPCO), pp. 56–60. IEEE, Piscataway (2018)
5. Dian, R., Li, S., Fang, L.: Learning a low tensor-train rank representation for hyperspectral image super-resolution. IEEE Trans. Neural Netw. Learn. Syst. **30**, 1–12 (2019). https://doi.org/10.1109/TNNLS.2018.2885616
6. Ermiş, B., Acar, E., Cemgil, A.T.: Link prediction in heterogeneous data via generalized coupled tensor factorization. Data Min. Knowl. Discov. **29**(1), 203–236 (2015)
7. Kanatsoulis, C.I., Fu, X., Sidiropoulos, N.D., Ma, W.K.: Hyperspectral super-resolution: a coupled tensor factorization approach. IEEE Trans. Signal Process. **66**(24), 6503–6517 (2018)
8. Li, S., Dian, R., Fang, L., Bioucas-Dias, J.M.: Fusing hyperspectral and multispectral images via coupled sparse tensor factorization. IEEE Trans. Image Process. **27**(8), 4118–4130 (2018)
9. Lika, B., Kolomvatsos, K., Hadjiefthymiades, S.: Facing the cold start problem in recommender systems. Expert Syst. Appl. **41**(4), 2065–2073 (2014)

10. Prévost, C., Usevich, K., Comon, P., Brie, D.: Coupled tensor low-rank multilinear approximation for hyperspectral super-resolution. In: ICASSP 2019-2019 IEEE International Conference on Acoustics, Speech and Signal Processing (ICASSP), pp. 5536–5540. IEEE, Piscataway (2019)
11. Simões, M., Bioucas-Dias, J., Almeida, L.B., Chanussot, J.: A convex formulation for hyperspectral image superresolution via subspace-based regularization. IEEE Trans. Geosci. Remote Sens. **53**(6), 3373–3388 (2014)
12. Şimşekli, U., Yılmaz, Y.K., Cemgil, A.T.: Score guided audio restoration via generalised coupled tensor factorisation. In: 2012 IEEE International Conference on Acoustics, Speech and Signal Processing (ICASSP), pp. 5369–5372. IEEE, Piscataway (2012)
13. Sorber, L., Van Barel, M., De Lathauwer, L.: Structured data fusion. IEEE J. Select.Topics Signal Process. **9**(4), 586–600 (2015)
14. Sørensen, M., De Lathauwer, L.: Coupled tensor decompositions for applications in array signal processing. In: 2013 5th IEEE International Workshop on Computational Advances in Multi-Sensor Adaptive Processing (CAMSAP), pp. 228–231. IEEE, Piscataway (2013)
15. Sørensen, M., De Lathauwer, L.D.: Coupled canonical polyadic decompositions and (coupled) decompositions in multilinear rank-(l_r,n,l_r,n,1) terms—part i: Uniqueness. SIAM J. Matrix Analy. Appl. **36**(2), 496–522 (2015)
16. Wei, Q., Dobigeon, N., Tourneret, J.Y.: Fast fusion of multi-band images based on solving a Sylvester equation. IEEE Trans. Image Process. **24**(11), 4109–4121 (2015)
17. Wimalawarne, K., Mamitsuka, H.: Efficient convex completion of coupled tensors using coupled nuclear norms. In: Advances in Neural Information Processing Systems, pp. 6902–6910 (2018)
18. Wimalawarne, K., Yamada, M., Mamitsuka, H.: Convex coupled matrix and tensor completion. Neural Comput. **30**(11), 3095–3127 (2018)
19. Yılmaz, K.Y., Cemgil, A.T., Simsekli, U.: Generalised coupled tensor factorisation. In: Advances in Neural Information Processing Systems, pp. 2151–2159 (2011)
20. Yokoya, N., Yairi, T., Iwasaki, A.: Coupled nonnegative matrix factorization unmixing for hyperspectral and multispectral data fusion. IEEE Trans. Geosci. Remote Sens. **50**(2), 528–537 (2011)
21. Zhang, K., Wang, M., Yang, S., Jiao, L.: Spatial–spectral-graph-regularized low-rank tensor decomposition for multispectral and hyperspectral image fusion. IEEE J. Sel. Top. Appl. Earth Obs. Remote Sens. **11**(4), 1030–1040 (2018)
22. Zhang, G., Fu, X., Huang, K., Wang, J.: Hyperspectral super-resolution: A coupled nonnegative block-term tensor decomposition approach. In: 2019 IEEE 8th International Workshop on Computational Advances in Multi-Sensor Adaptive Processing (CAMSAP), pp. 470–474 (2019). https://doi.org/10.1109/CAMSAP45676.2019.9022476
23. Zhang, S., Wang, X., Liu, A., Zhao, C., Wan, J., Escalera, S., Shi, H., Wang, Z., Li, S.Z.: A dataset and benchmark for large-scale multi-modal face anti-spoofing. In: Proceedings of the IEEE Conference on Computer Vision and Pattern Recognition, pp. 919–928 (2019)
24. Zheng, V.W., Zheng, Y., Xie, X., Yang, Q.: Towards mobile intelligence: Learning from GPS history data for collaborative recommendation. Artif. Intell. **184**, 17–37 (2012)

Chapter 6
Robust Principal Tensor Component Analysis

6.1 Principal Component Analysis: From Matrix to Tensor

In 1901, principal component analysis (PCA) was first formulated in statistics by Karl Pearson [39]. As a classic dimensionality reduction tool, PCA plays a vital role in data mining, bioinformatics, image processing, etc. It exploits the low-rank structure to effectively approximate a matrix by keeping its first few components in singular value decomposition (SVD).

One of PCA's main problems is the sensitivity to sparse noise or outliers. To deal with this issue, many improved versions have been proposed [8, 12, 16, 20, 24]. Among them, robust principal component analysis (RPCA) [4] proposed by Candès et al. in 2009 is the most prevalent one, which is the first polynomial-time algorithm with strong performance guarantees.

Given the observations in the matrix $\mathbf{X} \in \mathbb{R}^{I_1 \times I_2}$, RPCA can decompose it into two additive terms as follows:

$$\mathbf{X} = \mathbf{L} + \mathbf{E}, \tag{6.1}$$

where $\mathbf{L}$ represents a low-rank matrix and the sparse matrix is denoted by $\mathbf{E}$. Figure 6.1 provides an illustration for RPCA.

To separate the principal component from sparse corruption, the optimization model can be formulated as follows:

$$\min_{\mathbf{L},\,\mathbf{E}} \operatorname{rank}(\mathbf{L}) + \lambda \|\mathbf{E}\|_0 \text{ s.t. } \mathbf{X} = \mathbf{L} + \mathbf{E}, \tag{6.2}$$

where $\operatorname{rank}(\mathbf{L})$ is rank of the matrix $\mathbf{L}$, $\|\mathbf{E}\|_0$ is the ℓ_0 pseudo-norm which counts the nonzero entries, and λ is a parameter to make balance of sparsity and low-rank terms.

Y. Liu et al., *Tensor Computation for Data Analysis*,
https://doi.org/10.1007/978-3-030-74386-4_6

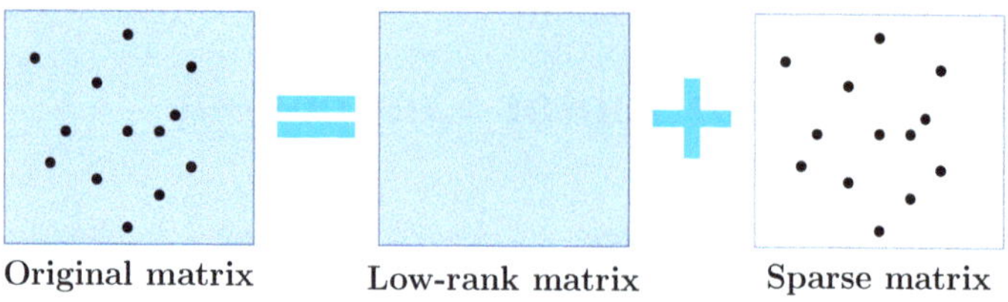

Fig. 6.1 Illustration of RPCA of matrix data

Problem (6.2) is highly nonconvex and hard to solve. By replacing the nonconvex rank and ℓ_0 norm with the convex matrix nuclear norm and ℓ_1 norm, it can be relaxed into a tractable convex optimization model as follows:

$$\min_{\mathbf{L},\,\mathbf{E}} \|\mathbf{L}\|_* + \lambda\|\mathbf{E}\|_1 \text{ s.t. } \mathbf{X} = \mathbf{L} + \mathbf{E}, \tag{6.3}$$

where $\|\cdot\|_*$ denotes the nuclear norm which is the sum of singular values of the low-rank component $\mathbf{L}$, $\|\cdot\|_1$ denotes the ℓ_1 norm which represents the sum of absolute values of all elements in the sparse component $\mathbf{E}$, and parameter λ is the weighting factor to balance the low-rank and sparse component, which is suggested to be set as $1/\sqrt{\max(I_1, I_2)}$ for good performance. It has been proven that under some incoherent conditions, $\mathbf{L}$ and $\mathbf{E}$ can be perfectly recovered.

The major shortcoming of RPCA is that it can only handle matrix data. In real-world applications, a lot of data are featured by its higher order, such as grayscale video, color images, and hyperspectral images. The classic RPCA needs to transform the high-dimensional data into a matrix, which inevitably leads to loss of structural information.

In order to better extract the low-rank structure from high-dimensional data, classical RPCA has been extended into tensor versions. A given tensor $\mathcal{X} \in \mathbb{R}^{I_1\times I_2\times I_3}$ can be decomposed as two additive tensor terms as follows:

$$\mathcal{X} = \mathcal{L} + \mathcal{E}, \tag{6.4}$$

where $\mathcal{L}$ and $\mathcal{E}$ represent a low-rank tensor and a sparse tensor, respectively.

Based on different tensor decomposition frameworks, there will be diverse definitions of tensor rank, which results in numerous robust principal tensor component analysis (RPTCA) methods.

6.2 RPTCA Methods Based on Different Decompositions

A variety of tensor decompositions have emerged over the years as introduced in Chap. 2, and corresponding RPTCA methods have been studied. In this section, the tensor singular value decomposition (t-SVD)-related methods are summarized in

Sects. 6.2.1, and 6.2.3 presents the other methods based on Tucker and tensor train decompositions.

6.2.1 *t-SVD-Based RPTCA*

Recently a tensor decomposition called tensor singular value decomposition (t-SVD) has been proposed for RPTCA. Different from others like Tucker decomposition, t-SVD factorizes a third-order tensor into three structural tensors connected by tensor products. It has lower computational complexity and possesses similar properties to those of matrix SVD. The tensor multi-rank and tubal rank defined based on t-SVD can well characterize the intrinsic multilinear structure of a tensor directly.

In this subsection, we first introduce the classic RPTCA model and algorithm and then present some sparse constraints for robustness in RPTCA for different applications [32, 33, 51, 57]. The improved methods with tensor nuclear norm (TNN) based on t-SVD are presented afterward.

6.2.1.1 Classical RPTCA Model and Algorithm

Given a tensor $\mathcal{A} \in \mathbb{R}^{I_1 \times I_2 \times I_3}$, its t-SVD is formulated as $\mathcal{A} = \mathcal{U} * \mathcal{S} * \mathcal{V}^{\mathrm{T}}$. In 2014, Zhang et al. first defined the tubal rank to be the number of nonzero singular tubes in $\mathcal{S}$ in the t-SVD framework. Afterward, Lu et al. defined the tensor nuclear norm (TNN) (see Definition 6.1) in 2016 [32]. In 2019, Lu et al. further proposed the tensor average rank (see Definition 6.2) and declared that the TNN is the convex envelope of the tensor average rank within the domain of the unit ball of the tensor spectral norm [33].

Definition 6.1 (Tensor Nuclear Norm [32, 33]) Given a tensor $\mathcal{A} \in \mathbb{R}^{I_1 \times I_2 \times I_3}$ and its t-SVD $\mathcal{A} = \mathcal{U} * \mathcal{S} * \mathcal{V}^{\mathrm{T}}$, when the tubal rank is R, the tensor nuclear norm (TNN) is defined as

$$\|\mathcal{A}\|_* = \frac{1}{I_3}\|\hat{\mathcal{A}}\|_* = \frac{1}{I_3}\sum_{i_3=1}^{I_3}\|\hat{\mathcal{A}}^{(i_3)}\|_* = \frac{1}{I_3}\sum_{i_3=1}^{I_3}\sum_{r=1}^{R}\hat{\mathcal{S}}(r, r, i_3), \tag{6.5}$$

where $\hat{\mathcal{A}}$, $\hat{\mathcal{S}}$ represent the result of FFT along the third mode of $\mathcal{A}$ and $\mathcal{S}$, respectively.

Definition 6.2 (Tensor Average Rank [33]) The tensor average rank of $\mathcal{A} \in \mathbb{R}^{I_1 \times I_2 \times I_3}$, denoted as $\mathrm{rank_a}(\mathcal{A})$, is defined as

$$\mathrm{rank_a}(\mathcal{A}) = \frac{1}{I_3}\mathrm{rank}(\mathrm{circ}(\mathcal{A})). \tag{6.6}$$

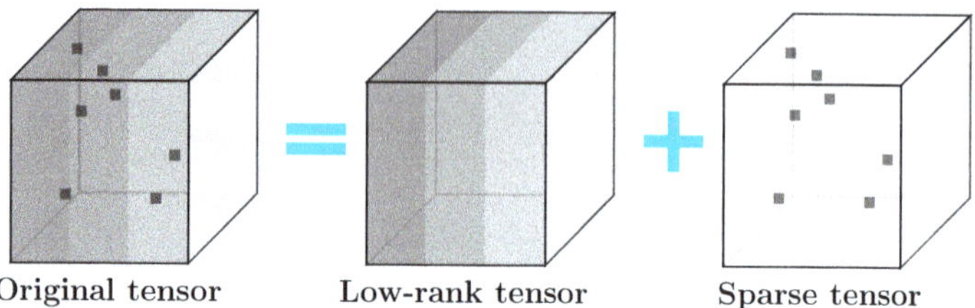

Fig. 6.2 Illustration of classical RPTCA method

For a 3rd-order tensor, equipped with the TNN and the classical ℓ_1 norm constraint, RPTCA [32, 33] convex optimization model can be formulated as

$$\min_{\mathcal{L},\,\mathcal{E}} \|\mathcal{L}\|_{\text{TNN}} + \lambda\|\mathcal{E}\|_1, \text{ s.t. } \mathcal{X} = \mathcal{L} + \mathcal{E}, \tag{6.7}$$

where λ is a regularization parameter and suggested to be set as $1/\sqrt{\max(I_1, I_2)I_3}$, $\|\mathcal{L}\|_{\text{TNN}}$ denotes the TNN of low-rank tensor component $\mathcal{L}$, and $\|\mathcal{E}\|_1$ is the ℓ_1 norm for the sparse tensor. Figure 6.2 gives an illustration of the classical RPTCA.

It has been analyzed that an exact recovery can be obtained for both the low-rank component and the sparse component under certain conditions. Firstly, we need to assume that the low-rank component is not sparse. In order to guarantee successful low-rank component extraction, the low-rank component $\mathcal{L}$ should satisfy the tensor incoherence conditions.

Definition 6.3 (Tensor Incoherence Conditions [33]) For the low-rank component $\mathcal{L} \in \mathbb{R}^{I_1\times I_2\times I_3}$, we assume that the corresponding tubal rank $\text{rank}_\text{t}(\mathcal{L}) = R$ and it has the t-SVD $\mathcal{L} = \mathcal{U} * \mathcal{S} * \mathcal{V}^\text{T}$, where $\mathcal{U} \in \mathbb{R}^{I_1\times R\times I_3}$ and $\mathcal{V} \in \mathbb{R}^{I_2\times R\times I_3}$ are orthogonal tensors and $\mathcal{S} \in \mathbb{R}^{R\times R\times I_3}$ is an f-diagonal tensor. Then $\mathcal{L}$ should satisfy the following tensor incoherence conditions with parameter δ:

$$\max_{i_1=1,\ldots,I_1} \|\mathcal{U}^\text{T} * \mathring{\mathbf{e}}_{i_1}\|_\text{F} \leqslant \sqrt{\frac{\delta R}{I_1 I_3}}, \tag{6.8}$$

$$\max_{i_2=1,\ldots,I_2} \|\mathcal{V}^\text{T} * \mathring{\mathbf{e}}_{i_2}\|_\text{F} \leqslant \sqrt{\frac{\delta R}{I_2 I_3}}, \tag{6.9}$$

and

$$\|\mathcal{U} * \mathcal{V}^\text{T}\|_\infty \leqslant \sqrt{\frac{\delta R}{I_1 I_2 {I_3}^2}}, \tag{6.10}$$

where $\mathring{e}_{i_1} \in \mathbb{R}^{I_1 \times 1 \times I_3}$ is the tensor column basis whose $(i_1, 1, 1)$-th element is 1 and the rest is 0. And $\mathring{e}_{i_2} \in \mathbb{R}^{I_2 \times 1 \times I_3}$ is the one with $(i_2, 1, 1)$-th entry equaling 1 and the rest equaling 0.

We should note that the sparse component is assumed not to be of low tubal rank. The RPTCA problem can be solved easily and efficiently by alternating direction method of multipliers (ADMM) [3]. The augmented Lagrangian function from the optimization model (6.7) is

$$\mathfrak{L}(\mathcal{L}, \mathcal{E}, \mathcal{Y}, \mu) = \|\mathcal{L}\|_{\text{TNN}} + \lambda\|\mathcal{E}\|_1 + \langle \mathcal{Y}, \mathcal{X} - \mathcal{L} - \mathcal{E} \rangle + \frac{\mu}{2}\|\mathcal{X} - \mathcal{L} - \mathcal{E}\|_{\text{F}}^2, \tag{6.11}$$

where $\mu > 0$ is a penalty parameter and $\mathcal{Y}$ is a Lagrangian multiplier. Each variable can be optimized alternatively while fixing the others. The key steps of this solution are the updates of low-rank and sparse tensors. Specifically, when the iteration index is $k + 1$, we have two main subproblems as follows:

(1) *Low-rank component approximation*: All terms with respect to low-rank component are extracted from the augmented Lagrangian function (6.11) to formulate the update as follows:

$$\mathcal{L}_{k+1} = \underset{\mathcal{L}}{\operatorname{argmin}} \|\mathcal{L}\|_{\text{TNN}} + \langle \mathcal{Y}_k, \mathcal{X} - \mathcal{L} - \mathcal{E}_k \rangle + \frac{\mu_k}{2}\|\mathcal{X} - \mathcal{L} - \mathcal{E}_k\|_{\text{F}}^2, \tag{6.12}$$

which is equivalent to

$$\mathcal{L}_{k+1} = \underset{\mathcal{L}}{\operatorname{argmin}} \|\mathcal{L}\|_{\text{TNN}} + \frac{\mu_k}{2}\|\mathcal{L} - \mathcal{X} + \mathcal{E}_k - \frac{\mathcal{Y}_k}{\mu_k}\|_{\text{F}}^2, \tag{6.13}$$

It can be solved by the tensor singular value thresholding (t-SVT) operator defined as follows.

Definition 6.4 (Tensor Singular Value Thresholding [33]) Given a tensor $\mathcal{A} \in \mathbb{R}^{I_1 \times I_2 \times I_3}$, the t-SVD is $\mathcal{A} = \mathcal{U} * \mathcal{S} * \mathcal{V}^{\text{T}}$. The tensor singular value thresholding (t-SVT) operator is defined as follows:

$$\text{tsvt}_\tau(\mathcal{A}) = \mathcal{U} * \mathcal{S}_\tau * \mathcal{V}^{\text{T}}, \tag{6.14}$$

where

$$\mathcal{S}_\tau = \text{ifft}((\hat{\mathcal{S}} - \tau)_+). \tag{6.15}$$

where $\tau > 0$, t_+ denotes the positive part of t, i.e., $t_+ = \max(t, 0)$, and $\hat{\mathcal{S}}$ is the fast Fourier Transform (FFT) of $\mathcal{S}$ along the third mode. Problem (6.14) will be solved by performing the soft thresholding operator to each frontal slice $\hat{\mathcal{S}}(:, :, i)$.

Therefore, we can update the low-rank component as follows:

$$\mathcal{L}_{k+1} := \text{tsvt}_{\frac{1}{\mu_k}}\left(\mathcal{X} - \mathcal{E}_k + \frac{\mathcal{Y}_k}{\mu_k}\right). \tag{6.16}$$

(2) *Sparse component approximation*: All terms containing the sparse component are extracted from the augmented Lagrangian function (6.11) for the update as follows:

$$\mathcal{E}_{k+1} = \underset{\mathcal{E}}{\text{argmin}}\ \lambda\|\mathcal{E}\|_1 + \langle \mathcal{Y}_k, \mathcal{X} - \mathcal{L}_k - \mathcal{E}\rangle + \frac{\mu_k}{2}\|\mathcal{X} - \mathcal{L}_k - \mathcal{E}\|_{\text{F}}^2, \tag{6.17}$$

which can be transformed into

$$\mathcal{E}_{k+1} = \underset{\mathcal{E}}{\text{argmin}}\ \lambda\|\mathcal{E}\|_1 + \frac{\mu_k}{2}\|\mathcal{L}_{k+1} - \mathcal{X} + \mathcal{E} - \frac{\mathcal{Y}_k}{\mu_k}\|_{\text{F}}^2. \tag{6.18}$$

It can be solved by

$$\mathcal{E}_{k+1} := \text{sth}_{\frac{\lambda}{\mu_k}}\left(\mathcal{X} - \mathcal{L}_{k+1} + \frac{\mathcal{Y}_k}{\mu_k}\right), \tag{6.19}$$

where $\text{sth}_\tau(\mathbf{X})$ and $\text{sth}_\tau(\mathcal{X})$ denote the entry-wise soft thresholding operators for matrix and tensor. For any element x in a matrix or a tensor, we have

$$\text{sth}_\tau(x) = \text{sgn}(x)\max(|x| - \tau). \tag{6.20}$$

where $\text{sgn}(\cdot)$ extracts the sign of a real number.

Algorithm 34: ADMM for RPTCA

Input: The observed tensor $\mathcal{X} \in \mathbb{R}^{I_1 \times I_2 \times I_3}$, parameter λ.
Output: $\mathcal{L}$, $\mathcal{E}$.
Initialize $\mathcal{L} = \mathcal{E} = \mathcal{Y} = 0$, μ, ρ, $\mu_{\max}$, ϵ.
while not converged **do**
1. Update the low-rank component: $\mathcal{L}_{k+1} = \text{tsvt}_{\frac{1}{\mu_k}}(\mathcal{X} - \mathcal{E}_k - \frac{\mathcal{Y}_k}{\mu_k})$;
2. Update the sparse component: $\mathcal{E}_{k+1} = \text{sth}_{\frac{\lambda}{\mu_k}}(\mathcal{X} - \mathcal{L}_{k+1} - \frac{\mathcal{Y}_k}{\mu_k})$;
3. Update $\mathcal{Y}_{k+1} = \mathcal{Y}_k + \mu_k(\mathcal{L}_{k+1} + \mathcal{E}_{k+1} - \mathcal{X})$;
4. Update $\mu_{k+1} = \min(\rho\mu_k, \mu_{\max})$;
5. Check the convergence conditions :

$$\|\mathcal{L}_{k+1} - \mathcal{L}_k\|_\infty \leq \epsilon,\ \|\mathcal{E}_{k+1} - \mathcal{E}_k\|_\infty \leq \epsilon,\ \|\mathcal{L}_{k+1} + \mathcal{E}_{k+1} - \mathcal{X}\|_\infty \leq \epsilon.$$

end while

Algorithm 34 presents the details of the RPTCA problem solved by ADMM. For this algorithm, parallel computation can be adopted to improve computational efficiency, as the calculation of matrix SVD in t-SVD for each iteration is independent. The computational complexity for updating low-rank component and sparse component in each iteration is $\mathrm{O}\left(I_1 I_2 I_3(\log I_3 + 1) + \lceil \frac{I_3+1}{2} \rceil I_1 I_2 I_{\min}\right)$, where $I_{\min} = \min(I_1, I_2)$.

6.2.1.2 RPTCA with Different Sparse Constraints

In addition to the classic ℓ_1 norm, there are several sparsity constraints with respect to different applications that have been developed.

Consider the situation where some tubes on a video data have heavy noises. In order to better process such pixels for the video recovery, another convex optimization model of RPTCA, called tubal-PTCA, was studied in [51] as follows:

$$\min_{\mathcal{L},\,\mathcal{E}} \|\mathcal{L}\|_{\mathrm{TNN}} + \lambda \|\mathcal{E}\|_{1,1,2}, \quad \text{s. t. } \mathcal{X} = \mathcal{L} + \mathcal{E}, \tag{6.21}$$

where $\|\mathcal{E}\|_{1,1,2}$ is defined as the sum of all Frobenius norms of the tubal fibers in the third mode, i.e. , $\|\mathcal{E}\|_{1,1,2} = \sum_{i,j} \|\mathcal{E}(i, j, :)\|_{\mathrm{F}}$, and λ is the regularization parameter to balance the low-rank and sparse components. The optimal value of λ is $1/\sqrt{\max(I_1, I_2)}$ for given tensor $\mathcal{X} \in \mathbb{R}^{I_1 \times I_2 \times I_3}$.

As the noises exist on the tubal fiber along the third mode, $\|\mathcal{E}\|_{1,1,2}$ can well characterize such kind of sparse components. Figure 6.3 shows the original tensor data and structurally sparse noise for this model. For this optimization model, it can remove the Gaussian noise on tubes while the moving objects are not considered as sparse component.

On the other hand, outliers or sample-specific corruptions are common in real data. When each sample is arranged as a lateral slice of a third-order tensor, an outlier-robust model for tensor PCA (OR-PTCA) has been proposed in [57] as follows:

$$\min_{\mathcal{L},\,\mathcal{E}} \|\mathcal{L}\|_{\mathrm{TNN}} + \lambda \|\mathcal{E}\|_{2,1} \quad , \text{ s. t. } \mathcal{X} = \mathcal{L} + \mathcal{E}, \tag{6.22}$$

where $\|\mathcal{E}\|_{2,1}$ is defined as the sum of all the Frobenius norms of the lateral slices, i.e., $\|\mathcal{E}\|_{2,1} = \sum_{i_2=1}^{I_2} \|\mathcal{E}(:, i_2, :)\|_{\mathrm{F}}$.

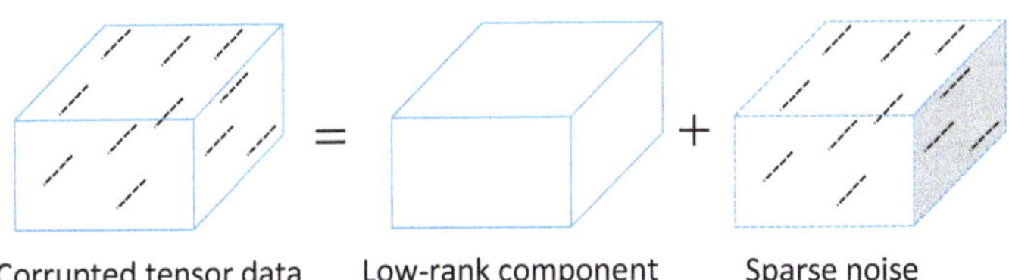

Fig. 6.3 Illustration of the tubal-RPTCA with sparse noise on tubes

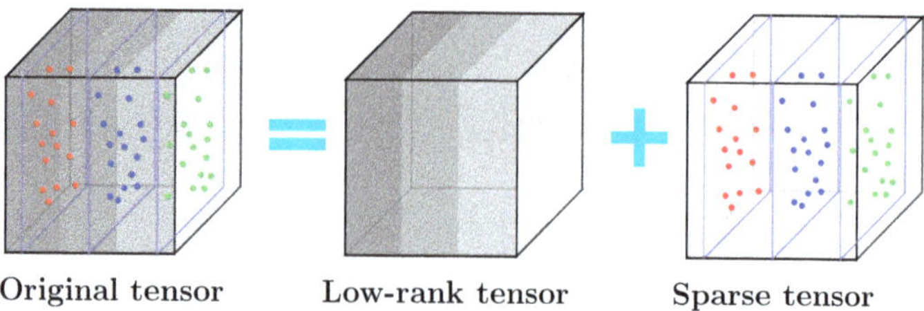

Fig. 6.4 Illustration of OR-PTCA method

OR-PTCA is the first method that has provable performance guarantee for exact tensor subspace recovery and outlier detection under mild conditions. Figure 6.4 shows the original tensor data and sparse noise on lateral slices for the measurement model assumed for OR-PTCA. Experimental results on four tasks including outlier detection, clustering, semi-supervised, and supervised learning have demonstrated the advantages of OR-PTCA method. In addition, it has been also applied into hyperspectral image classification in [42].

6.2.1.3 Improved RPTCA with Different Scales

The abovementioned works introduced in Sect. 6.2.1.2 mainly focus on application of different sparse constraints to the optimization models of RPTCA for different applications. In practical applications, the sparse tensor serves to extract outliers, noise, and sparse moving objects. Therefore, suitable sparse constraint can better characterize corresponding component and improve the performance in practical applications.

Similarly, low-rank term takes charge of extraction of principal components, and it can be of various forms for different applications too. Considering that the given tensor data are usually processed in the original scale, a block version of RPTCA has been proposed in [5, 11] as follows:

$$\begin{aligned} \min_{\mathcal{L}_p,\mathcal{E}_p} & \sum_{p=1}^{P}(\|\mathcal{L}_p\|_* + \lambda\|\mathcal{E}_p\|_1) \\ \text{s. t. } & \mathcal{X} = \mathcal{L}_1 \boxplus \cdots \boxplus \mathcal{L}_P + \mathcal{E}_1 \boxplus \cdots \boxplus \mathcal{E}_P, \end{aligned} \tag{6.23}$$

where "$\boxplus$" denotes the concatenation operator of block tensors; P denotes the number of blocks of the whole tensor, which is equal to $\lceil \frac{I_1}{n} \rceil \lceil \frac{I_2}{n} \rceil$ when the size of block tensor is $n \times n \times I_3$ and the size of whole tensor is $I_1 \times I_2 \times I_3$; $\mathcal{L}_p$, $p = 1, 2, \ldots, P$ denotes the block low-rank component; and $\mathcal{E}_p$, $p = 1, 2, \ldots, P$ represents the block sparse component. For convenience, it has been defined that the sparse component $\mathcal{E}_{\text{sum}} = \mathcal{E}_1 \boxplus \mathcal{E}_2 \boxplus \cdots \boxplus \mathcal{E}_P$. Figure 6.5 illustrates the data

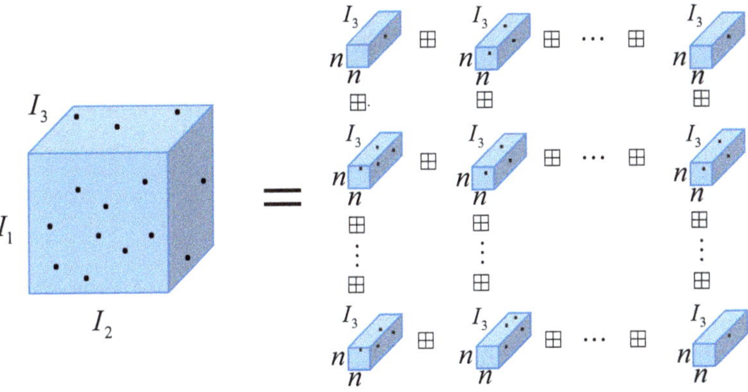

Fig. 6.5 Illustration of the data model for RBPTCA

model for robust block principal tensor component analysis (RBPTCA). We also call it block-RPTCA.

As the ℓ_1 norm is defined as the sum of the absolute values of all elements, problem (6.23) is equivalent to the following one:

$$\min_{\mathcal{L}_p,\mathcal{E}} \sum_{p=1}^{P} \|\mathcal{L}_p\|_* + \lambda\|\mathcal{E}\|_1 \\ \text{s. t. } \mathcal{X} = \mathcal{L}_1 \boxplus \mathcal{L}_2 \boxplus \cdots \boxplus \mathcal{L}_P + \mathcal{E}. \tag{6.24}$$

As shown in the above optimization model, the extraction of the sparse component is not influenced by the block decomposition. Taking advantage of the local similarity of pixels with different scale in analysis, the RBPTCA method can choose a suitable block size to better extract details.

Note that the blocks are not overlapping. In fact, it may exhibit different performance for a same scale with overlapping blocks. Correspondingly, the standard ℓ_1 norm-based sparse constraints for overlapping blocks would result in a weighted ℓ_1 norm for the estimate tensor and thus provide different sparse denoising performance.

6.2.1.4 Improved RPTCA with Respect to Rotation Invariance

As introduced in Chap. 2, t-SVD enjoys a considerably low computational complexity. However, it leads to performance degeneration due to its drawback on rotation variance. The t-SVD-based RPTCA can be sensitive to the choice of rotation on indexes. In fact, the low-rank structure in the first and the second dimensions can be well extracted by matrix SVD. But since the third mode is processed by the discrete Fourier transform (DFT), the low-rank information cannot be fully

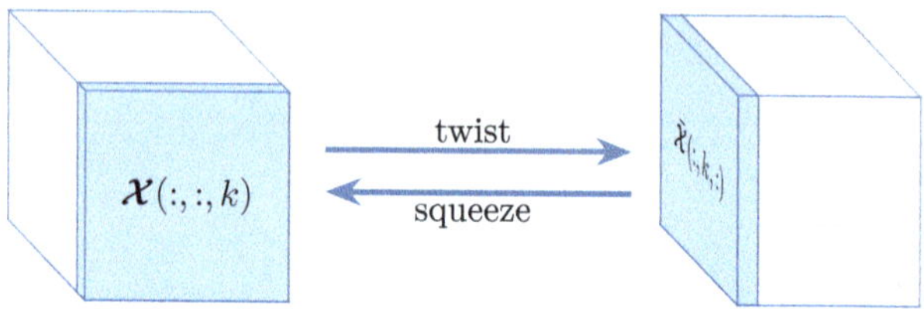

Fig. 6.6 Transformations between $\mathcal{X}$ and $\bar{\mathcal{X}}$

exploited. Therefore, the classical tensor nuclear norm (TNN) may not be a good choice under some circumstances.

In [17], a twist tensor nuclear norm (t-TNN) is proposed, and it applies a permutation operator on the block circulant matrices of original tensor. Its definition can be formulated as follows:

$$\|\mathcal{L}\|_{\text{t-TNN}} = \|\bar{\mathcal{L}}\|_{\text{TNN}} = \|\mathbf{P}_1 \operatorname{bcirc}(\mathcal{L})\mathbf{P}_2\|_{\text{TNN}}, \tag{6.25}$$

where $\mathbf{P}_i,\ i = 1,2$ denote so-called stride permutations which are invertible matrices. The illustration of the transformation between $\mathcal{X}$ and $\bar{\mathcal{X}}$ is shown in Fig. 6.6.

Interoperating this t-TNN, a new optimization model for RPTCA is formulated as follows:

$$\min_{\mathcal{L},\ \mathcal{E}} \|\mathcal{L}\|_{\text{t-TNN}} + \lambda\|\mathcal{E}\|_1 \quad \text{s. t. } \mathcal{X} = \mathcal{L} + \mathcal{E}. \tag{6.26}$$

We can see that this model is different from the classical one (6.7) in the definition of tensor nuclear norm. In this case, the t-TNN mainly exploits the linear relationship between frames, which is more suitable for describing a scene with motions. Indeed, t-TNN provides a new viewpoint in the low-rank structure via rotated tensor, and it makes t-SVD more flexible when dealing with complex applications.

6.2.1.5 Improved RPTCA with Low-Rank Core Matrix

In [6, 29], an improved robust tensor PCA model has been proposed. The authors find that low-rank information still exists in the core tensor for many visual data, and they define an improved tensor nuclear norm (ITNN) to further exploit the low-rank structures in multiway tensor data, which can be formulated as

$$\|\mathcal{X}\|_{\text{ITNN}} = \|\mathcal{X}\|_{\text{TNN}} + \gamma\|\bar{\mathbf{S}}\|_*, \tag{6.27}$$

where γ is set to balance the classical TNN and the nuclear norm of the core matrix. When we define $\mathcal{S} \in \mathbb{R}^{I_1 \times I_2 \times I_3}$ and $I = \min(I_1, I_2)$, the core matrix $\bar{\mathbf{S}} \in \mathbb{R}^{I \times I_3}$

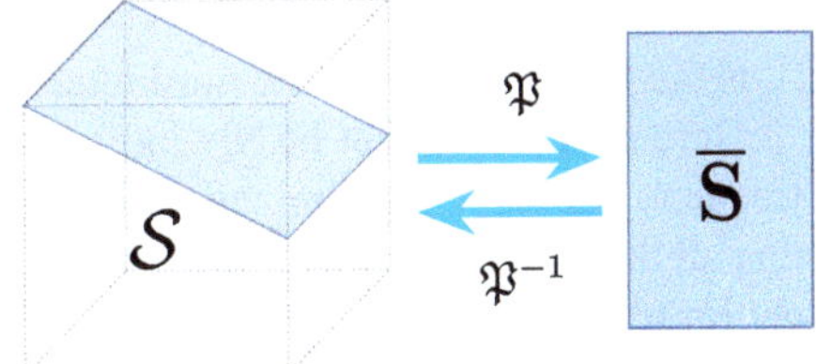

Fig. 6.7 Transforms between core tensor $\mathcal{S} \in \mathbb{R}^{I_1 \times I_2 \times I_3}$ and core matrix $\bar{\mathbf{S}} \in \mathbb{R}^{\min\{I_1, I_2\} \times I_3}$

satisfies $\bar{\mathbf{S}}(i, :) = \mathcal{S}(i, i\ :)$. Figure 6.7 shows the transformation between the core tensor $\mathcal{S}$ and the core matrix $\bar{\mathbf{S}}$.

Equipped with the newly defined ITNN (6.27), the convex optimization model for the improved robust principal tensor component analysis (IRPTCA or improved PTCA) is formulated as follows:

$$\min_{\mathcal{L},\, \mathcal{E}} \|\mathcal{L}\|_{\text{ITNN}} + \lambda \|\mathcal{E}\|_1, \quad \text{s. t. } \mathcal{X} = \mathcal{L} + \mathcal{E}, \tag{6.28}$$

where λ is set to balance the low-rank tensor and the sparse tensor.

In classical RPTCA based on t-SVD, the low-rank structure in the third mode is not fully utilized. Adding low-rank component extraction of the core matrix, ITNN can characterize the low-rank structure of the tensor data sufficiently. For example, in background modeling, the low-rank structure of a surveillance video mainly lies in the third dimension due to the correlation between frames. In this case, the additional low-rank extraction in the IRPTCA method can improve the performance effectively.

6.2.1.6 Improved TNN with Frequency Component Analysis

In [46], the authors have proposed a frequency-weighted tensor nuclear norm (FTNN) which can utilize the prior knowledge well in frequency domain. When minimizing the classical TNN, all the frontal matrix in core tensor $\hat{\mathcal{S}}$ is minimized equally which is inappropriate in some applications.

To explore the frequency prior knowledge of data, the authors combine frequency component analysis with principal component analysis and define the FTNN as follows:

$$\begin{aligned}
\|\mathcal{X}\|_{\text{FTNN}} &= \frac{1}{I_3} \|\text{bdiag}(\hat{\mathcal{X}})\|_{\text{FTNN}} \\
&= \frac{1}{I_3} \left\| \begin{bmatrix} \alpha_1 \hat{\mathcal{X}}^{(1)} & & & \\ & \alpha_2 \hat{\mathcal{X}}^{(2)} & & \\ & & \ddots & \\ & & & \alpha_{I_3} \hat{\mathcal{X}}^{(I_3)} \end{bmatrix} \right\|_* \\
&= \frac{1}{I_3} \sum_{i_3=1}^{I_3} \alpha_{i_3} \|\hat{\mathcal{X}}^{(i_3)}\|_*,
\end{aligned} \tag{6.29}$$

where $\alpha_{i_3} \geq 0, i_3 = 1, \ldots, I_3$ is called as frequency weight or filtering coefficient, $\hat{\mathcal{X}}$ is the FFT of $\mathcal{X}$ along the third mode, and $\hat{\mathcal{X}}^{(i_3)}$ is the i_3-th frontal slice of $\hat{\mathcal{X}}$.

Here we can see that FTNN just puts a series of weights on the frontal slices in frequency domain, thus implementing the frequency filtering at the same time. Note that all the weights are predefined by the prior knowledge of the data, which we can get by frequency component analysis. With FTNN, the frequency-weighted robust principal tensor component analysis model can be represented as follows:

$$\min_{\mathcal{L}, \mathcal{E}} \|\mathcal{L}\|_{\text{FTNN}} + \lambda \|\mathcal{E}\|_1, \text{ s. t. } \mathcal{X} = \mathcal{L} + \mathcal{E}. \tag{6.30}$$

6.2.1.7 Nonconvex RPTCA

Although the convex relaxation can easily get the optimal solution, the nonconvex approximation tends to yield more sparse or lower-rank local solutions. In [41], instead of using convex TNN, they use nonconvex surrogate functions to approximate the tensor tubal rank and propose a tensor-based iteratively reweighted nuclear norm solver.

Motivated by the convex TNN, given a tensor $\mathcal{X} \in \mathbb{R}^{I_1 \times I_2 \times I_3}$ and the tubal rank M, we can define the following nonconvex penalty function:

$$\mathfrak{R}(\mathcal{X}) = \frac{1}{I_3} \sum_{i_3=1}^{I_3} \sum_{m=1}^{M} \mathrm{g}_\lambda \left(\sigma_m \left(\hat{\mathcal{X}}^{(i_3)} \right) \right), \tag{6.31}$$

where g_λ represents a type of nonconvex functions and $\sigma_m(\hat{\mathcal{X}}^{(i_3)})$ denotes the m-th largest singular value of $\hat{\mathcal{X}}^{(i_3)}$. If $\mathrm{g}_\lambda(x) = x$, it is equivalent to the TNN.

Based on the nonconvex function (6.31), the following nonconvex low-rank tensor optimization problem is proposed:

$$\begin{aligned} \min_{\mathcal{X}} \mathfrak{F}(\mathcal{X}) &= \mathfrak{R}(\mathcal{X}) + \mathrm{f}(\mathcal{X}) \\ &= \frac{1}{I_3} \sum_{i_3=1}^{I_3} \sum_{m=1}^{M} \mathrm{g}_\lambda \left(\sigma_m \left(\hat{\mathcal{X}}^{(i_3)} \right) \right) + \mathrm{f}(\mathcal{X}), \end{aligned} \tag{6.32}$$

where f denotes some loss functions, which can be nonconvex. In general, it should be nontrivial. The convergence analysis of the new solver has been further provided.

In case of matrix-based RPCA, some inevitable biases might exist when minimizing the nuclear norm for extracting low-rank component [36, 37]. In fact, a similar problem may exist for tensor data. In [21], a novel nonconvex low-rank constraint called the partial sum of the tubal nuclear norm (PSTNN) has been proposed, which only consists of some small singular values. The PSTNN minimization directly shrinks the small singular values but does nothing on the large ones.

Given a third-order tensor $\mathcal{X} \in \mathbb{R}^{I_1 \times I_2 \times I_3}$, PSTNN is defined as

$$\|\mathcal{X}\|_{\text{PSTNN}} = \sum_{i_3=1}^{I_3} \left\| \hat{\mathcal{X}}^{(i_3)} \right\|_{p=N}, \tag{6.33}$$

where $\hat{\mathcal{X}}^{(i_3)}$ is the frontal slices of tensor $\hat{\mathcal{X}}$ and $\|\cdot\|_{p=N}$ denotes the partial sum of singular values (PSSV). PSSV is defined as $\|\mathbf{X}\|_{p=N} = \sum_{m=N+1}^{\min(I_1,I_2)} \sigma_m(\mathbf{X})$ for a matrix $\mathbf{X} \in \mathbb{R}^{I_1 \times I_2}$. The corresponding optimization model for RPTCA can be formulated as follows:

$$\min_{\mathcal{L},\, \mathcal{E}} \|\mathcal{L}\|_{\text{PSTNN}} + \lambda \|\mathcal{E}\|_1, \quad \text{s. t. } \mathcal{X} = \mathcal{L} + \mathcal{E}. \tag{6.34}$$

6.2.1.8 RPTCA with Transformed Domain

Rather than using discrete Fourier transform (DFT) for t-SVD, other forms have also been adopted such as discrete cosine transform (DCT) [31, 48] and unitary transform [40]. Besides, the quantum version of t-SVD and atomic-norm regularization can also refer to [9] and [44].

6.2.2 *Higher-Order t-SVD-Based RPTCA*

On the other hand, the development of higher-order t-SVD is impending to utilize the higher-dimensional structure information. In [25], the tensor-tensor product is proposed via any invertible linear transform. In [27], based on any multidimensional discrete transform, such as DCT and discrete wavelet transform (DWT), L-SVD for fourth-order tensors have been defined, and an efficient algorithm for decomposition is proposed. Furthermore, the tensor L-QR decomposition and a Householder QR algorithm have also been proposed to avoid the catastrophic cancellation problem associated with the conventional Gram-Schmidt process.

In order to compensate for the rotation inconsistency of t-SVD and develop Nth order ($N \geq 3$) versions of t-SVD, several TNNs have been proposed in [45, 56]. A corresponding tensor unfolding operator called mode-k_1k_2 tensor unfolding has been proposed in [55, 56], which is a three-way extension of the well-known mode-k unfolding.

Definition 6.5 (Mode-k_1k_2 Unfolding) For a multiway tensor $\mathcal{X} \in \mathbb{R}^{I_1 \times I_2 \times \cdots \times I_D}$, its mode-$k_1k_2$ unfolding tensor is represented as $\mathcal{X}_{(k_1k_2)} \in \mathbb{R}^{I_{k_1} \times I_{k_2} \times \prod_{s \neq k_1, k_2} I_s}$, whose frontal slices are the lexicographic ordering of the mode-k_1k_2 slices of $\mathcal{X}$. Mathematically, the mapping relation of $(i_1, i_2, \ldots, i_N)$-th entry of $\mathcal{A}$ and

(i_{k_1}, i_{k_2}, j)-th entry of $\mathcal{X}_{(k_1 k_2)}$ is as follows:

$$j = 1 + \sum_{\substack{s=1 \\ s \neq k_1, s \neq k_2}}^{D} (i_s - 1) J_s \text{ with } J_s = \prod_{\substack{m=1 \\ m \neq k_1, m \neq k_2}}^{s-1} i_m. \tag{6.35}$$

Based on this definition, the weighted sum of tensor nuclear norm (WSTNN) of tensor $\mathcal{X} \in \mathbb{R}^{I_1 \times I_2 \times \cdots \times I_D}$ is proposed as follows:

$$\|\mathcal{X}\|_{\text{WSTNN}} := \sum_{1 \leq k_1 < k_2 \leq N} \alpha_{k_1 k_2} \left\| \mathcal{X}_{(k_1 k_2)} \right\|_{\text{TNN}}, \tag{6.36}$$

where $\alpha_{k_1 k_2} \geq 0 \, (1 \leq k_1 < k_2 \leq D)$ and $\sum_{1 \leq k_1 < k_2 \leq D} \alpha_{k_1 k_2} = 1$. The corresponding optimization model for RPTCA can be formulated as follows:

$$\min_{\mathcal{L},\, \mathcal{E}} \|\mathcal{L}\|_{\text{WSTNN}} + \lambda \|\mathcal{E}\|_1, \text{ s. t. } \mathcal{X} = \mathcal{L} + \mathcal{E}, \tag{6.37}$$

which can handle multiway tensor and better model different correlations along different modes.

6.2.3 *RTPCA Based on Other Tensor Decompositions*

In addition to the t-SVD framework, some other tensor decompositions have also been applied to RTPCA recently. And an advanced model with additional total variation constraint is used for compressed sensing of dynamic MRI [30]. The PCA-based Tucker decomposition is firstly proposed by Huang in 2014 [18]. In addition, recently the tensor train-based RPTCA has been discussed by Zhang in 2019 [53] and by Yang in 2020 [49].

As a higher-order generalization of matrix SVD, the Tucker decomposition decomposes a tensor into a core tensor multiplied by factor matrix along each mode. Given the observed tensor $\mathcal{X} \in \mathbb{R}^{I_1 \times I_2 \times \cdots \times I_N}$, the corresponding Tucker rank denoted by $\text{rank}_n(\mathcal{X})$ is defined as a vector, the k-th entry of which is the rank of the mode-k unfolding matrix $\mathbf{X}_{(k)}$. Therefore, the tensor PCA optimization model (6.4) can be transformed into

$$\min_{\mathcal{L},\mathcal{E}} \text{rank}_n(\mathcal{L}) + \lambda \|\mathcal{E}\|_0, \text{ s. t. } \mathcal{X} = \mathcal{L} + \mathcal{E}. \tag{6.38}$$

Similar to the matrix RPCA model (6.2), problem (6.38) is also nonconvex. Since the matrix nuclear norm is the convex envelope of the matrix rank, the sum of nuclear norms (SNN) has been firstly proposed to solve the Tucker rank minimization problem in 2012, which is successfully applied to tensor completion problem [28]. In 2014, a robust tensor PCA model based on SNN minimization is firstly proposed by Huang [18] as follows:

$$\min_{\mathcal{L},\, \mathcal{E}} \sum_{n=1}^{N} \lambda_n \|\mathbf{L}_{(n)}\|_* + \|\mathcal{E}\|_1, \text{ s. t. } \mathcal{X} = \mathcal{L} + \mathcal{E}, \tag{6.39}$$

where $\mathbf{L}_{(n)} \in \mathbb{R}^{I_n \times \prod_{k \neq n}^{N} I_k}$ is the mode-n matricization of the low-rank tensor $\mathcal{L}$. For example, when we have a third-order tensor, it uses the combination of three matrix nuclear norms to solve the Tucker rank minimization problem. The theoretical guarantee has been given for successful recovery of low-rank and sparse components from corrupted tensor $\mathcal{X}$. However, the Tucker decomposition has to perform SVD over all the unfolding matrices, which would cause high computational costs. Meanwhile, the resulting unfolding matrices in Tucker decomposition are often unbalanced so that the global correlation of the tensor cannot be adequately captured.

It is shown that tensor train (TT) can provide a more compact representation for multidimensional data than traditional decomposition methods like Tucker decomposition [2]. In 2019, Zhang et al. firstly apply TT decomposition to recover noisy color images [53] with the optimization model as follows:

$$\min_{\mathbf{U}_n, \mathbf{V}_n, \mathcal{X}} \sum_{n=1}^{N-1} \frac{\alpha_n}{2} \|\mathbf{U}_n \mathbf{V}_n - \mathbf{X}_{<n>}\|_{\mathrm{F}}^2, \tag{6.40}$$

where $\mathbf{U}_n \in \mathbb{R}^{\prod_{j=1}^{n} I_j \times R_{n+1}}$, $\mathbf{V}_n \in \mathbb{R}^{R_{n+1} \times \prod_{j=n+1}^{N} I_j}$, and $\mathbf{X}_{<n>} \in \mathbb{R}^{\prod_{j=1}^{n} I_j \times \prod_{j=n+1}^{N} I_j}$ is the n-unfolding matrix of tensor $\mathcal{X}$, $[R_1, \ldots, R_N, R_{N+1}]$, $R_1 = R_{N+1} = 1$ is the tensor train rank. Model (6.40) can be solved by block coordinate descent (BCD) method, which is similar to the algorithm in [2]. It has been shown that the proposed method can remove the Gaussian noise efficiently.

Based on TT decomposition, Yang et al. defined a tensor train nuclear norm (TTNN) and proposed a TT-based RPTCA [49], which can be formulated as follows:

$$\min_{\mathcal{L},\, \mathcal{E}} \sum_{n=1}^{N-1} \lambda_n \|\mathbf{L}_{<n>}\|_* + \|\mathcal{E}\|_1, \text{ s. t. } \mathcal{X} = \mathcal{L} + \mathcal{E}, \tag{6.41}$$

where $\mathbf{L}_{<n>} \in \mathbb{R}^{\prod_{j=1}^{n} I_j \times \prod_{j=n+1}^{N} I_j}$ is the n-unfolding matrix of low-rank tensor $\mathcal{L}$. This TTNN-based RPTCA aims to recover a low-rank tensor corrupted by sparse noise, which can be solved by the classical ADMM algorithm.

6.3 Online RPTCA

The principal component analysis methods described above are implemented in a batch manner. They need to memorize all data samples, which leads to high storage complexity. In such a case, when a new sample is obtained, the above methods re-optimize all samples, which is very time-consuming.

In order to mitigate this issue, their online versions have been developed, which process only one sample per time. The matrix-based online methods have been studied in [10]. As for the tensor data, Zhang et al. [52] first proposed an online robust principal tensor component analysis (ORPTCA) method in 2016, and the convergence has been analyzed in [47].

As introduced, the convex optimization model for RPTCA can be formulated as follows:

$$\min_{\mathcal{X},\mathcal{E}} \frac{1}{2}\|\mathcal{Z}-\mathcal{X}-\mathcal{E}\|_{\mathrm{F}}^2+\lambda_1\|\mathcal{X}\|_{\mathrm{TNN}}+\lambda_2\|\mathcal{E}\|_1, \tag{6.42}$$

where $\mathcal{Z}$ is the observation, $\mathcal{X}$ denotes the low-rank tensor, and $\mathcal{E}$ is the sparse tensor. We introduce Lemma 6.1 with respect to the TNN.

Lemma 6.1 ([52]) *For a three-way tensor* $\mathcal{X} \in \mathbb{R}^{I_1\times T\times I_3}$ *with the tubal rank* $\mathrm{rank}_{\mathrm{t}}(\mathcal{X}) \leq R$, *we have*

$$\|\mathcal{X}\|_{\mathrm{TNN}} = \inf\left\{\frac{I_3}{2}\left(\|\mathcal{L}\|_F^2+\|\mathcal{R}\|_F^2\right) : \mathcal{X}=\mathcal{L}*\mathcal{R}^T\right\}. \tag{6.43}$$

where $\mathcal{L} \in \mathbb{R}^{I_1\times R\times I_3}$ and $\mathcal{R} \in \mathbb{R}^{T\times R\times I_3}$.

Then problem (6.42) can be reformulated as follows:

$$\min_{\mathcal{L},\mathcal{R},\mathcal{E}} \frac{1}{2}\|\mathcal{Z}-\mathcal{L}*\mathcal{R}^{\mathrm{T}}-\mathcal{E}\|_{\mathrm{F}}^2+\frac{I_3\lambda_1}{2}\left(\|\mathcal{L}\|_{\mathrm{F}}^2+\|\mathcal{R}\|_{\mathrm{F}}^2\right)+\lambda_2\|\mathcal{E}\|_1. \tag{6.44}$$

where $\mathcal{L} \in \mathbb{R}^{I_1\times R\times I_3}$, $\mathcal{R} \in \mathbb{R}^{T\times R\times I_3}$, $\mathcal{E} \in \mathbb{R}^{I_1\times T\times I_3}$. Considering the t-th sample $\mathcal{Z}(:,t,:)$, we abbreviate it as $\vec{\mathcal{Z}}_t \in \mathbb{R}^{I_1\times 1\times I_3}$. Then the loss function with respect to the t-th sample is represented as

$$\ell\left(\vec{\mathcal{Z}}_t,\mathcal{L}\right)=\min_{\vec{\mathcal{R}},\vec{\mathcal{E}}} \frac{1}{2}\left\|\vec{\mathcal{Z}}_t-\mathcal{L}*\vec{\mathcal{R}}^{\mathrm{T}}-\vec{\mathcal{E}}\right\|_{\mathrm{F}}^2+\frac{I_3\lambda_1}{2}\|\vec{\mathcal{R}}\|_{\mathrm{F}}^2+\lambda_2\|\vec{\mathcal{E}}\|_1. \tag{6.45}$$

where $\vec{\mathcal{R}} \in \mathbb{R}^{1\times R\times I_3}$ and $\vec{\mathcal{E}} \in \mathbb{R}^{I_1\times 1\times I_3}$ is the t-th component of $\mathcal{R}$ and $\mathcal{E}$, respectively. The total cumulative loss for the $t \leq T$ samples can be obtained by

$$\mathrm{f}_t(\mathcal{L})=\frac{1}{t}\sum_{\tau=1}^{t}\ell\left(\vec{\mathcal{Z}}_\tau,\mathcal{L}\right)+\frac{I_3\lambda_1}{2t}\|\mathcal{L}\|_{\mathrm{F}}^2. \tag{6.46}$$

The main idea of ORPTCA is to develop an algorithm for minimizing the total cumulative loss $\mathrm{f}_t(\mathcal{L})$ and achieve subspace estimation in an online manner. We define a surrogate function of the total cumulative loss $\mathrm{f}_t(\mathcal{L})$, which is represented as follows:

$$\mathrm{g}_t(\mathcal{L}) = \frac{1}{t}\sum_{\tau=1}^{t}\left(\frac{1}{2}\left\|\vec{\mathcal{Z}}_\tau - \mathcal{L} * \vec{\mathcal{R}}_\tau^{\mathrm{T}} - \vec{\mathcal{E}}_\tau\right\|_{\mathrm{F}}^2 + \frac{I_3\lambda_1}{2}\|\vec{\mathcal{R}}_\tau\|_{\mathrm{F}}^2 + \lambda_2\|\vec{\mathcal{E}}_\tau\|_1\right)$$
$$+ \frac{I_3\lambda_1}{2t}\|\mathcal{L}\|_{\mathrm{F}}^2. \tag{6.47}$$

This is the "non-optimized" version of $\mathrm{f}_t(\mathcal{L})$. In other words, for all $\mathcal{L}$, we have $\mathrm{g}_t(\mathcal{L}) \geq \mathrm{f}_t(\mathcal{L})$. It has been proved that minimizing $\mathrm{g}_t(\mathcal{L})$ is asymptotically equivalent to minimizing $\mathrm{f}_t(\mathcal{L})$ in [47].

At time t, we get a new sample $\vec{\mathcal{Z}}_t$; the coefficient $\vec{\mathcal{R}}_t$, sparse component $\vec{\mathcal{E}}_t$, and spanning basis $\mathcal{L}$ will be alternatively optimized. Based on the previous subspace estimation $\mathcal{L}_{t-1}$, $\vec{\mathcal{R}}_t$ and $\vec{\mathcal{E}}_t$ are optimized by minimizing the loss function with respect to $\vec{\mathcal{Z}}_t$. Based on these two newly acquired factors, we minimize $\mathrm{g}_t(\mathcal{L})$ to update the spanning basis $\mathcal{L}_t$. Therefore, the following two subproblems need to be solved:

(1) *Subproblem with respect to* $\mathcal{R}$ and $\mathcal{E}$:

$$\{\vec{\mathcal{R}}_t, \vec{\mathcal{E}}_t\} = \underset{\vec{\mathcal{R}}, \vec{\mathcal{E}}}{\operatorname{argmin}} \frac{1}{2}\left\|\vec{\mathcal{Z}}_t - \mathcal{L}_{t-1} * \vec{\mathcal{R}}^{\mathrm{T}} - \vec{\mathcal{E}}\right\|_{\mathrm{F}}^2 + \frac{I_3\lambda_1}{2}\|\vec{\mathcal{R}}\|_{\mathrm{F}}^2 + \lambda_2\|\vec{\mathcal{E}}\|_1. \tag{6.48}$$

For convenience, we define $\hat{\mathcal{Z}}_t$ as the result of FFT along the third mode of $\vec{\mathcal{Z}}_t$, i.e., $\hat{\mathcal{Z}}_t = \mathrm{fft}\left(\vec{\mathcal{Z}}_t, [], 3\right)$. In the same way, we can obtain $\vec{\mathcal{Z}}_t = \mathrm{ifft}\left(\hat{\mathcal{Z}}_t, [], 3\right)$. Then the solution of problem (6.48) is presented in Algorithm 35.

(2) *Subproblem with respect to* $\mathcal{L}$:

For convenience, given a tensor in the Fourier domain $\widehat{\mathcal{A}} \in \mathbb{R}^{I_1\times I_2\times I_3}$, the block diagonal matrix is denoted by $\overline{\mathbf{A}} \in \mathbb{R}^{I_1I_3\times I_2I_3}$ as follows:

$$\overline{\mathbf{A}} = \mathrm{bdiag}(\widehat{\mathcal{A}}) = \begin{bmatrix} \widehat{\mathcal{A}}^{(1)} & & & \\ & \widehat{\mathcal{A}}^{(2)} & & \\ & & \ddots & \\ & & & \widehat{\mathcal{A}}^{(I_3)} \end{bmatrix}. \tag{6.49}$$

Algorithm 35: Update the coefficients and sparse component

Input: $\mathcal{L}_{t-1} \in \mathbb{R}^{I_1 \times R \times I_3}$, $\vec{\mathcal{Z}}_t \in \mathbb{R}^{I_1 \times 1 \times I_3}$, parameter λ_1, λ_2.
Output: $\vec{\mathcal{R}}_t$, $\vec{\mathcal{E}}_t$.
Initialize $\vec{\mathcal{E}}_t \in \mathbb{R}^{I_1 \times 1 \times I_3}$, $\hat{\mathcal{R}}_t \in \mathbb{R}^{R \times 1 \times I_3}$.
1. $\hat{\mathcal{Z}}_t = \text{fft}\left(\vec{\mathcal{Z}}_t, [], 3\right)$, $\widehat{\mathcal{L}}_{t-1} = \text{fft}\left(\mathcal{L}_{t-1}, [], 3\right)$.
2. **While** not converge **do**
3. $\hat{\mathcal{E}}_t = \text{fft}\left(\vec{\mathcal{E}}_t, [], 3\right)$.
4. for $i = 1, 2, \cdots, I_3$ do
5. $\hat{\mathcal{R}}_t^{(i)} = \left(\left(\widehat{\mathcal{L}}_{t-1}^{(i)}\right)^{\mathrm{T}} \widehat{\mathcal{L}}_{t-1}^{(i)} + \lambda_1 \mathbf{I}\right)^{-1} \left(\widehat{\mathcal{L}}_{t-1}^{(i)}\right)^{\mathrm{T}} \left(\hat{\mathcal{Z}}_t^{(i)} - \hat{\mathcal{E}}_t^{(i)}\right)$.
6. end for
7. $\vec{\mathcal{R}}_t = \text{ifft}\left(\hat{\mathcal{R}}_t, [], 3\right)$.
8. $\vec{\mathcal{E}}_t = \text{sth}_{\lambda_2}(\vec{\mathcal{Z}}_t - \mathcal{L}_{t-1} * \vec{\mathcal{R}}_t^{\mathrm{T}})$.
9. **end while**

Then the spanning basis is updated by

$$\begin{aligned}
\mathcal{L}_t &= \underset{\mathcal{L}}{\operatorname{argmin}} \sum_{i=1}^{t} \left(\frac{1}{2}\left\|\vec{\mathcal{Z}}_i - \mathcal{L} * \vec{\mathcal{R}}_i^{\mathrm{T}} - \vec{\mathcal{E}}_i\right\|_{\mathrm{F}}^2\right) + \frac{I_3\lambda_1}{2}\|\mathcal{L}\|_{\mathrm{F}}^2 \\
&= \underset{\mathcal{L}}{\operatorname{argmin}} \sum_{i=1}^{t} \left(\frac{1}{2}\left\|\overline{\mathbf{Z}}_i - \overline{\mathbf{E}}_i - \overline{\mathbf{L}}\,\overline{\mathbf{R}}_i^{\mathrm{T}}\right\|_{\mathrm{F}}^2\right) + \frac{I_3\lambda_1}{2}\|\overline{\mathbf{L}}\|_{\mathrm{F}}^2 \\
&= \underset{\mathcal{L}}{\operatorname{argmin}} \frac{1}{2} \operatorname{tr}\left(\overline{\mathbf{L}}\left(\sum_{i=1}^{t}\left(\overline{\mathbf{R}}_i^{\mathrm{T}}\overline{\mathbf{R}}_i\right) + I_3\lambda_1\mathbf{I}\right)\overline{\mathbf{L}}^{\mathrm{T}}\right) \\
&\quad - \operatorname{tr}\left(\overline{\mathbf{L}}^{\mathrm{T}}\left(\sum_{i=1}^{t}\left(\overline{\mathbf{Z}}_i - \overline{\mathbf{E}}_i\right)\overline{\mathbf{R}}_i\right)\right).
\end{aligned} \tag{6.50}$$

Letting $\mathcal{A}_t = \mathcal{A}_{t-1} + \vec{\mathcal{R}}_t^{\mathrm{T}} * \vec{\mathcal{R}}_t$ and $\mathcal{B}_t = \mathcal{B}_{t-1} + (\vec{\mathcal{Z}}_t - \vec{\mathcal{E}}_t) * \vec{\mathcal{R}}_t$, we update the spanning basis in the Fourier domain as follows:

$$\overline{\mathbf{L}}_t = \underset{\overline{\mathbf{L}}}{\operatorname{argmin}} \frac{1}{2} \operatorname{tr}\left[\overline{\mathbf{L}}^{\mathrm{T}}\left(\overline{\mathbf{A}}_t + I_3\lambda_1\mathbf{I}\right)\overline{\mathbf{L}}\right] - \operatorname{tr}\left(\overline{\mathbf{L}}^{\mathrm{T}}\overline{\mathbf{B}}_t\right), \tag{6.51}$$

which can be solved by the block coordinate descent algorithm, and the details can refer to [52].

Finally, the whole process of ORPTCA method is summarized in Algorithm 36. For the batch RPTCA method, the total samples $\{\mathcal{Z}_t\}_{t=1}^{\mathrm{T}}$ need to be memorized, and the storage requirement is $I_1 I_2 T$. As for the ORPTCA, we can only need to save $\mathcal{L}_T \in \mathbb{R}^{I_1 \times R \times I_3}$, $\mathcal{R}_T \in \mathbb{R}^{T \times R \times I_3}$, $\mathcal{B}_T \in \mathbb{R}^{I_1 \times R \times I_3}$, and the storage requirement is $I_3 R T + 2 I_1 I_3 R$. When $R \ll T$, the storage requirement can be greatly reduced.

Algorithm 36: Online Robust Principal Tensor Component Analysis

Input: The observed tensor $\mathcal{Z} \in \mathbb{R}^{I_1 \times T \times I_3}$, parameter λ_1, λ_2.
Output: $\mathcal{X}_T,\ \mathcal{E}_T$.
Initialize $\mathcal{L}_0 \in \mathbb{R}^{I_1 \times R \times I_3}$, $\mathcal{R}_0 \in \mathbb{R}^{1 \times R \times I_3}$, $\mathcal{E} \in \mathbb{R}^{I_1 \times 1 \times I_3}$.
1. **for** $t = 1$ to T **do**
2. Reveal the sample $\vec{\mathcal{Z}}_t$.
3. Update the coefficient and the sparse component$\{\vec{\mathcal{R}}_t,\ \vec{\mathcal{E}}_t\}$ by Algorithm 35.
4. Compute $\mathcal{A}_t = \mathcal{A}_{t-1} + \vec{\mathcal{R}}_t^{\mathrm{T}} * \vec{\mathcal{R}}_t,\ \mathcal{B}_t = \mathcal{B}_{t-1} + (\vec{\mathcal{Z}}_t - \vec{\mathcal{E}}_t) * \vec{\mathcal{R}}_t$.
5. Update the spanning basis $\mathcal{L}_t$ by (6.51).
6. **end for**
7. $\mathcal{R}_T(t,:,:) = \vec{\mathcal{R}}_t, \mathcal{E}_T(:,t,:) = \vec{\mathcal{E}}_t,\ t = 1, \cdots, T$.
8. **Compute** $\mathcal{X} = \mathcal{L}_T * \mathcal{R}_T^{\mathrm{T}}$.

6.4 RTPCA with Missing Entries

The observed data may be not only partially missing but also contaminated by noise in some scenes. In this situation, robust matrix completion for matrix data has been developed in [4]. Extending to tensor field, robust principal tensor component analysis with missing entries, also called robust tensor completion (RTC), have been considered in [18, 19, 43]. In fact, it simultaneously deals with the robust principal tensor component analysis and the missing tensor component analysis.

Given the observed tensor $\mathcal{X} \in \mathbb{R}^{I_1 \times I_2 \times I_3}$, the RTC based on TNN (RTC-TNN) related to t-SVD [43] is formulated as follows:

$$\min_{\mathcal{L},\ \mathcal{E}} \lambda \|\mathcal{L}\|_{\mathrm{TNN}} + \|\mathcal{E}\|_1, ,\ \text{s. t. } \mathscr{P}_{\mathbb{O}}(\mathcal{L} + \mathcal{E}) = \mathscr{P}_{\mathbb{O}}(\mathcal{X}), \tag{6.52}$$

which is limited to the third-order tensors. The operator $\mathscr{P}_{\mathbb{O}}(\mathcal{X})$ means the projection of tensor $\mathcal{X}$ on the observation indices set $\mathbb{O}$. If $\{i_1, i_2, i_3\} \in \mathbb{O}$, then $\mathcal{X}(i_1, i_2, i_3)$ is observed. Otherwise, the entry $\mathcal{X}(i_1, i_2, i_3)$ is missing. For N-th order tensor $\mathcal{X} \in \mathbb{R}^{I_1 \times I_2 \times \cdots \times I_N}$, [18] proposed a RTC method based on SNN (RTC-SNN) related to Tucker decomposition as follows:

$$\min_{\mathcal{L},\ \mathcal{E}} \sum_{n=1}^{N} \lambda_n \|\mathbf{L}_{(n)}\|_* + \|\mathcal{E}\|_1,\ \text{s. t. } \mathscr{P}_{\mathbb{O}}(\mathcal{L} + \mathcal{E}) = \mathscr{P}_{\mathbb{O}}(\mathcal{X}), \tag{6.53}$$

where $\mathbf{L}_{(n)}$ is the mode-n unfolding matrix of the low-rank component $\mathcal{L}$. In addition, another RTC model based on TR decomposition (RTC-TR) has also been studied in [19] with the optimization model as follows:

$$\min_{\mathcal{L},\mathcal{E}} \sum_{n=1}^{N} \lambda_n \left\|\mathbf{L}_{\langle n,L\rangle}\right\|_* + \|\mathcal{E}\|_1, ,\ \text{s. t. } \mathscr{P}_{\mathbb{O}}(\mathcal{L} + \mathcal{E}) = \mathscr{P}_{\mathbb{O}}(\mathcal{X}), \tag{6.54}$$

where $\mathbf{L}_{\langle n,L\rangle}$ is the n-shifting L-unfolding matrix of $\mathcal{L}$, which relates to the tensor ring rank of $\mathcal{L}$. These models can be easily solved by classic ADMM.

6.5 Applications

RPTCA can extract low-rank component and sparse component in tensor data. In this section, we briefly demonstrate RPTCA for different applications, including illumination normalization for face images, image denoising, background extraction, video rain streaks removal, and infrared small target detection.

6.5.1 Illumination Normalization for Face Images

As we all know, face recognition algorithms are easily affected by the shadow of the face image. Therefore, it is important to remove the disturbance on face images and further improve the accuracy of the face recognition. The low-rank plus sparse model has been successfully used for shadow removal from face images.

We choose the face images in size of 192×168 with 64 different lighting conditions from Yale B face database [15], and construct the tensor data as $\mathcal{X} \in \mathbb{R}^{192\times 168\times 64}$. Matrix-based RPCA [4] and multi-scale low-rank decomposition [38] are chosen as the baselines, which reshape the tensor into the matrix $\mathbf{X} \in \mathbb{R}^{322256\times 64}$. To show the performance enhancement by tensor methods, we choose TNN-RPTCA [33] and block-RPTCA [11] for comparison. All the parameters are set as recommended in their papers. The recovery results of several RPTCA methods are shown in Fig. 6.8. We can see that the shadows can be removed efficiently, especially by the block-RPTCA method.

Moreover, we consider the situation that face images may be not only corrupted by noise but also partially missing. We apply robust tensor completion (RTC) methods mentioned in Sect. 6.4 for shadow removal on face images. From the YaleB face dataset,[1] we choose the first human subject for experiments. The clean and complete faces can be regarded as low-rank component when the shadows is sparse. We randomly choose 50% pixels from the original images as the observation. The comparison results are showed in Fig. 6.9. We can see that the RTRC method can remove most shadows.

[1] http://vision.ucsd.edu/content/yale-face-database.

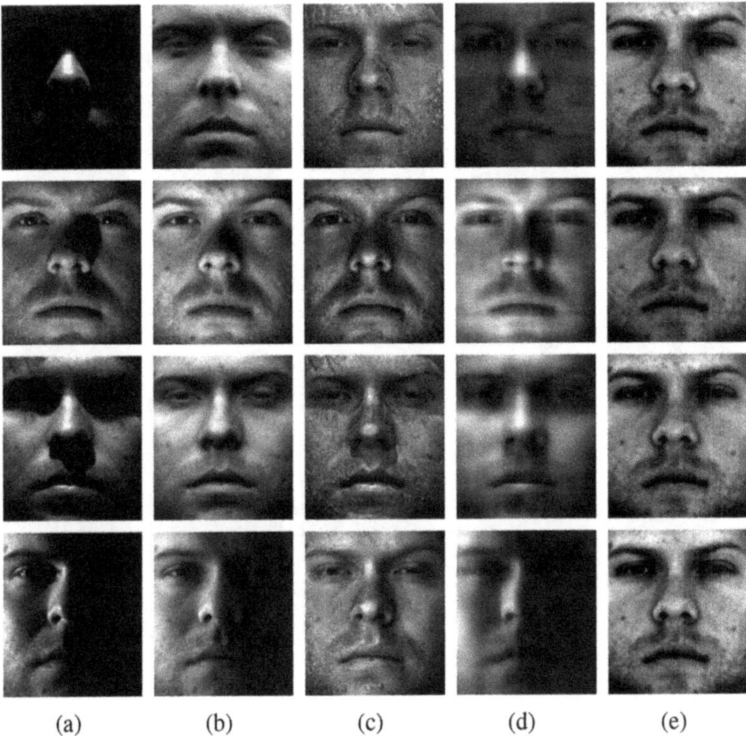

Fig. 6.8 Four methods for removing shadows on face images. (**a**) Original faces with shadows; (**b**) RPCA [4]; (**c**) multi-scale low-rank decomposition [38]; (**d**) TNN-RPTCA[33]; (**e**) block-RPTCA[11]

6.5.2 Image Denoising

In this section, we focus on image denoising. It is one of the most fundamental and classical tasks in image processing and computer vision. Natural images are always corrupted by various noise in imaging process. Since most information of a gray image is preserved by its first few singulars values, an image can be approximated by its low-rank approximation. If the noise ratio is not too high, the image denoising can be modeled as sparsely corrupted low-rank data which can be processed by robust principal component analysis.

A color image with the size of $I_1 \times I_2$ and three color channels can be represented as a third-order tensor $\mathcal{X} \in \mathbb{R}^{I_1 \times I_2 \times 3}$. Since the texture information in these three channels are very similar, the low-rank structure also exists among these channels. Classical RPCA has to process the color image on each channel independently, which neglects the strong correlations in the third dimension. RTPCA can fully utilize the low-rank structure between channels and deal with this problem better.

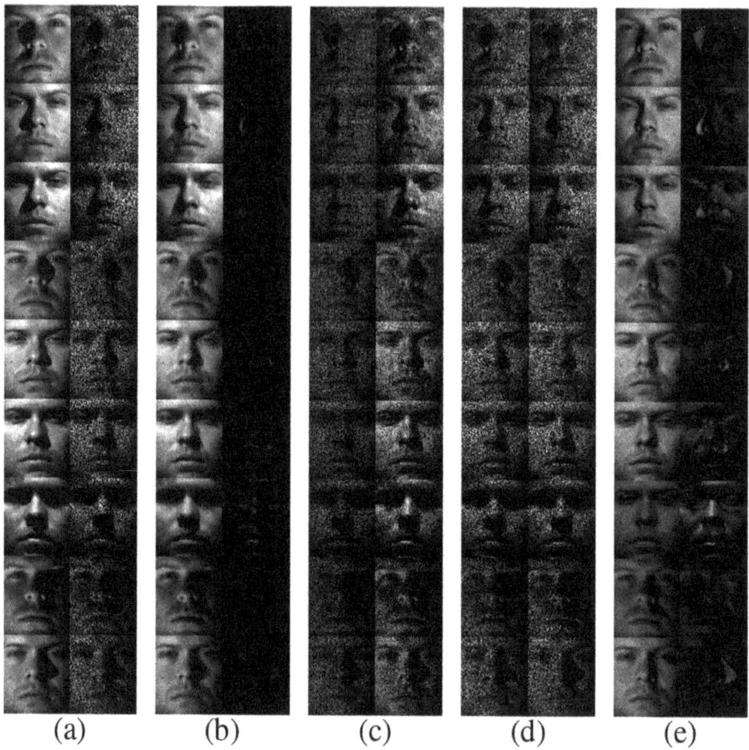

Fig. 6.9 Four methods for removing shadows on face images. (**a**) Original and observed; (**b**) RTRC; (**c**) RTC-TNN; (**d**) RTC-SNN; (**e**) RMC. For (**b**), (**c**), (**d**), (**e**), left, recovery results; right, absolute differences with respect to original images

We perform numerical experiments on the images from Berkeley Segmentation Dataset [35]. For each image, we randomly choose 10% of pixels and randomly set their values to [0, 255]. Matrix-based RPCA [4] is chosen as the baseline. We conduct experiments on each channel separately by RPCA and combine the results into a color image. For tensor-based methods, we choose SNN-RPTCA [18], TNN-RPTCA [33], block-RPTCA [11], improved RPTCA [29], and tubal-RPTCA [51] methods for comparison. All the parameters are set as recommended in their papers.

The experimental results by different methods are shown in Fig. 6.10. It can be seen that tensor-based methods including tubal-RPTCA [51], block-RPTCA [11], and improved RPTCA [29] perform better than others with more details of the images. The results by RPCA [4] tend to be vague in these examples. In summary, we conclude that tensor-based RTPCA is more suitable for color image denoising, and these tensor versions have better results in terms of accuracy.

Fig. 6.10 Several results of image denoising on five sample images. (**a**) Original images; (**b**) noisy images; (**c**) RPCA[4]; (**d**) SNN-PTCA[18]; (**e**) TNN-PTCA[33]; (**f**) tubal-PTCA[51]; (**g**) block-RPTCA[11]; (**h**) improved PTCA[29]

6.5.3 Background Extraction

The key task for background extraction is to model a clear background of the video frames and accurately separate the moving objects from the video. Generally, it is also called background modeling, background and foreground separation in different papers. Due to spatiotemporal correlations between the video frames, background can be regarded as a low-rank component. The moving foreground objects only occupy a fraction of pixels in each frame and can be treated as a sparse component.

Traditional matrix-based RPCA has to vectorize all the frames when dealing with this problem, which would lead to a spatial structure information loss. Driven by this, tensor-based RPCA methods are proposed and applied to solve such problems for enhanced performance.

We perform numerical experiments on the video sequences from SBI dataset [34]. Five color image sequences are selected with the size of $320 \times 240 \times 3$. We choose matrix-based RPCA [4] as the baseline. For comparison, we choose tensor-based methods including BRTF [54], TNN-RPTCA [33], and improved RPTCA [29]. The results of the background and foreground separation are shown in Fig. 6.11. We can observe that in most cases tensor-based methods outperform matrix-based one.

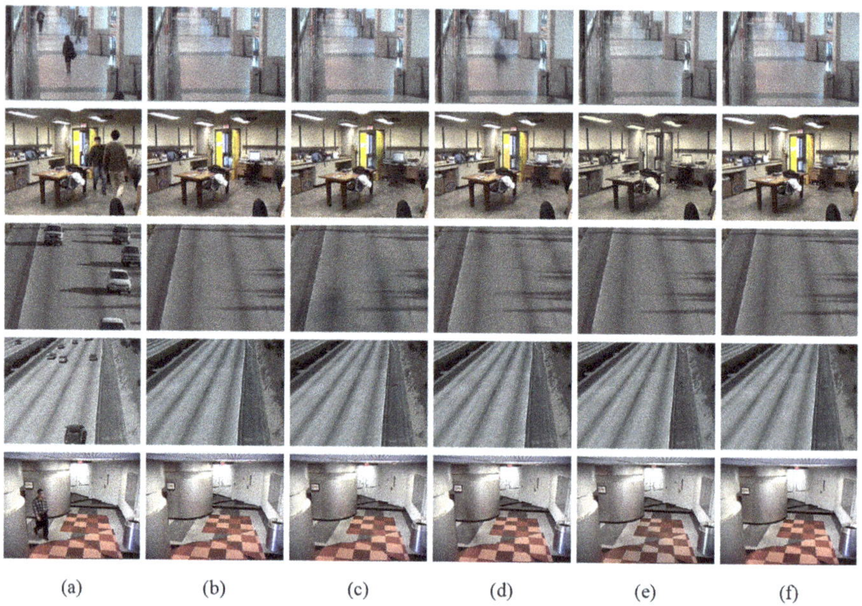

Fig. 6.11 Recovered background images of five example sequences. (**a**) Original; (**b**) ground truth; (**c**) RPCA [4]; (**d**) TNN-PTCA [33]; (**e**) BRTF [54]; (**f**) improved PTCA [29]

6.5.4 *Video Rain Streaks Removal*

Unlike background modeling, which is a basic technique in video processing, video rain streaks removal is a higher-level problem in the outdoor vision system and has been recently investigated extensively. In some video surveillance applications, both the moving objects and the rain streaks exist in the video. It is a hard problem to divide the original observed video into the background components, the foreground components (moving objects), and rain streaks. In this part, we only consider the rain streaks as our interesting point and leave the moving objects removal problem to be solved by RPTCA model as introduced in Sect. 6.5.3.

In general, the rainy video data can be modeled as $\mathcal{O} = \mathcal{B} + \mathcal{R}$, where $\mathcal{O}, \mathcal{B}, \mathcal{R} \in \mathbb{R}^{I_1 \times I_2 \times I_3}$ are third-order tensors for the rainy video, rain-free video, and rain streaks. In [1, 26], the authors apply robust principle component analysis to solve this problem. In [22, 23], a tensor-based RPCA method fully considers the prior knowledge of the rainy video model. The prior knowledge used in the video deraining model can be summarized as follows:

- **Sparsity of rain streaks:** When the rain is light, the drops only take a fraction of pixels of each frame. In most cases, the rain tends to be sparse in the rainy video, and the ℓ_1 norm can be applied for sparsity of the rain component in video.
- **Smoothness along the rain-perpendicular direction:** Normally the rain streaks are distributed from top to down, and each frame is smooth enough except for the

rain component along the rain-perpendicular direction. Therefore, the ℓ_1 norm of unidirectional total variation (TV) regularizer can be applied as the constraint for the rain-free component.

- **Peculiarity of the rain along rainy direction:** Since the pixels of rain streaks are very similar, the rain streaks blocks tend to be smooth, and the unidirectional TV regularizer can be used along the rainy direction for the rain component.
- **Correlation along time direction:** Because of the strong correlations along temporal direction, the rain-free video can be modeled as a low-rank data which can be approximated by the sum of nuclear norms (SNN) defined in Tucker decomposition for minimization. The unidirectional TV regularizer along the time direction can improve the smoothness of the rain-free video.

In summary, the optimization model of video rain streaks removal can be formulated as follows:

$$\begin{aligned} &\min_{\mathcal{B},\mathcal{R}} \alpha_1\|\nabla_1\mathcal{R}\|_1 + \alpha_2\|\mathcal{R}\|_1 + \alpha_3\|\nabla_2\mathcal{B}\|_1 + \alpha_4\|\nabla_t\mathcal{B}\|_1 + \|\mathcal{B}\|_{\text{SNN}}, \\ &\quad \text{s. t. } \mathcal{O} = \mathcal{B} + \mathcal{R}, \end{aligned} \tag{6.55}$$

where $\alpha_i, i = 1, 2, 3, 4$ is set to balance the four terms and ∇_1, ∇_2, and ∇_t are the vertical, horizontal, and temporal TV operators, respectively. In this model, the first item $\|\nabla_1\mathcal{R}\|_1$ indicates the smoothness priors of the rain streaks in the perpendicular direction. The second item $\|\mathcal{R}\|_1$ indicates the sparsity of the rain streaks. And the items $\|\nabla_2\mathcal{B}\|_1$ and $\|\nabla_t\mathcal{B}\|_1$ are imposed to regularize the smoothness of the background video. Finally, the whole rain-free video is approximated by a low-rank tensor model which is regularized by SNN. More details can be found in [22, 23].

We perform experiments on a selected color video named as "foreman"[2] and add the generated rain on it. Figure 6.12 shows two frames of the deraining results. We can see the rain component can be effectively removed in these experimental results.

6.5.5 *Infrared Small Target Detection*

Infrared search and track (IRST) plays a crucial role in some applications such as remote sensing and surveillance. The fundamental function of IRST system is the infrared small target detection. The small targets obtained by infrared imaging technology usually lack the texture and structure information, due to the influence by long distance and complex background. In general, the infrared small target likes a light spot, which is extremely difficult to be detected. Therefore, effective detection for infrared small target is really a challenge.

[2]http://trace.eas.asu.edu/yuv/.

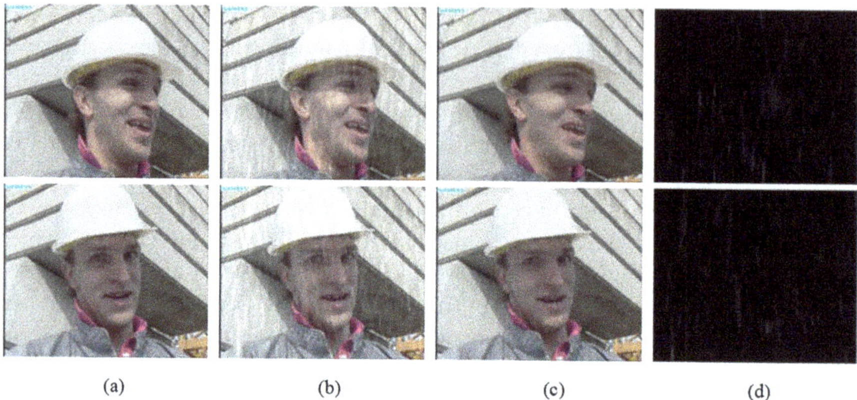

Fig. 6.12 Rain streaks removal from video data. From top to bottom are the 11th frame and 110th frame of the video. (**a**) Groundtruth. (**b**) Rainy image. (**c**) Recovered. (**d**) Rain removal

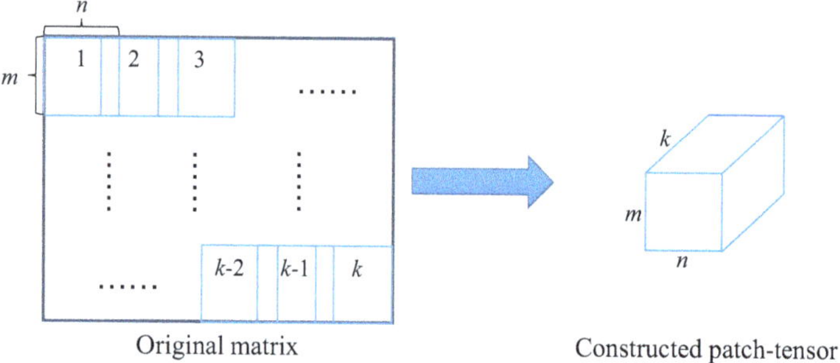

Fig. 6.13 The transformation between an original matrix data and a constructed tensor

The classical infrared patch image (IPI) model [14] regards the complex background as a low-rank component while the target as an outlier. Therefore, the conventional target detection problem is consistent with the RPCA optimization model. In [7], in order to make use of more correlations between different patches, Dai et al. set up an infrared patch-tensor (IPT) model. The transformation between the original matrix data and a constructed tensor is shown in Fig. 6.13.

In [50], the authors have applied the partial sum of the tubal nuclear norm (PSTNN) to detect the infrared small target. Based on the IPT model, they combine PSTNN with the weighted ℓ_1 norm to efficiently suppress the background and preserve the small target. The optimization model is as follows:

$$\min_{\mathcal{L},\mathcal{E}} \|\mathcal{L}\|_{\text{PSTNN}} + \lambda\|\mathcal{E} \circledast \mathcal{W}\|_1 \text{ s. t. } \mathcal{X} = \mathcal{L} + \mathcal{E}. \tag{6.56}$$

where $\circledast$ denotes the Hadamard product and $\mathcal{W}$ is a weighting tensor for sparse term.

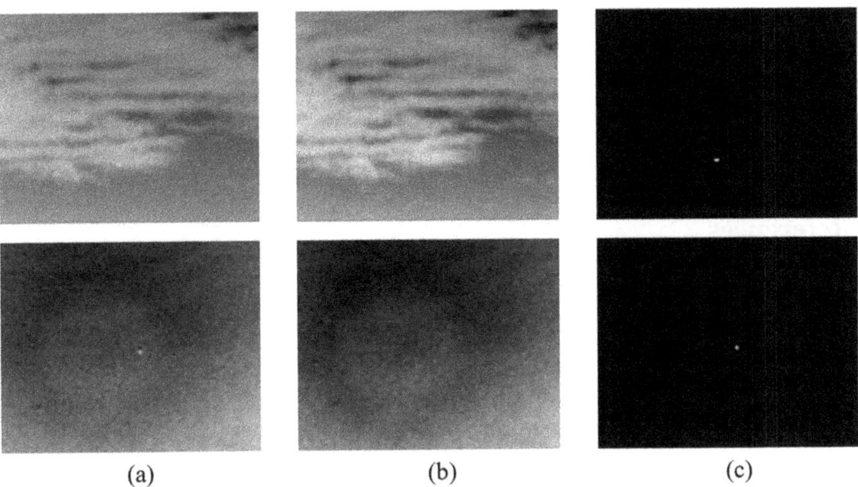

Fig. 6.14 Infrared small target detection results by Zhang and Peng [50]. (**a**) Original image; (**b**) background; (**c**) target

Two test images in [13, 14] are used to show the performance. Figure 6.14 shows the infrared small target detection results by Zhang and Peng [50]. We can see that the small target can be extracted very well.

6.6 Summary

In this section, we first introduce the basic principal component analysis model of the matrix data, which is then extended to the tensor version. We mainly overview the optimization models, algorithms, and applications of RPTCA. The tensor optimization model is set up by modeling the separation of low-rank component and sparse component in high-dimensional data. Several sparse constraints have been summarized according to different applications. In addition, a number of low-rank constraints based on t-SVD, Tucker decomposition, tensor train, and tensor ring decomposition have been discussed. In particular, a variety of low-rank tensor constraints-based t-SVD have been outlined with respect to different scales, rotation invariance, frequency component analysis, nonconvex versions, different transforms, and higher-order cases. Besides, online RPTCA and RPTCA with missing entries are briefly introduced. Finally, the effectiveness of RPTCA is illustrated in five image processing applications with experimental results, including illumination normalization for face images, image denoising, background extraction, video rain streaks removal, and infrared small target detection.

In the future, nonlocal similarity may be incorporated into the multi-scale RPTCA method for performance improvement. Meanwhile, the multidimensional tensor network decomposition via t-product can be considered to obtain a more compact representation. More regularization terms such as omnidirectional total variation can be applied for different applications. In addition, online versions based on various tensor decompositions can be explored.

References

1. Abdel-Hakim, A.E.: A novel approach for rain removal from videos using low-rank recovery. In: 2014 5th International Conference on Intelligent Systems, Modelling and Simulation, pp. 351–356. IEEE, Piscataway (2014)
2. Bengua, J.A., Phien, H.N., Tuan, H.D., Do, M.N.: Efficient tensor completion for color image and video recovery: low-rank tensor train. IEEE Trans. Image Proc. **26**(5), 2466–2479 (2017)
3. Boyd, S., Parikh, N., Chu, E., Peleato, B., Eckstein, J., et al.: Distributed optimization and statistical learning via the alternating direction method of multipliers. Found. Trends Mach. Learn. **3**(1), 1–122 (2011)
4. Candès, E.J., Li, X., Ma, Y., Wright, J.: Robust principal component analysis? J. ACM **58**(3), 1–37 (2011)
5. Chen, L., Liu, Y., Zhu, C.: Iterative block tensor singular value thresholding for extraction of low rank component of image data. In: 2017 IEEE International Conference on Acoustics, Speech and Signal Processing (ICASSP), pp. 1862–1866. IEEE, Piscataway (2017)
6. Chen, L., Liu, Y., Zhu, C.: Robust tensor principal component analysis in all modes. In: 2018 IEEE International Conference on Multimedia and Expo (ICME), pp. 1–6. IEEE, Piscataway (2018)
7. Dai, Y., Wu, Y.: Reweighted infrared patch-tensor model with both nonlocal and local priors for single-frame small target detection. IEEE J. Sel. Top. Appl. Earth Observ. Remote Sens. **10**(8), 3752–3767 (2017)
8. De La Torre, F., Black, M.J.: A framework for robust subspace learning. Int. J. Comput. Vision **54**(1–3), 117–142 (2003)
9. Driggs, D., Becker, S., Boyd-Graber, J.: Tensor robust principal component analysis: Better recovery with atomic norm regularization (2019). Preprint arXiv:1901.10991
10. Feng, J., Xu, H., Yan, S.: Online robust PCA via stochastic optimization. In: Advances in Neural Information Processing Systems, pp. 404–412 (2013)
11. Feng, L., Liu, Y., Chen, L., Zhang, X., Zhu, C.: Robust block tensor principal component analysis. Signal Process. **166**, 107271 (2020)
12. Fischler, M.A., Bolles, R.C.: Random sample consensus: a paradigm for model fitting with applications to image analysis and automated cartography. Commun. ACM **24**(6), 381–395 (1981)
13. Gao, C., Zhang, T., Li, Q.: Small infrared target detection using sparse ring representation. IEEE Aerosp. Electron. Syst. Mag. **27**(3), 21–30 (2012)
14. Gao, C., Meng, D., Yang, Y., Wang, Y., Zhou, X., Hauptmann, A.G.: Infrared patch-image model for small target detection in a single image. IEEE Trans. Image Process. **22**(12), 4996–5009 (2013)
15. Georghiades, A.S., Belhumeur, P.N., Kriegman, D.J.: From few to many: illumination cone models for face recognition under variable lighting and pose. IEEE Trans. Pattern Analy. Mach. Intell. **23**(6), 643–660 (2001)
16. Gnanadesikan, R., Kettenring, J.: Robust estimates, residuals, and outlier detection with multiresponse data. Biometrics **28**, 81–124 (1972)
17. Hu, W., Tao, D., Zhang, W., Xie, Y., Yang, Y.: The twist tensor nuclear norm for video completion. IEEE Trans. Neural Netw. Learn. Syst. **28**(12), 2961–2973 (2016)

18. Huang, B., Mu, C., Goldfarb, D., Wright, J.: Provable low-rank tensor recovery. Optimization-Online **4252**(2) (2014)
19. Huang, H., Liu, Y., Long, Z., Zhu, C.: Robust low-rank tensor ring completion. IEEE Trans. Comput. Imaging **6**, 1117–1126 (2020)
20. Huber, P.J.: Robust statistics, vol. 523. John Wiley & Sons, New York (2004)
21. Jiang, T.X., Huang, T.Z., Zhao, X.L., Deng, L.J.: A novel nonconvex approach to recover the low-tubal-rank tensor data: when t-SVD meets PSSV (2017). Preprint arXiv:1712.05870
22. Jiang, T.X., Huang, T.Z., Zhao, X.L., Deng, L.J., Wang, Y.: A novel tensor-based video rain streaks removal approach via utilizing discriminatively intrinsic priors. In: The IEEE Conference on Computer Vision and Pattern Recognition (CVPR), pp. 2818–2827 (2017). https://doi.org/10.1109/CVPR.2017.301
23. Jiang, T.X., Huang, T.Z., Zhao, X.L., Deng, L.J., Wang, Y.: Fastderain: A novel video rain streak removal method using directional gradient priors. IEEE Trans. Image Process. **28**(4), 2089–2102 (2018)
24. Ke, Q., Kanade, T.: Robust L1 norm factorization in the presence of outliers and missing data by alternative convex programming. In: 2005 IEEE Computer Society Conference on Computer Vision and Pattern Recognition (CVPR'05), vol. 1, pp. 739–746. IEEE, Piscataway (2005)
25. Kernfeld, E., Kilmer, M., Aeron, S.: Tensor–tensor products with invertible linear transforms. Linear Algebra Appl. **485**, 545–570 (2015)
26. Kim, J.H., Sim, J.Y., Kim, C.S.: Video deraining and desnowing using temporal correlation and low-rank matrix completion. IEEE Trans. Image Process. **24**(9), 2658–2670 (2015)
27. Liu, X.Y., Wang, X.: Fourth-order tensors with multidimensional discrete transforms (2017). Preprint arXiv:1705.01576
28. Liu, J., Musialski, P., Wonka, P., Ye, J.: Tensor completion for estimating missing values in visual data. IEEE Trans. Pattern Analy. Mach. Intell. **35**(1), 208–220 (2012)
29. Liu, Y., Chen, L., Zhu, C.: Improved robust tensor principal component analysis via low-rank core matrix. IEEE J. Sel. Topics Signal Process. **12**(6), 1378–1389 (2018)
30. Liu, Y., Liu, T., Liu, J., Zhu, C.: Smooth robust tensor principal component analysis for compressed sensing of dynamic MRI. Pattern Recogn. **102**, 107252 (2020)
31. Lu, C., Zhou, P.: Exact recovery of tensor robust principal component analysis under linear transforms (2019). Preprint arXiv:1907.08288
32. Lu, C., Feng, J., Chen, Y., Liu, W., Lin, Z., Yan, S.: Tensor robust principal component analysis: Exact recovery of corrupted low-rank tensors via convex optimization. In: Proceedings of the IEEE Conference on Computer Vision and Pattern Recognition, pp. 5249–5257 (2016)
33. Lu, C., Feng, J., Chen, Y., Liu, W., Lin, Z., Yan, S.: Tensor robust principal component analysis with a new tensor nuclear norm. IEEE Trans. Pattern Analy. Mach. Intell. **42**(4), 925–938 (2019)
34. Maddalena, L., Petrosino, A.: Towards benchmarking scene background initialization. In: International Conference on Image Analysis and Processing, pp. 469–476. Springer, Berlin (2015)
35. Martin, D., Fowlkes, C., Tal, D., Malik, J.: A database of human segmented natural images and its application to evaluating segmentation algorithms and measuring ecological statistics. In: Proceedings Eighth IEEE International Conference on Computer Vision. ICCV 2001, vol. 2, pp. 416–423. IEEE, Piscataway (2001)
36. Oh, T.H., Kim, H., Tai, Y.W., Bazin, J.C., So Kweon, I.: Partial sum minimization of singular values in RPCA for low-level vision. In: Proceedings of the IEEE International Conference on Computer Vision, pp. 145–152 (2013)
37. Oh, T.H., Tai, Y.W., Bazin, J.C., Kim, H., Kweon, I.S.: Partial sum minimization of singular values in robust PCA: Algorithm and applications. IEEE Trans. Pattern Analy. Mach. Intell. **38**(4), 744–758 (2015)
38. Ong, F., Lustig, M.: Beyond low rank + sparse: Multiscale low rank matrix decomposition. IEEE J. Sel. Topics Signal Process. **10**(4), 672–687 (2015)

39. Pearson, K.: LIII. On lines and planes of closest fit to systems of points in space. London, Edinburgh Dublin Philos. Mag. J. Sci. **2**(11), 559–572 (1901)
40. Song, G., Ng, M.K., Zhang, X.: Robust tensor completion using transformed tensor SVD (2019). Preprint arXiv:1907.01113
41. Su, Y., Wu, X., Liu, G.: Nonconvex low tubal rank tensor minimization. IEEE Access **7**, 170831–170843 (2019)
42. Sun, W., Yang, G., Peng, J., Du, Q.: Lateral-slice sparse tensor robust principal component analysis for hyperspectral image classification. IEEE Geosci. Remote Sens. Lett. **17**(1), 107–111 (2019)
43. Wang, A., Song, X., Wu, X., Lai, Z., Jin, Z.: Robust low-tubal-rank tensor completion. In: ICASSP 2019–2019 IEEE International Conference on Acoustics, Speech and Signal Processing (ICASSP), pp. 3432–3436. IEEE, Piscataway (2019)
44. Wang, X., Gu, L., Lee, H.w.J., Zhang, G.: Quantum tensor singular value decomposition with applications to recommendation systems (2019). Preprint arXiv:1910.01262
45. Wang, A., Li, C., Jin, Z., Zhao, Q.: Robust tensor decomposition via orientation invariant tubal nuclear norms. In: Proceedings of the AAAI Conference on Artificial Intelligence, vol. 34, No. 04, pp. 6102–6109 (2020)
46. Wang, S., Liu, Y., Feng, L., Zhu, C.: Frequency-weighted robust tensor principal component analysis (2020). Preprint arXiv:2004.10068
47. Wijnen, M., et al.: Online tensor robust principal component analysis. Technical Report, The Australian National University (2018)
48. Xu, W.H., Zhao, X.L., Ng, M.: A fast algorithm for cosine transform based tensor singular value decomposition (2019). Preprint arXiv:1902.03070
49. Yang, J.H., Zhao, X.L., Ji, T.Y., Ma, T.H., Huang, T.Z.: Low-rank tensor train for tensor robust principal component analysis. Appl. Math. Comput. **367**, 124783 (2020)
50. Zhang, L., Peng, Z.: Infrared small target detection based on partial sum of the tensor nuclear norm. Remote Sens. **11**(4), 382 (2019)
51. Zhang, Z., Ely, G., Aeron, S., Hao, N., Kilmer, M.: Novel methods for multilinear data completion and de-noising based on tensor-SVD. In: Proceedings of the IEEE Conference on Computer Vision and Pattern Recognition, pp. 3842–3849 (2014)
52. Zhang, Z., Liu, D., Aeron, S., Vetro, A.: An online tensor robust PCA algorithm for sequential 2D data. In: 2016 IEEE International Conference on Acoustics, Speech and Signal Processing (ICASSP), pp. 2434–2438. IEEE, Piscataway (2016)
53. Zhang, Y., Han, Z., Tang, Y.: Color image denoising based on low-rank tensor train. In: Tenth International Conference on Graphics and Image Processing (ICGIP 2018), vol. 11069, p. 110692P. International Society for Optics and Photonics, Bellingham (2019)
54. Zhao, Q., Zhang, L., Cichocki, A.: Bayesian CP factorization of incomplete tensors with automatic rank determination. IEEE Trans. Pattern Analy. Mach. Intell. **37**(9), 1751–1763 (2015)
55. Zheng, Y.B., Huang, T.Z., Zhao, X.L., Jiang, T.X., Ma, T.H., Ji, T.Y.: Mixed noise removal in hyperspectral image via low-fibered-rank regularization. IEEE Trans. Geosci. Remote Sens. **58**(1), 734–749 (2019)
56. Zheng, Y.B., Huang, T.Z., Zhao, X.L., Jiang, T.X., Ji, T.Y., Ma, T.H.: Tensor N-tubal rank and its convex relaxation for low-rank tensor recovery. Inf. Sci. **532**, 170–189 (2020)
57. Zhou, P., Feng, J.: Outlier-robust tensor PCA. In: Proceedings of the IEEE Conference on Computer Vision and Pattern Recognition, pp. 2263–2271 (2017)

Chapter 7
Tensor Regression

7.1 High-Dimensional Data-Related Regression Tasks

With the emergence of high-dimensional data, not only the data representation and information extraction but also the casual analysis and system modeling are very noteworthy research areas. Specifically, it is a critical research direction to model the time-varying networks in order to predict the possible climate state in the future or other missing values of adjacent locations in climatology [3, 35, 47], to explore the user relationship within social networks [15, 37, 50], to identify the effective features of specific economical activities [4]. However, such network data are usually indexed in multiple dimensions, such as time, 3D space, different features, or variables. The multidirectional relatedness between the multiway predictor and multiway response brings a big challenge to the traditional regression models. The main problem of utilizing the traditional regression models for such problems is the loss of the inherent structural information and heavy computational and storage burden, absence of uniqueness, sensitivity to noise, and difficulty of interpretation brought by the huge amount of model parameters [8, 55]. To address these issues, different tensor regression frameworks are developed to better explore the multiway dependencies along the high-dimensional input and output and help the robust modeling of the complex networks varying with multiple indexes.

As shown in Fig. 7.1, tensor regression can be applied in a wide range of fields. For example, facial images or high-dimensional features extracted from these images can be used for human age estimation [12], human facial expression attributes estimation [32], etc. The unique low-rank and smooth characteristics of the image itself can be well exploited by the tensor low-rank approximation method and smooth constraints, which can greatly benefit the modeling of the complex regression systems. Besides, in computer vision, motion reconstruction or tracking based on video sequences can also be regarded as multiway regression problems [31]. Neuroimaging analysis, including the medical diagnosis [14, 26, 27, 55],

Y. Liu et al., *Tensor Computation for Data Analysis*,
https://doi.org/10.1007/978-3-030-74386-4_7

Fig. 7.1 The high-dimensional data-related regression tasks

neuron decoding [1, 40, 53], and brain connectivity analysis [21, 41], is also a dominant research field, which inspires the development in sparse tensor regression methods. Furthermore, it also becomes important than ever to explore the relationship between manufacturing process parameters and the shape of manufactured parts [48], which can reduce a lot of time for process optimization and help achieve smart manufacturing.

7.2 Tensor Regression Framework

Given N observations $\{\mathcal{X}_n \in \mathbb{R}^{P_1 \times \cdots \times P_L}, \mathcal{Y}_n \in \mathbb{R}^{Q_1 \times \cdots \times Q_M}\}, n = 1, \ldots, N$, the common regression framework can be formulated as the following optimization problem:

$$\begin{aligned} \min \quad & \mathrm{L}(\mathcal{Y}_n, \mathrm{g}(\mathrm{f}(\mathcal{X}_n))) \\ \text{s. t.} \quad & \mathcal{Y}_n = \mathrm{g}(\mathrm{f}(\mathcal{X}_n)) + \mathcal{E}_n, \end{aligned} \tag{7.1}$$

where f is a function representing the feature extraction process, g is either a linear or nonlinear function which represents the relationship between the extracted features and the response, $\mathcal{E}_n$ is the model bias of the n-th sample, and L represents the loss function. With different assumptions over the functions f, different feature selection strategies will be employed, such as tensor decomposition [25, 43] or ensemble learning [47]. In this chapter, we focus on the concept and algorithms for regression analysis. So the preprocessing techniques are not illustrated in detail here; interested readers can refer to the related papers for more details.

Specifically, tensor regression can model the system between multiway predictor and the multiway response directly without preprocessing by optimizing the following problem:

$$\begin{aligned} \min \quad & \mathrm{L}(\mathcal{Y}_n, \mathrm{g}(\mathcal{X}_n)) \\ \text{s.t.} \quad & \mathcal{Y}_n = \mathrm{g}(\mathcal{X}_n) + \mathcal{E}_n. \end{aligned} \tag{7.2}$$

By selection of the function g, we categorize these approaches into two groups, linear tensor regression models and the nonlinear tensor regression models. Figure 7.2 gives a taxonomy for popular tensor regression models, which will be discussed in details in the following parts of this chapter.

Besides, aiming at different data forms, the regression models can also be classified into three groups, i.e., vector-on-tensor regression, tensor-on-vector regression, and tensor-on-tensor regression. Specifically, the vector-on-tensor regression is dedicated to regress a vector-based response from a multiway predictor, like the disease diagnosis task based on multidimensional medical images or human age estimation based on several facial images.

7.3 Linear Tensor Regression Models

In the years of research on regression analysis, linear regression has occupied a very important position. Although linear regression is difficult to well approximate some complex systems, its simplicity and efficiency make it widely used in industry, especially when the exact accuracy is not of paramount importance. Therefore, the linear tensor regression models have become the primary research direction for addressing multidimensional data prediction and analysis tasks. It assumes a simple

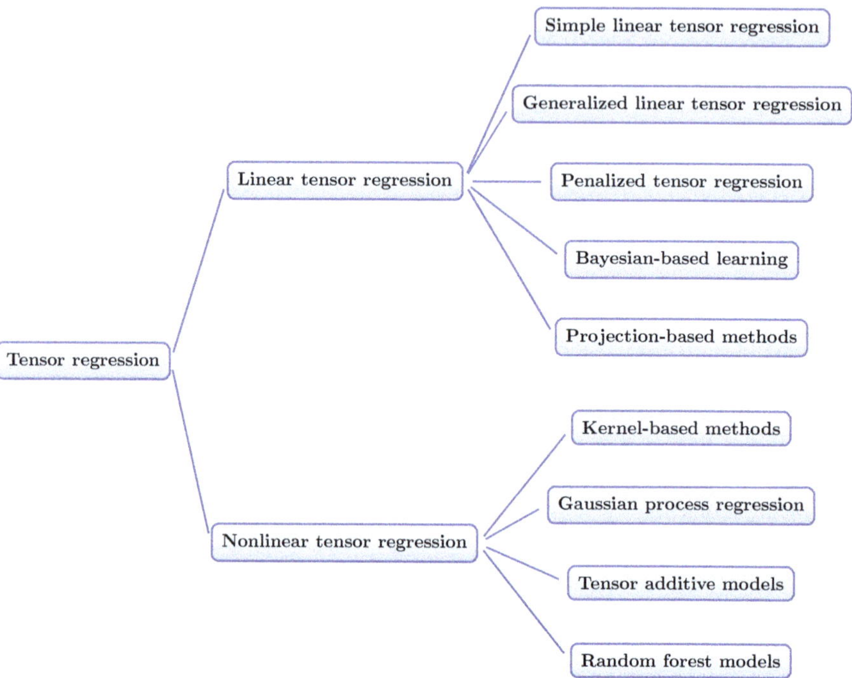

Fig. 7.2 A taxonomy of the tensor regression learning algorithms

linear relationship between the multiway predictor $\mathcal{X}_n$ and the multiway response $\mathcal{Y}_n$, as formulated by

$$\begin{aligned} \min_{\mathcal{B}} \quad & \mathrm{L}(\mathcal{Y}_n, << \mathcal{X}_n, \mathcal{B} >>) \\ \text{s. t.} \quad & \mathcal{Y}_n = << \mathcal{X}_n, \mathcal{B} >> + \mathcal{E}_n, \end{aligned} \tag{7.3}$$

where $\mathcal{B} \in \mathbb{R}^{P_1 \times \cdots \times P_L \times Q_1 \times \cdots \times Q_M}$ is the regression coefficient tensor which maps the predictor into the response by linear projections and $<< \mathcal{X}_n, \mathcal{B} >>$ is set to be a specific tensor product, such as tensor inner product and tensor contraction product. Model (7.3) can vary with different problems in terms of data forms, the adopted tensor products, assumptions over model parameters, and observation errors.

As shown in model (7.3), the model coefficient tensor is in much higher order compared with the tensor input and output. It brings a large amount of parameters to estimate during the training process, which inevitably leads to computational and storage burden. Moreover, it is easy for the number of model parameters to exceed the sample size, which brings a lot of numerical problems. Therefore, it is necessary to consider introducing assumptions over the model parameters or exploring the characteristics of the original data for improving the model identifiability and robustness, which is the key problem for linear regression.

Based on the correlation of the predictor or the response along different modes, a basic assumption over the high-dimensional regression coefficient tensor is low rankness. It can be taken as a generalization of the reduced rank regression, which solves the multivariate multiple outcome regression problems by

$$\min_{\mathbf{B}} \|\mathbf{y}_n - \mathbf{x}_n^{\mathrm{T}}\mathbf{B}\|_{\mathrm{F}}^2 \qquad (7.4)$$
$$\text{s. t.} \quad \text{rank}(\mathbf{B}) < R,$$

where $\mathbf{x}_n \in \mathbb{R}^P$, $\mathbf{y}_n \in \mathbb{R}^Q$, $\mathbf{B} \in \mathbb{R}^{P\times Q}$ is the coefficient matrix assumed to be low-rank and R gives an upper bound of the matrix rank. In fact, (7.4) can be rewritten as

$$\min_{\mathbf{B}} \| y_{n,q} - \mathbf{x}_n^{\mathrm{T}}\mathbf{b}_q \|_{\mathrm{F}}^2$$
$$\text{s. t.} \quad \text{rank}(\mathbf{B}) < R, \qquad (7.5)$$

where $y_{n,q}$ is q-th response in $\mathbf{y}_n$ and $\mathbf{b}_q$ is the q-th column of $\mathbf{B}$ which characterizes the relationship between the predictor $\mathbf{x}_n$ and the response $y_{n,q}$. The main reason for the low-rankness assumption over $\mathbf{B}$ here is to regress these outcomes $\{y_{n,q}\}_{q=1}^{Q}$ simultaneously and explore the correlations between these responses.

Extending it into tensor field, the optimization problem is transformed into

$$\min_{\mathcal{B}} \quad \mathrm{L}(\mathcal{Y}_n, << \mathcal{X}_n, \mathcal{B} >>) \qquad (7.6)$$
$$\text{s. t.} \quad \mathcal{Y}_n =<< \mathcal{X}_n, \mathcal{B} >> +\mathcal{E}_n$$
$$\text{rank}(\mathcal{B}) < R,$$

where R donates an upper bound of specific tensor rank. In fact, there are two main advantages of assuming the regression coefficient matrix or tensor to be low-rank rather than using the original least squares method. On the one hand, the low-rank assumption can be regarded as a regularization term, which improves the robustness of the model and reduces the model parameters. On the other hand, the low-rank assumption of the regression coefficients makes good use of the similarities between different inputs and outputs, which can play the role of data exploration during the training process.

7.3.1 *Simple Tensor Linear Regression*

Simple tensor linear regression model is the simplest regression model to illustrate the statistical relationship between two multidimensional variables. It directly solves the minimization problem in (7.6) when the observational error $\mathcal{E}$ is assumed to obey Gaussian distribution. Up to now, there are four different optimization methods used to tackle this tensor regression model, including rank minimization method, projected gradient descent, greedy low-rank learning, and alternating least squares method.

7.3.1.1 Rank Minimization Method

In [37], the low-rank tensor learning model is firstly developed for multitask learning, namely, multilinear multitask learning (MLMTL).

For a set of T linear regression tasks, M_t observations $\{\mathbf{x}_{\{m_t,t\}}, y_{\{m_t,t\}}\}$ are obtained for each task t, $t = 1, \ldots, T$, $m_t = 1, \ldots, M_t$. For multitask learning, the datasets may be defined by multiple indices, i.e., $T = T_1 \times \cdots \times T_N$. For example, to build a restaurant recommender system, we need to regress T_1 users' T_2 rating scores based on their D descriptive attributes of different restaurants. It can be regarded as a set of $T_1 \times T_2$ regression tasks. For each task t, one needs to regress t_1-th user's t_2-th type of scores $y_{\{m_t,t\}}$ based on the t_1-th user's descriptive attributes $\mathbf{x}_{\{m_t,t\}}$ which are in size of D, i.e.,

$$y_{\{m_t,t\}} =< \mathbf{x}_{\{m_t,t\}}, \mathbf{b}_t > +\varepsilon_{\{m_t,t\}},$$

$m_t = 1, \ldots, M_t$, $t = \overline{t_1, t_2}$.

Extending it into N different indices, the optimization problem for MLMTL is as follows:

$$\min_{\mathcal{B}} \sum_{t}^{T} \sum_{m_t}^{M_t} \mathrm{L}(y_{\{m_t,t\}}, < \mathbf{x}_{\{m_t,t\}}, \mathbf{b}_t >) + \lambda \sum_{n}^{N+1} \mathrm{rank}_n(\mathcal{B}), \tag{7.7}$$

where $\mathcal{B} \in \mathbb{R}^{D \times T_1 \times \cdots \times T_N}$ is the tensorization of the concatenation of the coefficient vector $\mathbf{b}_t \in \mathbb{R}^D$ as $[\mathbf{b}_1; \cdots; \mathbf{b}_T] \in \mathbb{R}^{D \times T}$ and $\mathrm{rank}_n(\mathcal{B})$ is the mode-n rank defined in Sect. 2.3 of Chap. 2.

The key challenge for rank minimization problem is how to find a good convex relaxation of specific tensor rank forms. As in [37], the overlapped trace norm is used to approximate the Tucker rank, which helps to transform the optimization problem into

$$\min_{\mathcal{B}, \mathcal{W}_n} \sum_{t}^{T} \sum_{m_t}^{M_t} \mathrm{L}(y_{\{m_t,t\}}, < \mathbf{x}_{\{m_t,t\}}, \mathbf{b}_t >) + \lambda \sum_{n=1}^{N+1} \|(\mathbf{W}_n)_{(n)}\|_{\mathrm{tr}} \tag{7.8}$$

$$\text{s.t.} \quad \mathcal{B} = \mathcal{W}_n, n = 1, \ldots, N+1,$$

where overlapped trace norm of tensor $\mathcal{B}$ is defined as

$$\|\mathcal{B}\|_{\mathrm{tr}} = \frac{1}{N+1} \sum_{n=1}^{N+1} \|\mathbf{B}_{(n)}\|_{\mathrm{tr}}, \tag{7.9}$$

where $\mathbf{B}_{(n)}$ is the mode-n unfolding matrix of $\mathcal{B}$, and where $(\mathbf{W}_n)_{(n)}$ is the mode-n unfolding matrix of $\mathcal{W}_n$.

Introducing a set of Lagrange multipliers $\mathcal{O}_n$, the resulting augmented Lagrangian function is given by

$$\mathrm{L}(\mathcal{B}, \mathcal{W}_n, \mathcal{O}_n) = \sum_t^T \sum_{m_t}^{M_t} \mathrm{L}(y_{\{m_t,t\}}, < \mathbf{x}_{\{m_t,t\}}, \mathbf{b}_t >) \quad (7.10)$$

$$+ \sum_{n=1}^{N+1} (\lambda \| (\mathbf{W}_n)_{(n)} \|_{\mathrm{tr}} - < \mathcal{O}_n, \mathcal{B} - \mathcal{W}_n > + \frac{\rho}{2} \| \mathcal{B} - \mathcal{W}_n \|_{\mathrm{F}}^2)$$

for some $\rho > 0$. The popular optimization algorithm ADMM can be employed to minimize the augmented Lagrangian function (7.10), and the detailed updating procedures are concluded in Algorithm 37.

Algorithm 37: Multilinear multitask learning

Input: Observations $\{\mathbf{x}_{\{m_t,t\}}, y_{\{m_t,t\}}\}$ for T tasks and λ, ρ
Output: the coefficient tensor $\mathcal{B}$
Initialize $\mathcal{B}, \mathcal{W}_n, \mathcal{O}_n$ for $n = 1, \ldots, N+1$
Iterate until convergence:
  Update $\mathcal{B}$:
    $\mathcal{B}^{k+1} = \mathrm{argmin}_{\mathcal{B}}\, \mathrm{L}(\mathcal{B}, \mathcal{W}_n^k, \mathcal{O}_n^k)$
  For $n = 1, \ldots, N+1$:
    Update $\mathcal{W}_n$:
      $\mathcal{W}_n^{k+1} = \mathrm{argmin}_{\mathcal{W}_n}\, \mathrm{L}(\mathcal{B}^{k+1}, \mathcal{W}_n, \mathcal{O}_n^k)$
    Update $\mathcal{O}_n$:
      $\mathcal{O}_n^{k+1} = \mathcal{O}_n^k - (\rho \mathcal{B}^{k+1} - \mathcal{W}_n^{k+1})$
  End for

Besides the overlapped trace norm, some other norms are also exploited for enhanced prediction performance, such as the latent trace norm [46] and the scaled latent trace norm [45].

7.3.1.2 Alternating Least Squares Method

Another group of approaches for low-rank learning is based on alternating minimization. The main concept is to replace the high-dimensional coefficient tensor with a set of low-order tensors or matrices and update these factors alternatively. Specifically, in [15], the bilinear regression model is extended into a more general form as follows:

$$\mathcal{Y} = \mathcal{X} \times_1 \mathbf{B}^{(1)} \cdots \times_M \mathbf{B}^{(M)} \times_{M+1} \mathbf{I}_N + \mathcal{E}, \quad (7.11)$$

where $\mathcal{Y} \in \mathbb{R}^{Q_1 \times \cdots \times Q_M \times N}$, $\mathcal{X} \in \mathbb{R}^{P_1 \times \cdots \times P_M \times N}$, $\mathbf{I}_N$ is an $N \times N$ diagonal matrix and $\mathbf{B}^{(m)} \in \mathbb{R}^{P_m \times Q_m}$, for $m = 1, \ldots, M$, is a coefficient matrix. But as shown in (7.11), there is a clear drawback for this model that the predictor and the response must be in the same order, which is not always the case in practice.

To get the model coefficients $\mathbf{B}^{(m)}$, $m = 1, \ldots, M$, the corresponding optimization can be given as follows:

$$\min_{\mathbf{B}^{(1)},\ldots,\mathbf{B}^{(M)}} \|\mathcal{Y} - \mathcal{X} \times_1 \mathbf{B}^{(1)} \cdots \times_M \mathbf{B}^{(M)} \times_{M+1} \mathbf{I}_N\|_{\mathrm{F}}^2. \tag{7.12}$$

Similar to the ALS-based algorithm for tensor decomposition, this problem can be solved by updating each of the model parameters $\mathbf{B}^{(1)}, \ldots, \mathbf{B}^{(M)}$ iteratively while others are fixed. Specifically, with respect to each factor $\mathbf{B}^{(m)}$, problem (7.12) boils down to the following subproblem:

$$\min_{\mathbf{B}^{(m)}} \|\mathbf{Y}_{(m)} - \mathbf{B}^{(m)} \tilde{\mathbf{X}}^{(m)}\|_{\mathrm{F}}^2, \tag{7.13}$$

which can be easily solved by the least squares method, where $\tilde{\mathbf{X}}^{(m)} = \mathbf{X}_{(m)}(\mathbf{B}^{(M)} \otimes \cdots \otimes \mathbf{B}^{(m+1)} \otimes \mathbf{B}^{(m-1)} \otimes \cdots \otimes \mathbf{B}^{(1)})^{\mathrm{T}}$. Algorithm 38 provides a summary of the updating procedures.

For ALS-based approaches, the main computational complexity comes from the construction of the matrix $\tilde{\mathbf{X}}^{(m)}$ and its inverse. In addition, for large-scale datasets, the storage of the intermediate variables is also a big challenge.

Algorithm 38: Multilinear tensor regression

Input: predictor $\mathcal{X}$, response $\mathcal{Y}$
Output: $\mathbf{B}^m$, $m = 1, \cdots, M$
Initialize $\mathbf{B}^m$, $m = 1, \cdots, M$
Iterate until convergence:
  For $m = 1, \cdots, M$:
    $\tilde{\mathbf{X}}^{(m)} = \mathbf{X}_{(m)}(\mathbf{B}^{(M)} \otimes \cdots \otimes \mathbf{B}^{(m+1)} \otimes \mathbf{B}^{(m-1)} \otimes \cdots \otimes \mathbf{B}^{(1)})^{\mathrm{T}}$
    $\mathbf{B}^{(m)} = \mathbf{Y}_{(m)}(\tilde{\mathbf{X}}^{(m)})^{\mathrm{T}}(\tilde{\mathbf{X}}^{(m)}(\tilde{\mathbf{X}}^{(m)})^{\mathrm{T}})^{-1}$
  End for

7.3.1.3 Greedy Low-Rank Learning

In [3], a unified framework is proposed for the cokriging and forecasting task in spatiotemporal data analysis into as follows:

$$\begin{aligned} \min_{\mathcal{B}} \quad & \mathrm{L}(\mathcal{B}; \mathcal{X}, \mathcal{Y}) \\ \text{s. t.} \quad & \mathrm{rank}(\mathcal{B}) \leq R, \end{aligned} \tag{7.14}$$

where $\mathcal{X} \in \mathbb{R}^{N \times I_1 \times I_3}$, $\mathcal{Y} \in \mathbb{R}^{N \times I_2 \times I_3}$, $\mathcal{B} \in \mathbb{R}^{I_1 \times I_2 \times I_3}$ is the coefficient tensor,

$$\mathrm{L}(\mathcal{B}; \mathcal{X}, \mathcal{Y}) = \sum_{i_3=1}^{I_3} \|\mathcal{Y}_{:,:,i_3} - \mathcal{X}_{:,:,i_3}\mathcal{B}_{:,:,i_3} + \mathcal{E}_{:,:,i_3}\|_{\mathrm{F}}^2, \tag{7.15}$$

and rank($\mathcal{B}$) is a sum of all the mode-n ranks of the tensor $\mathcal{B}$. Rather than using the rank minimization methods or ALS-based methods, [3] propose a greedy low n-rank tensor learning method which searches a best rank-1 estimation at a time until the stopping criteria are reached. At each iteration, the best rank-1 approximation is obtained by finding the best rank-1 approximation of all the unfolding matrices and selecting the mode which gives the maximum decrease of the objective function, as concluded in Algorithm 39. The operation refold means the inverse operation of the corresponding unfoldings, which satisfies refold$(\mathbf{B}_{(d)}, d) = \mathcal{B}$.

Algorithm 39: Greedy low-rank tensor learning

Input: predictor $\mathcal{X}$, response $\mathcal{Y}$
Output: the coefficient tensor $\mathcal{B}$
Initialize $\mathcal{B}$ as a zero tensor
Iterate until convergence:
  For $d = 1, \cdots, 3$:
    $\mathbf{W}_d \leftarrow \mathrm{argmin}_{\mathbf{W}_d : \mathrm{rank}(\mathbf{W}_d)=1} \mathrm{L}(\mathrm{refold}(\mathbf{B}_{(d)} + \mathbf{W}_d, d); \mathcal{X}, \mathcal{Y})$
    $\Delta_d \leftarrow \mathrm{L}(\mathcal{B}; \mathcal{X}, \mathcal{Y}) - \mathrm{L}\left(\mathrm{refold}\left(\mathbf{B}_{(d)} + \mathbf{W}_d, d\right); \mathcal{X}, \mathcal{Y}\right)$
  End for
  $d^* \leftarrow \mathrm{argmax}_d\, \Delta_d$
  if $\Delta_{d^*} \geq \eta$ **then**
    $\mathcal{B} \leftarrow \mathcal{B} + \mathrm{refold}(\mathbf{W}_{d^*}, d^*)$
  end if

7.3.1.4 Projected Gradient Descent Method

For the rank minimization methods and the ALS-based methods, they all suffer from high storage and computational cost for large-scale high-dimensional datasets. In addition, rank minimization methods usually converge slowly, while ALS-based methods often lead to suboptimal solutions. Therefore, an efficient algorithm based on projected gradient descent is proposed for spatiotemporal data analysis and multitask learning in [50], namely, tensor projected gradient (TPG).

Given the predictor $\mathcal{X} \in \mathbb{R}^{N \times I_1 \times I_3}$ and its corresponding response $\mathcal{Y} \in \mathbb{R}^{N \times I_2 \times I_3}$, the target problem is as follows:

$$\begin{aligned} \min_{\mathcal{B}} \quad & \mathrm{L}(\mathcal{B}; \mathcal{X}, \mathcal{Y}) \\ \text{s. t.} \quad & \mathrm{rank}(\mathcal{B}) \leq R, \end{aligned} \tag{7.16}$$

where $\mathcal{B} \in \mathbb{R}^{I_1 \times I_2 \times I_3}$ is the coefficient tensor and $\mathrm{L}(\mathcal{B}; \mathcal{X}, \mathcal{Y})$ is the fitting loss as it does in (7.15).

Treating problem (7.16) as an unconstrained optimization problem with only the loss term (7.15), TPG updates the coefficient tensor along the gradient descent direction, as shown in Algorithm 40. Then a projection procedure for rank constraint is performed over the estimation. Specifically, the projection of the estimated coefficient tensor can be formulated as follows:

$$\begin{aligned} &\min_{\mathcal{B}} \|\mathcal{B}^k - \mathcal{B}\|_{\mathrm{F}}^2 \\ &\mathcal{B} \in \{\mathcal{B} : \mathrm{rank}(\mathcal{B}) \leq R\}, \end{aligned} \tag{7.17}$$

where $\mathcal{B}^k$ is the update of the coefficient tensor in k-th iteration. Straightforward solutions for this problem can be obtained by the decomposition algorithms defined in Chap. 2. But it suffers from high computational complexity due to the requirement of full SVD over the unfolding matrices from the coefficient tensor. As shown in Algorithm 41, the tensor power iteration method can be employed to find the leading singular vectors at a time, which avoids performing SVD over large matrix and is especially efficient for low-rank coefficient tensors.

Algorithm 40: Tensor projected gradient method

Input: predictor $\mathcal{X}$, response $\mathcal{Y}$, rank R
Output: the coefficient tensor $\mathcal{B}$
Initialize $\mathcal{B}$ as zero tensor
Iterate until convergence:
$\bar{\mathcal{B}}^{k+1} = \mathcal{B}^k - \eta \nabla \mathrm{L}(\mathcal{B}^k; \mathcal{X}, \mathcal{Y})$
$\mathcal{B}^{k+1} = \mathrm{ITP}(\bar{\mathcal{B}}^{k+1})$

7.3.2 *Generalized Tensor Linear Regression*

As an extension of the generalized linear regression models (GLM), generalized tensor linear regression still aims at single response regression tasks but with tensor predictors. In GLM, the response y_n is commonly assumed to obey an exponential family distribution

$$\mathrm{Prob}(y_n | \mathcal{X}_n, \mathbf{z}_n) = \exp\left(\frac{y_n \theta_n - \mathrm{b}(\theta_n)}{\mathrm{a}(\phi_n)} + \mathrm{c}(y_n, \phi_n)\right), \tag{7.18}$$

where θ_n and ϕ_n are natural and dispersion parameters, $\mathrm{a}(\cdot)$, $\mathrm{b}(\cdot)$, and $\mathrm{c}(\cdot)$, are various functions according to different distributions, such as Poisson distribution and gamma distribution. Unlike simple linear regression which maps the predictor directly with the response, GLM link the linear model and the response by a link

Algorithm 41: Iterative tensor projection (ITP)

Input: coefficient tensor $\bar{\mathcal{B}}$, input $\mathcal{X}$, output $\mathcal{Y}$, rank R, stopping criteria parameter ϵ
Output: Projected coefficient tensor $\mathcal{B}$
Initialize $\mathbf{U}_n$ with R left singular vectors of $\bar{\mathbf{B}}_{(n)}$
while $r \leq R$ **do**
  repeat
  $\mathbf{u}_1^{k+1} = \frac{\bar{\mathcal{B}}\times_2 \mathbf{u}_2^{k\mathrm{T}} \times_3 \mathbf{u}_3^{k\mathrm{T}}}{\|\bar{\mathcal{B}}\times_2 \mathbf{u}_2^{k\mathrm{T}} \times_3 \mathbf{u}_3^{k\mathrm{T}}\|_2}$
  $\mathbf{u}_2^{k+1} = \frac{\bar{\mathcal{B}}\times_1 \mathbf{u}_1^{k\mathrm{T}} \times_3 \mathbf{u}_3^{k\mathrm{T}}}{\|\bar{\mathcal{B}}\times_1 \mathbf{u}_1^{k\mathrm{T}} \times_3 \mathbf{u}_3^{k\mathrm{T}}\|_2}$
  $\mathbf{u}_3^{k+1} = \frac{\bar{\mathcal{B}}\times_1 \mathbf{u}_1^{k\mathrm{T}} \times_2 \mathbf{u}_2^{k\mathrm{T}}}{\|\bar{\mathcal{B}}\times_1 \mathbf{u}_1^{k\mathrm{T}} \times_2 \mathbf{u}_2^{k\mathrm{T}}\|_2}$
  until convergence to $\{\mathbf{u}_1, \mathbf{u}_2, \mathbf{u}_3\}$
  $\mathbf{U}_n(:, r) = \mathbf{u}_n$ for $n = 1, 2, 3$
  $\mathcal{B} \leftarrow \bar{\mathcal{B}} \times_1 \mathbf{U}_1\mathbf{U}_1^{\mathrm{T}} \times_2 \mathbf{U}_2\mathbf{U}_2^{\mathrm{T}} \times_3 \mathbf{U}_3\mathbf{U}_3^{\mathrm{T}}$
  if $\mathrm{L}(\mathcal{B}; \mathcal{X}, \mathcal{Y}) \leq \epsilon$ **then**
   RETURN
  end if
end while

function. As in [55], given N observations $\{\mathcal{X}_n, \mathbf{z}_n, y_n\}$, the mean value μ_n of the distribution (7.18) is assumed to satisfy the following relationship:

$$
\begin{aligned}
\mathrm{g}(\mu_n) &= \alpha_n + \boldsymbol{\gamma}^{\mathrm{T}}\mathbf{z}_n + \langle \mathcal{B}, \mathcal{X}_n \rangle \\
&= \alpha_n + \boldsymbol{\gamma}^{\mathrm{T}}\mathbf{z}_n + \left\langle \sum_{r=1}^{R} \mathbf{u}_r^{(1)} \circ \cdots \circ \mathbf{u}_r^{(L)}, \mathcal{X}_n \right\rangle \qquad (7.19)\\
&= \alpha_n + \boldsymbol{\gamma}^{\mathrm{T}}\mathbf{z}_n + \left\langle \left(\mathbf{U}^{(L)} \odot \cdots \odot \mathbf{U}^{(1)}\right)\mathbf{1}_R, \mathrm{vec}(\mathcal{X}_n) \right\rangle, \qquad (7.20)
\end{aligned}
$$

where $\mathcal{X}_n \in \mathbb{R}^{P_1 \times \cdots \times P_L}$ is the tensor input, $\mathbf{z}_n$ is the vector input, α_n is the intercept, γ is the vector coefficients, and $\mathcal{B}$ is the coefficient tensor which represents the impact of the tensor input $\mathcal{X}_n$ over the response y_n. $\mathrm{g}(\mu_n)$ indicates a link function of the mean value $\mu_n = \mathrm{E}(y_n|\mathcal{X}_n, \mathbf{z}_n)$. For example, if the response y_n obeys the binomial distribution, then the link function should be the logit function in order to map the binary output into a continuous variable. For data exploration and dimension reduction, the CP decomposition form is employed, where $\mathbf{U}^{(l)} = [\mathbf{u}_1^{(l)}; \cdots ; \mathbf{u}_R^{(l)}] \in \mathbb{R}^{P_l \times R}$. By using the CP factorization form, the number of model parameters will be decreased from $\mathrm{O}(P_1 \cdots P_L)$ to $\mathrm{O}(\sum_{l=1}^{L} P_l R)$.

The model parameters can be estimated by maximizing the log likelihood function as follows:

$$
\mathrm{L}\left(\alpha_1, \ldots, \alpha_N, \gamma, \mathbf{U}^{(1)}, \ldots, \mathbf{U}^{(L)}\right) = \sum_{n=1}^{N} \log \mathrm{Prob}(y_n|\mathcal{X}_n, \mathbf{z}_n).
$$

And based on alternating minimization, subproblem with respect to each factor $\mathbf{U}^{(l)}$ is a simple GLM problem. The updating procedure is summarized in Algorithm 42.

Algorithm 42: Block relaxation algorithm for CP-based generalized linear tensor regression models

Input: predictor $\mathcal{X}_n$ and $\mathbf{z}_n$, response $\mathcal{Y}_n$ for $n = 1, \cdots, N$, rank R
Output: the coefficient tensor $\mathcal{B}$
Initialize α_n, γ and $\mathbf{U}^{(l)}$ for $n = 1, \cdots, N, l = 1, \cdots, L$
Iterate until convergence:
  for $l = 1, \cdots, L$ **do**
    $(\mathbf{U}^{(l)})^{k+1} = \arg\max_{\mathbf{U}^{(l)}} \mathrm{L}(\alpha_1^k, \cdots, \alpha_N^k, \gamma^k, (\mathbf{U}^{(1)})^{k+1}, \cdots, (\mathbf{U}^{(l-1)})^{k+1}, \mathbf{U}^{(l)}, (\mathbf{U}^{(l+1)})^k, \cdots, (\mathbf{U}^{(L)})^k)$
  end for
  $(\alpha_1^{k+1}, \cdots, \alpha_N^{k+1}, \gamma^{k+1}) = \arg\max_{\alpha_1, \cdots, \alpha_N, \gamma} \mathrm{L}(\alpha_1, \cdots, \alpha_N, \gamma, (\mathbf{U}^{(1)})^{k+1}, \cdots, (\mathbf{U}^{(L)})^{k+1})$
$\mathcal{B} = [\![\mathbf{U}^{(1)}, \cdots, \mathbf{U}^{(L)}]\!]$

Moreover, the popular Tucker decomposition and HT decomposition methods are also exploited for generalized linear tensor regression models [16, 25]. Compared with CP, those methods can explore the multilinear rank and thus be able to express the regression system with less parameters. Besides, rather than replacing the CP decomposition with other flexible form, [34] develop a soft version of the CP factorization which allows row-specific contributions. The experiments show that the soft CP improves the model flexibility of CP decomposition and provides enhanced prediction performance in brain connectomics analysis.

7.3.3 *Penalized Tensor Regression*

As described before, for most regression tasks with high-dimensional datasets, it is likely that the number of required model parameters is much larger than the sample size. Especially in neuroimaging analysis, the sample size is commonly small because of the difficulty of medical image collection and the existence of data islands. Therefore, it is essential to exploit the prior knowledge within the datasets for improving the model identifiability and robustness, such as the ridge regression [12, 32, 35], sparse regression [1, 24, 26, 27, 36, 40, 42, 46], and other regularization terms like smoothness [48], nonnegativity, or orthogonality [21].

7.3.3.1 Tensor Ridge Regression

Tensor ridge regression adds a Frobenius norm constraint over the regression coefficient tensor for controlling the model complexity [12, 31, 32, 35]. The corresponding optimization can be written as

$$\begin{aligned} \min_{\mathcal{B}} \quad & \mathrm{L}(\mathcal{B}; \mathcal{X}, \mathcal{Y}) + \lambda \|\mathcal{B}\|_{\mathrm{F}}^2 \\ \text{s. t.} \quad & \operatorname{rank}(\mathcal{B}) \leq R, \end{aligned} \tag{7.21}$$

where the form of the loss function $\mathrm{L}(\mathcal{B}; \mathcal{X}, \mathcal{Y})$ varies with different developed models, as the vector-on-tensor regression model $y_n = < \mathcal{X}_n, \mathcal{B} > + \epsilon_n$ for N observations $\{\mathcal{X}_n \in \mathbb{R}^{P_1 \times \cdots \times P_L}, y_n \in \mathbb{R}\}$ in [12]; the tensor-on-vector regression model $\mathcal{Y} = \mathcal{B} \times_1 \mathbf{X} + \mathcal{E}$ for $\{\mathbf{X} \in \mathbb{R}^{N \times D}, \mathcal{Y} \in \mathbb{R}^{N \times Q_1 \times \cdots \times Q_M}\}$ in [35]; and the tensor-on-tensor regression model $\mathcal{Y} = < \mathcal{X}, \mathcal{B} >_L + \mathcal{E}$ for $\{\mathcal{X} \in \mathbb{R}^{N \times P_1 \times \cdots \times P_L}, \mathcal{Y} \in \mathbb{R}^{N \times Q_1 \times \cdots \times Q_M}\}$ in [30–32].

To tackle problem (7.21), one way is to incorporate the regularization term into the data fitting term as follows:

$$\begin{aligned} \min_{\mathcal{B}} \quad & \mathrm{L}(\mathcal{B}; \breve{\mathcal{X}}, \breve{\mathcal{Y}}) \\ \text{s. t.} \quad & \operatorname{rank}(\mathcal{B}) \leq R, \end{aligned} \tag{7.22}$$

where $\breve{\mathbf{X}}_{(1)} = [\mathbf{X}_{(1)}, \lambda \mathbf{I}]^{\mathrm{T}}$, $\breve{\mathbf{Y}}_{(1)} = [\mathbf{Y}_{(1)}, \lambda \mathbf{0}]^{\mathrm{T}}$. Then the algorithms mentioned before can be directly used for addressing this unregularized optimization problem.

Besides imposing the Frobenius norm constraint over the original coefficient tensor $\mathcal{B}$, it is also reasonable to enforce the latent factors under specific factorization forms like CP decomposition as follows:

$$\Omega(\mathcal{B}) = \sum_{m=1}^{M} \|\mathbf{U}^{(m)}\|_{\mathrm{F}}^2, \tag{7.23}$$

where $\mathcal{B}$ is assumed to be of low rank in CP factorization form $\mathcal{B} = [\![\mathbf{U}^{(1)}, \ldots, \mathbf{U}^{(M)}]\!]$.

7.3.3.2 Tensor Sparse Regression

Mainly motivated by the neuroimaging analysis, tensor sparse regression assumes the relatedness between the multiway neuroimages and the medical assessment outcomes is sparse. In other words, with respect to specific behavior or diseases, there is only part of the brain neurons important in predicting the outcomes.

To this end, a natural method is to introduce the element-wise sparsity constraint over the coefficient tensor as follows:

$$\begin{aligned} \min_{\mathcal{B}} \quad & \mathrm{L}(\mathcal{B}; \mathcal{X}, \mathcal{Y}) + \lambda \|\mathcal{B}\|_1 \\ \text{s. t.} \quad & \operatorname{rank}(\mathcal{B}) \leq R, \end{aligned} \tag{7.24}$$

which has been studied with the vector-on-tensor regression model $y_n =< \mathcal{X}_n, \mathcal{B} > +\epsilon_n$ for N observations $\{\mathcal{X}_n \in \mathbb{R}^{P_1 \times \cdots \times P_L}, y_n \in \mathbb{R}\}$ in [1, 27, 40].

To tackle problem (7.24), three different ways can be used, where two are based on the previously mentioned rank minimization method and projected gradient descent method, respectively.

Rank Minimization Approaches Specifically, the optimization problem proposed in [40] is given by

$$\min_{\mathcal{B}} \quad \mathrm{L}(\mathcal{B}; \mathcal{X}, \mathcal{Y}) + \tau \|\mathcal{B}\|_* + \lambda \|\mathcal{B}\|_1, \tag{7.25}$$

which uses the convex tensor nuclear norm to replace the rank constraint, as defined in the following equation:

$$\|\mathcal{B}\|_* = \frac{1}{D} \sum_{d=1}^{D} \|\mathbf{B}_{(d)}\|_*, \tag{7.26}$$

where D is the order of tensor $\mathcal{B}$.

Introducing necessary auxiliary variables, problem (7.25) can be solved by the ADMM. The subproblem with the tensor nuclear norm and ℓ_1 norm can be tackled by the proximity operators of the nuclear norm and ℓ_1 norm, respectively. In addition, a variant based on tubal rank regularization, called sparse tubal-regularized multilinear regression, is proposed in [27] and solved by the convex relaxation in the same way.

Projected Gradient Descent Approach Aiming at the vector-on-tensor regression model, projected gradient descent approach is used to solve the following problem:

$$\min_{\mathcal{B} \in \mathbb{C}} \frac{1}{2} \sum_{n=1}^{N} \|y_n - < \mathcal{B}, \mathcal{X}_n > \|_{\mathrm{F}}^2, \tag{7.27}$$

where $\mathbb{C}$ is a set of R-rank and S-sparse tensors as follows

$$\mathbb{C} = \{\mathcal{G} \times_1 \mathbf{U}^{(1)} \times \cdots \times_L \mathbf{U}^{(L)} : \mathcal{G} \in \mathbb{R}^{R_1 \times \cdots \times R_L}, \mathbf{U}^{(l)} \in \mathbb{R}^{P_l \times R_l}, \tag{7.28}$$

$$\|\mathbf{U}^{(l)}(:, r_l)\|_0 \leq S_l, l = 1, \ldots, L, r_l = 1, \ldots, R_l\},$$

where S_l controls the sparsity level of $\mathbf{U}^{(l)}$.

Different from the TPG method in [50], which only considers the low rankness of the coefficient tensor, [1] projects the estimation of $\mathcal{B}$ in each iteration to be a R-rank and S-sparse tensor by the sparse higher-order SVD [2]. In fact, for projected gradient descent methods, the projection procedure is essentially a representation problem. In that sense, the introduction of the regularization term over the regression coefficient tensor just changes the projection procedure into a constrained low-rank approximation problem. And existing tensor representation methods can be directly used for projecting the estimation into the desired subspace.

Greedy Low-Rank Tensor Learning In [14], a fast unit-rank tensor factorization method is proposed, which tries to sequentially find unit-rank tensor for the following problem:

$$\begin{aligned} \min_{\mathcal{B}_r} \frac{1}{N} \sum_{n=1}^{N} \left(y_{n,r} - \langle \mathcal{X}_n, \mathcal{B}_r \rangle\right)^2 + \lambda_r \|\mathcal{B}_r\|_1 + \alpha \|\mathcal{B}_r\|_{\mathrm{F}}^2, \\ \text{s. t.} \quad \operatorname{rank}(\mathcal{B}_r) \leq 1 \end{aligned} \tag{7.29}$$

where r is the sequential number of the unit-rank tensors and $y_{n,r}$ is the remaining residuals after $r-1$ approximations. Until the stopping condition is reached, the full estimation is given by $\mathcal{B} = \sum_r \mathcal{B}_r$.

The key problem for [14] becomes the unit-rank tensor factorization problem, which can be divided into several regression subproblems with elastic net penalty terms by the alternating search approach. And each subproblem can be solved using the stagewise regression algorithms.

Other Variants Besides, instead of simply enforcing the whole coefficient tensor to be sparse, the sparsity constraint can also be imposed over the latent factors. For example, the adopted regression model in [42] maps the vector input $\mathbf{x}_n$ with the multidimensional response $\mathcal{Y}_n$ by $\mathcal{Y}_n = \mathcal{B} \times_{M+1} \mathbf{x}_n + \epsilon_n$. Imposing CP decomposition over $\mathcal{B}$, it can be rewritten as

$$\mathcal{Y}_n = \sum_r w_r \mathbf{u}_r^{(1)} \circ \cdots \circ \mathbf{u}_r^{(M)} \circ \mathbf{u}_r^{(M+1)} \times_{M+1} \mathbf{x}_n + \epsilon_n. \tag{7.30}$$

Considering the sparsity constraint over the latent factors, we can get the optimization problem as

$$\begin{aligned} \min_{w_r, \mathbf{u}_r^{(1)}, \ldots, \mathbf{u}_r^{(M+1)}} \frac{1}{N} \sum_{n=1}^{N} \|\mathcal{Y}_n - \sum_r w_r \left(\left(\mathbf{u}_r^{(M+1)}\right)^{\mathrm{T}} \mathbf{x}_n\right) \mathbf{u}_r^{(1)} \circ \cdots \circ \mathbf{u}_r^{(M)} \|_{\mathrm{F}}^2 \\ \text{s. t.} \quad \|\mathbf{u}_r^{(m)}\|_2 = 1, \quad \|\mathbf{u}_r^{(m)}\|_0 \leq S_m. \end{aligned} \tag{7.31}$$

In this way, based on alternating minimization, the subproblem with respect to each latent factor can be solved by efficient sparse tensor decomposition methods.

However, the element-wise sparsity cannot utilize the structural information of the data. As for multiresponse regression model $y_{n,q} =< \mathcal{X}_n, \mathcal{B}_q > +\epsilon_n$ for N observations $\{\mathcal{X}_n \in \mathbb{R}^{P_1 \times \cdots \times P_L}, y_{n,q} \in \mathbb{R}\}$, simply imposing the coefficient tensor to be sparse means to fit each response with the sparse tensor regression model individually. It is not in line with the actual situation that the important features for potentially correlated response are similar. Therefore, when $\mathcal{B}_q$ admits CP decomposition as $\mathcal{B}_q = [\![\mathcal{U}_q^{(1)}, \ldots, \mathcal{U}_q^{(L)}]\!]$ for $q = 1, \ldots, Q$, [26] proposed a group lasso penalty as follows:

$$\Omega(\mathcal{B}_1, \ldots, \mathcal{B}_Q) = \sum_{l=1}^{L} \sum_{r=1}^{R} \sum_{p_l=1}^{P_l} \left(\sum_{q=1}^{Q} \mathbf{U}_q^{(l)}(p_l, r) \right)^{1/2}, \tag{7.32}$$

which regards the same position of $\{\mathcal{B}_1, \ldots, \mathcal{B}_Q\}$ as a group and imposes sparsity constraint over these groups. In this way, if a subregion is not related with any response, that region will be dropped out from the model.

Besides, group sparsity constraints can be used for rank estimation. It has been proved in [10, 51] that the rank minimization problem can be regarded as a decomposition with sparse constraints. And derived by this consideration, [12] have developed a group sparsity term to estimate the CP rank during the training process as follows:

$$\Omega(\mathcal{B}) = \sum_{r=1}^{R} \left(\sum_{l=1}^{L} \|\mathbf{U}^{(l)}(:, r)\|_2^2 \right)^{1/2}, \tag{7.33}$$

where $\mathcal{B} = \sum_{r=1}^{R} \mathbf{U}^{(1)}(:, r) \circ \cdots \circ \mathbf{U}^{(L)}(:, r)$. Since $\mathcal{B}$ is the sum of R rank-1 tensors, treating each rank-1 tensor as a group and enforcing these components to be sparse can make unnecessary component to be jointly zero and thus reduces the CP rank. In addition, a variant which aims to estimate the appropriate tensor ring rank during the training process is proposed [29].

7.3.4 Bayesian Approaches for Tensor Linear Regression

The key to Bayesian learning is the utilization of prior knowledge. The parameters in Bayesian learning can be tuned according to the data during the training process. Unlike the penalized regression, the weighting factor of the penalty term is directly specified. In other words, penalized regression introduces the prior information based on experience but not the data, and the weighting factor needs to be selected by a large number of cross-validation experiments. Bayesian regression considers all the possibilities of the parameters and tunes them in the estimation process according to the training data. In addition, Bayesian regression can give a probability distribution of predicted values, not just a point estimate.

For tensor linear regression, the key challenge for Bayesian learning is the design of multiway shrinkage prior for the tensor coefficients. For example, in [11], a novel class of multiway shrinkage priors for the coefficient tensor, called multiway Dirichlet generalized double Pareto (M-DGDP) prior, is proposed for the model

$y_n = \alpha + \mathbf{z}_n^{\mathrm{T}} \gamma + \langle \mathcal{X}_n, \mathcal{B} \rangle + \varepsilon_n$. The Bayesian model can be formulated as follows:

$$\begin{aligned}
&y_n | \gamma, \mathcal{B}, \sigma \sim \mathrm{N}\left(\mathbf{z}_n^{\mathrm{T}} \boldsymbol{\gamma} + \langle \mathcal{X}_n, \mathcal{B} \rangle, \sigma^2\right) \\
&\mathcal{B} = \sum_{r=1}^{R} \mathcal{U}_r, \mathcal{U}_r = \mathbf{u}_r^{(1)} \circ \cdots \circ \mathbf{u}_r^{(L)} \\
&\sigma^2 \sim \pi_\sigma, \ \gamma \sim \pi_\gamma, \ \mathbf{u}_r^{(l)} \sim \pi_{\mathbf{U}}
\end{aligned} \tag{7.34}$$

where the shrinkage prior $\pi_{\mathbf{U}}$ can be represented as

$$\mathbf{u}_r^{(l)} \sim \mathrm{N}\left(0, (\phi_r \tau) \mathbf{W}_{lr}\right), \tag{7.35}$$

which can be illustrated by the hierarchical graph in Fig. 7.3. As shown in Fig. 7.3, the distribution of the latent factor relays on the distribution of τ, $\mathbf{\Phi}$, and $\mathbf{W}_{lr}$. And these three factors play different roles, where $\tau \sim \mathrm{Ga}(a_\tau, b_\tau)$ is a global scale parameter which fits in each component by $\tau_r = \phi_r \tau$ for $r = 1, \ldots, R$, $\mathbf{\Phi} = [\phi_1, \ldots, \phi_R] \sim \mathrm{Dirichlet}(\alpha_1, \ldots, \alpha_R)$ encourages the latent factors to be low-rank under the tensor decomposition, and $\mathbf{W}_{lr} = \mathrm{diag}(w_{lr,1}, \ldots, w_{lr,P_l})$, $l = 1, \ldots, L$ are margin-specific scale parameters which encourage shrinkage at local scale, where $w_{lr,p_l} \sim \mathrm{Exp}\left(\lambda_{lr}^2/2\right)$, $\lambda_{lr} \sim \mathrm{Ga}\left(a_\lambda, b_\lambda\right)$ for $p_l = 1, \ldots, P_L$, $r = 1, \ldots, R, l = 1, \ldots, L$.

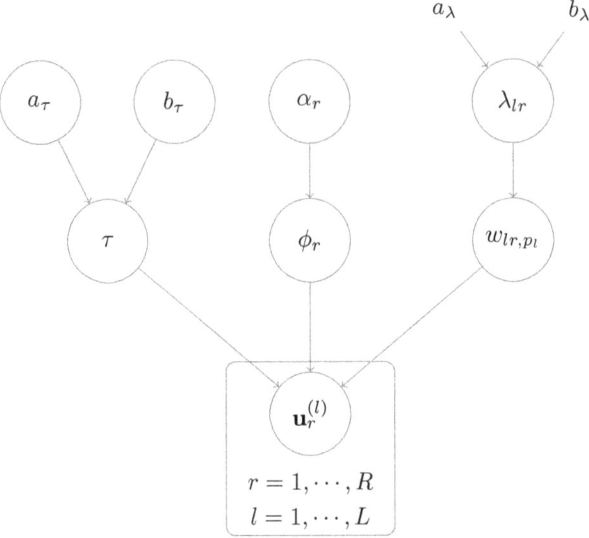

Fig. 7.3 The diagram of the hierarchical shrinkage prior for Bayesian dynamic tensor regression [4]

With the similar shrinkage prior, the Bayesian learning approaches are extended to tensor autoregression models for analyzing time-varying networks in economics [4]. And an application to brain activation and connectivity analysis can refer to [41].

Since the computation of the posterior distribution is complex, the posterior distribution is usually calculated using an approximate method. Approximate methods are mainly divided into two categories: the first class simplifies the posterior distribution, such as variational inference, or Laplace approximation, and the second group uses sampling methods, such as Gibbs sampling and HMC sampling. Most Bayesian learning approaches for tensor linear regression employ the Gibbs sampling for computing the posterior distribution of the coefficient tensor. Although the sampling methods can achieve high accuracy, it is too simple and costs a lot of time for training. Although the derivation process of variational inference is more difficult, the speed of its calculation makes its promotion very meaningful.

7.3.5 Projection-Based Tensor Regression

As an important role in multivariate statistical data analysis, partial least squares (PLS) method is shown to be effective to address collinear datasets and conduct regression analysis with small sample size or identify system information from even nonrandom noise. With these advantages, PLS is extended into tensor field under different assumptions, such as multilinear PLS [5], higher-order PLS (HOPLS) [53], and tensor envelope PLS (TEPLS) [52].

7.3.5.1 Partial Least Squares Regression

The main idea of PLS is to first project the response and the predictor into vector space and then find the relationship between the latent factors of them. Specifically, given input and output matrices $\mathbf{X} \in \mathbb{R}^{N\times I}$ and $\mathbf{Y} \in \mathbb{R}^{N\times J}$, we assume there exists a common latent factor in their matrix factorization forms as follows:

$$\mathbf{X} = \mathbf{T}\mathbf{U}^{\mathrm{T}} + \mathbf{E} = \sum_{r=1}^{R} \mathbf{t}_r \mathbf{u}_r^{\mathrm{T}} + \mathbf{E}$$

$$\mathbf{Y} = \mathbf{T}\mathbf{D}\mathbf{V}^{\mathrm{T}} + \mathbf{F} = \sum_{r=1}^{R} d_r \mathbf{t}_r \mathbf{v}_r^{\mathrm{T}} + \mathbf{F}, \tag{7.36}$$

where $\mathbf{T} \in \mathbb{R}^{N\times R}$ is the shared common latent factor, $\mathbf{U} \in \mathbb{R}^{I\times R}$ and $\mathbf{V} \in \mathbb{R}^{J\times R}$ are the loading matrices of $\mathbf{X}$ and $\mathbf{Y}$, $\mathbf{D} = \mathrm{diag}(d_1, \ldots, d_R)$ is a diagonal matrix, and $\mathbf{E} \in \mathbb{R}^{N\times I}$ and $\mathbf{F} \in \mathbb{R}^{N\times J}$ are residuals. Figure 7.4 provides a graphical illustration for the PLS.

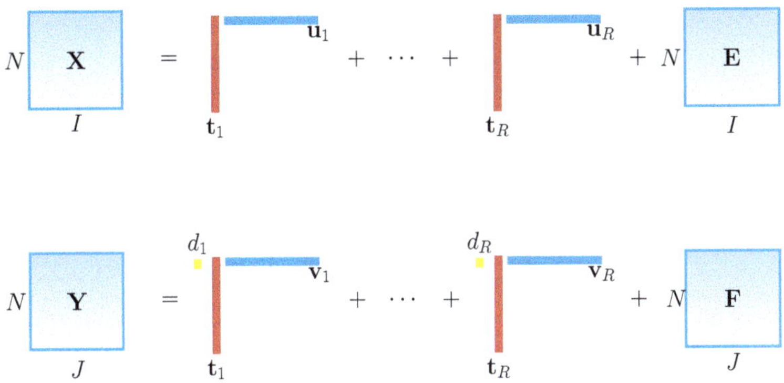

Fig. 7.4 The diagram of PLS

The solution for PLS regression is commonly obtained based on the greedy learning. It is to say at each iteration, PLS only searches one set of latent factors which corresponds to one desired rank-1 component. The corresponding problem at r-th iteration is formulated as follows:

$$\max_{\mathbf{u}_r, \mathbf{v}_r} \mathbf{u}_r^{\mathrm{T}} \mathbf{X}_r^{\mathrm{T}} \mathbf{Y}_r \mathbf{v}_r \tag{7.37}$$

$$\text{s.t. } \mathbf{u}_r^{\mathrm{T}} \mathbf{u}_r = 1, \mathbf{v}_r^{\mathrm{T}} \mathbf{v}_r = 1, \tag{7.38}$$

where $\mathbf{t}_r = \mathbf{X}_r \mathbf{u}_r$, $\mathbf{c}_r = \mathbf{Y}_r \mathbf{v}_r$, $d_r = \mathbf{c}_r^{\mathrm{T}} \mathbf{t}_r / \mathbf{t}_r^{\mathrm{T}} \mathbf{t}_r$, $\mathbf{X}_r = \mathbf{X} - \sum_{k=1}^{r-1} \mathbf{t}_k \mathbf{u}_k^{\mathrm{T}}$, and $\mathbf{Y}_r = \mathbf{Y} - \sum_{k=1}^{r-1} d_k \mathbf{t}_k \mathbf{v}_k^{\mathrm{T}}$ are the remaining parts after previous $r-1$ approximations. After getting all the latent factors, we can provide the prediction for new predictor $\mathbf{X}'$ as

$$\mathbf{Y}' = \mathbf{T}' \mathbf{D} \mathbf{V}^{\mathrm{T}} \tag{7.39}$$

with $\mathbf{T}' = \mathbf{X}' \mathbf{U}$.

7.3.5.2 Tensor Partial Least Squares Regression

For example, based on CP decomposition, multilinear PLS first decomposes the multiway predictor, and response into its CP factorization forms as follows:

$$\begin{aligned} \mathcal{X} &= \sum_{r=1}^{R} \mathbf{t}_r \circ \mathbf{u}_r^{(1)} \circ \mathbf{u}_r^{(2)} + \mathcal{E} \\ \mathcal{Y} &= \sum_{r=1}^{R} d_r \mathbf{t}_r \circ \mathbf{v}_r^{(1)} \circ \mathbf{v}_r^{(2)} + \mathcal{F} \end{aligned} \tag{7.40}$$

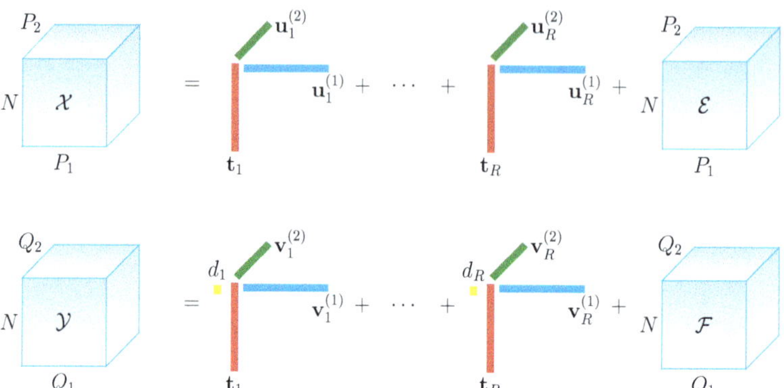

Fig. 7.5 The diagram of multilinear PLS

where $\mathbf{t}_r, r = 1, \ldots, R$ are the common latent vectors of $\mathcal{X}$ and $\mathcal{Y}$, $\mathbf{u}_r^{(1)}$ and $\mathbf{u}_r^{(2)}$ are loading vectors for $\mathcal{X}$, while $\mathbf{v}_r^{(1)}$ and $\mathbf{v}_r^{(2)}$ are loading vectors of $\mathcal{Y}$, $r = 1, \ldots, R$, as shown in Fig. 7.5.

Then multilinear PLS obtains the model parameters by sequentially maximizing the covariance of the r-th common latent factor as follows:

$$\left\{\mathbf{u}_r^{(1)}, \mathbf{u}_r^{(2)}, \mathbf{v}_r^{(1)}, \mathbf{v}_r^{(2)}\right\} = \arg \max_{\mathbf{u}_r^{(1)}, \mathbf{u}_r^{(2)}, \mathbf{v}_r^{(1)}, \mathbf{v}_r^{(2)}} \operatorname{Cov}(\mathbf{t}_r, \mathbf{c}_r) \tag{7.41}$$

$$\text{s.t.} \quad \mathbf{t}_r = \mathcal{X}_r \times_2 \mathbf{u}_r^{(1)} \times_3 \mathbf{u}_r^{(2)}, \quad \mathbf{c}_r = \mathcal{Y}_r \times_2 \mathbf{v}_r^{(1)} \times_3 \mathbf{v}_r^{(2)},$$

$$\|\mathbf{u}_r^{(1)}\|_2^2 = \|\mathbf{u}_r^{(2)}\|_2^2 = \|\mathbf{v}_r^{(1)}\|_2^2 = \|\mathbf{v}_r^{(2)}\|_2^2 = 1.$$

where $\mathcal{X}_r$ and $\mathcal{Y}_r$ are the remaining parts after $r - 1$ approximations. The detailed updating procedure is concluded in Algorithm 43, where $\mathbf{T} = [\mathbf{t}_1; \cdots; \mathbf{t}_R]$, $\mathbf{C} = [\mathbf{c}_1; \cdots; \mathbf{c}_R]$. The linear relationship between $\mathbf{T}$ and $\mathbf{C}$ is characterized by $\mathbf{D} = (\mathbf{T}^{\mathrm{T}}\mathbf{T})^{-1}\mathbf{T}^{\mathrm{T}}\mathbf{C}$. After getting all the parameters, the prediction can be achieved for new predictors $\mathcal{X}'$ by

$$\mathbf{Y}'_{(1)} = \mathbf{X}'_{(1)}\big(\mathbf{U}^{(2)} \odot \mathbf{U}^{(1)}\big)\mathbf{D}\big(\mathbf{V}^{(2)} \odot \mathbf{V}^{(1)}\big)^{\mathrm{T}}, \tag{7.42}$$

where $\mathbf{X}'_{(1)}$ is the mode-1 unfolding matrix of $\mathcal{X}'$, $\mathbf{U}^{(1)} = [\mathbf{u}_1^{(1)}, \ldots, \mathbf{u}_R^{(1)}]$, $\mathbf{U}^{(2)} = [\mathbf{u}_1^{(2)}, \ldots, \mathbf{u}_R^{(2)}]$, $\mathbf{V}^{(1)} = [\mathbf{v}_1^{(1)}, \ldots, \mathbf{v}_R^{(1)}]$, $\mathbf{V}^{(2)} = [\mathbf{v}_1^{(2)}, \ldots, and \mathbf{v}_R^{(2)}]$. And a Bayesian framework for model learning of N-way PLS is drawn in [6], which can determine the number of latent components automatically.

In addition, based on the block term decomposition, [53] propose a PLS-based method for both high-order tensor inputs and outputs. The HOPLS model can be

Algorithm 43: Multilinear PLS

Input: predictor $\mathcal{X}$, response $\mathcal{Y}$
Output: $\mathbf{t}_r, \mathbf{c}_r, \mathbf{u}_r^{(1)}, \mathbf{u}_r^{(2)}, \mathbf{v}_r^{(1)}, \mathbf{v}_r^{(2)}, \; r = 1, \ldots, R, \mathbf{D}$
Initialization: $\mathcal{E}_1 = \mathcal{X}, \mathcal{F}_1 = \mathcal{Y}$
for $r \leq R$
  $\mathbf{c}_r \leftarrow$ the first leading left singular vector of $(\mathbf{F}_r)_{(1)}$
  Iterate until convergence:
    calculate matrix $\mathbf{Z} = \mathcal{E}_r \times_1 \mathbf{c}_r$
    obtain $\mathbf{u}_r^{(1)}$ and $\mathbf{u}_r^{(2)}$ by performing SVD over $\mathbf{Z}$
    $\mathbf{t}_r = \mathcal{E}_r \times_2 \mathbf{u}_r^{(1)} \times_3 \mathbf{u}_r^{(2)}$
    calculate matrix $\mathbf{H} = \mathcal{F}_r \times_1 \mathbf{t}_r$
    obtain $\mathbf{v}_r^{(1)}$ and $\mathbf{v}_r^{(2)}$ by performing SVD over $\mathbf{H}$
    $\mathbf{c}_r = \mathcal{F}_r \times_2 \mathbf{v}_r^{(1)} \times_3 \mathbf{v}_r^{(2)}$
  $\mathbf{T}(:, r) = \mathbf{t}_r, \mathbf{C}(:, r) = \mathbf{c}_r$
  $\mathbf{D} = (\mathbf{T}^{\mathrm{T}}\mathbf{T})^{-1}\mathbf{T}^{\mathrm{T}}\mathbf{C}$
  $\mathcal{E}_{r+1} = \mathcal{E}_r - \mathbf{t}_r \circ \mathbf{u}_r^{(1)} \circ \mathbf{u}_r^{(2)}$
  $\mathbf{G}(:, r) = \mathrm{vec}(\mathbf{v}_r^{(1)} \circ \mathbf{v}_r^{(2)})$
  $\mathcal{F}_{r+1} = \mathcal{Y} - \mathbf{T}\mathbf{D}\mathbf{G}^{\mathrm{T}}$
end

expressed as

$$\mathcal{X} = \sum_{r=1}^{R} \mathcal{G}_{xr} \times_1 \mathbf{t}_r \times_2 \mathbf{U}_r^{(1)} \cdots \times_{L+1} \mathbf{U}_r^{(L)} + \mathcal{E} \tag{7.43}$$

$$\mathcal{Y} = \sum_{r=1}^{R} \mathcal{G}_{yr} \times_1 \mathbf{t}_r \times_2 \mathbf{V}_r^{(1)} \cdots \times_{M+1} \mathbf{V}_r^{(M)} + \mathcal{F} \tag{7.44}$$

where the predictor $\mathcal{X}$ is decomposed as a sum of R rank-$(1, I_1, \ldots, I_L)$ subcomponents and the residuals $\mathcal{E}$ and the response $\mathcal{Y}$ is decomposed as the sum of R rank-$(1, K_1, \ldots, K_M)$ subcomponents and the residuals $\mathcal{F}$. The variable $\mathbf{t}_r$ is the r-th common latent factor between the predictor $\mathcal{X}$ and the response $\mathcal{Y}$ along the sample mode and $\{\mathbf{U}_r^{(l)}\}_{l=1}^{L}$ and $\{\mathbf{V}_r^{(m)}\}_{m=1}^{M}$ are the loading matrices of $\mathcal{X}$ and $\mathcal{Y}$ along specific modes, respectively. The solution process of HOPLS is similar to that of multilinear PLS.

Furthermore, some online learning methods for PLS are also studied, such as recursive multilinear PLS [9], online incremental higher-order partial least squares [17], and recursive higher-order partial least squares [18]. The main idea for these frameworks is to first add the newly arrived data to the previous data by appending operations, extract the new factors from the joint data, and finally compress all the model factors to produce a new set of factors. In this way, these is no need to retrain the regression model when new datasets are arrived to achieve real-time prediction.

7.4 Nonlinear Tensor Regression

The previous section presents linear tensor regression models and some popular solutions. However, in real data analysis, the relationship between system input and output is complex and affected by many factors. The simple linear assumption may not simulate complex regression systems very well, and the resulting regression model can be difficult to give ideal prediction performance. In this section, we will introduce some prevalent nonlinear methods which are known and studied for their powerful fitting capabilities.

7.4.1 Kernel-Based Learning

The simplest idea may be how to linearize the nonlinear model. In that case, the existing linear methods can be exploited for subsequent processing. One example is the kernel technique, which projects the low-dimensional features into high-dimensional space, and the data features can be classified by a linear function, as stated in Fig. 7.6.

Given samples $\{\mathbf{x}_n, y_n\}_{n=1}^N$, the mathematical formulation is as follows:

$$y_n = < \phi(\mathbf{x}_n), \mathbf{b} > + \epsilon_n, \tag{7.45}$$

where $\phi(\cdot)$ projects the input features into a higher dimensional space which make the given samples linear separable. Regarding $\phi(\mathbf{x}_n)$ as the extracted feature, $\mathbf{b}$ characterizes the relationship between $\phi(\mathbf{x}_n)$ and y_n. For example, the optimization problem for kernel ridge regression is

$$\min_{\mathbf{b}} \|\mathbf{y} - < \phi(\mathbf{X}), \mathbf{b} > \|_F^2 + \lambda \mathbf{b}^{\mathrm{T}} \mathbf{b}, \tag{7.46}$$

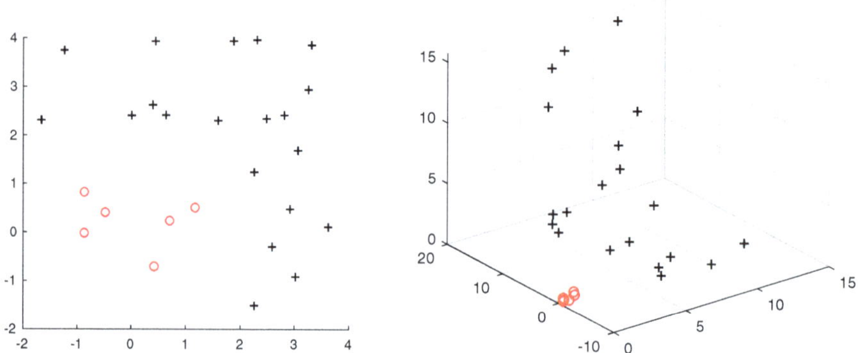

Fig. 7.6 An example of the nonlinear projection

where $\mathbf{X} = [\mathbf{x}_1, \ldots, \mathbf{x}_N]^{\mathrm{T}}$, $\mathbf{y} = [y_1, \ldots, y_N]^{\mathrm{T}}$. The corresponding solution is given by

$$\mathbf{b} = (\phi(\mathbf{X})^{\mathrm{T}}\phi(\mathbf{X}) + \lambda\mathbf{I})^{-1}\phi(\mathbf{X})^{\mathrm{T}}\mathbf{y}. \tag{7.47}$$

Thus, the prediction of new sample $\mathbf{x}'$ can be obtained as

$$y' = \mathbf{b}^{\mathrm{T}}\phi(\mathbf{x}') = \mathbf{y}^{\mathrm{T}}\left(\phi(\mathbf{X})\phi(\mathbf{X})^{\mathrm{T}} + \lambda\mathbf{I}\right)^{-1}\phi(\mathbf{X})\phi(\mathbf{x}')^{\mathrm{T}}, \tag{7.48}$$

which can be rewritten by the kernel functions as

$$y' = \mathbf{b}^{\mathrm{T}}\phi(\mathbf{x}') = \mathbf{y}^{\mathrm{T}}(\mathrm{k}(\mathbf{X}, \mathbf{X}) + \lambda\mathbf{I})^{-1}\mathrm{k}(\mathbf{X}, \mathbf{x}'). \tag{7.49}$$

In this way, only the kernel functions need to be defined for kernel-based regression with no need to know the projection function $\phi(\cdot)$.

There may be a question on why we use the kernel methods for regression analysis. Taken from other points of view, it is hard to extract the similarity of two features in low-dimensional space, as shown in Fig. 7.6 But it becomes clear that these two features belong to the same group when projected into a high-dimensional space. Therefore, kernel function is actually a better measurement metric of the similarity of two features.

Derived by the superiority of kernel methods, there are some kernel-based tensor regression models that have been developed as extensions to kernel spaces, such as the kernel-based higher-order linear ridge regression (KHOLRR) [35] and kernel-based tensor PLS [54]. The main challenge here is how to design the tensor-based kernel functions. As implemented in [35], the simplest way is to vectorize the tensor predictors and use the existing kernel functions, such as the linear kernel and Gaussian RBF kernel. But in such a way, the inherent structure information within the original tensor inputs is ignored. Therefore, some kernel functions defined in tensor field are proposed, such as the product kernel [38, 39] measuring the similarity of all the mode-d unfolding matrices of two predictor tensors $\mathcal{X}$ and $\mathcal{X}$

$$\mathrm{k}(\mathcal{X}, \mathcal{X}') = \sum_{d=1}^{D} \mathrm{k}\left(\mathbf{X}_{(d)}, \mathbf{X}'_{(d)}\right) \tag{7.50}$$

and the probabilistic product kernels [54]

$$\mathrm{k}(\mathcal{X}, \mathcal{X}') = \alpha^2 \sum_{d=1}^{D} \exp\left(-\frac{1}{2\beta_d^2} \mathrm{S}_d(\mathcal{X}||\mathcal{X}')\right), \tag{7.51}$$

where $\mathrm{S}_d(\mathcal{X}||\mathcal{X}'))$ represents the similarity of $\mathcal{X}$ and $\mathcal{X}'$ in mode-d, which is measured by the information divergence between the distributions of $\mathcal{X}$ and $\mathcal{X}'$, α is a magnitude parameter, and β_d denotes length-scales parameter.

Besides, it is also reasonable to decompose the original tensor into its factorization forms and measure their similarity through the inner product of kernel function performed over the latent factors. For example, applying SVD over the mode-d unfolding matrices of $\mathcal{X}$, the Chordal distance-based kernel [38] for tensor data is given by

$$\mathrm{k}(\mathcal{X}, \mathcal{X}') = \sum_{d=1}^{D} \exp\left(-\frac{1}{2\beta_d^2}\|\mathbf{V}_{\mathcal{X}}^{(d)}\left(\mathbf{V}_{\mathcal{X}}^{(d)}\right)^{\mathrm{T}} - \mathbf{V}_{\mathcal{X}'}^{(d)}\left(\mathbf{V}_{\mathcal{X}'}^{(d)}\right)^{\mathrm{T}}\|_{\mathrm{F}}^2\right), \tag{7.52}$$

where $\mathbf{X}_{(d)} = \mathbf{U}_{\mathcal{X}}^{(d)}\Sigma_{\mathcal{X}}^{(d)}(\mathbf{V}_{\mathcal{X}}^{(d)})^{\mathrm{T}}$, $\mathcal{X}'_{(d)} = \mathbf{U}_{\mathcal{X}'}^{(d)}\Sigma_{\mathcal{X}'}^{(d)}(\mathbf{V}_{\mathcal{X}'}^{(d)})^{\mathrm{T}}$. Meanwhile, [7] proposed the TT-based kernel functions by measuring the similarity of two tensors through kernel functions over their core factors under TT decomposition as follows:

$$\mathrm{k}(\mathcal{X}, \mathcal{X}') = \sum_{d=1}^{D}\sum_{r_d=1}^{R_d}\sum_{r_{d+1}=1}^{R_{d+1}} \mathrm{k}\left(\mathcal{G}_{\mathcal{X}}^{(d)}(r_d, :, r_{d+1}), \mathcal{G}_{\mathcal{X}'}^{(d)}(r_d, :, r_{d+1})\right), \tag{7.53}$$

where $\mathcal{G}_{\mathcal{X}}^{(d)}$ and $\mathcal{G}_{\mathcal{X}'}^{(d)}$ are the d-th core factor of tensor $\mathcal{X}$ and $\mathcal{X}'$, respectively, and $\{R_1, \ldots, R_{D+1}\}$ is the tensor train rank.

7.4.2 Gaussian Process Regression

Gaussian process (GP) regression is a typical nonparametric regression model based on Bayesian inference. The essence of Gaussian process regression is the exploitation of unknown functions, which fits the overall distribution of the system output based on existing observations.

In [54], the GP regression is considered with tensor inputs. The observations $\mathbb{D} = \{\mathcal{X}_n, y_n\}_{n=1}^N$ are assumed to be generated from the generative model

$$y_n = \mathrm{f}(\mathcal{X}_n) + \epsilon, \tag{7.54}$$

where the latent function $\mathrm{f}(\mathcal{X})$ is modeled by the GP

$$\mathrm{f}(\mathcal{X}) \sim \mathrm{GP}(\mathrm{m}(\mathcal{X}), \mathrm{k}(\mathcal{X}, \mathcal{X}')|\boldsymbol{\theta}), \tag{7.55}$$

where $\mathrm{m}(\mathcal{X})$ is the mean function of $\mathcal{X}$ (usually set to zero), $\mathrm{k}(\mathcal{X}, \mathcal{X}')$ is the covariance function like kernel function, and $\boldsymbol{\theta}$ is the collection of hyperparameters. The noise distribution is $\epsilon \sim \mathcal{N}(0, \sigma)$.

After determining the distribution of the GP prior and hyperparameters in model (7.54), the posterior distribution of the latent function can be obtained, and the prediction can be performed based on Bayesian inference. Given new predictor

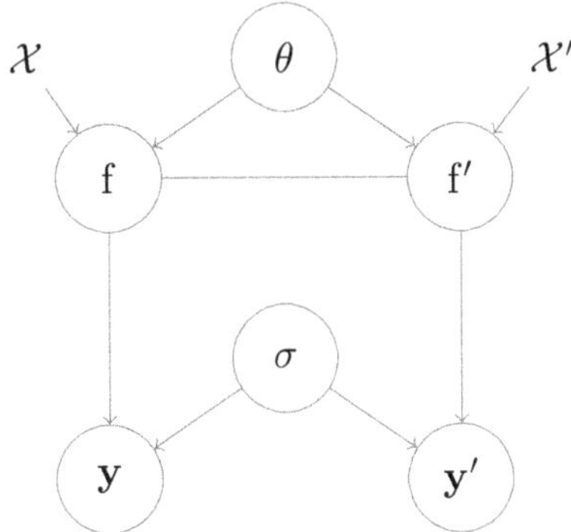

Fig. 7.7 The graphical model of the tensor-based Gaussian process regression

$\mathcal{X}'$, the prediction of the response can be made by

$$\mathbf{y}'|\mathcal{X}', \mathcal{X}, \mathbf{y}, \theta, \sigma \sim \mathcal{N}(\overline{y'}, \text{cov}(y')), \tag{7.56}$$

where

$$\overline{y'} = \text{k}(\mathcal{X}', \mathcal{X})\big(\text{k}(\mathcal{X}, \mathcal{X}) + \sigma^2\mathbf{I}\big)^{-1}\mathbf{y},$$
$$\text{cov}(y') = \text{k}(\mathcal{X}', \mathcal{X}') - \text{k}(\mathcal{X}', \mathcal{X})\big(\text{k}(\mathcal{X}, \mathcal{X}) + \sigma^2\mathbf{I}\big)^{-1}\text{k}(\mathcal{X}, \mathcal{X}').$$

Figure 7.7 provides a graphical model of the tensor-based Gaussian process regression.

In fact, the key difference between the typical GP regression and the tensor-based one is the design of kernel functions, which is demonstrated in Sect. 7.4.1. The subsequent processing of tensor GP regression is similar as the typical GP regression.

Moreover, [20] analyzed the statistical convergence rate of the Gaussian process tensor estimator and proved the derived convergence rate is minimax optimal. It also discusses the relationship between the linear tensor regression models and nonparametric models based on the generative models, while the connections between the tensor regression and Gaussian process models are further analyzed in [49]. In addition, a nonparametric extension of generalized linear regression models is proposed and used for multitask learning in neuroimaging data analysis [23].

7.4.3 Tensor Additive Models

Additive model, as a useful nonlinear regression model in statistics, is a flexible extension of the linear models retaining most of the interpretability. The familiar tools for modeling and inferring in linear models are also suitable for additive models.

Specifically, the sparse tensor additive model is expressed as follows:

$$y_n = \mathrm{T}(\mathcal{X}_n) + \epsilon_n = \sum_{l=1}^{L} \sum_{p_l=1}^{P_l} \mathrm{f}_{p_1,\ldots,p_L}(\mathcal{X}_n(p_1, \ldots, p_L)) + \epsilon_n, \tag{7.57}$$

where $\mathrm{f}^*_{p_1,\ldots,p_L}$ is nonparametric additive function which belongs to a smooth function class. Approximating function $\mathrm{f}_{p_1,\ldots,p_L}$ with H basis functions, $\mathrm{T}(\mathcal{X}_n)$ can be rewritten as

$$\mathrm{T}(\mathcal{X}_n) = \sum_{l=1}^{L} \sum_{p_l=1}^{P_l} \sum_{h=1}^{H} \beta_{p_1,\ldots,p_L,h} \psi_{p_1,\ldots,p_L,h}(\mathcal{X}_n(p_1, \ldots, p_L)), \tag{7.58}$$

where $\psi_{p_1,\ldots,p_L,h}(\cdot)$ belongs to basis functions, such as polynomial splines [13] or Fourier series [44], and $\beta_{p_1,\ldots,p_L,h}$ represents the weight of basis function $\psi_{p_1,\ldots,p_L,h}(\cdot)$ in function $\mathrm{f}_{p_1,\ldots,p_L}(\cdot)$.

Letting $\mathcal{B}_h \in \mathbb{R}^{P_1 \times \cdots \times P_L}$ with $\mathcal{B}_h(p_1, \ldots, p_L) = \beta_{p_1,\ldots,p_L,h}$ and $\Psi_h(\mathcal{X}_n) \in \mathbb{R}^{P_1 \times \cdots \times P_L}$ with $p_1, \ldots, p_L$-th element equaling to $\psi_{p_1,\ldots,p_L,h}(\mathcal{X}_n(p_1, \ldots, p_L))$, we can reformulate the regression model in (7.58) as

$$\mathrm{T}(\mathcal{X}_n) = \sum_{h=1}^{H} < \mathcal{B}_h, \Psi_h(\mathcal{X}_n) > . \tag{7.59}$$

Then the estimation of the model becomes to the estimation of the coefficients $\mathcal{B}^*_h$. In this way, the algorithms developed for linear tensor regression models can be employed directly for parameter estimation.

However, the additive model has limitations for applications in large-scale data mining or tasks with few samples. For additive model, all the predictors are included and fitted, but most predictors are not feasible or necessary in real applications.

7.4.4 Random Forest-Based Tensor Regression

Random forest is a classic ensemble learning algorithm. It trains multiple decision trees and packs to form a powerful model. The multiple subsets are obtained by random subsampling, and each decision tree is constructed according to the subset. For typical Random forest, all the optimizations are performed on vectors or matrices. For training datasets with high-dimensionality, the traditional methods may suffer from either high computational burden or poor accuracy performance.

To exploit the structural information within the multiway data, tensor-based random forest methods are proposed. For example, in [22], the linear tensor regression model based on CP decomposition is used to fit the response and the features arrived at the leaf nodes. Figure 7.8 gives an illustration of the training

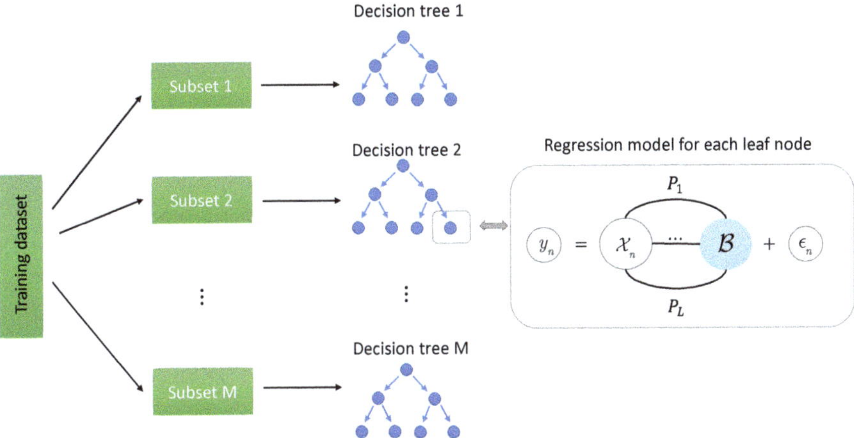

Fig. 7.8 An illustration of the training process of random forest based on tensor regression

process of random forest based on tensor regression. Compared with the traditional random forest method, the tensor-based method employs the tensor regression models instead of multivariate Gaussian distribution models within the leaf nodes. Recently, a tensor basis random forest method is proposed in [19], where not only the leaf nodes but also other child nodes are fitted using regression models. It should be noted that [19] and [22] predict based on the average of the existing observations in the leaf nodes, which means that the prediction can be inferred to be values out of the observation range.

7.5 Applications

Tensor regression has been applied in a wide range of fields, such as sociology, climatology, computer vision, and neuroimaging analysis. In this section, we will go through some examples for the applications of tensor regression.

7.5.1 Vector-on-Tensor Regression

Vector-on-tensor regression aims to explore the relationship between multiway predictor and a scalar response or multiple response, such as the disease diagnosis predicting the clinical assessment outcomes from medical images [1, 14, 16, 25–27, 34, 40, 55], human age or other attributes estimation from facial images [12, 32], and head pose estimation from video sequences [12].

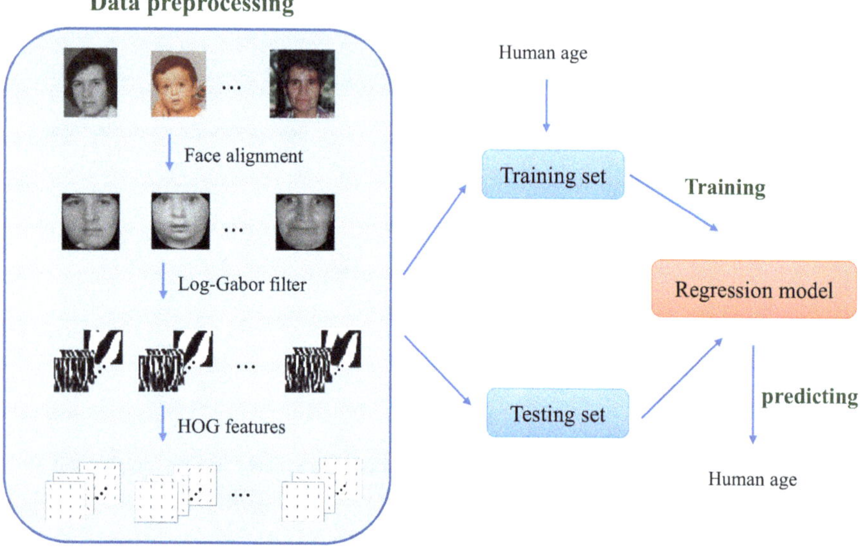

Fig. 7.9 The experimental process of the age estimation task

Here we provide a practical example for the human age estimation task from corresponding facial images. The dataset used in this experiment is the FG-NET dataset.[1] The experimental procedure for human age estimation is illustrated in Fig. 7.9. The facial images are first aligned and downsized into the same size. Then a Log-Gabor filter with four scales and eight orientations is employed to process each image. After obtaining the filtered images, a set of histograms are calculated for each blocks. In this experiment, the block size is set to be 8×8, and the number of orientation histogram bins is set to be 59. The extracted feature tensor of each image is in the size of $944 \times 4 \times 8$, which is used to build the relationship between the facial images and human age. The preprocessed dataset is divided into training set and testing set to train the desired regression model and validate the performance of the algorithm.

We compare some tensor regression methods mentioned in terms of estimation error and training time, including higher rank tensor ridge regression (hrTRR), optimal rank tensor ridge regression (orSTR), higher rank support tensor regression (hrSTR) [12], regularized multilinear regression and selection (Remurs) [40], sparse tubal-regularized multilinear regression (Sturm) [27], and fast stagewise unit-rank tensor factorization (SURF) [14]. Table 7.1 reports the experimental results of these algorithms. To visualize the difference between different age ranges, we evaluate the performance of these algorithms in five age ranges, namely, 0–6, 7–12, 13–17,

[1] https://yanweifu.github.io/FG_NET_data/index.html.

Table 7.1 The performance comparison for age estimation on FG-NET dataset

Model	RMSE	Q^2	Correlation	CPU time
hrTRR	10.80	0.4068	0.3964	257.64 ± 238.04
hrSTR	8.14	0.6630	0.5520	75.11 ± 1.71
orSTR	9.38	0.5532	0.5447	188.96 ± 18.91
Remurs	9.73	0.5184	0.5470	25.34 ± 0.28
Sturm	9.85	0.5064	0.5412	26.22 ± 0.20
SURF	10.35	0.4558	0.4752	280.71 ± 1.65

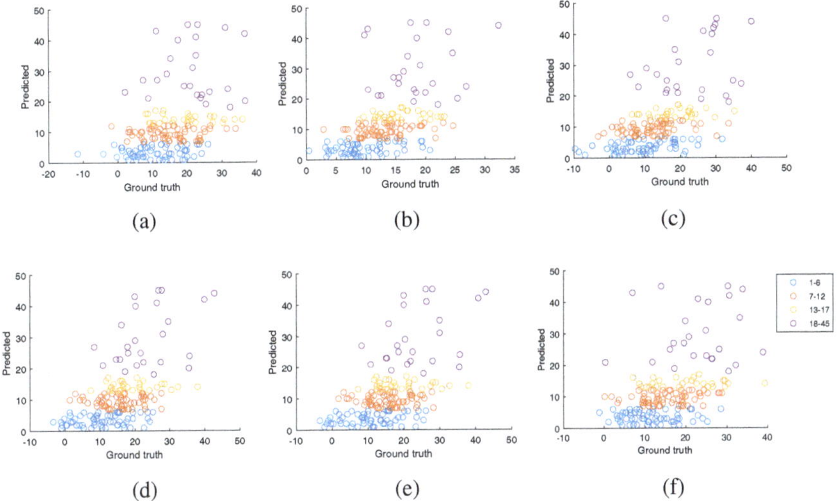

Fig. 7.10 The visualization of the predicted values vs. the ground truth of the human age estimation task. (**a**) hrTRR. (**b**) hrSTR. (**c**) orSTR. (**d**) Remurs. (**e**) Sturm. (**f**) SURF

and 18–45. Figure 7.10 shows the predicted values and the given values for the test data over all employed algorithms and five age ranges.

7.5.2 *Tensor-on-Vector Regression*

Tensor-on-vector regression is mainly derived by the attention that what is the difference between the brain activation regions or connectivity patterns with respect to specific behaviors or different subjects [24, 42]. The comparison between different subjects is commonly performed on the rest-sate fMRI images, while the comparison with specific task is also analyzed through the task-related fMRI images. Besides, the process optimization in mechanical manufacturing can be formulated as 3D shape analysis based on different process parameters [48].

To illustrate how the tensor-on-vector regression methods work, we conduct a group of experiments on EEG comparison tasks between the normal subjects and the alcoholic individuals. The employed EEG dataset is publicly available.[2]In this experiment, only part of the dataset is used. The chosen part is average of all the trails on same condition. After preprocessing, the resulting EEG signal is in the size of 64 time points $\times$ 64 channels. Mapping the type of the subjects (alcoholic individuals or controls) with the corresponding EEG signals, we can obtain the coefficient tensors using several classical tensor-on-vector regression methods, such as the sparse tensor response regression (STORE) [42], tensor envelope tensor response regression (TETRR) [24], and HOLRR [35]. The OLS method is taken as a baseline method too. Figure 7.11 presents the estimated coefficient tensor using OLS, STORE, TETRR, HOLRR, and the corresponding p-value maps.

7.5.3 *Tensor-on-Tensor Regression*

Tensor-on-tensor regression is a generalization of two regression models described before, which tends to explore the relationship between multiway predictor and multiway response. A common example is the autoregression problem of multiway datasets, such as the forecasting or cokriging task in sociology or climatology [3, 50] and casual analysis in brain connectivity [21].

Besides of the autoregression problem, mapping multiway predictor into another multiway tensor is also an important task, such as the multitask learning [37, 50], neuron decoding of the ECoG signals [53], and the human motion reconstruction from the video sequences.

Regarding this group, we take the classical forecasting task in climatology as an example. The dataset, namely, Meteo-UK [35], which records the monthly measurements of weather information across the UK, is employed. We select the complete data of 5 variables across 16 stations in the UK from 1960 to 2000. Using the observations in past three time points $t-3, t-2, t-1$ to predict the future five time points $t, t+1, \ldots, t+4$, the resulting predictor is in the size of *sample size* $\times$ 16 $\times$ 5 $\times$ 3. The corresponding response is in the size of *sample size* $\times 16 \times 5 \times 5$. In this experiment, 80% of the samples are used for training, and 20% are used for testing. The performance of several classical tensor-based regression methods, including nonconvex multilinear multitask learning (MLMTL-NC) [37], convex multilinear multitask learning (MLMTL-C) [37], greedy tensor regression solution based on orthogonal matching pursuit (greedy) [3], TPG [50], HOPLS [53], HOLRR [35], tensor-on-tensor regression based on CP decomposition (TTR-CP) [32], and tensor-on-tensor regression based on TT decomposition (TTR-TT) [31], is given in Table 7.2. We visualize the predicted waves by different algorithms in Fig. 7.12.

[2]https://archive.ics.uci.edu/ml/datasets/EEG+Database.

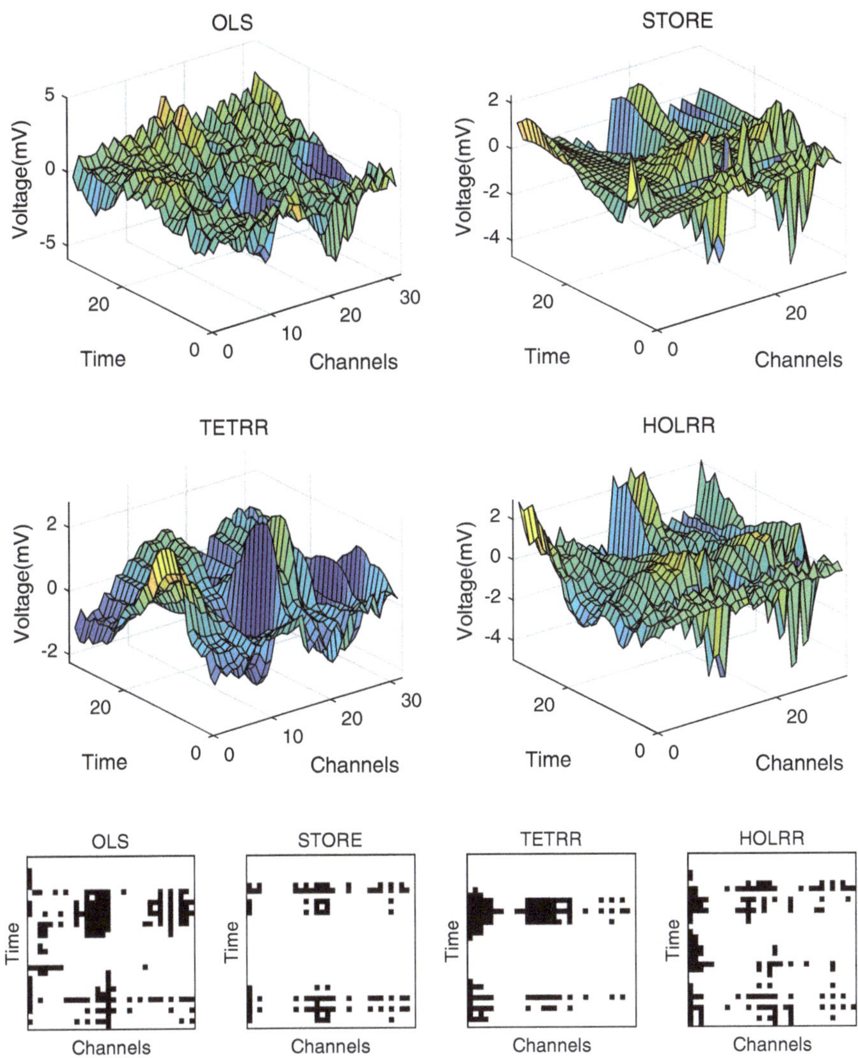

Fig. 7.11 Performance comparison of EEG data analysis, where the top two rows are the estimated coefficient tensor by the OLS, STORE, TETRR, and HOLRR and the bottom line shows the corresponding p-value maps, thresholded at 0.05

7.6 Summary

In this section, we overview the motivations, popular solutions, and applications of tensor regression. In decades of research, linear tensor regression has always been the focus of research, especially regression models with penalty terms. The core

Table 7.2 The performance comparison for forecasting task on Meteo-UK dataset

Model	RMSE	Q^2	Correlation	CPU time
Greedy	0.6739	0.5072	0.7126	1.83
TPG	0.6694	0.5138	0.7175	7.19
HOPLS	0.6222	0.5621	0.7570	0.82
HOLRR	0.6084	0.5813	0.7695	0.8
TTR-CP	0.6133	0.5745	0.7656	129.35
TTR-TT	0.5979	0.5957	0.7752	0.33

of the linear tensor regression model lies in the approximation of the regression coefficient tensor. There are three main research directions:

1. How to use fewer parameters to better represent the multidimensional correlation, or multidimensional function? It is also the key problem in low-rank tensor learning.
2. How to use the data prior to improve the model's identifiability and robustness?
3. Noise modeling is also one of the main problems in regression model, such as the research on sparse noise [28] and mixed noise [33]. But there is few literature on robust tensor regression.

The linear model is simple with strong generalization ability, but naturally it cannot fit the local characteristics well. Nonlinear models have strong fitting capabilities and have received widespread attention in recent years. For example, for kernel learning, how to design a suitable tensor kernel function, which can not only use the multidimensional structure of the data but also simplify the calculation, is an open problem. In addition, support tensor regression and tensor-based multilayer regression network are also promising research directions.

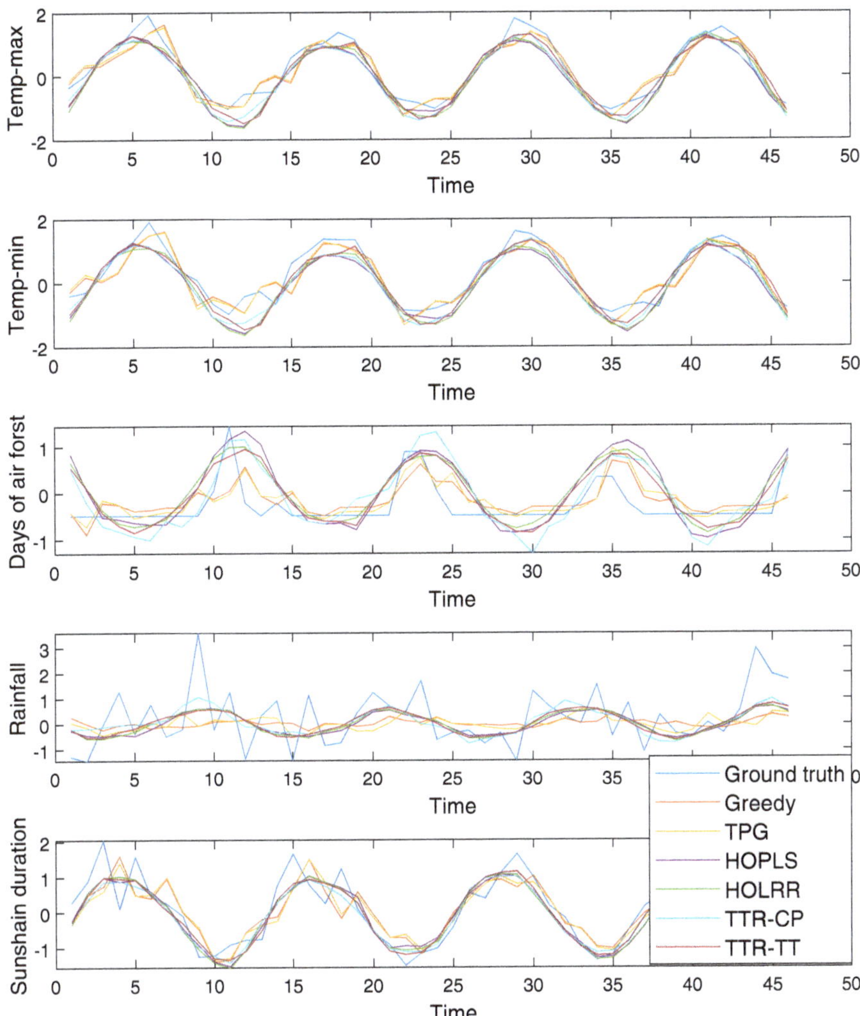

Fig. 7.12 The visualization of the prediction of five variables in one station, including the mean daily maximum temperature (temp-max), mean daily minimum temperature (temp-min), days of air forest, total rainfall, and total sunshine duration, by different algorithms

References

1. Ahmed, T., Raja, H., Bajwa, W.U.: Tensor regression using low-rank and sparse Tucker decompositions. SIAM J. Math. Data Sci. **2**(4), 944–966 (2020)
2. Allen, G.: Sparse higher-order principal components analysis. In: Artificial Intelligence and Statistics, pp. 27–36 (2012)
3. Bahadori, M.T., Yu, Q.R., Liu, Y.: Fast multivariate spatio-temporal analysis via low rank tensor learning. In: NIPS, pp. 3491–3499. Citeseer (2014)

4. Billio, M., Casarin, R., Kaufmann, S., Iacopini, M.: Bayesian dynamic tensor regression. University Ca'Foscari of Venice, Dept. of Economics Research Paper Series No, 13 (2018)
5. Bro, R.: Multiway calibration. multilinear PLS. J. Chemom. **10**(1), 47–61 (1996)
6. Camarrone, F., Van Hulle, M.M.: Fully bayesian tensor-based regression. In: 2016 IEEE 26th International Workshop on Machine Learning for Signal Processing (MLSP), pp. 1–6. IEEE, Piscataway (2016)
7. Chen, C., Batselier, K., Yu, W., Wong, N.: Kernelized support tensor train machines (2020). Preprint arXiv:2001.00360
8. Cichocki, A., Phan, A.H., Zhao, Q., Lee, N., Oseledets, I., Sugiyama, M., Mandic, D.P., et al.: Tensor networks for dimensionality reduction and large-scale optimization: part 2 applications and future perspectives. Found Trends Mach. Learn. **9**(6), 431–673 (2017)
9. Eliseyev, A., Aksenova, T.: Recursive N-way partial least squares for brain-computer interface. PloS One **8**(7), e69962 (2013)
10. Fan, J., Ding, L., Chen, Y., Udell, M.: Factor group-sparse regularization for efficient low-rank matrix recovery. In: Advances in Neural Information Processing Systems, pp. 5105–5115 (2019)
11. Guhaniyogi, R., Qamar, S., Dunson, D.B.: Bayesian tensor regression. J. Machine Learn. Res. **18**(1), 2733–2763 (2017)
12. Guo, W., Kotsia, I., Patras, I.: Tensor learning for regression. IEEE Trans. Image Proc. **21**(2), 816–827 (2012)
13. Hao, B., Wang, B., Wang, P., Zhang, J., Yang, J., Sun, W.W.: Sparse tensor additive regression (2019). Preprint arXiv:1904.00479
14. He, L., Chen, K., Xu, W., Zhou, J., Wang, F.: Boosted sparse and low-rank tensor regression. In: Advances in Neural Information Processing Systems, pp. 1009–1018 (2018)
15. Hoff, P.D.: Multilinear tensor regression for longitudinal relational data. Ann. Appl. Stat. **9**(3), 1169 (2015)
16. Hou, M., Chaib-Draa, B.: Hierarchical Tucker tensor regression: Application to brain imaging data analysis. In: 2015 IEEE International Conference on Image Processing (ICIP), pp. 1344–1348. IEEE, Piscataway (2015)
17. Hou, M., Chaib-draa, B.: Online incremental higher-order partial least squares regression for fast reconstruction of motion trajectories from tensor streams. In: 2016 IEEE International Conference on Acoustics, Speech and Signal Processing (ICASSP), pp. 6205–6209. IEEE, Piscataway (2016)
18. Hou, M., Chaib-draa, B.: Fast recursive low-rank tensor learning for regression. In: Proceedings of the Twenty-Sixth International Joint Conference on Artificial Intelligence, pp. 1851–1857 (2017)
19. Kaandorp, M.L., Dwight, R.P.: Data-driven modelling of the reynolds stress tensor using random forests with invariance. Comput. Fluids **202**, 104497 (2020)
20. Kanagawa, H., Suzuki, T., Kobayashi, H., Shimizu, N., Tagami, Y.: Gaussian process nonparametric tensor estimator and its minimax optimality. In: International Conference on Machine Learning, pp. 1632–1641. PMLR (2016)
21. Karahan, E., Rojas-Lopez, P.A., Bringas-Vega, M.L., Valdes-Hernandez, P.A., Valdes-Sosa, P.A.: Tensor analysis and fusion of multimodal brain images. Proc. IEEE **103**(9), 1531–1559 (2015)
22. Kaymak, S., Patras, I.: Multimodal random forest based tensor regression. IET Comput. Vision **8**(6), 650–657 (2014)
23. Kia, S.M., Beckmann, C.F., Marquand, A.F.: Scalable multi-task gaussian process tensor regression for normative modeling of structured variation in neuroimaging data (2018). Preprint arXiv:1808.00036
24. Li, L., Zhang, X.: Parsimonious tensor response regression. J. Amer. Stat. Assoc. **112**, 1–16 (2017)
25. Li, X., Xu, D., Zhou, H., Li, L.: Tucker tensor regression and neuroimaging analysis. Stat. Biosci. **10**, 1–26 (2013)

26. Li, Z., Suk, H.I., Shen, D., Li, L.: Sparse multi-response tensor regression for Alzheimer's disease study with multivariate clinical assessments. IEEE Trans. Med. Imag. **35**(8), 1927–1936 (2016)
27. Li, W., Lou, J., Zhou, S., Lu, H.: Sturm: Sparse tubal-regularized multilinear regression for fmri. In: International Workshop on Machine Learning in Medical Imaging, pp. 256–264. Springer, Berlin (2019)
28. Liu, J., Cosman, P.C., Rao, B.D.: Robust linear regression via ℓ_0 regularization. IEEE Trans. Signal Process. **66**(3), 698–713 (2017)
29. Liu, J., Zhu, C., Liu, Y.: Smooth compact tensor ring regression. IEEE Trans. Knowl. Data Eng. (2020). https://doi.org/10.1109/TKDE.2020.3037131
30. Liu, J., Zhu, C., Long, Z., Huang, H., Liu, Y.: Low-rank tensor ring learning for multi-linear regression. Pattern Recognit. **113**, 107753 (2020)
31. Liu, Y., Liu, J., Zhu, C.: Low-rank tensor train coefficient array estimation for tensor-on-tensor regression. IEEE Trans. Neural Netw. Learn. Syst. **31**, 5402–5411 (2020)
32. Lock, E.F.: Tensor-on-tensor regression. J. Comput. Graph. Stat. **27**(3), 638–647 (2018)
33. Luo, L., Yang, J., Qian, J., Tai, Y.: Nuclear-L1 norm joint regression for face reconstruction and recognition with mixed noise. Pattern Recognit. **48**(12), 3811–3824 (2015)
34. Papadogeorgou, G., Zhang, Z., Dunson, D.B.: Soft tensor regression (2019). Preprint arXiv:1910.09699
35. Rabusseau, G., Kadri, H.: Low-rank regression with tensor responses. In: Advances in Neural Information Processing Systems, pp. 1867–1875 (2016)
36. Raskutti, G., Yuan, M.: Convex regularization for high-dimensional tensor regression (2015). Preprint arXiv:1512.01215 **639**
37. Romera-Paredes, B., Aung, H., Bianchi-Berthouze, N., Pontil, M.: Multilinear multitask learning. In: International Conference on Machine Learning, pp. 1444–1452 (2013)
38. Signoretto, M., De Lathauwer, L., Suykens, J.A.: A kernel-based framework to tensorial data analysis. Neural Netw. **24**(8), 861–874 (2011)
39. Signoretto, M., Olivetti, E., De Lathauwer, L., Suykens, J.A.: Classification of multichannel signals with cumulant-based kernels. IEEE Trans. Signal Process. **60**(5), 2304–2314 (2012)
40. Song, X., Lu, H.: Multilinear regression for embedded feature selection with application to fmri analysis. In: Proceedings of the AAAI Conference on Artificial Intelligence, vol. 31 (2017)
41. Spencer, D., Guhaniyogi, R., Prado, R.: Bayesian mixed effect sparse tensor response regression model with joint estimation of activation and connectivity (2019). Preprint arXiv:1904.00148
42. Sun, W.W., Li, L.: STORE: sparse tensor response regression and neuroimaging analysis. J. Mach. Learn. Res. **18**(1), 4908–4944 (2017)
43. Tang, X., Bi, X., Qu, A.: Individualized multilayer tensor learning with an application in imaging analysis. J. Amer. Stat. Assoc. **115**, 1–26 (2019)
44. Wahls, S., Koivunen, V., Poor, H.V., Verhaegen, M.: Learning multidimensional fourier series with tensor trains. In: 2014 IEEE Global Conference on Signal and Information Processing (GlobalSIP), pp. 394–398. IEEE, Piscataway (2014)
45. Wimalawarne, K., Sugiyama, M., Tomioka, R.: Multitask learning meets tensor factorization: task imputation via convex optimization. In: Advances in Neural Information Processing Systems, pp. 2825–2833 (2014)
46. Wimalawarne, K., Tomioka, R., Sugiyama, M.: Theoretical and experimental analyses of tensor-based regression and classification. Neural Comput. **28**(4), 686–715 (2016)
47. Xu, J., Zhou, J., Tan, P.N., Liu, X., Luo, L.: Spatio-temporal multi-task learning via tensor decomposition. IEEE Trans. Knowl. Data Eng. **33**, 2764–2775 (2019)
48. Yan, H., Paynabar, K., Pacella, M.: Structured point cloud data analysis via regularized tensor regression for process modeling and optimization. Technometrics **61**(3), 385–395 (2019)
49. Yu, R., Li, G., Liu, Y.: Tensor regression meets gaussian processes. In: International Conference on Artificial Intelligence and Statistics, pp. 482–490 (2018)
50. Yu, R., Liu, Y.: Learning from multiway data: Simple and efficient tensor regression. In: International Conference on Machine Learning, pp. 373–381 (2016)

51. Zha, Z., Yuan, X., Wen, B., Zhou, J., Zhang, J., Zhu, C.: A benchmark for sparse coding: When group sparsity meets rank minimization. IEEE Trans. Image Process. **29**, 5094–5109 (2020)
52. Zhang, X., Li, L.: Tensor envelope partial least-squares regression. Technometrics **59**, 1–11 (2017)
53. Zhao, Q., Caiafa, C.F., Mandic, D.P., Chao, Z.C., Nagasaka, Y., Fujii, N., Zhang, L., Cichocki, A.: Higher order partial least squares (HOPLS): a generalized multilinear regression method. IEEE Trans. Pattern Analy. Mach. Intell. **35**(7), 1660–1673 (2013)
54. Zhao, Q., Zhou, G., Adali, T., Zhang, L., Cichocki, A.: Kernelization of tensor-based models for multiway data analysis: Processing of multidimensional structured data. IEEE Signal Process. Mag. **30**(4), 137–148 (2013)
55. Zhou, H., Li, L., Zhu, H.: Tensor regression with applications in neuroimaging data analysis. J. Amer. Stat. Asso. **108**(502), 540–552 (2013)

Chapter 8
Statistical Tensor Classification

8.1 Introduction

Classification is a main type of supervised learning that aims to classify input samples into specific groups. Compared with regression tasks with continuous responses, the output of classification tasks is usually discrete values. For example, common classification tasks include spam identification, disease diagnosis, traffic sign detection, image classification, etc.

With the emergence of many multidimensional data such as visual videos and medical images, classification tasks involving high-dimensional data seem to be increasing. Using traditional classification methods to solve classification tasks related to multidimensional data faces some challenges. First, it is necessary to vectorize multidimensional data for traditional matrix computation-based classification methods, which inevitably causes the loss of multidimensional structural information; second, the amount of parameters required for modeling can be very large, and both storage and calculation burden can be very heavy.

Tensor classification has been proposed as the extension of traditional matrix computation-based classification. It follows a similar procedure as it does in vector-based classification, such as preprocessing, training, and predicting. However, compared with original vector- or matrix-based classification, tensor-based one can explore the multidimensional structure information by either extracting useful features by tensor decomposition or utilizing tensor operations to map the multiway input into binary output. The common framework of performing classification tasks based on tensor analysis can be illustrated as in Fig. 8.1.

In the rest of this chapter, we first give a basic framework for the statistical tensor classification. Then some popular tensor-based classification algorithms are introduced, including the logistic tensor classification; several tensor versions of support vector machine, including support tensor machine, higher-rank support

Y. Liu et al., *Tensor Computation for Data Analysis*,
https://doi.org/10.1007/978-3-030-74386-4_8

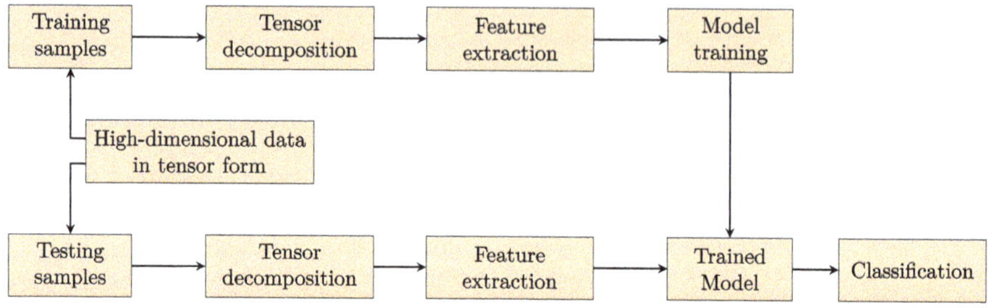

Fig. 8.1 Framework of tensor classification

tensor machine, and support Tucker machine; and tensor Fisher discriminant analysis. Finally, some practical applications on tensor classification for digital image classification and neuroimaging analysis are given.

8.2 Tensor Classification Basics

Tensor-based classification aims to extend the vector-based classification to deal with tensor data. Specifically, given training samples $\{\mathbf{x}_n, y_n \in \{+1, -1\}\}_{n=1}^N$, here $\mathbf{x}_n \in \mathbb{R}^I$, a vector-based classification model can be represented as

$$\begin{aligned} \min_{\mathbf{w}, b, \boldsymbol{\xi}} \quad & \mathrm{f}(\mathbf{w}, b, \boldsymbol{\xi}) \\ \text{s. t.} \quad & y_n \, \mathrm{c}_n \left(\mathbf{w}^{\mathrm{T}} \mathbf{x}_n + b\right) \geq \xi_n, \ 1 \leq n \leq N, \end{aligned} \tag{8.1}$$

where $\mathrm{f} : \mathbb{R}^{I+N+1} \to \mathbb{R}$ is an optimization criterion for classification tasks, such as the maximum likelihood function for the logistic regression; $\mathrm{c}_n : \mathbb{R} \to \mathbb{R}, 1 \leq n \leq N$ represents a convex function; and $\boldsymbol{\xi} = [\xi_1; \xi_2; \cdots; \xi_N]$ are slack variables.

From the perspective of optimization theory, the optimization model (8.1) is an optimization problem with inequality constraints. Denote the sign function $\mathrm{sign}(x)$ as

$$\mathrm{sign}(x) = \begin{cases} +1 & x \geq 0, \\ -1 & x \leq 0, \end{cases} \tag{8.2}$$

the estimated variables $\mathbf{w}$ and b determine a decision hyperplane like $\hat{y} = \mathrm{sign}(\mathbf{w}^{\mathrm{T}}\mathbf{x} + b)$, which is used to separate the negative measurements from the positive measurements. With different objective functions and convex constraints, we can derive several learning machines, such as support vector machine, logistic regression, and Fisher discriminant analysis.

To extend the vector-based learning machines into their tensor forms which can accept tensors as inputs, the key problem is how to model the relationship between a multidimensional array and its corresponding label. In vector-based classification model, the decision function maps the vector input into its corresponding label by a vector inner product of the vector measurements $\mathbf{x}_n \in \mathbb{R}^I$ and projection vector $\mathbf{w} \in \mathbb{R}^I$. Therefore, it is nature to consider tensor products to map the multiway tensors into its corresponding labels.

Here we consider N training measurements which are represented as tensors $\mathcal{X}_n \in \mathbb{R}^{I_1 \times I_2 \times \cdots \times I_K}$, $1 \leq n \leq N$, and their corresponding class labels are $y_n \in \{+1, -1\}$. By employing tensor product of tensor inputs $\mathcal{X}_n \in \mathbb{R}^{I_1 \times I_2 \times \cdots \times I_K}$ and multiway projection tensor $\mathcal{W} \in \mathbb{R}^{I_1 \times I_2 \times \cdots \times I_K}$, we can get the basic tensor learning models as follows:

$$\begin{aligned} \min_{\mathcal{W}, b, \boldsymbol{\xi}} \quad & \mathrm{f}(\mathcal{W}, b, \boldsymbol{\xi}) \\ \text{s. t.} \quad & y_n \, \mathrm{c}_n \left(\langle \mathcal{X}_n, \mathcal{W} \rangle + b \right) \geq \xi_n, \; 1 \leq n \leq N, \end{aligned} \tag{8.3}$$

where $\langle \mathcal{X}_n, \mathcal{W} \rangle$ refers to one specific tensor product, such as tensor inner product. In this way, the decision functions are determined by the tensor plane $\langle \mathcal{X}, \mathcal{W} \rangle + b = 0$.

However, an obvious shortcoming is that the high dimensionality of the coefficient tensor inevitably requires numerous training samples and causes huge computational and storage costs. In machine learning, when the dimension of training samples is much larger than the number of samples, it is easy to cause the problem of overfitting of classifier. Dimensionality reduction for this high-order data classification is important. The commonly used assumptions for dimensionality reduction of coefficient tensor include low rankness and sparsity.

Compared with the low-rank tensor approximation for data recovery, the difference is that the common tensor approximation task is to approximate real data, such as images and videos, while the approximation target in regression or classification is to approximate the multidimensional coefficient tensor which reflects the relationship between multi-features of data.

In the following, we will give detailed illustration for popular statistical tensor classification algorithms from the perspective of its development from matrix versions to tensor versions, such as from logistic regression to logistic tensor regression, from support vector machine to support tensor machine, and from Fisher discriminant analysis (FDA) to tensor FDA. In fact, all these approaches can be regarded as a special case of the framework in (8.3) with different objective functions and constraints. In order to make this chapter self-contained and let readers better understand the evolution of tensor classification algorithms, before introducing all tensor machines, we have given a basic introduction to the corresponding vector version.

8.3 Logistic Tensor Regression

8.3.1 Logistic Regression

Logistic regression is a basic and widely used binary classification method which is derived from linear regression. Given a training set $\mathbf{x}_n \in \mathbb{R}^I$, $1 \leq n \leq N$ and their corresponding class labels $y_n \in \{0, 1\}$, the decision function of logistic regression is given by

$$y(\mathbf{x}) = \phi\left(\mathbf{w}^{\mathrm{T}}\mathbf{x} + b\right), \tag{8.4}$$

where $\mathbf{w}$ is the weighting vector, b is the bias, and $\phi(\cdot)$ is logistic function which is a kind of sigmoid function as

$$\phi(x) = \frac{e^x}{1 + e^x}. \tag{8.5}$$

The decision function uses a nonlinear function $\phi(x)$ to project the value of linear function $\mathbf{w}^{\mathrm{T}}\mathbf{x} + b$ to the range [0, 1]. Equation (8.5) can be transformed into

$$\log \frac{y}{1 - y} = \mathbf{w}^{\mathrm{T}}\mathbf{x} + b, \tag{8.6}$$

where the mapped value can be interpreted as the corresponding probability as

$$\mathrm{Prob}(y = 1|\mathbf{x}) = \frac{1}{1 + e^{-(\mathbf{w}^{\mathrm{T}}\mathbf{x}+b)}}, \quad \mathrm{Prob}(y = 0|\mathbf{x}) = 1 - \mathrm{Prob}(y = 1|\mathbf{x}). \tag{8.7}$$

The likelihood function is as follows:

$$\mathrm{L}(\mathbf{w}, b) = \prod_{n=1}^{N} [\mathrm{Prob}(y = 1|\mathbf{x}_n)]^{y_n} [\mathrm{Prob}(y = 0|\mathbf{x}_n)]^{1-y_n}, \tag{8.8}$$

and $\mathbf{w}$ and b can be estimated by minimizing the corresponding loss function as

$$\min_{\mathbf{w},b} \quad -\log \ \mathrm{L}(\mathbf{w}, b), \tag{8.9}$$

which is obtained as

$$\min_{\mathbf{w},b} \quad -\sum_{n=1}^{N} \left[y_n \left(\mathbf{w}^{\mathrm{T}}\mathbf{x}_n + b \right) - \log\left(1 + e^{\mathbf{w}^{\mathrm{T}}\mathbf{x}_n + b} \right) \right]. \tag{8.10}$$

The performance of this basic classifier would be deteriorated when dealing with large-scale higher-order data. To solve the problem, some regularization terms are applied to utilize the possible priors over the weighted vector. For example, the ℓ_0 norm regularizer is commonly used to overcome the overfitting and makes the model more sparse. Considering the ℓ_0 optimization is NP-hard, the ℓ_1 norm regularizer is often used as a substitution; the resulted new convex optimization model is as follows:

$$\min_{\mathbf{w},b} \sum_{n=1}^{N} \left[\log\left(1 + e^{\mathbf{w}^{\mathrm{T}}\mathbf{x}_n + b}\right) - y_n\left(\mathbf{w}^{\mathrm{T}}\mathbf{x}_n + b\right)\right] + \lambda \sum_i |\mathbf{w}_i| . \tag{8.11}$$

8.3.2 *Logistic Tensor Regression*

The classical logistic regression is based on the vector form and would lead to an information loss when dealing with the multiway data. To avoid such a problem, we discuss the way to extend the logistic regression to the tensor version. The training set of logistic tensor regression are denoted as tensors $\mathcal{X}_n \in \mathbb{R}^{I_1 \times I_2 \times \cdots \times I_M},\ 1 \le n \le N$ with their class labels $y_n \in \{+1, -1\}$. One decision function for logistic tensor regression can be selected as follows:

$$y(\mathcal{X}) = \phi(< \mathcal{W}, \mathcal{X} > +b), \tag{8.12}$$

where $\mathcal{W} \in \mathbb{R}^{I_1 \times I_2 \times \cdots \times I_M}$ denotes the weighting tensor and b is the bias. Therefore, the maximum likelihood method optimization can be formulated as follows:

$$\min_{\mathcal{W},b} \sum_{n=1}^{N} \left[\log\left(1 + e^{<\mathcal{W},\mathcal{X}_n> + b}\right) - y_n\left(< \mathcal{W}, \mathcal{X}_n > +b\right)\right]. \tag{8.13}$$

In order to reduce the model parameters, we model the weighting tensor as a low-rank tensor which is assumed to be a sum of R rank-1 tensors as follows:

$$\mathcal{W} = \sum_{r=1}^{R} \mathbf{u}_r^{(1)} \circ \mathbf{u}_r^{(2)} \circ \cdots \circ \mathbf{u}_r^{(M)} = [\![\mathbf{U}^{(1)}, \ldots, \mathbf{U}^{(M)}]\!],$$

where $\mathbf{u}_r^{(m)} \in \mathbb{R}^{I_m}, m = 1, 2, \ldots, M$ is the projection vector, $\mathbf{U}^{(m)} = [\mathbf{u}_1^{(m)}; \cdots; \mathbf{u}_R^{(m)}]$, and R is the CP decomposition-based rank of the weighting tensor. By substituting the CP form of $\mathcal{W}$ into (8.13), we can get the following

Algorithm 44: Logistic tensor regression [19]

Input: The training measurements $\mathcal{X}_n \in \mathbb{R}^{I_1 \times I_2 \times \cdots \times I_M}$, $1 \leq n \leq N$, the corresponding class labels $y_n \in \{+1, -1\}$, iterations number $T_{\max}$

Output: The weighting tensor $\mathcal{W}$ and the bias b.

Step 1: Initialize randomly $\{\mathbf{U}^{(1)}, \ldots, \mathbf{U}^{(M)}\}|_{(0)}$;

Step 2: Repeat;

Step 3: $t \leftarrow t + 1$

Step 4: For $m = 1$ to M do
update $\mathbf{U}^{(m)}|_{(t)}$ by using BBR with $(\text{vec}(\tilde{\mathbf{X}}_{n(m)}), y_n)$ as input

Step 5: End For

Step 5: Until $\|\mathcal{W}^{(t)} - \mathcal{W}^{(t-1)}\|_F / \|\mathcal{W}^{(t-1)}\|_F$ or $t \geq T_{\max}$

problem:

$$\begin{aligned} \min_{\mathbf{U}^{(m)}, m=1,\ldots,M,b} \quad & \sum_{n=1}^{N} \log\left(1 + e^{<[\![\mathbf{U}^{(1)},\ldots,\mathbf{U}^{(M)}]\!],\mathcal{X}_n>+b}\right) \\ & - y_n(< [\![\mathbf{U}^{(1)}, \ldots, \mathbf{U}^{(M)}]\!], \mathcal{X}_n > +b). \end{aligned}$$

Inspired by the alternative minimization, the problem can be solved with respect to one of the variables $\{\mathbf{U}^{(m)}, m = 1, \ldots, M, b\}$ by keeping the others fixed.

Specifically, the subproblem with respect to $\mathbf{U}^{(m)}$ is rewritten as

$$\min_{\mathbf{U}^{(m)}} \sum_{n=1}^{N} \log\left(1 + e^{(\text{tr}(\mathbf{U}^{(m)}(\mathbf{U}^{(-m)})^{\text{T}}\mathbf{X}_{n(m)}^{\text{T}})+b)}\right) - y_n\left(\text{tr}\left(\mathbf{U}^{(m)}\left(\mathbf{U}^{(-m)}\right)^{\text{T}}\mathbf{X}_{n(m)}^{\text{T}}\right) + b\right),$$

where $\mathbf{U}^{(-m)} = \mathbf{U}^{(M)} \odot \cdots \odot \mathbf{U}^{(m+1)} \odot \mathbf{U}^{(m-1)} \odot \cdots \odot \mathbf{U}^{(1)}$, $\mathbf{X}_{n(m)}$ is mode-m unfolding matrix of tensor $\mathcal{X}_n$. Letting $\tilde{\mathbf{X}}_{n(m)} = \mathbf{X}_{n(m)}\mathbf{U}^{(-m)}$, since

$$\text{tr}\left(\mathbf{U}^{(m)}\tilde{\mathbf{X}}_{n(m)}^{\text{T}}\right) = \left[\text{vec}\left(\mathbf{U}^{(m)}\right)\right]^{\text{T}}\left[\text{vec}\left(\tilde{\mathbf{X}}_{n(m)}\right)\right],$$

we can vectorize the subproblem as

$$\min_{\mathbf{U}^{(m)}} \sum_{n=1}^{N} \log\left(1 + e^{[\text{vec}(\mathbf{U}^{(m)})]^{\text{T}}[\text{vec}(\tilde{\mathbf{X}}_{n(m)})]+b}\right) - y_n\left(\left[\text{vec}\left(\mathbf{U}^{(m)}\right)\right]^{\text{T}}\left[\text{vec}\left(\tilde{\mathbf{X}}_{n(m)}\right)\right]+b\right).$$

It is clear that the logistic tensor regression problem can be transformed into several vector logistic regression subproblems and easily solved in this way. The details of logistic tensor regression can be found in Algorithm 44. In addition, the ℓ_1 norm regularizer can be applied to make the solution sparse and interpretable, which can be well solved by the Bayesian binary regression (BBR) software introduced in [9].

8.4 Support Tensor Machine

8.4.1 *Support Vector Machine*

Support vector machine (SVM) [6] is a popular supervised learning model for classification problem. It finds an optimal classification hyperplane by maximizing the margin between the positive measurements and the negative measurements. Figure 8.2 provides an illustration of the support vector machine when the samples are linearly separable.

Mathematically, given N training measurements $\mathbf{x}_n \in \mathbb{R}^I (1 \leq n \leq N)$ associated with the class labels $y_n \in \{+1, -1\}$, the standard SVM finds a projection vector $\mathbf{w}$ and a bias b by a constrained optimization model as follows:

$$\begin{aligned} \min_{\mathbf{w},b,\xi} \quad & \mathrm{J}_{\mathrm{SVM}}(\mathbf{w}, b, \xi) = \frac{1}{2}\|\mathbf{w}\|_{\mathrm{F}}^2 + \lambda \sum_{n=1}^{N} \xi_n \\ \text{s. t.} \quad & y_n\left(\mathbf{w}^{\mathrm{T}}\mathbf{x}_n + b\right) \geq 1 - \xi_n, \ 1 \leq n \leq N, \xi_n \geq 0, \end{aligned} \tag{8.14}$$

where $\boldsymbol{\xi} = [\xi_1; \xi_2; \cdots; \xi_M]$ is a vector which contains all the slack variables, and it is used to deal with the linearly non-separable problem. ξ_n is also called the marginal error for the n-th measurement, and the margin is $2/\|\mathbf{w}\|_2^2$. When the problem is

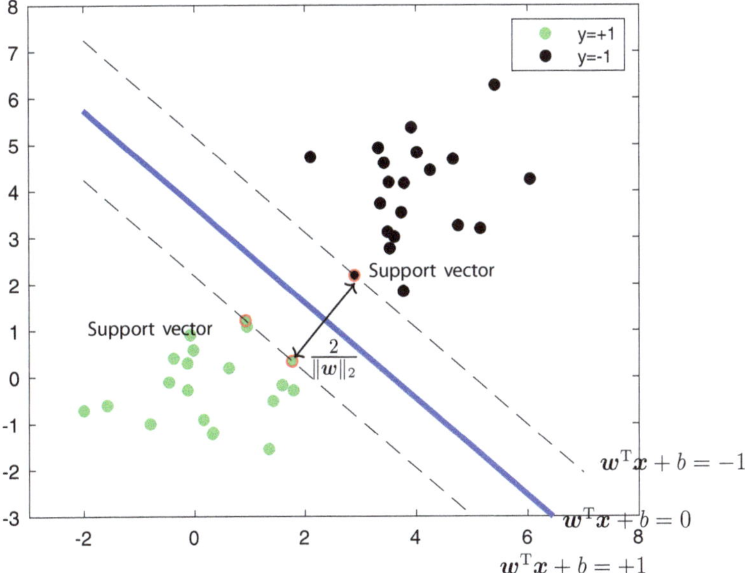

Fig. 8.2 An illustration of the support vector machine when the samples are linearly separable

linearly separable, we can set $\boldsymbol{\xi} = \mathbf{0}$. Under this condition the decision function is $y(\mathbf{x}_n) = \text{sign}(\mathbf{w}^{\mathrm{T}}\mathbf{x}_n + b)$.

To solve this problem, we give the Lagrangian function of the optimization model (8.14) as follows:

$$\begin{aligned} \mathrm{L}(\mathbf{w}, b, \boldsymbol{\xi}, \boldsymbol{\alpha}, \boldsymbol{\kappa}) = & \frac{1}{2}\|\mathbf{w}\|_{\mathrm{F}}^2 + \lambda \sum_{n=1}^{N} \xi_n \\ & - \sum_{n=1}^{N} \alpha_n \left(y_n \left(\mathbf{w}^{\mathrm{T}} \mathbf{x}_n + b \right) - 1 + \xi_n \right) - \sum_{n=1}^{N} \kappa_n \xi_n, \end{aligned} \tag{8.15}$$

where $\alpha_n \geq 0$, $\kappa_n \geq 0$, $1 \leq n \leq N$ are Lagrangian multipliers. And the solution is determined by the saddle point of the Lagrangian:

$$\max_{\boldsymbol{\alpha}, \boldsymbol{\kappa}} \min_{\mathbf{w}, b, \boldsymbol{\xi}} \ \mathrm{L}(\mathbf{w}, b, \boldsymbol{\xi}, \boldsymbol{\alpha}, \boldsymbol{\kappa}), \tag{8.16}$$

which can be solved by

$$\begin{aligned} \partial_{\mathbf{w}} \mathrm{L} = 0 & \Rightarrow \mathbf{w} = \sum_{n=1}^{N} \alpha_n y_n \mathbf{x}_n \\ \partial_b \mathrm{L} = 0 & \Rightarrow \sum_{n=1}^{N} \alpha_n y_n = 0 \\ \partial_{\boldsymbol{\xi}} \mathrm{L} = 0 & \Rightarrow \lambda - \boldsymbol{\alpha} - \boldsymbol{\kappa} = 0. \end{aligned} \tag{8.17}$$

Based on Eq. (8.17), the dual problem of optimization model (8.14) can be formulated as follows:

$$\begin{aligned} \max_{\boldsymbol{\alpha}} \quad & \mathrm{J}_D(\boldsymbol{\alpha}) = -\frac{1}{2} \sum_{n_1=1}^{N} \sum_{n_2=1}^{N} y_{n_1} y_{n_2} \mathbf{x}_{n_1} \mathbf{x}_{n_2} \alpha_{n_1} \alpha_{n_2} + \sum_{n=1}^{N} \alpha_n \\ \text{s. t.} \quad & \sum_{n=1}^{N} \alpha_n y_n = 0, 0 \leq \boldsymbol{\alpha} \leq \lambda. \end{aligned} \tag{8.18}$$

Here we can see that the dual problem (8.18) of the original problem (8.14) is a quadratic programming problem, which has some efficient solvers.

8.4.2 Support Tensor Machine

Considering a specific data processing problem, SVM may suffer from overfitting and result in a bad performance when the number of training measurements is limited or the measurements are tensors. Driven by the success of tensor analysis in many learning techniques, it is expected to design a tensor extension of SVM, namely, support tensor machine (STM) [1, 10, 20].

We consider that there are N training measurements denoted as M-th order tensors $\mathcal{X}_n \in \mathbb{R}^{I_1 \times I_2 \times \cdots \times I_M}$, $1 \leq n \leq N$ with their corresponding class labels $y_n \in \{+1, -1\}$. The STM aims to find M projection vectors $\mathbf{w}_m \in \mathbf{R}^{I_m}$, $1 \leq m \leq M$ and a bias b to formulate a decision function $y(\mathcal{X}_n) = \text{sign}(\mathcal{X}_n \prod_{m=1}^{M} \times_m \mathbf{w}_m + b)$. The desired parameters can be obtained by the following minimization problem:

$$\begin{aligned} \min_{\mathbf{w}_m|_{m=1}^M, b, \xi} \quad & \mathrm{J}_{\mathrm{STM}}\left(\mathbf{w}_m|_{m=1}^M, b, \xi\right) = \frac{1}{2}\|\circ_{m=1}^M \mathbf{w}_m\|_{\mathrm{F}}^2 + \lambda \sum_{n=1}^N \xi_n \\ \text{s. t.} \quad & y_n\left(\mathcal{X}_n \prod_{m=1}^M \times_m \mathbf{w}_m + b\right) \geq 1 - \xi_n, 1 \leq n \leq N, \xi_n \geq 0, \end{aligned} \tag{8.19}$$

where λ is a balance parameter. Here $\circ_{m=1}^M \mathbf{w}_m$ is short for $\mathbf{w}_1 \circ \mathbf{w}_2 \circ \cdots \circ \mathbf{w}_M$, and $\mathcal{X}_n \prod_{m=1}^M \times_m \mathbf{w}_m$ is short for $\mathcal{X}_n \times_1 \mathbf{w}_1 \times_2 \mathbf{w}_2 \times \cdots \times_M \mathbf{w}_M$.

Unlike the classical soft-margin SVM, the soft-margin STM does not have a closed-form solution. To solve this problem, we apply an alternating projection method [20] which is usually used to deal with tensor approximation problems. The details of this algorithm can refer to Algorithm 45. It just replaces Step 4 in Algorithm 45 with solution of the following optimization problem:

$$\begin{aligned} \min_{\mathbf{w}_m, b, \boldsymbol{\xi}} \quad & J_{\mathrm{SVM}}(\mathbf{w}_m, b, \boldsymbol{\xi}) = \frac{\eta}{2}\|\mathbf{w}_m\|_{\mathrm{F}}^2 + \lambda \sum_{n=1}^N \xi_n \\ \text{s. t.} \quad & y_n\left(\mathbf{w}_m^{\mathrm{T}}\left(\mathcal{X}_n \prod_{1 \leq k \leq M}^{k \neq m} \times_k \mathbf{w}_k\right) + b\right) \geq 1 - \xi_n, 1 \leq n \leq N, \end{aligned} \tag{8.20}$$

where $\eta = \prod_{1 \leq k \leq M}^{k \neq m} \|\mathbf{w}_k\|_{\mathrm{F}}^2$.

Algorithm 45: Alternating projection method for tensor optimization[20]

Input: The training measurements $\mathcal{X}_n \in \mathbb{R}^{I_1 \times I_2 \times \cdots \times I_M}$, $1 \leq n \leq N$, the corresponding class labels $y_n \in \{+1, -1\}$
Output: The projection vector of the tensorplane $\mathbf{w}_m \in \mathbb{R}^{I_m}$, $1 \leq m \leq M$ and $b \in \mathbb{R}$.
Step 1: Set the $\mathbf{w}_m$ as the random unit vectors in $\mathbb{R}^{I_m}$;
Step 2: Do steps 3–5 until convergence;
Step 3: For $m = 1$ to M:
Step 4: Update $\mathbf{w}_m$ by minimizing the SVM problem

$$\min_{\mathbf{w}_m, b, \xi} \ \mathrm{f}(\mathbf{w}_m, b, \xi)$$

$$\text{s. t.} \quad y_n c_n (\mathbf{w}_m^{\mathrm{T}} (\mathcal{X}_n \prod_{1 \leq k \leq M}^{k \neq m} \times_k \mathbf{w}_k) + b) \geq \xi_n, \ 1 \leq n \leq N$$

Step 5: End For
Step 6: Convergence checking:

$$\text{if} \ \sum_{m=1}^{M} [\mathbf{w}_{m,t}^{\mathrm{T}} \mathbf{w}_{m,t-1} (\|\mathbf{w}_{m,t}\|_{\mathrm{F}}^{-2}) - 1] \leq \epsilon.$$

the calculate $\mathbf{w}_m$, $m = 1, \ldots, M$ have converged. Here $\mathbf{w}_{m,t}$ is the current projection vector and $\mathbf{w}_{m,t-1}$ is the previous projection vector.
Step 7: End

8.4.3 *Higher-Rank Support Tensor Machine*

Considering the decision function $y = \text{sign}(\mathcal{X} \prod_{m=1}^{M} \times_m \mathbf{w}_m + b)$ from the classical STM, we can see that the weighting parameters $\mathbf{w}_m|_{m=1}^{M}$ form a rank-1 CP tensor $\mathcal{W} = \circ_{m=1}^{M} \mathbf{w}_m$. Each mode of this weighting tensor is represented by a vector, which could lead to poor representation ability. That is because real tensor data may not be very low-rank and a rank-1 weighting tensor cannot represent the true structure of original data sufficiently. The higher-rank support tensor machine (HRSTM) estimates the weighting tensor as the sum of several rank-1 tensors using CP decomposition [14]. It can still be regarded as a low-rank tensor but not strictly rank-1 tensor. The optimization problem for HRSTM is formulated as follows:

$$\begin{aligned} \min_{\mathcal{W}, b, \xi} \quad & \frac{1}{2} < \mathcal{W}, \mathcal{W} > + C \sum_{n=1}^{N} \xi_i \\ \text{s.t.} \quad & y_n (< \mathcal{W}, \mathcal{X}_n > + b) \geq 1 - \xi_n, \quad 1 \leq n \leq N, \quad \xi \geq 0 \\ & \mathcal{W} = [\![\mathbf{U}^{(1)}, \ldots, \mathbf{U}^{(M)}]\!]. \end{aligned} \tag{8.21}$$

Unlike STM, which use multiple projections through mode-m products $\mathcal{X}\prod_{m=1}^{M}\times_m \mathbf{w}_m$ to map the input tensor to a scalar, HRSTM applies tensor inner product $<\mathcal{W},\mathcal{X}_n>$ to map the input tensor. Here all the parameters are the same as STM except for the weighting tensor $\mathcal{W}$.

Based on the definition of tensor inner product and CP decomposition in Chaps. 1 and 2, we can get the following:

$$<\mathcal{W},\mathcal{W}> = \mathrm{tr}\big(\mathbf{W}_{(m)}\mathbf{W}_{(m)}^{\mathrm{T}}\big) = \mathrm{tr}\big(\mathbf{U}^{(m)}\big(\mathbf{U}^{(-m)}\big)^{\mathrm{T}}\mathbf{U}^{(-m)}\big(\mathbf{U}^{(m)}\big)^{\mathrm{T}}\big), \quad (8.22)$$

$$<\mathcal{W},\mathcal{X}_n> = \mathrm{tr}\big(\mathbf{W}_{(m)}\mathbf{X}_{n(m)}^{\mathrm{T}}\big) = \mathrm{tr}\big(\mathbf{U}^{(m)}\big(\mathbf{U}^{(-m)}\big)^{\mathrm{T}}\mathbf{X}_{n(m)}^{\mathrm{T}}\big), \quad (8.23)$$

where $\mathbf{W}_{(m)}$ is the mode-m unfolding matrix of tensor $\mathcal{W}$ and $\mathbf{X}_{n(m)}$ is mode-m unfolding matrix of tensor $\mathcal{X}_n$, $\mathbf{U}^{(-m)} = \mathbf{U}^{(M)}\odot\cdots\odot\mathbf{U}^{(m+1)}\odot\mathbf{U}^{(m-1)}\odot\cdots\odot\mathbf{U}^{(1)}$.

Substituting (8.22) and (8.23) into (8.21), the problem becomes

$$\begin{aligned}\min_{\mathbf{U}^{(m)},b,\boldsymbol{\xi}} \quad & \frac{1}{2}\mathrm{tr}\left(\mathbf{U}^{(m)}(\mathbf{U}^{(-m)})^{\mathrm{T}}\mathbf{U}^{(-m)}(\mathbf{U}^{(m)})^{\mathrm{T}}\right) + C\sum_{n=1}^{N}\xi_n \\ \text{s. t.} \quad & y_n\left(\mathrm{tr}\left(\mathbf{U}^{(m)}(\mathbf{U}^{(-m)})^{\mathrm{T}}\mathbf{X}_{n(m)}^{\mathrm{T}}\right)+b\right) \geq 1-\xi_n,\ \ 1\leq n\leq N,\ \ \xi\geq 0.\end{aligned} \quad (8.24)$$

To solve this problem, we define $\mathbf{A} = \mathbf{U}^{(-m)^{\mathrm{T}}}\mathbf{U}^{(-m)}$, which is a positive definite matrix, and let $\tilde{\mathbf{U}}^{(m)} = \mathbf{U}^{(m)}\mathbf{A}^{\frac{1}{2}}$; then we have:

$$\mathrm{tr}\left(\mathbf{U}^{(m)}\big(\mathbf{U}^{(-m)}\big)^{\mathrm{T}}\mathbf{U}^{(-m)}\big(\mathbf{U}^{(m)}\big)^{\mathrm{T}}\right) = \mathrm{tr}\left(\tilde{\mathbf{U}}^{(m)}\big(\tilde{\mathbf{U}}^{(m)}\big)^{\mathrm{T}}\right) = \mathrm{vec}\big(\tilde{\mathbf{U}}^{(m)}\big)^{\mathrm{T}}\mathrm{vec}\big(\tilde{\mathbf{U}}^{(m)}\big),$$

$$\mathrm{tr}\left(\mathbf{U}^{(m)}(\mathbf{U}^{(-m)})^{\mathrm{T}}\mathbf{X}_{n(m)}^{\mathrm{T}}\right) = \mathrm{tr}\left(\tilde{\mathbf{U}}^{(m)}\tilde{\mathbf{X}}_{n(m)}^{\mathrm{T}}\right) = \mathrm{vec}(\tilde{\mathbf{U}}^{(m)})^{\mathrm{T}}\mathrm{vec}(\tilde{\mathbf{X}}_{n(m)}),$$

where $\tilde{\mathbf{X}}_{n(m)} = \mathbf{X}_{n(m)}\mathbf{U}^{(-m)}\mathbf{A}^{-\frac{1}{2}}$.

Finally, the original optimization problem (8.21) can be simplified as

$$\begin{aligned}\min_{\tilde{\mathbf{U}}^{(m)},b,\boldsymbol{\xi}} \quad & \frac{1}{2}\mathrm{vec}(\tilde{\mathbf{U}}^{(m)})^{\mathrm{T}}\mathrm{vec}(\tilde{\mathbf{U}}^{(m)}) + C\sum_{n=1}^{N}\xi_n \\ \text{s. t.} \quad & y_n\big(\mathrm{vec}\big(\tilde{\mathbf{U}}^{(m)}\big)^{\mathrm{T}}\mathrm{vec}\big(\tilde{\mathbf{X}}_{n(m)}\big)+b\big) \geq 1-\xi_n,\ \ 1\leq n\leq N,\ \ \xi\geq 0,\end{aligned} \quad (8.25)$$

which can be solved similarly by the alternative projection method.

8.4.4 Support Tucker Machine

The classical STM and HRSTM models assume that weighting tensor admits low-rank structure based on CP decomposition. It is well known that the best rank-R approximation in CP decomposition of a given data is NP-hard to obtain. In [13], a support Tucker machine (STuM) obtains low-rank weighting tensor by Tucker decomposition. Compared with STM and HRSTM, which formulate weighting tensor under CP decomposition, STuM is more general and can explore the multilinear rank especially for unbalanced datasets. Furthermore, since the Tucker decomposition of a given tensor is represented as a core tensor multiplied by a series of factor matrices along all modes, we can easily perform a low-rank tensor approximation for dimensionality reduction which is practical when dealing with complex real data.

Assuming that the weighting tensor is $\mathcal{W}$, the optimization problem of STuM is as follows:

$$\begin{aligned} \min_{\mathcal{W},b,\boldsymbol{\xi}} \quad & \frac{1}{2} < \mathcal{W}, \mathcal{W} > +C \sum_{n=1}^{N} \xi_n \\ \text{s. t.} \quad & y_n(< \mathcal{W}, \mathcal{X}_n > +b) \geq 1 - \xi_n, \quad 1 \leq n \leq N, \quad \xi \geq 0 \\ & \mathcal{W} = [\![\mathcal{G}, \mathbf{U}^{(1)}, \ldots, \mathbf{U}^{(M)}]\!]. \end{aligned} \tag{8.26}$$

Based on the Tucker decomposition theory, the mode-m matricization of tensor $\mathcal{W}$ can be written as

$$\begin{aligned} \mathbf{W}_{(m)} &= \mathbf{U}^{(m)} \mathbf{G}_{(m)} \left(\mathbf{U}^{(M)} \otimes \cdots \otimes \mathbf{U}^{(m+1)} \otimes \mathbf{U}^{(m-1)} \otimes \cdots \otimes \mathbf{U}^{(1)} \right)^{\mathrm{T}} \\ &= \mathbf{U}^{(m)} \mathbf{G}_{(m)} (\mathbf{U}^{(-m)})^{\mathrm{T}}, \end{aligned}$$

where $\mathbf{G}_{(m)}$ is mode-m matricization of the core tensor $\mathcal{G}$. The resulting subproblem with respect to $\mathbf{U}^{(m)}$ can be represented as follows:

$$\begin{aligned} \min_{\mathbf{U}^{(m)},b,\boldsymbol{\xi}} \quad & \frac{1}{2} \mathrm{tr} \left(\mathbf{U}^{(m)} \mathbf{P}^{(m)} (\mathbf{P}^{(m)})^{\mathrm{T}} (\mathbf{U}^{(m)})^{\mathrm{T}} \right) + C \sum_{n=1}^{N} \xi_n \\ \text{s. t.} \quad & y_n (\mathrm{tr} \left(\mathbf{U}^{(m)} \mathbf{P}^{(m)} \mathbf{X}_{n(m)}^{\mathrm{T}} \right) + b) \geq 1 - \xi_n, \quad 1 \leq n \leq N, \quad \xi \geq 0, \end{aligned} \tag{8.27}$$

where $\mathbf{P}^{(m)} = \mathbf{G}_{(m)}(\mathbf{U}^{(-m)})^{\mathrm{T}}$, $\mathbf{X}_{n(m)}$ is the mode-m matricization of $\mathcal{X}_n$. It can be solved similarly as it does for (8.24). The subproblem with respect to $\mathcal{G}$ is as follows:

$$\begin{aligned} \min_{\mathbf{G}_{(1)}, b, \boldsymbol{\xi}} \quad & \frac{1}{2} \left(\mathbf{U}\operatorname{vec}(\mathbf{G}_{(1)})\right)^{\mathrm{T}} \left(\mathbf{U}\operatorname{vec}(\mathbf{G}_{(1)})\right) + C\sum_{n=1}^{N} \xi_n \\ \text{s. t.} \quad & y_n\left(\left(\mathbf{U}\operatorname{vec}(\mathbf{G}_{(1)})\right)^{\mathrm{T}} \operatorname{vec}(\mathcal{X}) + b\right) \geq 1 - \xi_n, \ \ 1 \leq n \leq N, \ \ \boldsymbol{\xi} \geq 0, \end{aligned} \tag{8.28}$$

where $\mathbf{U} = \mathbf{U}^{(M)} \otimes \cdots \otimes \mathbf{U}^{(1)}$ and $\mathbf{G}_{(1)}$ is the mode-1 unfolding matrix of the core tensor $\mathcal{G}$.

As shown above, the optimization problem of STuM can also be transformed into several standard SVM problems. The difference of STuM from HRSTM is that STuM has an additional iteration for solving the subproblem of core tensor $\mathcal{G}$. More details can be found in [13].

In addition, another variant based on tensor train decomposition, namely, support tensor train machine (STTM), is also proposed due to the enhanced expression capability of tensor networks [4]. Besides, in order to deal with the complex case that the samples are not linearly separate, some kernelized tensor machines are proposed, including the dual structure-preserving kernel (DuSK) for supervised tensor learning [11] and the kernelized STTM [5].

8.5 Tensor Fisher Discriminant Analysis

8.5.1 Fisher Discriminant Analysis

Fisher discriminant analysis (FDA) [8, 12] has been widely applied for classification. It aims to find a direction which separates the means of class well while minimizing the variance of the total training measurements. The basic idea of FDA is to project the high-dimensional samples into a best vector space for classification feature extraction and dimensionality reduction. After projecting all the measurements onto a specific line $y = \mathbf{w}^{\mathrm{T}}\mathbf{x}$, FDA separates the positive measurements from the negative measurements by maximizing the generalized Rayleigh quotient[1] between the between-class scatter matrix and within-class scatter matrix.

Here we consider N training measurements $\mathbf{x}_n$, $1 \leq n \leq N$ and the corresponding class labels $y_n \in \{+1, -1\}$. Among these measurements, there are N_+ positive measurements with the mean $\mathbf{m}_+ = \frac{1}{N_+}\sum_{n=1}^{N}(\mathrm{I}(y_n + 1)\mathbf{x}_n)$ and N_- measurements

[1]For a real symmetric matrix $\mathbf{A}$, the Rayleigh quotient is defined as $\mathrm{R}(\mathbf{x}) = \frac{\mathbf{x}^{\mathrm{T}}\mathbf{A}\mathbf{x}}{\mathbf{x}^{\mathrm{T}}\mathbf{x}}$.

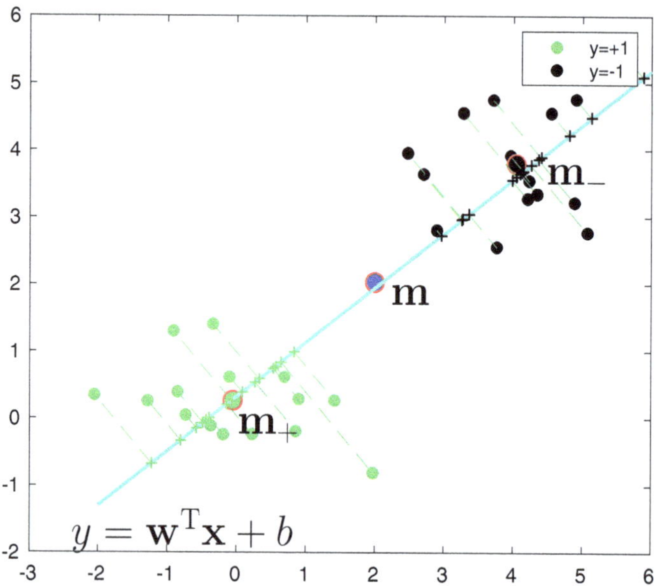

Fig. 8.3 An illustration of the Fisher discriminant analysis when the samples are linear separable

with the mean $\mathbf{m}_- = \frac{1}{N_-}\sum_{n=1}^{N}(\mathrm{I}(y_n - 1)\mathbf{x}_n)$. The mean of all the measurements is $\mathbf{m} = \frac{1}{N}\sum_{n=1}^{N}\mathbf{x}_n$, where the indicator function is defined as follows:

$$\mathrm{I}(x) = \begin{cases} 0 & x = 0, \\ 1 & x \neq 0. \end{cases}$$

Figure 8.3 gives an illustration of the FDA. The key problem is how to find the best projection vector in order to ensure that the samples have the largest inter-class distance and smallest intra-class distance in the new subspace. The projection vector **w** can be determined by solving the following problem:

$$\begin{aligned} \max_{\mathbf{w}} \quad J_{\mathrm{FDA}}(\mathbf{w}) &= \frac{\mathbf{w}^{\mathrm{T}}\mathbf{S}_{\mathrm{b}}\mathbf{w}}{\mathbf{w}^{\mathrm{T}}\mathbf{S}_{\mathrm{w}}\mathbf{w}} \\ &= \frac{\|\mathbf{w}^{\mathrm{T}}(\mathbf{m}_+ - \mathbf{m}_-)\|_2}{\sqrt{\mathbf{w}^{\mathrm{T}}\Sigma\mathbf{w}}}, \end{aligned} \tag{8.29}$$

where $\mathbf{S}_{\mathrm{b}} = (\mathbf{m}_+ - \mathbf{m}_-)(\mathbf{m}_+ - \mathbf{m}_-)^{\mathrm{T}}$ is the between-class scatter matrix and $\mathbf{S}_{\mathrm{w}} = \sum_{i=1}^{2}\mathbf{S}_{\mathrm{w}_i}$, where $\mathbf{S}_{\mathrm{w}_i}$ is the scatter matrix for the positive or negative class, is the within-class scatter matrix. This problem is equivalent to maximize the symmetric Kullback-Leibler divergence (KLD) [7] between the positive and the negative measurements with identical covariances. Therefore, we can separate the positive samples from the negative measurements by the decision function $y(\mathbf{x}) = \text{sign}(\mathbf{w}^T\mathbf{x} + b)$, where $\mathbf{w}$ is the eigenvector of $\Sigma^{-1}(\mathbf{m}_+ - \mathbf{m}_-)(\mathbf{m}_+ - \mathbf{m}_-)^T$ associated with the largest eigenvalue, and the bias is given by

$$b = \frac{N_- - N_+ - (N_+\mathbf{m}_+ + N_-\mathbf{m}_-)^T\mathbf{w}}{N_- + N_+}.$$

8.5.2 Tensor Fisher Discriminant Analysis

Here we consider N measurements represented as tensors $\mathcal{X}_n$, $1 \le n \le N$ and the corresponding class labels $y_n \in \{+1, -1\}$. Among these measurements, there are N_+ positive measurements with their mean $\mathcal{M}_+ = \frac{1}{N_+}\sum_{n=1}^{N}(\mathrm{I}(y_n + 1)\mathcal{X}_n)$ and N_- measurements with their mean $\mathcal{M}_- = \frac{1}{N_-}\sum_{n=1}^{N}(\mathrm{I}(y_n - 1)\mathcal{X}_n)$. The mean of all the measurements is $\mathcal{M} = \frac{1}{N}\sum_{n=1}^{N}\mathcal{X}_n$.

In tensor FDA (TFDA) [16, 17], the decision function is given by $y(\mathcal{X}) = \text{sign}(\mathcal{X}\prod_{m=1}^{M}\times_m \mathbf{w}_m + b)$, and the multiple projection vector $\{\mathbf{w}_m, m = 1, \ldots, M\}$ and b can be obtained by

$$\max_{\mathbf{w}_m|_{m=1}^{M}} \quad \mathrm{J_{TFDA}} = \frac{\|(\mathcal{M}_+ - \mathcal{M}_-)\prod_{m=1}^{M}\times_m \mathbf{w}_m\|_2^2}{\sum_{n=1}^{N}\|(\mathcal{X}_n - \mathcal{M})\prod_{m=1}^{M}\times_m \mathbf{w}_m\|_2^2}. \tag{8.30}$$

Just like STM in (8.19), there is no closed-form solution for TFDA. The alternating projection method can be applied to solve this problem, and one of the multiple projection vectors will be updated at each iteration. At each iteration, the solution of $\mathbf{w}_m$ can be updated by

$$\max_{\mathbf{w}_m} \quad \mathrm{J_{TFDA}} = \frac{\|\mathbf{w}_m^T((\mathcal{M}_+ - \mathcal{M}_-)\prod_{1\le k\le M}^{k\ne m}\times_k \mathbf{w}_k)\|_2^2}{\sum_{n=1}^{N}\|\mathbf{w}_m^T((\mathcal{X}_n - \mathcal{M})\prod_{1\le k\le M}^{k\ne m}\times_k \mathbf{w}_k)\|_2^2}. \tag{8.31}$$

To get the multiple projection vectors, we only need to replace the subproblem of Step 4 in Algorithm 45 with (8.31). And the solution to the bias is computed by

$$b = \frac{N_- - N_+ - (N_+\mathcal{M}_+ + N_-\mathcal{M}_-)\prod_{m=1}^{M}\times_m \mathbf{w}_m}{N_- + N_+}.$$

8.6 Applications

Generally, tensor classification methods are chosen to deal with classification tasks related to multiway data, such as network abnormal detection[18], pattern classification [21], and gait recognition [14]. To validate the effectiveness of tensor classification, two groups of experiments have been performed in this section.

8.6.1 Handwriting Digit Recognition

To briefly show the differences between the vector-based methods and the tensor-based methods, two classical methods STM and SVM are compared on the well-known MNIST dataset[15]. There are 60k training samples and 10k testing samples in this database. Each sample is a grayscale image of size 28×28, denoting a handwritten digit {"0", ..., "9"}. Since STM and SVM are naturally binary classifiers, we divide these 9 digits into 45 digit pairs and do classification task on each digit pair separately. Then all the classification accuracy of these 45 group experiments is calculated.

Table 8.1 shows the classification results on several selected digit pairs. As shown for these five pairs, STM gets higher accuracy than SVM. This can be possibly explained by that the spatial structure information is reserved in STM. The classification accuracy of SVM and STM for all the digit pairs are also given in Fig. 8.4. It can be seen that the performance of STM is better than SVM in most cases.

Table 8.1 Classification accuracy for different MNIST digit pairs

Digit pair	{'0','3'}	{'1','9'}	{'2','7'}	{'4','6'}	{'5','8'}
SVM	98.09%	95.85%	96.69%	98.25%	95.97%
STM	99.05%	99.58%	97.91%	97.68%	92.87%

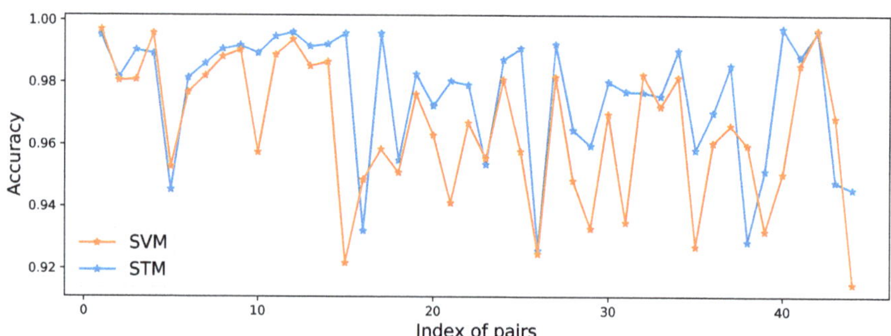

Fig. 8.4 Classification accuracy of SVM and STM for all the digit pairs

8.6.2 Biomedical Classification from fMRI Images

In order to make a comprehensive comparison of various tensor classification methods, we choose several tensor-based methods to classify the instantaneous cognitive states from fMRI data. The compared methods include:

- SVM [6]: classical support vector machine, which is achieved by the standard library for support vector machines (libsvm)[2].
- STM [1]: support tensor machine based on rank-1 CP decomposition
- STuM [13]: support tensor machine based on Tucker decomposition
- DuSK [11]: a kernelized support tensor machine based on CP decomposition
- TT classifier [3]: an advanced tensor classification method, which is trained as a polynomial classifier in tensor train format
- KSTTM [5]: kernelized support tensor train machine

We consider a higher order fMRI dataset, namely, the StarPlus fMRI dataset,[2] to evaluate the performance of different models mentioned above. StarPlus fMRI dataset contains the brain images of six human subjects. The data of each human subject are partitioned into trials, and each subject has 40 effective trials. During each effective trial, the subject is shown a sentence and a simple picture in sequence and then presses a button to indicate whether the sentence correctly describes the picture. In half of the trials, the picture is presented first, followed by the sentence. In the remaining trials, the sentence is presented first, followed by the picture. In either case, the first stimulus (sentence or picture) is presented for 4 s.

Here we only use the first 4 s of each trial since the subject is shown only one kind of stimulus (sentence or picture) during the whole period. The fMRI images are collected every 500 ms, and we can utilize eight fMRI images in each trial. Overall, we have 320 fMRI images: half for picture stimulus, half for sentence stimulus. Each fMRI image is in size of $64 \times 64 \times 8$, which contains 25–30 anatomically defined regions (called "regions of interest", or ROIs). To achieve a better classification accuracy, we only consider the following ROIs: "CALC," "LIPL," "LT," "LTRIA," "LOPER," "LIPS," "LDLPFC". After extracting those ROIs, we further normalize the data of each subject.

The main goal for this task is to train classifiers to distinguish whether the subject is viewing a sentence or a picture. We randomly select 140 images for training, 60 for validation, and the others for testing. It is worth noting that in neuroimaging task, it is very hard to achieve even moderate classification accuracy on StarPlus dataset since this dataset is extremely high dimensional (number of voxels $= 64 \times 64 \times 8 =$ 32,768) but with small sample size (number of samples 320).

The classification results by SVM, STM, STuM, DuSK, TT classifier, and KSTTM are presented in Table 8.2. Due to the very high dimensionality and sparsity of the data, SVM fails to find a good hyperparameter setting, thus performing

[2] http://www.cs.cmu.edu/afs/cs.cmu.edu/project/theo-81/www/.

Table 8.2 Classification accuracy of different methods for different subjects in StarPlus fMRI datasets

Subject	SVM	STM	STuM	DuSK	TT classifier	KSTTM
04799	36.67%	45.83%	52.50%	59.18%	57.50%	55.83%
04820	43.33%	44.12%	56.67%	58.33%	54.17%	70.00%
04847	38.33%	47.50%	49.17%	50.83%	61.67%	65.00%
05675	35.00%	37.50%	45.83%	57.50%	55.00%	60.00%
05680	38.33%	40.00%	58.33%	64.17%	60.83%	73.33%
05710	40.00%	43.33%	62.50%	59.17%	53.33%	59.17%

poorly on fMRI image classification task. Since fMRI data are very complicated, the STM which is based on rank-1 CP decomposition cannot achieve an acceptable classification accuracy. We attribute this to the fact that the rank-1 CP tensor is simple and cannot extract complex structure of an fMRI image. Compared with SVM and STM, the Tucker-based STuM and higher-rank CP-based DuSK give better results. Meanwhile, the TT classifier and KSTTM based on tensor train also give satisfactory results. Overall, we can conclude that tensor classification methods are more effective than their counterparts for high-dimensional data.

8.7 Summary

This chapter gives a brief introduction to the statistical tensor classification. First, we give an introduction for the concepts and the framework of tensor classification. Second, we give some examples and show how to convert vector classification models to tensor classification models. Specifically, the techniques of logistic tensor regression, support tensor machines, and tensor Fisher discriminant analysis are discussed in details. Finally, we apply tensor classification methods to two applications, i.e., handwriting digit recognition from grayscale images and biomedical classification from fMRI images. Two groups of experiments are performed on MNIST and fMRI datasets, respectively. The former one simply validates the effectiveness of tensor classification, while the latter provides a comprehensive comparison of existing tensor classifiers. Numerical results show that the classification methods based on complicated tensor structures, like TT format and higher-rank CP tensor, have stronger representation ability and perform better in practice.

Here are some interesting research directions in the future. Instead of the linear classifiers like SVM and the logistic regression, some nonlinear methods like Bayesian probability model and kernel tricks can be applied into tensor classification. Besides using tensor decomposition for observed multiway data for tensor classification, constructing tensor-based features is also an attractive way in the future.

References

1. Biswas, S.K., Milanfar, P.: Linear support tensor machine with LSK channels: Pedestrian detection in thermal infrared images. IEEE Trans. Image Process. **26**(9), 4229–4242 (2017)
2. Chang, C.C., Lin, C.J.: LIBSVM: A library for support vector machines. ACM Trans. Intell. Syst. Technol. **2**, 27:1–27:27 (2011). Software available at http://www.csie.ntu.edu.tw/~cjlin/libsvm
3. Chen, Z., Batselier, K., Suykens, J.A., Wong, N.: Parallelized tensor train learning of polynomial classifiers. IEEE Trans. Neural Netw. Learn. Syst. **29**(10), 4621–4632 (2017)
4. Chen, C., Batselier, K., Ko, C.Y., Wong, N.: A support tensor train machine. In: 2019 International Joint Conference on Neural Networks (IJCNN), pp. 1–8. IEEE, Piscataway (2019)
5. Chen, C., Batselier, K., Yu, W., Wong, N.: Kernelized support tensor train machines (2020). Preprint arXiv:2001.00360
6. Cortes, C., Vapnik, V.: Support-vector networks. Mach. Learn. **20**(3), 273–297 (1995)
7. Cover, T.M.: Elements of Information Theory. Wiley, Hoboken (1999)
8. Fisher, R.A.: The statistical utilization of multiple measurements. Ann. Eugen. **8**(4), 376–386 (1938)
9. Genkin, A., Lewis, D.D., Madigan, D.: Large-scale Bayesian logistic regression for text categorization. Technometrics **49**(3), 291–304 (2007)
10. Hao, Z., He, L., Chen, B., Yang, X.: A linear support higher-order tensor machine for classification. IEEE Trans. Image Process. **22**(7), 2911–2920 (2013)
11. He, L., Kong, X., Yu, P.S., Yang, X., Ragin, A.B., Hao, Z.: DuSk: A dual structure-preserving kernel for supervised tensor learning with applications to neuroimages. In: Proceedings of the 2014 SIAM International Conference on Data Mining, pp. 127–135. SIAM, Philadelphia (2014)
12. Kim, S.J., Magnani, A., Boyd, S.: Robust fisher discriminant analysis. In: Advances in Neural Information Processing Systems, pp. 659–666 (2006)
13. Kotsia, I., Patras, I.: Support tucker machines. In: CVPR 2011, pp. 633–640. IEEE, Piscataway (2011)
14. Kotsia, I., Guo, W., Patras, I.: Higher rank support tensor machines for visual recognition. Pattern Recognit. **45**(12), 4192–4203 (2012)
15. Deng, L.: The MNIST database of handwritten digit images for machine learning research [best of the web]. IEEE Signal Process. Mag. **29**(6), 141–142 (2012)
16. Li, Q., Schonfeld, D.: Multilinear discriminant analysis for higher-order tensor data classification. IEEE Trans. Pattern Analy. Mach. Intell. **36**(12), 2524–2537 (2014)
17. Phan, A.H., Cichocki, A.: Tensor decompositions for feature extraction and classification of high dimensional datasets. Nonlinear Theory Appl. **1**(1), 37–68 (2010)
18. Sun, J., Tao, D., Faloutsos, C.: Beyond streams and graphs: dynamic tensor analysis. In: Proceedings of the 12th ACM SIGKDD International Conference on Knowledge Discovery and Data Mining, pp. 374–383 (2006)
19. Tan, X., Zhang, Y., Tang, S., Shao, J., Wu, F., Zhuang, Y.: Logistic tensor regression for classification. In: International Conference on Intelligent Science and Intelligent Data Engineering, pp. 573–581. Springer, Berlin (2012)
20. Tao, D., Li, X., Hu, W., Maybank, S., Wu, X.: Supervised tensor learning. In: Fifth IEEE International Conference on Data Mining (ICDM'05), pp. 8–pp. IEEE, Piscataway (2005)
21. Tao, D., Li, X., Maybank, S.J., Wu, X.: Human carrying status in visual surveillance. In: 2006 IEEE Computer Society Conference on Computer Vision and Pattern Recognition (CVPR'06), vol. 2, pp. 1670–1677. IEEE, Piscataway (2006)

Chapter 9
Tensor Subspace Cluster

9.1 Background

As the scale and diversity of data increase, adaptations to existing algorithms are required to maintain cluster quality and speed. Many advances [43] in processing high-dimensional data have relied on the observation that, even though these data are high dimensional, their intrinsic dimension is often much smaller than the dimension of the ambient space. Similarly, subspace clustering is a technique about how to cluster those high-dimensional data within multiple low-dimensional subspaces. For example, suppose a group of data $\mathbf{X} = [\mathbf{x}_1, \mathbf{x}_2, \cdots, \mathbf{x}_N] \in \mathbb{R}^{D \times N}$, where N is the number of samples and D is the feature dimension. The subspace clustering seeks clusters by assuming that these samples are chosen from K unknown lower dimensional subspaces $\mathbb{S}_1, \mathbb{S}_2, \cdots, \mathbb{S}_K$. Exploring these useful subspaces and further grouping the observations according to these distinct subspaces are the main aims of subspace clustering. As illustrated in Fig. 9.1, the observed data can be clustered with three subspaces, namely, a plane and two lines.

Traditional matrix-based subspace clustering can be divided into four categories: statistical, algebraic, iterative, and spectral clustering-based methods [11]. Statistical approaches can be divided into three subclasses, iterative statistical approaches [41], robust statistical approaches [14], and information theoretic statistical approaches [34], which introduce different statistical distributions in their models. Algebraic methods contain two subclasses, factorization-based methods using matrix factorization [8, 15] and algebraic geometric methods using polynomials to fit the data, such as [29, 44]. The iterative approaches alternate two steps (i.e., allocate points to subspaces and fit subspaces to each cluster) to obtain the clustering results such as [42, 55]. Spectral clustering-based methods include two subclasses, one calculates the similarity matrix based on local information [16, 52], while the other employs global information [9, 10].

However, these subspace clustering approaches based on matrix operations often suffer from two limitations. The first one is that in most existing matrix-based works,

Y. Liu et al., *Tensor Computation for Data Analysis*,
https://doi.org/10.1007/978-3-030-74386-4_9

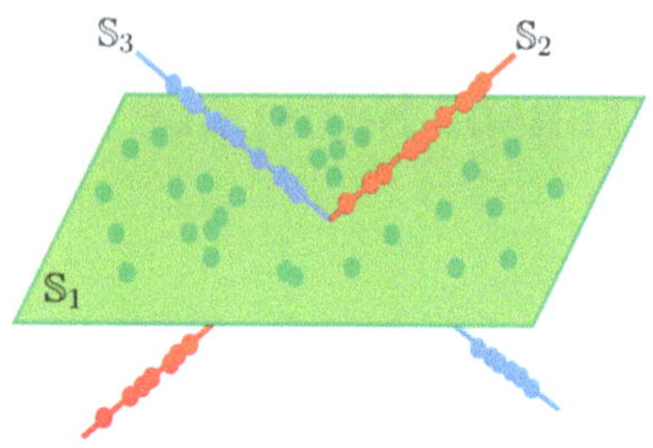

Fig. 9.1 Illustration of subspace clustering. In this figure, the samples in the dataset can be clustered according to three subspaces $\mathbb{S}_1$, $\mathbb{S}_2$, and $\mathbb{S}_3$

samples are simply considered to be independent and identically distributed. That is to say, samples are ordered arbitrarily. In this way, the spatial relationships of samples are ignored, resulting in poor clustering performance [47]. The other is that the structure information of samples is lost due to the vectorization or matricization of higher-order samples [26, 34].

In many applications, the high-dimensional data can be represented by tensors, which can ensure the integrity of the data to the greatest extent, and the spatial structures can be well preserved. But how to deal with these tensor data after introducing the tensor representation is an important issue.

Tensor subspace clustering is a generalization to the traditional subspace clustering, which can be applied to deal with complex datasets like heterogeneous data. In this section, we will mainly introduce two tensor subspace clustering models, including K-means and self-representation, and then illustrate several interesting applications and examples of tensor subspace clustering for better understanding.

9.2 Tensor Subspace Cluster Based on K-Means

9.2.1 Matrix-Based K-Means Clustering

K-means is an unsupervised clustering algorithm, which divides unlabeled data into a certain number (commonly denoted as K) of distinct groupings by minimizing Euclidean distances between them. It believes that the closer the distance between two points, the greater the similarity. For example, given N samples of size D $\mathbf{Z} \in \mathbb{R}^{D\times N}$, K-means clustering starts with randomly chosen K centroids as initial centroids $\mathbf{c}_k, k = 1, \cdots, K$, then proceeds to assign each of the rest samples to its nearest centroid based on the squared Euclidean distance, and recomputes the new cluster centroids in an iterative way. Its mathematical formulation is as follows:

$$\min_{\{\mathbf{c}_k\}} \sum_{n=1}^{N} \min_{1\leq k\leq K} \|\mathbf{z}_n - \mathbf{c}_k\|_2^2 = \sum_{k=1}^{K}\sum_{n=1}^{N} s_{kn} \|\mathbf{z}_n - \mathbf{c}_k\|_2^2, \tag{9.1}$$

where $\mathbf{z}_n \in \mathbb{R}^D$ is the n-th column of matrix $\mathbf{Z}$, $\mathbf{c}_k \in \mathbb{R}^D$ is a cluster centroid, and s_{kn} is a binary indicator variable such that

$$s_{kn} = \begin{cases} 1, & \text{if } \mathbf{z}_n \text{ belongs to } k\text{-th cluster,} \\ 0, & \text{otherwise .} \end{cases} \tag{9.2}$$

The traditional K-means clustering algorithm that classifies the columns of $\mathbf{Z} \in \mathbb{R}^{D\times N}$ into K clusters is illustrated in Algorithm 46. Furthermore, it is proved that the above formula (9.1) can be transformed into the form of matrix decomposition as follows [3]:

$$\begin{aligned} &\min_{\mathbf{C},\mathbf{S}} \|\mathbf{Z} - \mathbf{CS}\|_{\mathrm{F}}^2 \\ &\text{s.t. } \|\mathbf{S}(:, n)\|_0 = 1, \mathbf{S}(k, n) \in \{0, 1\}, \end{aligned} \tag{9.3}$$

where $\mathbf{C} \in \mathbb{R}^{D\times K}$ represents the centroids of the clusters, $\mathbf{S} \in \mathbb{R}^{K\times N}$ is the cluster indication matrix, and $\mathbf{S}(k, n) = 1$ means $\mathbf{Z}(:, n)$ belongs to k-th cluster.

Driven by the equivalence between the (9.1) and matrix factorization problem shown in (9.3), the K-means clustering can be naturally generalized into tensor space based on different tensor decompositions, such as classical canonical polyadic (CP) and Tucker decomposition.

Algorithm 46: Traditional K-means algorithm

Input: Data matrix $\mathbf{Z}$.
Randomly select K initial centroids;
Repeat:
Each data point is assigned to its nearest centroid based on the squared Euclidean distance;
Recompute the centroids, which is done by taking the mean value of all data points assigned to that centroid's cluster.
Until the centroids no longer change.
Output: $\mathbf{C} = \{\mathbf{c}_1, \cdots, \mathbf{c}_K\}$

9.2.2 *CP Decomposition-Based Clustering*

Applying tensor decomposition model into K-means clustering algorithm can accurately find the multilinear subspaces of data and cluster these high-order data at the same time. It finds the subspace of high-dimensional data by CP decomposition and clusters the latent factor along the sample mode into K groups. Specifically, considering a third-order tensor $\mathcal{X} \in \mathbb{R}^{N\times I_2\times I_3}$, its CPD factor matrices are $\mathbf{U}^{(1)} \in \mathbb{R}^{N\times R}$, $\mathbf{U}^{(2)} \in \mathbb{R}^{I_2\times R}$, and $\mathbf{U}^{(3)} \in \mathbb{R}^{I_3\times R}$, where R is the CP rank of $\mathcal{X}$. Supposing that the rows of $\mathbf{U}^{(1)}$ can be clustered into K categories, the problem that joints the non-negative tensor decomposition and K-means of mode-1 (JTKM) can be expressed as follows [53]:

$$\begin{aligned}\min_{\substack{\mathbf{U}^{(1)},\mathbf{U}^{(2)},\mathbf{U}^{(3)},\\ \mathbf{S},\mathbf{C},\{d_n\}_{n=1}^N}} & \|\mathbf{X}_{(1)} - \mathbf{D}\mathbf{U}^{(1)}(\mathbf{U}^{(3)} \odot \mathbf{U}^{(2)})^{\mathrm{T}}\|_{\mathrm{F}}^2 + \lambda\|\mathbf{U}^{(1)} - \mathbf{S}\mathbf{C}\|_{\mathrm{F}}^2 \\ & + \eta(\|\mathbf{U}^{(2)}\|_{\mathrm{F}}^2 + \|\mathbf{U}^{(3)}\|_{\mathrm{F}}^2) \\ & \text{s.t.}\ \ \mathbf{U}^{(1)}, \mathbf{U}^{(2)}, \mathbf{U}^{(3)} \geq 0,\ \|\mathbf{U}^{(1)}(n,:)\|_2 = 1, \\ & \mathbf{D} = \operatorname{diag}(d_1, \cdots, d_N),\ \|\mathbf{S}(n,:)\|_0 = 1,\ \mathbf{S}(n,k) \in \{0,1\},\end{aligned} \tag{9.4}$$

where $\mathbf{S}(n,k) = 1$ means $\mathbf{U}^{(1)}(n,:)$ belongs to k-th cluster, the second term in objective function $\|\mathbf{U}^{(1)} - \mathbf{S}\mathbf{C}\|_{\mathrm{F}}^2$ is an K-means penalty, the regularization terms $\|\mathbf{U}^{(2)}\|_{\mathrm{F}}^2$ and $\|\mathbf{U}^{(3)}\|_{\mathrm{F}}^2$ are used to control scaling, and the diagonal matrix $\mathbf{D}$ is applied to normalize $\mathbf{U}^{(1)}$ using their ℓ_2-norms in advance [53].

Introducing a variable $\mathbf{Z}$, we can get

$$\begin{aligned}\min_{\substack{\mathbf{U}^{(1)},\mathbf{U}^{(2)},\mathbf{U}^{(3)},\\ \mathbf{S},\mathbf{C},\mathbf{Z},\{d_n\}_{n=1}^N}} & \|\mathbf{X}_{(1)} - \mathbf{D}\mathbf{U}^{(1)}(\mathbf{U}^{(3)} \odot \mathbf{U}^{(2)})^{\mathrm{T}}\|_{\mathrm{F}}^2 + \lambda\|\mathbf{U}^{(1)} - \mathbf{S}\mathbf{C}\|_{\mathrm{F}}^2 \\ & + \eta(\|\mathbf{U}^{(2)}\|_{\mathrm{F}}^2 + \|\mathbf{U}^{(3)}\|_{\mathrm{F}}^2) + \mu\|\mathbf{U}^{(1)} - \mathbf{Z}\|_{\mathrm{F}}^2 \\ & \text{s.t.}\ \ \mathbf{U}^{(1)}, \mathbf{U}^{(2)}, \mathbf{U}^{(3)} \geq 0,\ \|\mathbf{Z}(n,:)\|_2 = 1, \\ & \mathbf{D} = \operatorname{diag}(d_1, \cdots, d_N),\ \|\mathbf{S}(n,:)\|_0 = 1,\ \mathbf{S}(n,k) \in \{0,1\},\end{aligned} \tag{9.5}$$

which can be solved by the ADMM algorithm. The corresponding subproblems for different variables are as follows.

Letting $\mathbf{W} = (\mathbf{U}^{(3)} \odot \mathbf{U}^{(2)})^{\mathrm{T}}$, the optimization model (9.5) with respect to $\mathbf{U}^{(1)}$ can be transformed into (9.6)

$$\mathbf{U}^{(1)} = \arg\min_{\mathbf{U}^{(1)} \geq 0} \|\mathbf{X}_{(1)} - \mathbf{D}\mathbf{U}^{(1)}\mathbf{W}\|_{\mathrm{F}}^2 + \lambda\|\mathbf{U}^{(1)} - \mathbf{S}\mathbf{C}\|_{\mathrm{F}}^2 + \mu\|\mathbf{U}^{(1)} - \mathbf{Z}\|_{\mathrm{F}}^2 \tag{9.6}$$

Next, the subproblem w.r.t. d_n admits closed-form solution, i.e.

$$d_n = \frac{\mathbf{b}_n(\mathbf{X}_{(1)}(n,:))^{\mathrm{T}}}{\left(\mathbf{b}_n\mathbf{b}_n^{\mathrm{T}}\right)}, \tag{9.7}$$

where $\mathbf{b}_n$ is the n-th row of $\mathbf{U}^{(1)}\mathbf{W} \in \mathbb{R}^{N \times I_2 I_3}$.

And the update of $\mathbf{Z}(n,:)$ is also in closed-form

$$\mathbf{Z}(n,:)=\frac{\mathbf{U}^{(1)}(n,:)}{\|\mathbf{U}^{(1)}(n,:)\|_2}. \tag{9.8}$$

Moreover, the update of $\mathbf{C}$ comes from the K-means algorithm. Let $\mathcal{J}_k=\{n \mid \mathbf{S}(n,k)=1, n=1,\cdots,N\}, k=1,\cdots,K$, then

$$\mathbf{C}(k,:)=\frac{\sum_{n\in\mathcal{J}_k}\mathbf{U}^{(1)}(n,:)}{|\mathcal{J}_k|}. \tag{9.9}$$

The update of $\mathbf{S}$ also comes from the K-means algorithm

$$\mathbf{S}(n,k)=\begin{cases}1, \ k=\arg\min_k \|\mathbf{U}^{(1)}(n,:)-\mathbf{C}(k,:)\|_2,\\ 0, \ \text{otherwise}.\end{cases} \tag{9.10}$$

Finally, to update $\mathbf{U}^{(2)}$ and $\mathbf{U}^{(3)}$, we make use of the other two matrix unfoldings as follows:

$$\mathbf{U}^{(2)}=\underset{\mathbf{U}^{(2)}\geq\mathbf{0}}{\operatorname{argmin}}\left\|\mathbf{X}_{(2)}-\mathbf{U}^{(2)}(\mathbf{U}^{(3)}\odot\mathbf{D}\mathbf{U}^{(1)})^{\mathrm{T}}\right\|_{\mathrm{F}}^2+\eta\|\mathbf{U}^{(2)}\|_{\mathrm{F}}^2, \tag{9.11}$$

$$\mathbf{U}^{(3)}=\underset{\mathbf{U}^{(3)}\geq\mathbf{0}}{\operatorname{argmin}}\left\|\mathbf{X}_{(3)}-\mathbf{U}^{(3)}(\mathbf{U}^{(2)}\odot\mathbf{D}\mathbf{U}^{(1)})^{\mathrm{T}}\right\|_{\mathrm{F}}^2+\eta\|\mathbf{U}^{(3)}\|_{\mathrm{F}}^2. \tag{9.12}$$

The overall algorithm alternates between updates (9.6)–(9.12) until it achieves the convergence condition, and its brief process has been illustrated in Algorithm 47.

Algorithm 47: JTKM algorithm

Input: Data set $\mathcal{X}$.
Repeat:
Update $\mathbf{U}^{(1)}$ by solving (9.6);
Update $\mathbf{D}, \mathbf{Z}, \mathbf{C}, \mathbf{S}$ by Eqs. (9.7), (9.8), (9.9), and (9.10) respectively;
Update $\mathbf{U}^{(2)}, \mathbf{U}^{(3)}$ by Eqs. (9.11) and (9.12);
Update λ, η, μ;
Until the algorithm converges, the loop ends.
Output: $\mathbf{C}, \mathbf{S}$

9.2.3 *Tucker Decomposition-Based Clustering*

Another tensor subspace clustering with K-means is based on Tucker decomposition, which is equivalent to simultaneous data compression (i.e., two-dimensional singular value decomposition (2DSVD), which is equivalent to Tucker2) and K-means clustering [18]. Given a dataset $\mathcal{X} = \{\mathbf{X}_1, \mathbf{X}_2, \cdots, \mathbf{X}_N\} \in \mathbb{R}^{I_1 \times I_2 \times N}$, where $\mathbf{X}_n$ is a two-dimensional matrix (e.g., an image), we first project the observations $\{\mathbf{X}_n, n = 1, \cdots, N\}$ with sharing singular vectors by 2DSVD as follows [25]:

$$\begin{aligned} \min_{\mathbf{U},\mathbf{V},\mathbf{Z}_n} & \sum_{n=1}^{N} \left\| \mathbf{X}_n - \mathbf{U}\mathbf{Z}_n\mathbf{V}^{\mathrm{T}} \right\|_{\mathrm{F}}^2 \\ \text{s.t.} \ & \mathbf{V}^{\mathrm{T}}\mathbf{V} = \mathbf{I}, \mathbf{U}^{\mathrm{T}}\mathbf{U} = \mathbf{I}. \end{aligned} \tag{9.13}$$

After obtaining $\{\mathbf{Z}_n\}_{n=1}^N$, we vectorize the diagonal elements of each matrix $\mathbf{Z}_n$ and stack them as the row vectors of a new matrix $\widetilde{\mathbf{Z}} \in \mathbb{R}^{N \times \min(I_1, I_2)}$. Therefore, based on the distance relationship, the K-means clustering (9.3) of the observations $\{\mathbf{X}_n, n = 1, \cdots, N\}$ can be transformed as the clustering of $\{\mathbf{Z}_n, n = 1, \cdots, N\}$, namely, the rows of $\widetilde{\mathbf{Z}}$:

$$\min_{\mathbf{S},\mathbf{C}} \|\widetilde{\mathbf{Z}} - \mathbf{S}\mathbf{C}\|_{\mathrm{F}}^2, \tag{9.14}$$

where $\mathbf{S} \in \mathbb{R}^{N \times K}$, $\mathbf{C} \in \mathbb{R}^{K \times \min(I_1, I_2)}$.

Moreover, Huang et al. [18] have proved that K-means clustering in $\{\mathbf{Z}_n\}_{n=1}^N$ is equivalent to Tucker decomposition with optimization problem as follows:

$$\begin{aligned} \min_{\mathbf{U},\mathbf{V},\mathbf{W},\mathcal{Y}} & \|\mathcal{X} - \mathcal{Y} \times_1 \mathbf{U} \times_2 \mathbf{V} \times_3 \mathbf{W}\|_{\mathrm{F}}^2 \\ \text{s.t.} \ & \mathbf{U}^{\mathrm{T}}\mathbf{U} = \mathbf{I}, \ \mathbf{V}^{\mathrm{T}}\mathbf{V} = \mathbf{I}, \ \mathbf{W}^{\mathrm{T}}\mathbf{W} = \mathbf{I}, \end{aligned} \tag{9.15}$$

where $\mathbf{U}, \mathbf{V}, \mathbf{W}$ are the factor matrices and $\mathcal{Y}$ is the core tensor. For better understanding the relationship between K-means and Tucker decomposition, we have

$$\begin{aligned} \min_{\mathbf{U},\mathbf{V},\mathbf{W},\mathbf{Y}_{(3)}} & \left\| \mathbf{X}_{(3)} - \mathbf{W}\mathbf{Y}_{(3)}(\mathbf{V} \otimes \mathbf{U})^{\mathrm{T}} \right\|_{\mathrm{F}}^2 \\ \text{s.t.} \ & \mathbf{U}^{\mathrm{T}}\mathbf{U} = \mathbf{I}, \ \mathbf{V}^{\mathrm{T}}\mathbf{V} = \mathbf{I}, \ \mathbf{W}^{\mathrm{T}}\mathbf{W} = \mathbf{I}, \end{aligned} \tag{9.16}$$

where $\mathbf{X}_{(3)}$ and $\mathbf{Y}_{(3)}$ are matrices along mode-3 of $\mathcal{X}$ and core tensor $\mathcal{Y}$, respectively. In this way, we can clearly see that $\mathbf{W}$ in (9.16) contains the information of cluster indicator in K-means clustering. $\mathbf{Y}_{(3)}(\mathbf{V} \otimes \mathbf{U})^{\mathrm{T}}$ contains the cluster centroid information, and $\mathbf{U}, \mathbf{V}$ in Eq. (9.13) are the same as $\mathbf{U}, \mathbf{V}$ in Tucker decomposition.

Therefore, Tucker decomposition is equivalent to simultaneous subspace selection (i.e., data compression) in Eq. (9.13) and K-means clustering [4] as shown in Eq. (9.14). And the final clustering indicator can be found by applying clustering method (e.g., K-means) on matrix $\mathbf{W}$.

Besides, instead of performing Tucker decomposition first and then clustering the observations according to the factor matrix $\mathbf{W}$, another algorithm called adaptive subspace iteration on tensor (ASI-T) [31] completes these two steps simultaneously. The corresponding mathematical optimization model can be written as

$$\begin{aligned} \min_{\mathbf{U},\mathbf{V},\mathbf{W},\mathcal{Y}} \ & \|\mathcal{X} - \mathcal{Y} \times_1 \mathbf{U} \times_2 \mathbf{V} \times_3 \mathbf{W}\|_{\mathrm{F}}^2 \\ \text{s.t.}\ & \mathbf{U}^{\mathrm{T}}\mathbf{U} = \mathbf{I},\ \mathbf{V}^{\mathrm{T}}\mathbf{V} = \mathbf{I}, \\ & \mathbf{W} \text{ is a binary matrix.} \end{aligned} \tag{9.17}$$

Different from Eq. (9.15), $\mathbf{W}$ is a binary matrix which is the cluster indication matrix in Eq. (9.17), and $(\mathcal{Y} \times_1 \mathbf{U} \times_2 \mathbf{V})$ is the cluster centroid matrix. And $\mathcal{Y}$, $\mathbf{W}, \mathbf{U}, \mathbf{V}$ in Eq. (9.17) can be optimized using the alternating least square algorithm which iteratively updates one variable while fixing the others until the convergence condition is reached.

9.3 Tensor Subspace Cluster Based on Self-Representation

In this section, we will introduce two classical self-representation clustering methods based on sparse structure and low rank structure in matrix forms first and then illustrate several tensor decomposition-based methods with examples.

9.3.1 Matrix Self-Representation-Based Clustering

Self-representation is a concept that each sample can be represented as a linear combination of all the samples [38, 45]. Specifically, given a dataset $\mathbf{X} \in \mathbb{R}^{D\times N}$, of which each column denotes a sample, the classic matrix self-expression of the data samples is

$$\mathbf{X} = \mathbf{XC}, \tag{9.18}$$

where $\mathbf{C} \in \mathbb{R}^{N\times N}$ denotes the representations of the data samples $\mathbf{X}$ and $\mathbf{C}_{i,j}$ can be considered as a measure for certain relationship, such as similarity and correlation, between the i-th and the j-th samples.

The self-representation matrix $\mathbf{C} \in \mathbb{R}^{N\times N}$ can reveal the structure of the data samples in the perspective of relationships [39]. However, $\mathbf{C}_{i,j}$ is not always the same as $\mathbf{C}_{j,i}$, and the pairwise affinity matrix $\mathbf{W} = |\mathbf{C}| + \left|\mathbf{C}^{\mathrm{T}}\right|$ is often applied [9, 11, 12]. Then the spectral clustering is performed over $\mathbf{W}$ to obtain the final results. Spectral clustering [37] is based on graph theory which uses information from the eigenvalues (spectrum) of special matrices that are built from the graph or the data set to find an optimal partition. Specifically, an undirected graph $\mathrm{G}(\mathbb{V}, \mathbb{E})$ can be defined by the affinity matrix $\mathbf{W}$, where $\mathbb{V}$ contains all the data samples; edge set $\mathbb{E}$ means there is a relationship between i-th and j-th points when $\mathbf{W}_{i,j} \neq 0$. By putting this graph $\mathrm{G}(\mathbb{V}, \mathbb{E})$ into spectral clustering algorithm, the optimal partition (i.e., clustering results) can be obtained.

However, the number of data points is often greater than its dimension, i.e., $N > D$, giving rise to infinite representations of each data point. As a result, certain extra constraints on the representation matrix $\mathbf{C}$ are required to ensure that the self-expression has a unique solution. In subspace clustering, some extra constraints are often implemented for some data structures, like sparsity and low rank, which can reveal the membership of data samples.

By imposing sparsity structure on the self-expression matrix $\mathbf{C}$ of the data samples $\mathbf{X}$, a sparse subspace clustering (SSC) algorithm can be formulated as follows [9]:

$$\min_{\mathbf{C}} \|\mathbf{C}\|_0 \quad \text{s.t. } \mathbf{X} = \mathbf{X}\mathbf{C},\ \operatorname{diag}(\mathbf{C}) = 0. \tag{9.19}$$

Since the (9.19) with ℓ_0-norm is a nondeterministic polynomial (NP) hard problem, minimizing the tightest convex relaxation of the ℓ_0-norm is considered:

$$\min_{\mathbf{C}} \|\mathbf{C}\|_1 \quad \text{s.t. } \mathbf{X} = \mathbf{X}\mathbf{C},\ \operatorname{diag}(\mathbf{C}) = 0. \tag{9.20}$$

Additionally, in the case of data contaminated with Gaussian noise, the optimization problems in (9.20) can be generalized as follows:

$$\begin{aligned} &\min_{\mathbf{C},\mathbf{E}} \|\mathbf{C}\|_1 + \lambda\|\mathbf{E}\|_F^2 \\ &\text{s.t. } \mathbf{X} = \mathbf{X}\mathbf{C} + \mathbf{E},\ \operatorname{diag}(\mathbf{C}) = 0. \end{aligned} \tag{9.21}$$

By enforcing $\mathbf{X}$ to be of low-rank structure, the low-rank representation (LRR)-based subspace clustering algorithm is another option [24, 27]. It is similar to sparse subspace clustering, except that it aims to find a low-rank representation instead of a sparse representation. Specifically, it aims to minimize $\operatorname{rank}(\mathbf{C})$ instead of $\|\mathbf{C}\|_1$. Since this rank-minimization problem is NP hard, the researchers replace the rank of self-expression matrix $\mathbf{C}$ by its nuclear norm $\|\mathbf{C}\|_*$. In the case of noise-free data drawn from linear (affine) subspaces, this leads to the problem (9.22)

$$\min_{\mathbf{C}} \|\mathbf{C}\|_* \quad \text{s.t. } \mathbf{X} = \mathbf{X}\mathbf{C}. \tag{9.22}$$

Furthermore, in the case of data contaminated with noise or outliers, the LRR algorithm solves the problem

$$\begin{aligned} &\min_{\mathbf{C},\mathbf{E}} \|\mathbf{C}\|_* + \lambda\|\mathbf{E}\|_{2,1} \\ &\text{s.t. } \mathbf{X} = \mathbf{X}\mathbf{C} + \mathbf{E}. \end{aligned} \tag{9.23}$$

which can be solved using an augmented Lagrangian method.

Based on that, given a dataset $\mathbf{X} = \{\mathbf{x}_1, \mathbf{x}_2, \cdots, \mathbf{x}_N\} \in \mathbb{R}^{D \times N}$, we can obtain the standard subspace model based on self-representation:

$$\begin{aligned} &\min_{\mathbf{C},\mathbf{E}} \mathrm{R}(\mathbf{C}) + \lambda\, \mathrm{L}(\mathbf{E}) \\ &\text{s.t. } \mathbf{X} = \mathbf{X}\mathbf{C} + \mathbf{E}, \end{aligned} \tag{9.24}$$

where $\mathbf{E}$ is the error matrix, $\mathrm{L}(\mathbf{E})$ and $\mathrm{R}(\mathbf{C})$ represent the fitting error and data self-representation regularizers, and λ is used to make balance of those two regularizers. After obtaining the matrix $\mathbf{C}$, we can apply the spectral clustering algorithm to obtain the final clustering results by calculating the affinity matrix $\mathbf{W}$ from $\mathbf{C}$.

9.3.2 Self-Representation in Tucker Decomposition Form

Since the self-representation clustering models in matrix form ignore the spatial relationships of data, it has been extended to tensor space to obtain the better clustering performance. Assume we have a tensor dataset $\mathcal{X} = \{\mathbf{X}_1, \mathbf{X}_2, \cdots, \mathbf{X}_N\} \in \mathbb{R}^{I_1 \times I_2 \times N}$, where $\mathbf{X}_n$ is a two-dimensional matrix, the self-representation model in tensor space can be formulated as follows:

$$\begin{aligned} &\min_{\mathcal{C},\mathcal{E}} \ \mathrm{R}(\mathcal{C}) + \lambda\, \mathrm{L}(\mathcal{E}) \\ &\text{s.t. } \ \mathcal{X} = \mathcal{X} * \mathcal{C} + \mathcal{E}, \end{aligned} \tag{9.25}$$

where $*$ is t-product based on circular convolution operation, $\mathcal{E}$ is the error tensor, and $\mathcal{C} \in \mathbb{R}^{I_2 \times I_2 \times N}$ is the learned subspace representation tensor.

By encouraging low-rank structure and block-sparse structure for self-representation tensor and error tensor [7], (9.25) can be transformed into the following form:

$$\begin{aligned} &\min_{\mathcal{C},\mathcal{E}} \|\mathcal{C}\|_* + \lambda \|\mathcal{E}\|_{2,1} \\ &\text{s.t. } \mathcal{X} = \mathcal{X} * \mathcal{C} + \mathcal{E}, \end{aligned} \tag{9.26}$$

where $\|\mathcal{C}\|_* = \sum_{n=1}^{3} \alpha_n \left\|\mathbf{C}_{(n)}\right\|_*$ is a weighted tensor nuclear norm based on Tucker decomposition with $\alpha_n \geq 0$, $\sum_{n=1}^{3} \alpha_n = 1$, $\mathcal{E}$ is the error tensor, and λ is a tradeoff parameter.

Tackling (9.26) by the classical alternating direction method of multipliers (ADMM) algorithm, we can obtain the self-representation tensor $\mathcal{C}$, and the corresponding affinity matrix can be obtained by $\mathbf{W} = \frac{1}{N} \sum_{n=1}^{N} |\mathbf{C}_n| + \left|\mathbf{C}_n^{\mathrm{T}}\right|$. Then the traditional clustering, such as K-means, spectral clustering, can be used to obtain the final clustering results.

Similarly in [7], a multi-view clustering learning method based on low-rank tensor representation called tRLMvC (tensor-based representation learning method for multi-view clustering) algorithm has been proposed. It constructs a self-representation tensor and applies the Tucker decomposition on it to obtain a low-dimensional multi-view representation. Given the data $\mathcal{X} \in \mathbb{R}^{I_1 \times N \times I_3}$, it can be written as

$$\min_{\mathcal{C},\mathcal{G},\mathbf{U},\mathbf{V},\mathbf{W}} \lambda\|\mathcal{C}\|_{\mathrm{FF1}} + \|\mathcal{X} - \mathcal{X} * \mathcal{C}\|_{\mathrm{F}}^2 + \beta \|\mathcal{C} - \mathcal{G} \times_1 \mathbf{U} \times_2 \mathbf{V} \times_3 \mathbf{W}\|_{\mathrm{F}}^2, \tag{9.27}$$

where $\mathcal{C} \in \mathbb{R}^{N \times N \times I_3}$, $\mathbf{U} \in \mathbb{R}^{N \times R_1}$, $\mathbf{V} \in \mathbb{R}^{N \times R_2}$, and $\mathbf{W}$ is a $(I_3 \times 1)$ matrix quantifying the contributions of different views, thus $\mathcal{G} \in \mathbb{R}^{R_1 \times R_2 \times 1}$. Besides, $\|\mathcal{C}\|_{\mathrm{FF1}} = \sum_i \|\mathcal{C}(i,:,:)\|_{\mathrm{F}}$, λ and β are parameters to trade off self-representation learning and low-dimensional representation learning. These two parts are iteratively performed so that the interaction between self-representation tensor learning and its factorization can be enhanced and the new representation can be effectively generated for clustering purpose.

Concatenating $\mathbf{U}$ and $\mathbf{V}$, i.e., $\mathbf{H} = [\mathbf{U}, \mathbf{V}] \in \mathbb{R}^{N \times R}$, $R = R_1 + R_2$, we will have the new representation $\mathbf{H}$ of multi-view data. Applying the traditional clustering methods, e.g., K-means or spectral clustering, on $\mathbf{H}$, we can obtain the membership of the original data.

9.3.3 *Self-Representation in t-SVD Form*

Applying t-SVD [20] to self-representation cluster model has the similar purpose as that of Tucker decomposition. It also aims at obtaining a low-dimensional representation tensor.

Assume a dataset $\mathcal{X} = \{\mathbf{X}_1, \mathbf{X}_2, \cdots, \mathbf{X}_N\} \in \mathbb{R}^{I_1 \times I_2 \times N}$, where $\mathbf{X}_n$ is a two-dimensional matrix. Based on a new type of low-rank tensor constraint (i.e., the t-SVD-based tensor nuclear norm (t-TNN)), a subspace clustering model called t-SVD-MSC, has been proposed [50]. The optimization model is as follows:

$$\begin{aligned} \min_{\mathbf{C}_n,\mathbf{E}_n} \ & \|\mathcal{C}\|_{\text{t-TNN}} + \lambda\|\mathbf{E}\|_{2,1} \\ \text{s.t. } & \mathbf{X}_n = \mathbf{X}_n\mathbf{C}_n + \mathbf{E}_n, \ n = 1, \cdots, N, \\ & \mathcal{C} = \phi(\mathbf{C}_1, \mathbf{C}_2, \cdots, \mathbf{C}_N), \\ & \mathbf{E} = [\mathbf{E}_1; \mathbf{E}_2; \cdots; \mathbf{E}_N], \end{aligned} \tag{9.28}$$

where $\phi(\cdot)$ forms different $\mathbf{C}_n$, $n = 1, 2, \cdots, N$ into a third-order tensor $\mathcal{C}$, $\mathbf{X}_n$ represents the features matrix of the n-th view, $\mathbf{E} = [\mathbf{E}_1; \mathbf{E}_2; \cdots; \mathbf{E}_N]$ is the vertical concatenation along the column of the all error matrices $\mathbf{E}_n$, and $\lambda > 0$

is a regularizer parameter. Additionally, $\|\mathcal{C}\|_{\text{t-TNN}}$ is the tensor nuclear norm based on t-SVD whose mathematical formulation is

$$\|\mathcal{C}\|_{\text{t-TNN}} = \sum_{i=1}^{\min(I_1,I_2)} \sum_{n=1}^{N} \left|\hat{\mathbf{S}}_n(i,i)\right|. \tag{9.29}$$

where $\hat{\mathcal{C}} = \text{fft}(\mathcal{C}, [], 3)$ denotes the fast Fourier transformation (FFT) of a tensor $\mathcal{C}$ along the third dimension. Besides, $\hat{\mathbf{S}}_n$ is computed by the SVD of n-th frontal slice of $\hat{\mathcal{C}}$ as $\hat{\mathbf{C}}_n = \hat{\mathbf{U}}_n\hat{\mathbf{S}}_n\hat{\mathbf{V}}_n^{\text{T}}$.

By introduction of the auxiliary tensor variable $\mathcal{G}$, the optimization problem (9.28) can be transformed and solved by augmented Lagrangian method [31] whose equation is as follows:

$$\begin{aligned} &\mathcal{L}(\mathbf{C}_1, \cdots, \mathbf{C}_N; \mathbf{E}_1, \cdots, \mathbf{E}_N; \mathcal{G}) \\ &= \|\mathcal{G}\|_{\text{t-TNN}} + \lambda\|\mathbf{E}\|_{2,1} + \sum_{n=1}^{N} (\langle \mathbf{Y}_n, \mathbf{X}_n - \mathbf{X}_n\mathbf{C}_n - \mathbf{E}_n\rangle) \\ &+\tfrac{\mu}{2}\|\mathbf{X}_n - \mathbf{X}_n\mathbf{C}_n - \mathbf{E}_n\|_{\text{F}}^2 + \langle \mathcal{W}, \mathcal{C} - \mathcal{G}\rangle + \tfrac{\rho}{2}\|\mathcal{C} - \mathcal{G}\|_{\text{F}}^2, \end{aligned} \tag{9.30}$$

where $\mathbf{Y}_n$ $(n = 1, \cdots, N)$ and $\mathcal{W}$ represent two Lagrangian multipliers and λ, μ, and ρ are penalty coefficients, which can be adjusted by using an adaptive update strategy [23]. Besides, we can update $\mathbf{C}_n$, $\mathbf{E}_n$, and $\mathcal{G}$, respectively, as follows.

Subproblem with respect to $\mathbf{C}_n$: Update $\mathbf{C}_n$ by solving the following subproblem:

$$\begin{aligned} \min_{\mathbf{C}_n} \ &\langle \mathbf{Y}_n, \mathbf{X}_n - \mathbf{X}_n\mathbf{C}_n - \mathbf{E}_n\rangle + \frac{\mu}{2}\|\mathbf{X}_n - \mathbf{X}_n\mathbf{C}_n - \mathbf{E}_n\|_{\text{F}}^2 \\ &+ \langle \mathbf{W}_n, \mathbf{C}_n - \mathbf{G}_n\rangle + \frac{\rho}{2}\|\mathbf{C}_n - \mathbf{G}_n\|_{\text{F}}^2, \end{aligned} \tag{9.31}$$

where $\phi_n^{-1}(\mathcal{W}) = \mathbf{W}_n,\ \phi_n^{-1}(\mathcal{G}) = \mathbf{G}_n$.

Subproblem with respect to $\mathbf{E}_n$: Then update $\mathbf{E}_n$ by solving

$$\begin{aligned} \mathbf{E}^* &= \arg\min_{\mathbf{E}} \ \lambda\|\mathbf{E}\|_{2,1} + \sum_{n=1}^{N} (\langle \mathbf{Y}_n, \mathbf{X}_n - \mathbf{X}_n\mathbf{Z}_n - \mathbf{E}_n\rangle) \\ &\quad + \frac{\mu}{2}\|\mathbf{X}_n - \mathbf{X}_n\mathbf{C}_n - \mathbf{E}_n\|_{\text{F}}^2 \\ &= \arg\min_{\mathbf{E}} \ \frac{\lambda}{\mu}\|\mathbf{E}\|_{2,1} + \frac{1}{2}\|\mathbf{E} - \mathbf{D}\|_{\text{F}}^2, \end{aligned} \tag{9.32}$$

where $\mathbf{D} = \mathbf{X}_n - \mathbf{X}_n\mathbf{C}_n - \frac{1}{\mu}\mathbf{Y}_n$.

Subproblem with respect to $\mathcal{G}$: Furthermore, $\mathcal{G}$ can be updated by

$$\mathcal{G}^* = \arg\min_{\mathcal{G}} \|\mathcal{G}\|_{\text{t-TNN}} + \frac{\rho}{2}\left\|\mathcal{G} - \left(\mathcal{C} + \frac{1}{\rho}\mathcal{W}\right)\right\|_{\text{F}}^2. \tag{9.33}$$

The Lagrange multipliers $\mathbf{Y}_n$ and $\mathcal{W}$ are updated as follows:

$$\mathbf{Y}_n^* = \mathbf{Y}_n + \mu\left(\mathbf{X}_n - \mathbf{X}_n\mathbf{C}_n - \mathbf{E}_n\right), n = 1, \cdots, N, \tag{9.34}$$

$$\mathcal{W}^* = \mathcal{W} + \rho(\mathcal{C} - \mathcal{G}). \tag{9.35}$$

Finally, when the algorithm converges, the loop ends, and the whole process is shown in Algorithm 48.

Algorithm 48: ALM algorithm of t-SVD-MSC model

Input: Multi-view data $\mathcal{X} = \{\mathbf{X}_1, \mathbf{X}_2, \cdots, \mathbf{X}_N\} \in \mathbb{R}^{I_1\times I_2\times N}$, λ and the cluster number K
Repeat:
Update $\mathbf{C}_n$ by solving (9.31);
Update $\mathbf{E}_n$ by solving (9.32);
Update $\mathcal{G}$ by solving (9.33);
Update $\mathbf{Y}_n, \mathcal{W}$ by Eqs. (9.34) and (9.35);
Until the algorithm converges, the loop ends.
Output: $\mathbf{C}_n$, $\mathbf{E}_n$

Furthermore, based on t-SVD-MSC, some researchers have proposed a hyper-Laplace regularized multilinear multi-view self-representation model [51], referred to as HLR $-$ M^{2}VS for short, to solve the representation problem of multi-view nonlinear subspace. Based on the optimization model (9.28), the hyper-Laplace matrix is added and leads to a new model as follows:

$$\begin{aligned}
&\min_{\mathbf{C}_n,\mathbf{E}_n} \|\mathcal{C}\|_{\text{t-TNN}} + \lambda_1\|\mathbf{C}\|_{2,1} + \lambda_2\sum_{n=1}^{N}\text{tr}\left(\mathbf{C}_n(\mathbf{L}_{\text{h}})_n\mathbf{C}_n^{\text{T}}\right)\\
&\text{s.t. } \mathbf{X}_n = \mathbf{X}_n\mathbf{C}_n + \mathbf{E}_n,\ n = 1, \cdots, N,\\
&\qquad \mathcal{C} = \phi\left(\mathbf{C}_1, \mathbf{C}_2, \cdots, \mathbf{C}_N\right),\\
&\qquad \mathbf{E} = [\mathbf{E}_1; \mathbf{E}_2; \cdots; \mathbf{E}_N],
\end{aligned} \tag{9.36}$$

where $(\mathbf{L}_{\text{h}})_n$ represents the view-specific hyper-Laplacian matrix built on the optimized subspace representation matrix $\mathbf{C}_n$, so that the local high-order geometrical structure would be discovered by using the hyper-Laplacian regularized term. The optimization problem in (9.36) can be solved by ADMM to get the optimal $\mathbf{C}_n$, $\mathbf{E}_n$. For more details about this hyper-Laplacian matrix $(\mathbf{L}_h)_n$, readers can refer to [51].

9.4 Applications

In this section, we briefly demonstrate some applications of tensor clustering in heterogeneous information networks, bioinformatics signals, and multi-view subspace clustering.

9.4.1 Heterogeneous Information Networks Clustering

Heterogeneous information networks are widely used in the real-world applications, such as bibliographic networks and social media networks. Generally, heterogeneous information networks consist of multiple types of objects and rich semantic relations among different object types.

An example of heterogeneous information networks is given in Fig. 9.2. Clustering analysis is essential for mining this kind of networks. However, most existing approaches are designed to analyze homogeneous information networks which are composed of only one type of objects and links [40]. In this way, we can use tensor model to represent this complicated and high-order network. Specifically, as Fig. 9.3 shows, each type of objects is represented by one mode of the tensor, and the semantic relations among different object types correspond to the elements in the tensor.

In [46], authors adopt a stochastic gradient descent algorithm to process the CP decomposition-based tensor clustering in heterogeneous information networks. Specifically, assume that we have a tensor representation $\mathcal{X} \in \mathbb{R}^{I_1 \times I_2 \times \cdots \times I_N}$ of the heterogeneous information network $\mathbb{G} = (\mathbb{V}, \mathbb{E})$, where $\mathbb{V}$ is a set of objects and $\mathbb{E}$ is a set of links and we have K clusters. Figure 9.4 shows the result of applying the CP

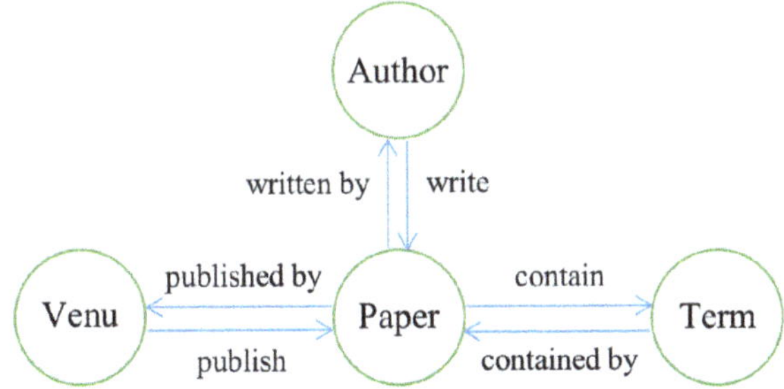

Fig. 9.2 An example of heterogeneous information network: a bibliographic network extracted from the DBLP database (http://www.informatik.uni-trier.de/ ley/db/), which contains four types of objects: author, paper, venue, and term. The links are labeled by "write" or "written by" between author and paper or labeled by "publish" or "published by" between paper and venue or labeled by "contain" or "contained in" between paper and term. Since it contains more than one type of objects and links, we cannot directly measure the similarity among the different types of objects and relations, and it is more difficult for us to achieve clustering in this heterogeneous network

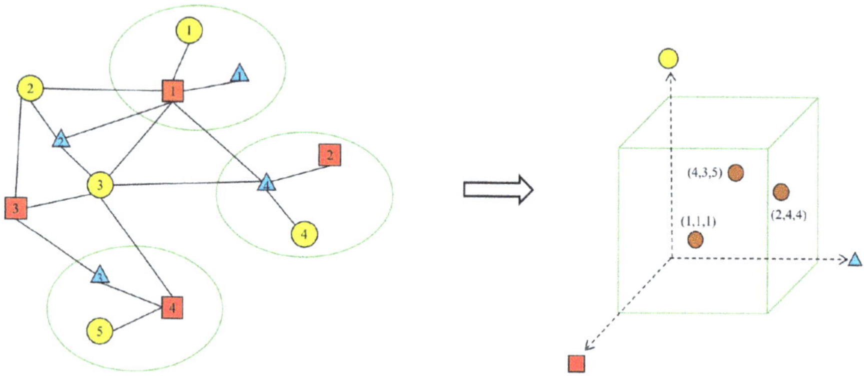

Fig. 9.3 An example of tensor model to represent heterogeneous information network. The left one is the original network with three types of objects (yellow, red, and blue), and the number within each object is the identifier of the object. Each element (brown dot) in the third-order tensor (right one) represents a gene network (i.e., the minimum subnetwork of heterogeneous information network) in the network (green circle in the left), and the coordinates represent the corresponding identifiers of the objects

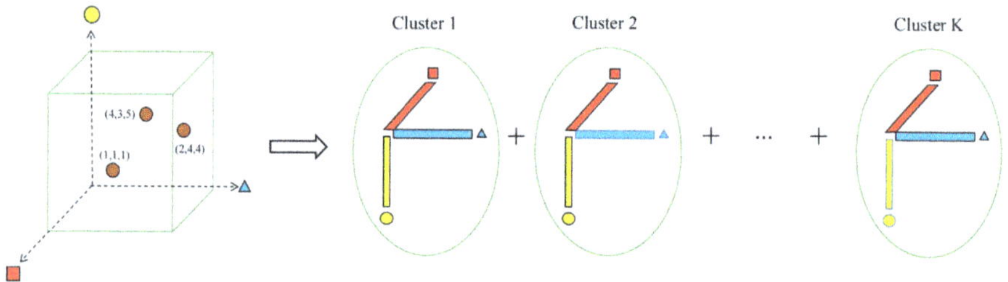

Fig. 9.4 An example of tensor CP decomposition for clustering heterogeneous information network, and each component (green circle in the right) shows one cluster of the original network

decomposition to the tensor that we construct in Fig. 9.3. The optimization model for CP decomposition can be written as

$$\min_{\mathbf{U}^{(1)},\mathbf{U}^{(2)},\cdots,\mathbf{U}^{(\mathrm{N})}} \|\mathcal{X} - [\![\mathbf{U}^{(1)}, \mathbf{U}^{(2)}, \cdots, \mathbf{U}^{(\mathrm{N})}]\!]\|_{\mathrm{F}}^2, \tag{9.37}$$

where $\mathbf{U}^{(n)} \in \mathbb{R}^{I_n \times K}$, $n = 1, \cdots, N$, is the cluster indication matrix of the n-th type of objects and each column $\mathbf{u}_k^{(n)} = \left[u_{k,1}^{(n)}, u_{k,2}^{(n)}, \cdots, u_{k,I_n}^{(n)}\right]^{\mathrm{T}}$ of the factor matrix $\mathbf{U}^{(n)}$ is the probability vector of each object in n-th type objects belonging to the k-th cluster.

It can be seen that the problem in (9.37) is an NP-hard nonconvex optimization problem [1]. Thus, in [46], authors adopt a stochastic gradient descent algorithm to process this CP decomposition-based tensor clustering problem with Tikhonov regularization-based loss function [30]. At the same time, they prove that the CP decomposition in (9.37) is equivalent to the K-means clustering for the n-th type of objects in heterogeneous information network $\mathbb{G} = (\mathbb{V}, \mathbb{E})$, which means that the CP decomposition of $\mathcal{X}$ obtains the clusters of all types of objects in the heterogeneous information network $\mathbb{G} = (\mathbb{V}, \mathbb{E})$ simultaneously.

9.4.2 Biomedical Signals Clustering

In bioinformatics, clinical 12-lead electrocardiogram (ECG) signals are now widely used in hospitals due to more detailed information about heart activities. The process of a heartbeat cycle can be investigated by 12 leads from different angles, which is helpful for a cardiologist to give an accurate cardiovascular disease diagnosis for the patients. However, comparing with the standard 2-lead ECG, the data volume generated from 12-lead ECG are large and its complexity is high.

As Fig. 9.5 illustrates, He et al. proposed a scheme to process and classify these signals using wavelet tensor decomposition and two-dimensional Gaussian spectral clustering (TGSC) [17]. Specifically, given a 12-lead ECG signals $\mathbf{X} \in \mathbb{R}^{D \times N_0}$, where $D = 12$, N_0 is the length of original signal. After filtering and segmentation, $\mathbf{X}$ can be divided into M heartbeats. Extracting features from each matrix $\mathbf{S}_m \in \mathbb{R}^{D \times N}$ by the discrete wavelet packet transform (DWPT), a new coefficient matrix $\mathbf{Z}_m \in \mathbb{R}^{D \times N}$ can be constructed.

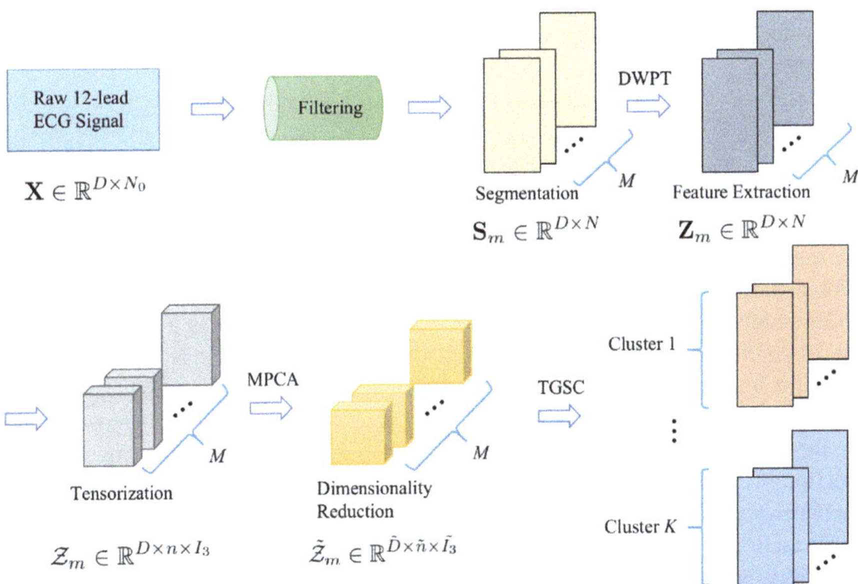

Fig. 9.5 Schematic diagram of unsupervised classification of 12-lead ECG signals using wavelet tensor spectral clustering

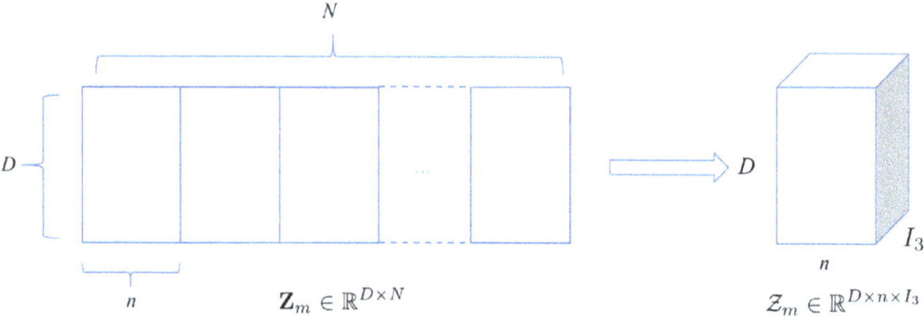

Fig. 9.6 Tensorization of ECG wavelet matrix, where $\mathbf{Z}_m \in \mathbb{R}^{D\times N}$ is one of the ECG wavelet matrices, n is the length of one wavelet frequency sub-band, D of $\mathcal{Z}_m$ represents the number of the measured ECG variables which equals to 12 in this case, and I_3 refers to the frequency sub-band (i.e., $I_3 = N/n$)

The operations above are for preprocessing. To further reduce the dimensionality of feature data for better pattern recognition results and less memory requirements, each $\mathbf{Z}_m$ is transformed into the tensor form $\mathcal{Z}_m \in \mathbb{R}^{D\times n\times I_3}$ as presented in Fig. 9.6, where D represents the number of the measured ECG variables which equals to 12 in this case, n is the sampling time period, and I_3 refers to the frequency sub-band. By applying the multilinear principal component analysis (MPCA) [28], the original tensor space can be mapped onto a tensor subspace. Finally, a new kernel function based on Euclidean distance and cosine distance, namely, two-dimensional Gaussian kernel function (TGSC), has been applied to spectral clustering algorithm to obtain the final clustering results, and each cluster contains all samples from one patient.

9.4.3 *Multi-View Subspace Clustering*

In real-world applications, many datasets are naturally comprised of multiple views. For example, images can be described by different kinds of features, such as color, edge, and texture. Each type of features is referred to as a particular view, and combining multiple views of dataset for data analysis has been a popular way for improving data processing performance. Since the high-dimensional data representation can ensure the integrity to the great extent, we can apply tensor model to the multi-view subspace clustering for further performance improvement. In this section, we use several multi-view datasets for clustering performance comparison between tensor-based models and matrix-based models. In Table 9.1, we give a brief introduction of several datasets in different applications, including object clustering, face clustering, scene clustering, and digit clustering, and then some samples of these datasets are presented in Fig. 9.7.

Table 9.1 Different datasets covering four applications, namely, generic object clustering, face clustering, scene clustering, and digit clustering

Dataset	Images	Clusters	Views
COIL-20	1440	20	3
ORL	400	40	3
Scene-15	4485	15	3
UCI-digits	2000	10	3

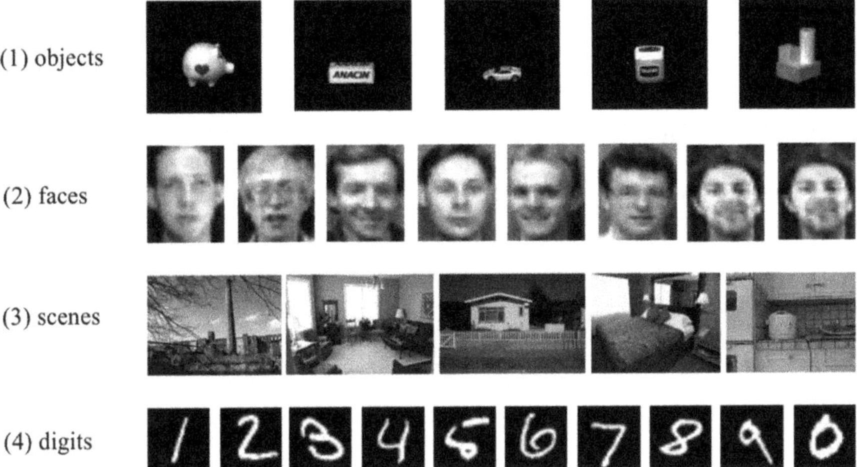

Fig. 9.7 Several examples from different datasets in various applications. (1) The COIL-20 dataset [11]; (2) the ORL dataset [50]; (3) the Scene-15 dataset [13]; (4) the UCI-Digits dataset [2]

We conclude the clustering results by applying six subspace clustering algorithms into our experiment. Here is a brief introduction to these methods.

LRR [27] can be seen as a subspace clustering method based on matrix, whose purpose is to find the joint lowest rank representation of all data.

Co-reg [21] is a co-regularization method for spectral clustering, which explores complementary information through the co-regularization hypothesis.

RMSC [49] takes the shared low-rank transfer probability matrix as input into the spectral clustering based on the Markov chain.

DiMSC [5] extends the subspace clustering to the multi-view fields and effectively solves the integrity problem of multi-view representation by using Hilbert-Schmidt Independent Criterion (HSIC).

t-SVD-MSC [50] imposes a new low-rank tensor constraint on the rotation tensor by introducing the t-SVD to ensure consensus among multiple views.

ETLMSC [48] focuses on a novel essential tensor learning method that explores the high-order correlations for multi-view representation for the Markov chain-based spectral clustering.

To better compare the performance of those methods, we divide them into two categories. LRR [27], Co-reg [21], DiMSC [5], and RMSC [49] can be regarded as the matrix-based clustering methods, and t-SVD-MSC [50] and ETLMSC [48] are tensor-based clustering methods.

Additionally, six commonly used evaluation metrics are employed, namely, F-score (F), precision (P), recall (R), normalized mutual information (NMI), adjusted rand index (ARI), and accuracy (ACC) [19, 36].

What we need to note is that these six metrics represent different properties in clustering scenes and the higher the value of these six commonly used clustering evaluation metrics we mentioned above, the better the clustering performance.

As shown in Table 9.2, t-SVD-MSC and ETLMSC, which apply the tensor model into clustering, achieve the best and second-best results in terms of almost all these different metrics in four applications. Since those tensor models ensure the integrity of the data to the greatest extent and preserve the spatial structures well, the tensor-based methods will obtain better performance than matrix-based clustering methods.

Table 9.2 The clustering results of four datasets, namely, UCI-Digits, ORL, COIL-20, and Scene-15 datasets. The best results are all highlighted in bold black

Datasets	Method	F-score	Precision	Recall	NMI	AR	ACC
UCI-Digits	LRR	0.316	0.292	0.344	0.422	0.234	0.356
	Co-reg	0.585	0.576	0.595	0.638	0.539	0.687
	RMSC	0.833	0.828	0.839	0.843	0.815	0.902
	DiMSC	0.341	0.299	0.397	0.424	0.257	0.530
	t-SVD-MSC	**0.936**	**0.935**	0.938	0.934	**0.929**	**0.967**
	ETLMSC	0.915	0.876	**0.963**	**0.957**	0.905	0.905
ORL	LRR	0.679	0.631	0.734	0.882	0.671	0.750
	Co-reg	0.535	0.494	0.583	0.813	0.523	0.642
	RMSC	0.623	0.586	0.665	0.853	0.614	0.700
	DiMSC	0.717	0.669	0.769	0.905	0.709	0.775
	t-SVD-MSC	**0.976**	**0.962**	**0.989**	**0.995**	**0.975**	**0.974**
	ETLMSC	0.904	0.856	0.958	0.977	0.901	0.900
COIL20	LRR	0.597	0.570	0.628	0.761	0.576	0.679
	Co-reg	0.597	0.577	0.619	0.765	0.576	0.627
	RMSC	0.664	0.637	0.693	0.812	0.645	0.692
	DiMSC	0.733	0.726	0.739	0.837	0.719	0.767
	t-SVD-MSC	0.747	0.725	0.770	0.853	0.734	0.784
	ETLMSC	**0.854**	**0.825**	**0.886**	**0.936**	**0.846**	**0.867**
Scene15	LRR	0.268	0.255	0.282	0.371	0.211	0.370
	Co-reg	0.344	0.346	0.341	0.425	0.295	0.451
	RMSC	0.294	0.295	0.292	0.410	0.241	0.396
	DiMSC	0.181	0.176	0.186	0.261	0.118	0.288
	t-SVD-MSC	0.692	0.671	0.714	0.776	0.668	0.756
	ETLMSC	**0.826**	**0.820**	**0.831**	**0.877**	**0.812**	**0.859**

9.5 Summary

In this chapter, we have introduced the background, motivation, and popular algorithms of tensor subspace clustering and illustrated its advantages. We mainly overview tensor subspace clustering based on K-means and self-expression starting from its matrix counterparts. Furthermore, to explore the spatial relationships of samples and ensure the integrity of the data, we introduce several standard models in tensor space about those two aspects. Several applications are also discussed to show its superior performance in compression with traditional ones for higher-order data.

Besides the discussed ones, tensor subspace clustering has also been applied in some other plication problems, such as user clustering in mobile Internet traffic interaction patterns [54], large scale urban reconstruction with tensor clustering [33], tensor-based recommender with clustering [22], and so forth. In the future, possible research may involve how to better construct higher-order similarities and integrate them into tensor clustering models [32]. There are several methods [6, 35, 51] which start to deal with data that come from nonlinear subspaces, but how to enhance their computational efficiency and decrease their storage requirements still deserve attention.

References

1. Acar, E., Dunlavy, D.M., Kolda, T.G.: A scalable optimization approach for fitting canonical tensor decompositions. J. Chemometrics **25**(2), 67–86 (2011)
2. Asuncion, A., Newman, D.: UCI machine learning repository (2007)
3. Bauckhage, C.: K-means clustering is matrix factorization (2015, preprint). arXiv:1512.07548
4. Bishop, C.M.: Pattern Recognition and Machine Learning. Springer, Berlin (2006)
5. Cao, X., Zhang, C., Fu, H., Liu, S., Zhang, H.: Diversity-induced multi-view subspace clustering. In: Proceedings of the IEEE Conference on Computer Vision and Pattern Recognition, pp. 586–594 (2015)
6. Chen, Y., Xiao, X., Zhou, Y.: Jointly learning kernel representation tensor and affinity matrix for multi-view clustering. IEEE Trans. Multimedia **22**(8), 1985–1997 (2019)
7. Cheng, M., Jing, L., Ng, M.K.: Tensor-based low-dimensional representation learning for multi-view clustering. IEEE Trans. Image Process. **28**(5), 2399–2414 (2018)
8. Costeira, J.P., Kanade, T.: A multibody factorization method for independently moving objects. Int. J. Comput. Visi. **29**(3), 159–179 (1998)
9. Elhamifar, E., Vidal, R.: Sparse subspace clustering. In: 2009 IEEE Conference on Computer Vision and Pattern Recognition, pp. 2790–2797. IEEE, Piscataway (2009)
10. Elhamifar, E., Vidal, R.: Clustering disjoint subspaces via sparse representation. In: 2010 IEEE International Conference on Acoustics, Speech and Signal Processing, pp. 1926–1929. IEEE, Piscataway (2010)
11. Elhamifar, E., Vidal, R.: Sparse subspace clustering: algorithm, theory, and applications. IEEE Trans. Pattern Anal. Mach. Intell. **35**(11), 2765–2781 (2013)
12. Favaro, P., Vidal, R., Ravichandran, A.: A closed form solution to robust subspace estimation and clustering. In: IEEE Conference on Computer Vision and Pattern Recognition (CVPR 2011), pp. 1801–1807. IEEE, Piscataway (2011)

13. Fei-Fei, L., Perona, P.: A Bayesian hierarchical model for learning natural scene categories. In: 2005 IEEE Computer Society Conference on Computer Vision and Pattern Recognition (CVPR'05), vol. 2, pp. 524–531. IEEE, Piscataway (2005)
14. Fischler, M.A., Bolles, R.C.: Random sample consensus: a paradigm for model fitting with applications to image analysis and automated cartography. Commun. ACM **24**(6), 381–395 (1981)
15. Gear, C.W.: Multibody grouping from motion images. Int. J. Comput. Vision **29**(2), 133–150 (1998)
16. Goh, A., Vidal, R.: Segmenting motions of different types by unsupervised manifold clustering. In: 2007 IEEE Conference on Computer Vision and Pattern Recognition, pp. 1–6. IEEE, Piscataway (2007)
17. He, H., Tan, Y., Xing, J.: Unsupervised classification of 12-lead ECG signals using wavelet tensor decomposition and two-dimensional gaussian spectral clustering. Knowl.-Based Syst. **163**, 392–403 (2019)
18. Huang, H., Ding, C., Luo, D., Li, T.: Simultaneous tensor subspace selection and clustering: the equivalence of high order SVD and k-means clustering. In: Proceedings of the 14th ACM SIGKDD International Conference on Knowledge Discovery and Data Mining, pp. 327–335 (2008)
19. Hubert, L., Arabie, P.: Comparing partitions. J. Classif. **2**(1), 193–218 (1985)
20. Kilmer, M.E., Martin, C.D.: Factorization strategies for third-order tensors. Linear Algebra Appl. **435**(3), 641–658 (2011)
21. Kumar, A., Rai, P., Daume, H.: Co-regularized multi-view spectral clustering. In: Advances in Neural Information Processing Systems, pp. 1413–1421 (2011)
22. Leginus, M., Dolog, P., Žemaitis, V.: Improving tensor based recommenders with clustering. In: International Conference on User Modeling, Adaptation, and Personalization, pp. 151–163. Springer, Berlin (2012)
23. Lin, Z., Liu, R., Su, Z.: Linearized alternating direction method with adaptive penalty for low-rank representation. In: Advances in Neural Information Processing Systems, pp. 612–620 (2011)
24. Liu, G., Lin, Z., Yu, Y.: Robust subspace segmentation by low-rank representation. In: Proceedings of the 27th International Conference on Machine Learning (ICML-10), pp. 663–670 (2010)
25. Liu, J., Liu, J., Wonka, P., Ye, J.: Sparse non-negative tensor factorization using columnwise coordinate descent. Pattern Recognit. **45**(1), 649–656 (2012)
26. Lu, C.Y., Min, H., Zhao, Z.Q., Zhu, L., Huang, D.S., Yan, S.: Robust and efficient subspace segmentation via least squares regression. In: European Conference on Computer Vision, pp. 347–360. Springer, Berlin (2012)
27. Liu, G., Lin, Z., Yan, S., Sun, J., Yu, Y., Ma, Y.: Robust recovery of subspace structures by low-rank representation. IEEE Trans. Pattern Anal. Mach. Intell. **35**(1), 171–184 (2012)
28. Lu, H., Plataniotis, K.N., Venetsanopoulos, A.N.: MPCA: multilinear principal component analysis of tensor objects. IEEE Trans. Neural Netw. **19**(1), 18–39 (2008)
29. Ma, Y., Yang, A.Y., Derksen, H., Fossum, R.: Estimation of subspace arrangements with applications in modeling and segmenting mixed data. SIAM Rev. **50**(3), 413–458 (2008)
30. Paatero, P.: A weighted non-negative least squares algorithm for three-way 'PARAFAC' factor analysis. Chemom. Intell. Lab. Syst. **38**(2), 223–242 (1997)
31. Peng, W., Li, T.: Tensor clustering via adaptive subspace iteration. Intell. Data Anal. **15**(5), 695–713 (2011)
32. Peng, H., Hu, Y., Chen, J., Haiyan, W., Li, Y., Cai, H.: Integrating tensor similarity to enhance clustering performance. IEEE Trans. Pattern Anal. Mach. Intell. (2020). https://doi.org/10.1109/TPAMI.2020.3040306
33. Poullis, C.: Large-scale urban reconstruction with tensor clustering and global boundary refinement. IEEE Trans. Pattern Anal. Mach. Intell. **42**(5), 1132–1145 (2019)

34. Rao, S., Tron, R., Vidal, R., Ma, Y.: Motion segmentation in the presence of outlying, incomplete, or corrupted trajectories. IEEE Trans. Pattern Anal. Mach. Intell. **32**(10), 1832–1845 (2009)
35. Ren, Z., Mukherjee, M., Bennis, M., Lloret, J.: Multikernel clustering via non-negative matrix factorization tailored graph tensor over distributed networks. IEEE J. Sel. Areas Commun. **39**(7), 1946–1956 (2021)
36. Schütze, H., Manning, C.D., Raghavan, P.: Introduction to Information Retrieval, vol. 39. Cambridge University Press, Cambridge (2008)
37. Shi, J., Malik, J.: Normalized cuts and image segmentation. IEEE Trans. Pattern Anal. Mach. Intell. **22**(8), 888–905 (2000)
38. Sui, Y., Zhao, X., Zhang, S., Yu, X., Zhao, S., Zhang, L.: Self-expressive tracking. Pattern Recognit. **48**(9), 2872–2884 (2015)
39. Sui, Y., Wang, G., Zhang, L.: Sparse subspace clustering via low-rank structure propagation. Pattern Recognit. **95**, 261–271 (2019)
40. Sun, Y., Han, J., Zhao, P., Yin, Z., Cheng, H., Wu, T.: Rankclus: integrating clustering with ranking for heterogeneous information network analysis. In: Proceedings of the 12th International Conference on Extending Database Technology: Advances in Database Technology, pp. 565–576 (2009)
41. Tipping, M.E., Bishop, C.M.: Mixtures of probabilistic principal component analyzers. Neural Comput. **11**(2), 443–482 (1999)
42. Tseng, P.: Nearest q-flat to m points. J. Optim. Theory Appl. **105**(1), 249–252 (2000)
43. Vidal, R.: Subspace clustering. IEEE Signal Process. Mag. **28**(2), 52–68 (2011)
44. Vidal, R., Ma, Y., Sastry, S.: Generalized principal component analysis (GPCA). IEEE Trans. Pattern Anal. Mach. Intell. **27**(12), 1945–1959 (2005)
45. Wright, J., Ma, Y., Mairal, J., Sapiro, G., Huang, T.S., Yan, S.: Sparse representation for computer vision and pattern recognition. Proc. IEEE **98**(6), 1031–1044 (2010)
46. Wu, J., Wang, Z., Wu, Y., Liu, L., Deng, S., Huang, H.: A tensor CP decomposition method for clustering heterogeneous information networks via stochastic gradient descent algorithms. Sci. Program. **2017**, 2803091 (2017)
47. Wu, T., Bajwa, W.U.: A low tensor-rank representation approach for clustering of imaging data. IEEE Signal Process. Lett. **25**(8), 1196–1200 (2018)
48. Wu, J., Lin, Z., Zha, H.: Essential tensor learning for multi-view spectral clustering. IEEE Trans. Image Proces. **28**(12), 5910–5922 (2019)
49. Xia, R., Pan, Y., Du, L., Yin, J.: Robust multi-view spectral clustering via low-rank and sparse decomposition. In: Proceedings of the Twenty-Eighth AAAI Conference on Artificial Intelligence, pp. 2149–2155 (2014)
50. Xie, Y., Tao, D., Zhang, W., Liu, Y., Zhang, L., Qu, Y.: On unifying multi-view self-representations for clustering by tensor multi-rank minimization. Int. J. Comput. Vis. **126**(11), 1157–1179 (2018)
51. Xie, Y., Zhang, W., Qu, Y., Dai, L., Tao, D.: Hyper-laplacian regularized multilinear multiview self-representations for clustering and semisupervised learning. IEEE Trans. Cybern. **50**(2), 572–586 (2018)
52. Yan, J., Pollefeys, M.: A general framework for motion segmentation: independent, articulated, rigid, non-rigid, degenerate and non-degenerate. In: European Conference on Computer Vision, pp. 94–106. Springer, Berlin (2006)
53. Yang, B., Fu, X., Sidiropoulos, N.D.: Learning from hidden traits: joint factor analysis and latent clustering. IEEE Trans. Signal Process. **65**(1), 256–269 (2016)
54. Yu, K., He, L., Philip, S.Y., Zhang, W., Liu, Y.: Coupled tensor decomposition for user clustering in mobile internet traffic interaction pattern. IEEE Access. **7**, 18113–18124 (2019)
55. Zhang, T., Szlam, A., Lerman, G.: Median k-flats for hybrid linear modeling with many outliers. In: 2009 IEEE 12th International Conference on Computer Vision Workshops, ICCV Workshops, pp. 234–241. IEEE, Piscataway (2009)

Chapter 10
Tensor Decomposition in Deep Networks

10.1 Introduction to Deep Learning

Inspired by information processing and distributed communication nodes in biological systems, deep learning can extract useful features with respect to specific tasks for data processing. Compared with other machine learning methods, deep learning has better performance on a number of applications, such as speech recognition [21, 24], computer vision [11, 16, 26], and natural language processing [6, 25]. The rise of deep learning has set off a new round of artificial intelligence.

Deep neural network (DNN) is a general model in deep learning to establish an end-to-end data mapping. It is composed of multiple nonlinear processing layers. Different deep learning models can be obtained with different operators in each layer and various connections between layers. Figure 10.1 gives a graphical illustration of a deep neural network. Among all the existing deep learning models, convolutional neural network (CNN) and recurrent neural network (RNN) are two classical ones.

As it shows in Fig. 10.1, $\mathbf{x}_0 \in \mathbb{R}^{N_0}$ is the input, $\mathbf{y} \in \mathbb{R}^{N_L}$ is the output, where we have $L = 3$ layers, and $\mathrm{f}_l(\cdot)$ is the forward propagation of the l-th dense layer which is denoted as

$$\mathrm{f}_l(\mathbf{x}_{l-1}) = \sigma(\mathbf{W}_l \mathbf{x}_{l-1} + b_l), \tag{10.1}$$

where $\mathbf{x}_{l-1} \in \mathbb{R}^{N_{l-1}}$, $\mathbf{W}_l \in \mathbb{R}^{N_l \times N_{l-1}}$, and $b_l \in \mathbb{R}^{N_l}$ denote the input, weight, and bias of l-th layer, respectively, and $\sigma(\cdot)$ is a nonlinear activation function. The nonlinearity of the activation function brings powerful representation capability to the DNN. Different activation functions may have different effects, and commonly used activation functions include ReLU, sigmoid, etc. The model parameters $\mathbf{W}_l$ and b_l are trainable and often achieved by minimizing a specific loss function by error backpropagation. Loss functions can be divided into two broad categories: classification and regression. Mean absolute error (MAE, ℓ_1 loss) and mean square error (MSE, ℓ_2 loss) are used for regression tasks, which are respectively represented

Y. Liu et al., *Tensor Computation for Data Analysis*,
https://doi.org/10.1007/978-3-030-74386-4_10

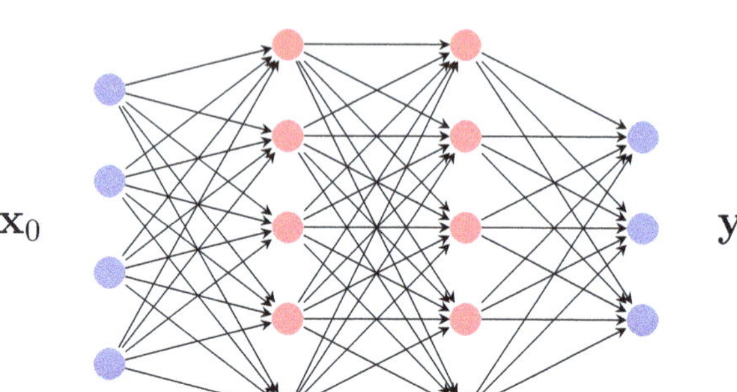

Fig. 10.1 Graphical illustration of a basic neural network model

as

$$\mathrm{MAE} = \frac{1}{\mathrm{N_L}} \sum^{\mathrm{N_L}} |\mathrm{y_n} - \bar{\mathrm{y}}_\mathrm{n}|$$

and

$$\mathrm{MSE} = \frac{1}{\mathrm{N_L}} \sum^{\mathrm{N_L}} (\mathrm{y_n} - \bar{\mathrm{y}}_\mathrm{n})^2,$$

where $\bar{y}_n$ is the n-th elements of the actual (goal) vector. For classification tasks, the cross-entropy loss function can not only measure the effect of the model well but also carry out the derivation calculation easily. The size of output N_L is set to the number of categories, and the cross-entropy error (CEE) of multi-classification is expressed as

$$\mathrm{CEE} = \sum_{\mathrm{n}=1}^{\mathrm{N_L}} \bar{\mathrm{y}}_\mathrm{n} \ln \mathrm{p_n},$$

where $\bar{\mathbf{y}} = [\bar{y}_1, \bar{y}_2, \cdots, \bar{y}_{N_L}]$ is the target (one-hot coding) and p_n is the probability of the n-th category. The probability is usually obtained by softmax function, which is represented as $p_n = \frac{e^{y_n}}{\sum e^{y_n}}$.

When dealing with multiway data, classical neural networks usually perform vectorization operation over the input data, which may cause loss of structural information. In order to explore the spatial structure of multiway data, CNN is proposed by introducing convolution operators [15]. Besides, RNN is designed for temporally related data, which can complete the tasks related to a continuous speech or text, such as automatic speech recognition (ASR) and text generation [24]. Figure 10.2 gives the illustration of core part of CNN, and Fig. 10.4 illustrates the main idea of RNN.

10.1.1 Convolutional Neural Network

A CNN usually includes convolutional layers, pooling layers, and fully connected (dense) layers, and the architecture is shown in Fig. 10.2. Convolutional layers, which can extract local features and need fewer parameters than dense layers, are the core of CNNs. Each convolutional layer is followed by a pooling layer which is used to reduce the size of the output of the convolutional layer to avoid overfitting. At the end of a CNN, there is one fully connected layer but is not limited to one, which can aggregate global information and easily output a vector in the size of the number of categories.

Suppose an input image $\mathcal{X} \in \mathbb{R}^{H\times W\times I}$, where H is height of the image, W is width of the image, and I is dimension of the image (normally $D = 3$ for image), and design a convolutional kernel $\mathcal{K} \in \mathbb{R}^{K\times K\times I\times O}$, where K is the size of the window and O is the number of output channel. The convolutional operation is mathematically expressed in element-wise as

$$\mathcal{Y}_{h',w',o} = \sum_{k_1,k_2=1}^{K} \sum_{i=1}^{I} \mathcal{X}_{h,w,i} \mathcal{K}_{k_1,k_2,i,o},$$

where $h = (h' - 1)S + k_1 \in \{1, 2, \cdots, H + 2P\}$, $w = (w' - 1)S + k_2 \in \{1, 2, \cdots, W + 2P\}$, S is stride size, P is zero-padding size, and O is the number of output channel. Output of the convolutional operation is $\mathcal{Y} \in \mathbb{R}^{H'\times W'\times O}$, where $H' = (H + 2P - K)/S + 1$, $W' = (W + 2P - K)/S + 1$.

Pooling layer is also named subsampling (downsampling) layer. The sampling operation usually includes taking the average and the maximum, which corresponds to average-pool and max-pool, respectively. As shown in Fig. 10.3, the max pool layer chooses a pool and takes the maximum of the values in the pool. Suppose the

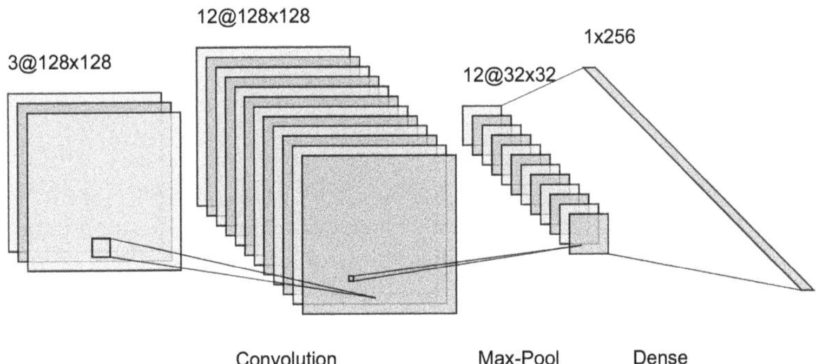

Fig. 10.2 A graphical illustration for CNN. 3@128 × 128 represents an image with three channels whose size is 128 × 128

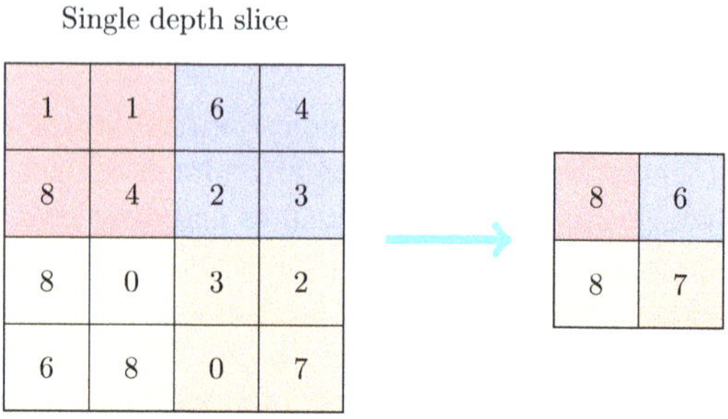

Fig. 10.3 Max pool with 2×2 filters and stride size 2

Single depth slice

1	1	6	4
8	4	2	3
8	0	3	2
6	8	0	7

8	6
8	7

size of input is $H \times W$ and the size of the pool is $P \times P$, then the output size is $\frac{H}{P} \times \frac{W}{P}$ (the stride size is usually set to P).

10.1.2 Recurrent Neural Network

The input data of the neural network includes not only images but also speech signals and video which are temporal (sequence) data. Although CNNs can handle these data types, the performance is not ideal. Since the relational information between the sequences will be ignored if the sequence data is put into the CNNs one by one, it will result in information redundancy if the whole sequence is trained at once. Thus RNNs are designed to handle temporal data, which can deal with the relationship between adjacent sequences more effectively.

The main feature of RNN is additional hidden state in each moment. Driven by the dynamic system, the t-th hidden state is the sum of linear transformation of the $(t-1)$-th hidden state and the linear transformation of the current input,

$$\mathbf{h}_t = \mathbf{W}\mathbf{h}_{t-1} + \mathbf{U}\mathbf{x}_t.$$

The t-th output is the linear transformation of the current hidden state,

$$\mathbf{o}_t = \mathbf{V}\mathbf{h}_t.$$

Fig. 10.4 shows the architecture of a basic RNN. The long short-term memory (LSTM) [9] is a popular RNN with enhanced capability to store longer-term temporal information, and it can deal with the problem of the gradient disappearing in RNN.

DNN is a powerful tool for computer vision tasks, but it may suffer from some limitations. One well-known shortcoming is the black box property. It is hard to understand how and why deep networks produce an output corresponding to the input and why the better performance can be achieved by deeper networks. Besides, DNN needs lots of neurons and hidden layers to improve performance. It means that memory and computing resources cost can be much higher than

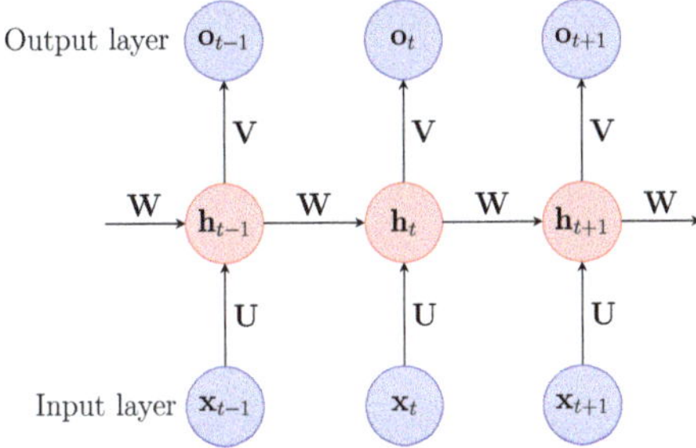

Fig. 10.4 A graphical illustration for RNN

that of traditional machine learning algorithms. To alleviate these problems, tensor decompositions are used to enhance the interpretability and compress DNNs to reduce the computational and storage memory.

In the following sections, we will introduce how tensor decompositions help to compress DNNs in Sect. 10.2 and to enhance the interpretability of DNNs in Sect. 10.3.

10.2 Network Compression by Low-Rank Tensor Approximation

In Sect. 10.1, we have briefly discussed some shortcomings of DNN. As the number of hidden layers increases, the accuracy performance can be improved, but its disadvantages are obvious. In order to learn from high-dimensional datasets, different strategies are explored to reduce the network parameters while keeping the multiway structures exploited. In this section, we will introduce these methods through two aspects, one is based on tensorizing specific layers and the other is to employ tensor low-rank approximation for reducing parameter redundancy in popular DNNs.

10.2.1 Network Structure Design

In either CNN or RNN, the weights of fully connected layers occupy the majority of memory [23]. In a fully connected layer, the input is flattened to a vector and converted to another output vector through a linear transformation based on matrix product. In this way, not only the flattening operation causes dimensionality information loss, but also the number of parameters required is huge. In this section, we will introduce how to design a new network structure by using tensor products, including t-product, mode-k product, and tensor contraction, instead of matrix product to compress the parameters or improve the performance of neural networks.

10.2.2 t-Product-Based DNNs

For example, suppose we have M samples of two-dimensional data in size of $N_0 \times K$, the forward propagation of the l-th dense layer (10.1) can be rewritten in matrix form as

$$\mathrm{f}_l(\mathbf{X}_{l-1}) = \sigma(\mathbf{W}_l \mathbf{X}_{l-1} + \mathbf{B}_l), \tag{10.2}$$

where $\mathbf{X}_{l-1} \in \mathbb{R}^{N_{l-1}K \times M}$, $\mathbf{W}_l \in \mathbb{R}^{N_l K \times N_{l-1}K}$, $\mathbf{B}_l$ is the bias. In [19], keeping the original multiway data representation and replacing the matrix product with t-product, we can get the layers in form of

$$\mathrm{f}_l(\mathcal{X}_{l-1}) = \sigma(\mathcal{W}_l * \mathcal{X}_{l-1} + \mathcal{B}_l), \tag{10.3}$$

where $\mathcal{W}_l \in \mathbb{R}^{N_l \times N_{l-1} \times K}$ denotes the weight tensor, $\mathcal{X}_{l-1} \in \mathbb{R}^{N_{l-1} \times M \times K}$ denotes the input tensor, $\mathrm{f}_l(\mathcal{X}_{l-1}) \in \mathbb{R}^{N_l \times M \times K}$ denotes the output tensor, and $\mathcal{B}_l$ denotes the bias tensor.

In Fig. 10.5, we can see the linear and multilinear combinations by matrix and tensor products, respectively. Notice that the matrix product in Fig. 10.5a requires N^2K^2 weight parameters, while t-product in Fig. 10.5b only requires N^2K if $N_l = N$ for $l = 1, \cdots, L$. Furthermore, according to the properties of t-product, the computation of t-product can be accelerated by fast Fourier transform as

$$\hat{\mathcal{X}}_l^{(k)} = \hat{\mathcal{W}}_l^{(k)} \hat{\mathcal{X}}_{l-1}^{(k)}, \tag{10.4}$$

where $\hat{\mathcal{X}}_l$ and $\hat{\mathcal{W}}_l$ are the fast Fourier transforms (FFTs) of $\mathcal{X}_l$ and $\mathcal{W}_l$ along their third modes, respectively, $\hat{\mathcal{X}}_l^{(k)}$ is the k-th frontal slice of $\hat{\mathcal{X}}_l$, and $k = 1, \cdots, K$. Then the inverse Fourier transform can be used to obtain $\mathcal{X}_{l+1}$, which greatly benefit the efficient implementation of DNNs. But there is a strict requirement that the input tensor $\mathcal{X}_{l-1}$ and the output tensor $\mathcal{X}_l$ must have two same dimensions and t-product is commonly used for third-order tensors.

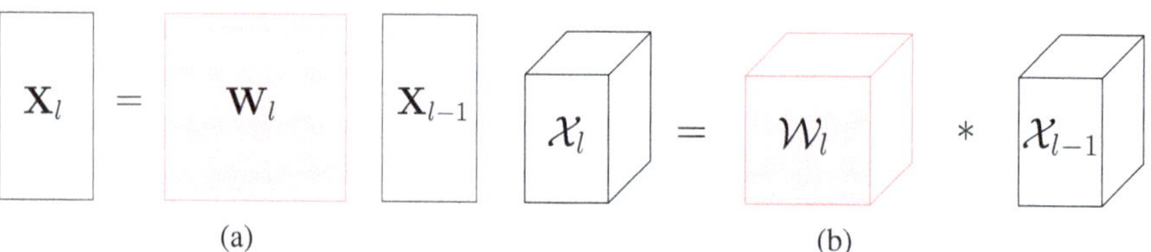

Fig. 10.5 Comparison of matrix and tensor product-based linear operators. (**a**) Matrix weighted connection. (**b**) Tensor weighted connection

10.2.3 Tensor Contraction Layer

Here we consider an N-way input tensor $\mathcal{X} \in \mathbb{R}^{I_1 \times I_2 \times \ldots \times I_N}$. In a fully connected layer of classical deep neural networks, the input tensor is unfolded to a one-way high-dimensional vector $\mathbf{x} \in \mathbb{R}^{I_1 I_2 \ldots I_N}$. If the number of neurons in next layer is I_{N+1}, then the fully connected layer is in form of

$$\mathrm{f}(\mathbf{x}) = \sigma(\mathbf{W}\mathbf{x} + \mathbf{b}), \tag{10.5}$$

where the corresponding weight matrix $\mathbf{W} \in \mathbb{R}^{I_1 I_2 \ldots I_N \times I_{N+1}}$. Because of the high dimensionality of the input vector, the weight matrix will be very large, occupy a lot of memory, and slow down the computation.

In [2], based on tensor contraction, the aforementioned layer is replaced with

$$\mathrm{f}(\mathcal{X}) = \sigma(< \mathcal{X}, \mathcal{W} >_N + \mathcal{B}), \tag{10.6}$$

where $\mathcal{W} \in \mathbb{R}^{I_1 \times \cdots \times I_N}$ is the coefficient tensor which usually admits low rankness. The tensor contraction layer can maintain the multiway structure of input and reduce the model parameters by low-rank tensor approximation.

Taking the Tucker decomposition as an example, i.e.

$$\mathcal{W} = \mathcal{G} \times_1 \mathbf{U}^{(1)} \times_2 \cdots \times_N \mathbf{U}^{(N)} \times_{N+1} \mathbf{U}^{(N+1)},$$

where $\mathcal{G} \in \mathbb{R}^{R_1 \times R_2 \times \ldots \times R_N \times R_{N+1}}$, $\mathbf{U}^{(n)} \in \mathbb{R}^{I_n \times R_n}$ for $n = 1, \cdots, N+1$. For easy understanding, we give a tensor graphical representation of the Tucker layer in Fig. 10.6.

In fact, when the original tensor weight $\mathcal{W}$ is decomposed according to different tensor decompositions, new neural networks with different structures can be obtained, as the TT layer based on tensor train (TT) decomposition shown in Fig. 10.7. In Fig. 10.7a, the weight tensor $\mathcal{W} \in \mathbb{R}^{I_1 \times \ldots \times I_N \times I_{N+1}}$ is decomposed to

$$\mathcal{W}(i_1, \cdots, i_N, i_{N+1}) = \mathcal{G}^{(1)}(:, i_1, :) \cdots \mathcal{G}^{(N)}(:, i_N, :)\mathcal{G}^{(N+1)}(:, i_{N+1}, :),$$

the core factors $\mathcal{G}^{(n)} \in \mathbb{R}^{R_n \times I_n \times R_{n+1}}$, $n = 1, \cdots, N+1$ ($R_1 = R_{N+2} = 1$). Specially, if the output of tensor contraction layer is a tensor in size of $J_1 \times \cdots J_N$, then the weight tensor $\mathcal{W} \in \mathbb{R}^{I_1 \times \ldots \times I_N \times J_1 \times \ldots \times J_N}$ is decomposed to

$$\mathcal{W}(i_1, \cdots, i_N, j_1, \cdots, j_N) = \mathcal{G}^{(1)}(:, i_1, j_1, :) \cdots \mathcal{G}^{(N)}(:, i_N, j_N, :),$$

where the core factors $\mathcal{G}^{(n)} \in \mathbb{R}^{R_n \times I_n \times J_n \times R_{n+1}}$, $n = 1, \cdots, N$ ($R_1 = R_{N+1} = 1$), as shown in Fig. 10.7b.

Table 10.1 gives the compression factor (CF) of these tensor contraction layers, where CF is the ratio of number of parameters for the fully connected layer and the tensorized fully connected layer to measure the compression efficiency of tensorized neural networks.

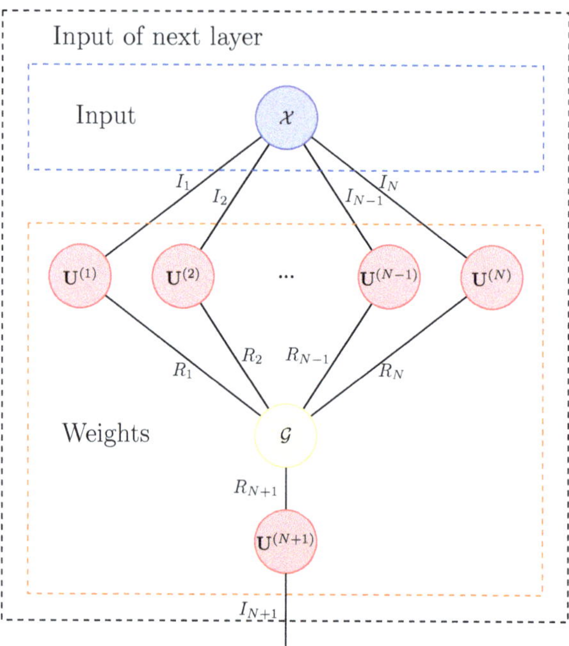

Fig. 10.6 A graphical illustration for the linear transformation in Tucker layer. Tucker layer is a hidden layer whose weight tensor is in Tucker format

Besides of above Tucker decomposition and TT decomposition, many tensor decomposition forms also can be applied to approximate the coefficient tensor, such as CP decomposition [14], TR decomposition [27], HT decomposition [31], and BT decomposition [30]. And these tensor decomposition forms can be applied to approximate the convolutional kernels [5, 14, 27].

10.2.4 *Deep Tensorized Neural Networks Based on Mode-n Product*

Besides the t-product and tensor contraction-based deep neural networks, some other works use mode-n product to replace matrix product [3, 12], resulting in deep tensorized neural networks, which have fewer parameters than classical networks.

For input tensor $\mathcal{X} \in \mathbb{R}^{I_1 \times \cdots \times I_N}$, one simple deep tensorized neural network based on mode-n product [12] aims to seek a output tensor in small size by

$$f(\mathcal{X}) = \sigma(\mathcal{X} \times_1 \mathbf{U}^{(1)} \times_2 \cdots \times_N \mathbf{U}^{(N)} + \mathcal{B}), \tag{10.7}$$

where $\mathbf{U}^{(n)} \in \mathbb{R}^{I_n \times R_n}$ for $n = 1, \cdots, N$. It can be either used as an additional layer after the convolutional layer or the pooling layer for dimension reduction or employed as a substitute of the fully connected layer to reducing the number of parameters required.

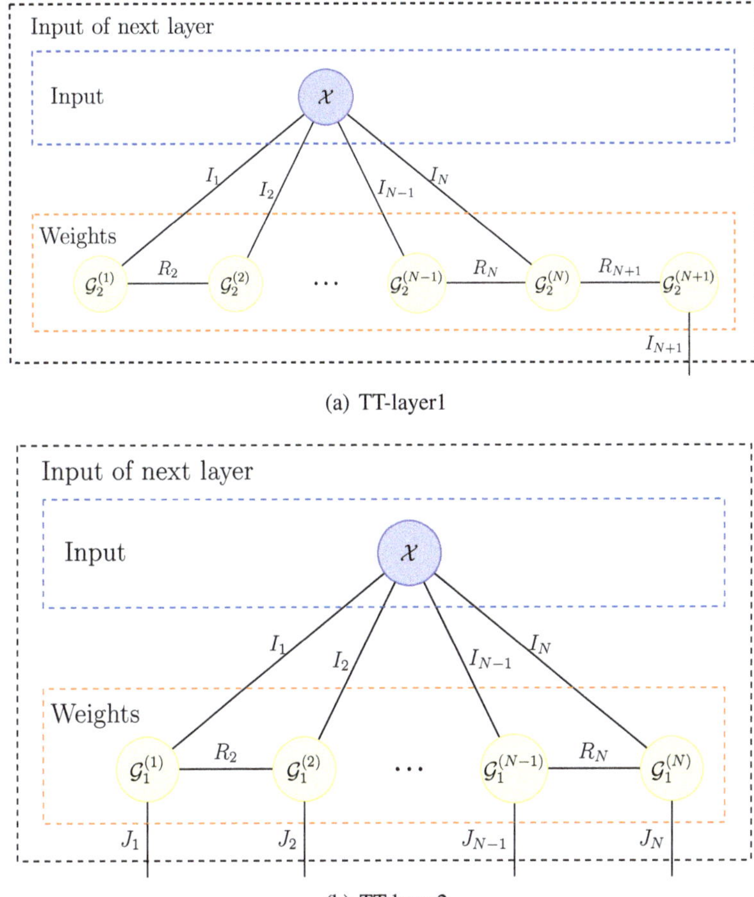

Fig. 10.7 A graphical illustration for the linear transformation in tensor train layer. Tensor train layer is a hidden one whose weight tensor is in tensor train format. There are two kinds of graphical structures: (**a**) the output is a vector [2]. It can be applied in last hidden layer for classification task. (**b**) The output is a tensor [20]. It can be applied in middle hidden layer and connected with next tensor train or Tucker layer

Table 10.1 The compression factor (CF) of tensor contraction layers

Model	Tucker-layer	TT-layer1	TT-layer2
CF	$\frac{\prod_{n=1}^{N+1} I_n}{\prod_{n=1}^{N+1} R_n+\sum_{n=1}^{N+1} I_n R_n}$	$\frac{\prod_{n=1}^{N} I_n J_n}{\sum_{n=1}^{N} R_n I_n J_n R_{n+1}}$	$\frac{\prod_{n=1}^{N+1} I_n}{\sum_{n=1}^{N+1} R_n I_n R_{n+1}}$

Generally, more hidden layers can improve performance of deep neural networks. In tensor factorization neural network (TFNN) [3], multiple factor layers and activation functions are used, as graphically illustrated in Fig. 10.8. With the sharing of weight factors along different ways, TFNN can efficiently capture

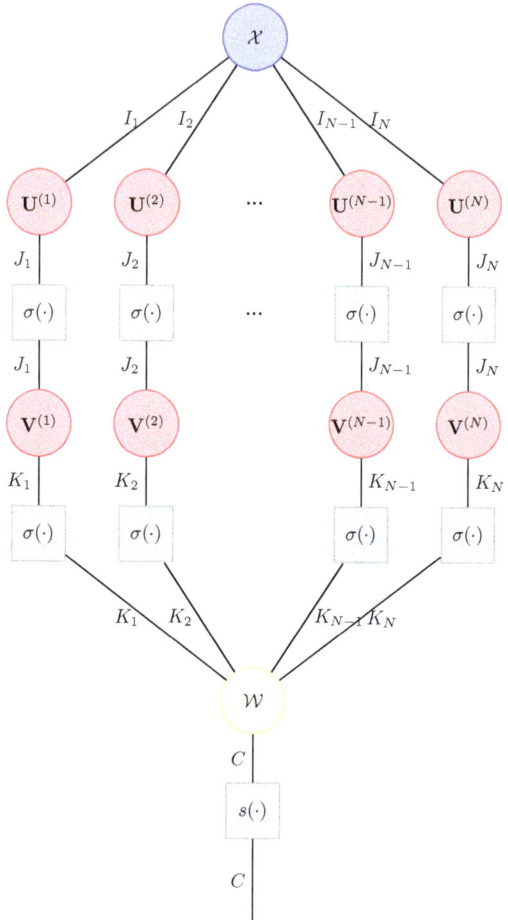

Fig. 10.8 A graphical illustration for the structure of TFNN

the data structure with much fewer parameters than the neural networks without factorization. In Fig. 10.8, the input is a tensor $\mathcal{X} \in \mathbb{R}^{I_1 \times I_2 \times \ldots \times I_N}$, and the weights are composed of $\mathbf{U}^{(n)} \in \mathbb{R}^{I_n \times J_n}, n = 1, \cdots, N$, $\mathbf{V}^{(n)} \in \mathbb{R}^{J_n \times K_n}, n = 1, \cdots, N$ and $\mathcal{W} \in \mathbb{R}^{K_1 \times \ldots \times K_N \times C}$. Following behind each factor layer is a nonlinear activation function $\sigma(\cdot)$. The softmax function s$(\cdot)$ is usually used in deep neural network for classification task.

Besides, abovementioned neural network structures compress DNN whose input is a tensor. Here, we additionally introduce another neural network whose input is a vector, named tensor neural network (TNN) [32]. TNN is more efficient for large vocabulary speech recognition tasks. Figure 10.9 gives a graphical illustration for the network structure of TNN, composed of a double projection layer and a tensor layer. The forward propagation in this model can be represented as follows:

$$\mathbf{a}_1 = \sigma(\mathbf{U}_1\mathbf{x}), \mathbf{a}_2 = \sigma(\mathbf{U}_2\mathbf{x}), \tag{10.8}$$

$$\mathbf{z} = \sigma(\mathcal{W} \times_1 \mathbf{a}_1 \times_2 \mathbf{a}_2), \tag{10.9}$$

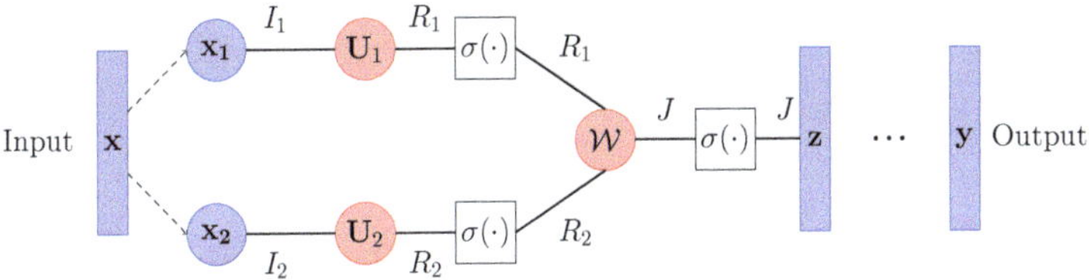

Fig. 10.9 A graphical illustration for the structure of TNN

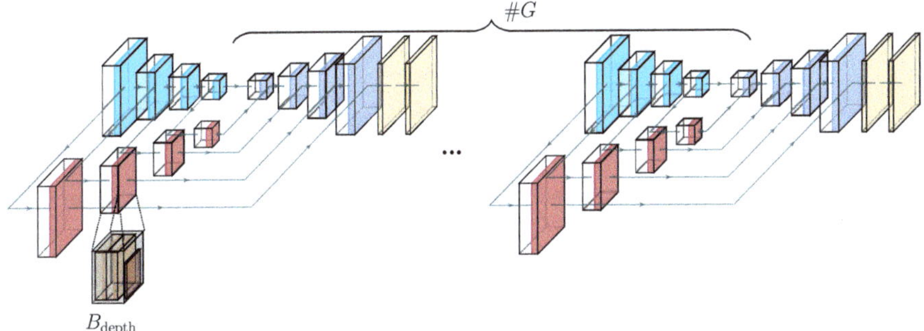

Fig. 10.10 A graphical illustration for the architecture of modified HG. They are three signal pathways. Cyan region means downsampling/encoder, blue region means upsampling/decoder, and red blocks represent skip connection. The transparent black block gives a detailed illustration for each basic block module, which is composed of B_{dept} convolutional layers followed by a pooling layer. The kernel of each convolutional layer is sized of $H \times W \times F_{\text{in}} \times F_{\text{out}}$

where $\mathbf{x} \in \mathbb{R}^I$ is the input vector, $\mathbf{U}_1 \in \mathbb{R}^{R_1 \times I}$ and $\mathbf{U}_2 \in \mathbb{R}^{R_2 \times I}$ are two affine matrices constituting the double projection layer, and $\mathcal{W} \in \mathbb{R}^{R_1 \times R_2 \times J}$ is the bilinear transformation tensor constituting the tensor layer, and $\sigma(\cdot)$ is nonlinear activation function. Note that $\mathbf{x}_1 = \mathbf{x}_2 = \mathbf{x}$ in Fig. 10.9.

10.2.5 Parameter Optimization

The abovementioned methods compress the network parameters by tensorizing individual layers only. They change the structure of traditional matrix operator-based neural networks into tensor-based ones and train these networks on their structures.

In fact, the whole parameters in deep neural networks may be redundant due to their over-parametrization, which can easily cause overfitting in learning too. In [13], a new approach considers the whole parameters of modified Hour-Glass (HG) networks [16] as a single high-order tensor and uses low-rank tensor approximation to reduce the number of parameters. A graphical illustration of the architecture of modified HourGlass (HG) networks [16] is given in Fig. 10.10.

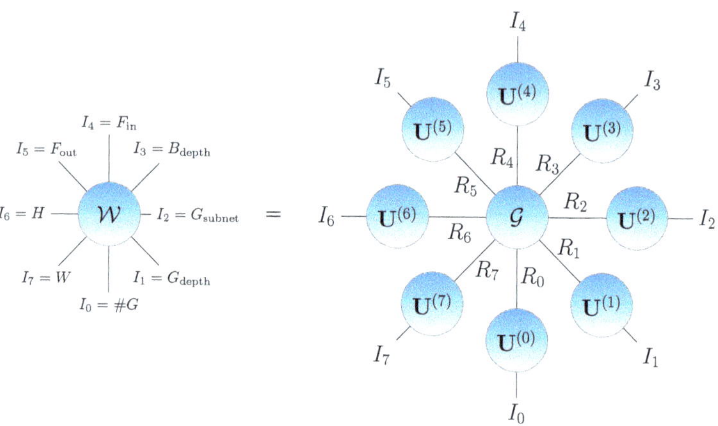

Fig. 10.11 A graphical illustration for Tucker decomposition of the whole parameters of modified HG networks

As shown in Fig. 10.10, HG network has several kinds of network design hyperparameters—the number of sub HG networks ($\#G$), the depth of each HG (G_{depth}), the three signal pathways (G_{subnet}), the number of convolutional layers per block (B_{depth}), the number of input features (F_{in}), the number of output features (F_{out}), and the height (H) and width (W) of each convolutional kernel. Therefore, all weights of the network can be collected as a single eighth-order tensor $\mathcal{W} \in \mathbb{R}^{I_0 \times I_1 \times \ldots \times I_7}$. Driven by the possibility of over-parametrization, tensor decompositions, including Tucker decomposition and TT decomposition, are used to compress the parameter tensor from $\mathrm{O}(\prod_{k=0}^{7} I_k)$ to $\mathrm{O}(\prod_{k=0}^{7} R_k + \sum_{k=0}^{7} I_k R_k)$ or $\mathrm{O}(\sum_{k=0}^{7} I_k R_k R_{k+1})$. A graphical illustration is given in Fig. 10.11 based on Tucker decomposition.

In fact, the tensor contraction layer mentioned in Fig. 10.6 can be seen as a layer-wise decomposition for specific layers in order to compress the network parameters.

10.3 Understanding Deep Learning with Tensor Decomposition

Most of deep neural networks are end-to-end models, and it is difficult to understand the network characteristics, just like "black boxes." One well-publicized proposition is that deep networks are more efficient compared with shallow ones. A series of literatures have demonstrated this phenomenon using various network architectures [4, 18]. However, they fail to explain why the performance of a deep network is better than that of a shallow network.

In [4], authors try to establish the ternary correspondence between functions, tensor decomposition, and neural networks and provide some explanations of depth efficiency theoretically. As we know, neural networks can express any functions, and its expressiveness varies in different network architectures. Generally, deep neural networks can simulate more complex functions than shallow ones, as shown in Fig. 10.12.

To verify this phenomenon, we consider an input image to the network, denoted $\mathbf{X}$, which is composed of N patches $\{\mathbf{x}_1, \cdots, \mathbf{x}_N\}$ with $\mathbf{x}_n \in \mathbb{R}^I$. For example, given a RGB image sized of 32×32, we construct the input data through a 5×5 selection window crossing the three color bands. Assuming a patch is taken for every pixel (boundaries padded), we have $N = 32 * 32 = 1024$ and $I = 5 * 5 * 3 = 75$. For image classification tasks, the output y belongs one of the categories $\{1, \cdots, K\}$. Then the hypotheses space of the required functions can be expressed as follows:

$$\mathrm{h}_y(\mathbf{x}_1, \ldots, \mathbf{x}_N) = \sum_{d_1,\ldots,d_N=1}^{D} \mathcal{A}^y_{d_1,\ldots,d_N} \prod_{n=1}^{N} \mathrm{f}_{\theta_{d_n}}(\mathbf{x}_n), \tag{10.10}$$

where $\mathcal{A}^y \in \mathbb{R}^{D\times\cdots\times D}$ is a Nth-order coefficient tensor and $\mathrm{f}_{\theta_1}, \cdots, \mathrm{f}_{\theta_D} :\in \mathbb{R}^I \to \mathbb{R}$ are representation functions commonly selected from wavelets, affine functions, or radial basis functions.

To realize the score functions $\mathrm{h}_y(\mathbf{x}_1, \ldots, \mathbf{x}_N)$ with a neural network, the representation functions can be achieved by performing different functions over each entries of the input data, namely, the representation layer. Then a fully connected layer parameterized by $\mathcal{A}^y$ can be used to give the final scores.

The function coefficients (i.e., neural network parameters) can be considered as a tensor. In [4], it is proposed that parameter tensor in a shallow neural network and a deep neural network can be represented by CP decomposition and hierarchical Tucker (HT) decomposition, respectively.

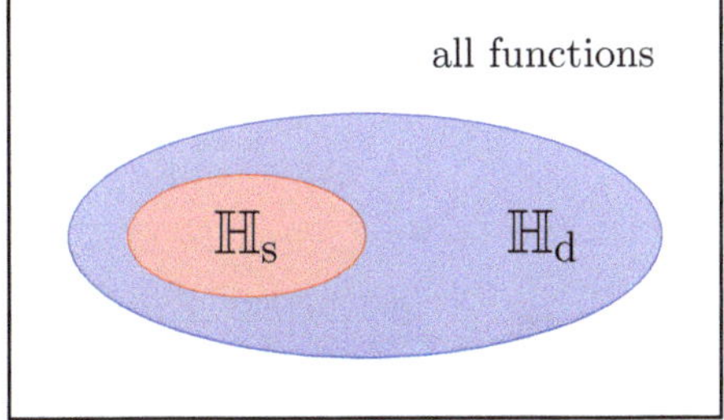

Fig. 10.12 A graphical illustration for the relationship between neural networks and functions. $\mathbb{H}_\mathrm{d}$ and $\mathbb{H}_\mathrm{s}$ are the sets of functions practically realizable by deep network and shallow network, respectively

Specifically, assuming the coefficient tensor admits CP decomposition as $\mathcal{A}^y = \sum_{r=1}^{R} a_r^y \mathbf{u}_r^{(1)} \circ \cdots \circ \mathbf{u}_r^{(N)}$

$$\mathrm{h}_y(\mathbf{x}_1, \ldots, \mathbf{x}_N) = \sum_{r=1}^{R} a_r^y \prod_{n=1}^{N} \sum_{j_n=1}^{J} \mathbf{u}_r^n(j_n)\, \mathrm{f}_{\theta_{j_n}}(\mathbf{x}_n), \tag{10.11}$$

where $\mathbf{a}^y \in \mathbb{R}^R = [a_1^y; \cdots, a_R^y]$. It can be achieved by a single hidden layer convolutional arithmetic circuit as shown in Fig. 10.13.

Based on hierarchical Tucker (HT) decomposition, we can get the expression of $\mathcal{A}^y$ as follows:

$$\phi_{r_1}^{(1,j)} = \sum_{r_0=1}^{R_0} a_{r_0}^{1,j,r_1} \mathbf{u}_{r_0}^{(0,2j-1)} \circ \mathbf{u}_{r_0}^{(0,2j)} \tag{10.12}$$

$$\cdots$$

$$\phi_{r_l}^{(l,j)} = \sum_{r_{l-1}=1}^{R_{l-1}} a_{r_{l-1}}^{l,j,r_l} \phi_{r_{l-1}}^{(l-1,2j-1)} \circ \phi_{r_{l-1}}^{(l-1,2j)} \tag{10.13}$$

$$\cdots$$

$$\phi_{r_{L-1}}^{(L-1,j)} = \sum_{r_{L-2}=1}^{R_{L-2}} a_{r_{L-2}}^{L-1,j,r_{L-1}} \phi_{r_{L-2}}^{(L-2,2j-1)} \circ \phi_{r_{L-2}}^{(L-2,2j)} \tag{10.14}$$

$$\mathcal{A}^y = \sum_{r_{L-1}=1}^{R_{L-1}} a_{r_{L-1}}^{L,y} \phi_{r_{L-1}}^{(L-1,1)} \circ \phi_{r_{L-1}}^{(L-1,2)}, \tag{10.15}$$

where $R_0, \ldots, R_{L-1}$ are HT ranks.

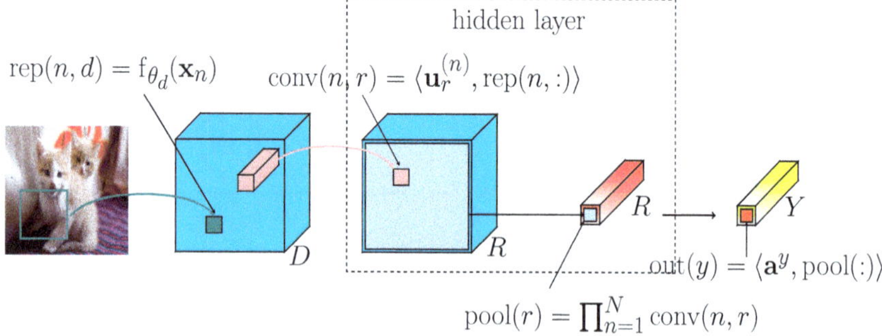

Fig. 10.13 A graphical illustration for single hidden layer convolutional arithmetic circuit architecture. d is the index for output channels in the first convolution layer, r is the index for output channels in hidden layer, and y is the index for output channels in last dense layer (output layer), $d = 1, \cdots, D, r = 1, \cdots, R, y = 1, \cdots, Y$

Denoting $\mathbf{a}^{L,y} = [a_1^{L,y}; \cdots; a_{R_{L-1}}^{L,y}]$, $\mathbf{a}^{l,j,r_l} = [a_1^{l,j,r_l}; \cdots; a_{R_{l-1}}^{l,j,r_l}]$, $\mathbf{A}^{l,j} \in \mathbb{R}^{R_{l-1} \times R_l}$ with $\mathbf{A}^{l,j}(:, r_l) = \mathbf{a}^{l,j,r_l}$, $\mathbf{U}^{(0,n)} \in \mathbb{R}^{D \times R_0}$ with $\mathbf{U}^{(0,n)}(:, r_0) = \mathbf{u}_{r_0}^{0,n}$, an illustration for special HT factorization is shown in Fig. 10.14 when $L = 2$. To construct the HT factorization of multiway tensors, parameters required to be stored include the first-level parameters $\mathbf{U}^{(0,n)}$ for $n = 1, \cdots, n$, intermediate levels' weights $\mathbf{A}^{l,j}$ for $l = 1, \cdots, L-1$, $j = 1, \cdots, N/2^l$, and the final level weights $\mathbf{a}^{L,y}$. Figure 10.15 presents the corresponding convolutional arithmetic circuit architecture with L hidden layers, where the size of pooling windows is 2. Thus, the number of hidden layers is $L = \log_2^N$, and in l-th hidden layer, the index $i = 1, 2, \ldots, I/2^l$.

In Figs. 10.13 and 10.15, we can see an initial convolutional layer (referred to as representation layer) followed by hidden layers, which in turn are followed by a dense (linear) output layer. Each hidden layer consists of a convolution and spatial pooling, and the receptive field of the convolution is 1×1. Another different point with traditional CNN is that the pooling operator is based on products ($\mathrm{P}(c_i) = \prod_i c_i$).

As shown in Figs. 10.13 and 10.15, the only difference between these two networks is the depth of the network. Specifically, the shallow network corresponds to the CP decomposition of the weighting tensors, and the deep network corresponds

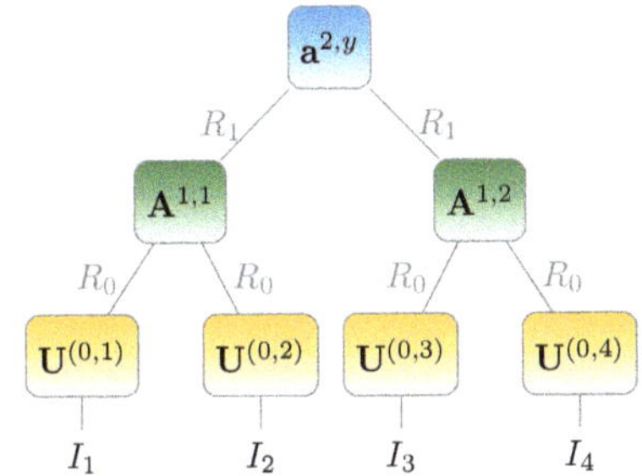

Fig. 10.14 A graphical illustration for HT decomposition of the coefficient tensor $\mathcal{A}^y$

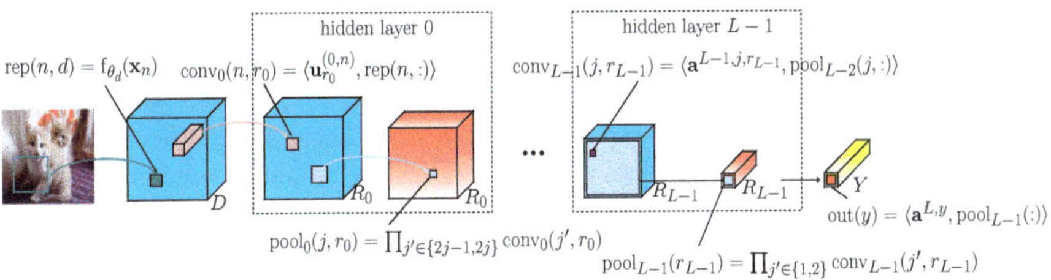

Fig. 10.15 A graphical illustration for convolutional arithmetic circuit architecture with L hidden layers. d is the index for output channels in the first convolution layer, r_l is the index for output channels in l-th hidden layer, and y is the index for output channels in the last dense layer (output layer)

to the HT decomposition. And it is well known that the generated tensor from HT decomposition can catch more information than that from CP decomposition since the number of HT ranks is more than 1. Therefore, to a certain extent, this provides an explanation for the depth efficiency of neural networks.

10.4 Applications

As a state-of-the-art machine learning technique, deep neural networks have been extensively applied to solve two major problems, i.e., classification and regression. In this section, we mainly focus on image classification. Different tensor decompositions are employed for deep network compression, in order to demonstrate the potentials by the introduction of tensor computation to deep learning. We will compare the performance of the uncompressed deep neural networks with that of corresponding compressed ones in terms of compressed rate and classification accuracy. Three commonly used datasets are employed in this section as follows.

- **MNIST dataset**: It collects 70,000 grayscale images in size of 28×28 for the handwritten digit recognition task, where 60,000 grayscale images are used for training and 10,000 are used for testing.
- **Fashion MNIST**: It is highly similar to MNIST dataset but depicts fashion products (t-shirt, trouser, pullover, dress, coat, sandal, shirt, sneaker, bag, and ankle boot).
- **CIFAR-10 dataset**: It consists of 60,000 images with size $32 \times 32 \times 3$ including 10 different classes: airplane, automobile, bird, cat, deer, dog, frog, horse, ship, and truck. The dataset contains 50,000 training and 10,000 testing images.

10.4.1 MNIST Dataset Classification

For MNIST dataset, 55,000 images in the training set are used to train, and the other 5000 images in the training set are used to validate. At last we get the result by applying the best model for testing. In addition, for all the neural network models compared here, the optimizer used is the popular stochastic gradient descent (SGD) algorithm with learning rate = 0.01, momentum = 0.95, and weight decay=0.0005, and the batch size is set to 500 and the epoch is 100. It should be noted that we pad original 28×28 images to 32×32 in this group of experiments for easy implementation of different TT networks to get more choices for compression factors.

The neural network models compared in this part include a single hidden layer dense neural network and its extensions in tensor field, including TT networks and Tucker networks. In fact, with different network design parameters, the resulting networks will be in different structures. We demonstrate the construction of these networks in detail as follows.

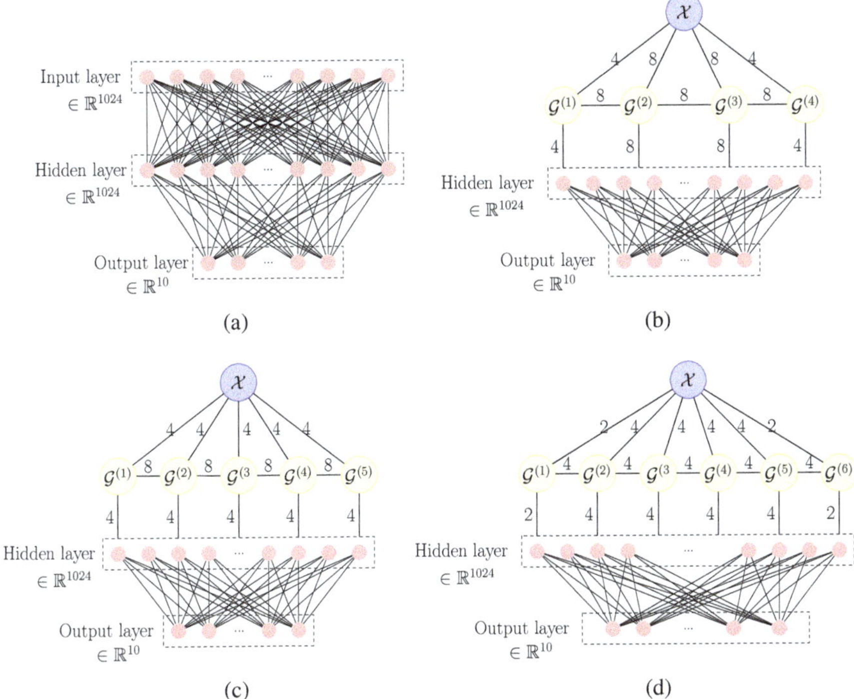

Fig. 10.16 A graphical illustration of the baseline network and its extension to different TT network structures. (**a**) Baseline. (**b**) TT network1. (**c**) TT network2. (**d**) TT network3

- **Baseline** The construction of the baseline network is shown in Fig. 10.16a, where the input data is a vector $\mathbf{x} \in \mathbb{R}^{1024}$, the weight matrix of the hidden layer is in size of 1024×1024, the activation function used after the hidden layer is ReLU, and the weight matrix of the output layer is in size of 1024×10, respectively.
- **TT network** Driven by the TT layer we introduced in Fig. 10.7, we can replace the hidden layer in Fig. 10.16a by a TT layer as shown in Fig. 10.16. Regrading the input data as multiway tensors $\mathcal{X} \in \mathbb{R}^{4\times8\times8\times4}$, $\mathcal{X} \in \mathbb{R}^{4\times4\times4\times4\times4}$, $\mathcal{X} \in \mathbb{R}^{2\times4\times4\times4\times4\times2}$, different TT layers can be obtained. The weights of the TT layer include $\mathcal{G}^{(n)}$, $n = 1, \cdots, N$; N is set to be 4, 5, and 6 in Fig. 10.16b, c, and d, respectively.
- **Tucker network** Similarly, based on the Fig. 10.6, we can substitute the hidden layer in Fig. 10.17a with a Tucker layer as shown in Fig. 10.17b. The input data is a matrix $\mathbf{X} \in \mathbb{R}^{32\times32}$; the weights of the Tucker-layer include $\mathcal{G} \in \mathbb{R}^{R_1\times R_2\times R_3}$ and three matrices $\mathbf{U}^{(1)}$, $\mathbf{U}^{(2)}$, and $\mathbf{U}^{(3)}$.

Because MNIST dataset is relatively easy, we also used fashion MNIST which is similar to MNIST. The classification results of the MNIST and fashion MNIST datasets by the baseline network, TT networks, and Tucker networks with different

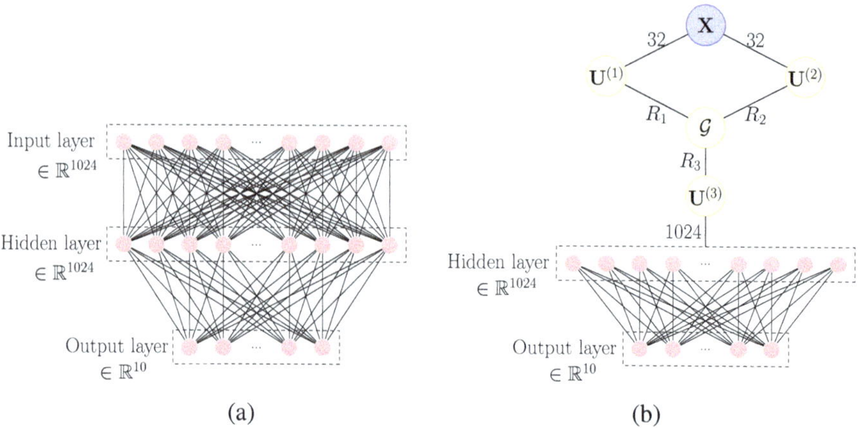

Fig. 10.17 A graphical comparison of the baseline network and its extension to Tucker network, where R_1, R_2, $and R_3$ are Tucker ranks. (**a**) Baseline. (**b**) Tucker network

Table 10.2 TT network and Tucker network performance

Dataset	Model	CF	Input data size	Ranks	Classification accuracy(%)
MNIST	Baseline	1	1024	/	98.23
	TT network	56.68	$4 \times 8 \times 8 \times 4$	1,8,8,8,1	97.59
		78.06	$4 \times 4 \times 4 \times 4 \times 4$	1,8,8,8,8,1	98.01
		95.09	$4 \times 4 \times 4 \times 4 \times 4$	1,4,4,4,4,1	97.90
		93.94	$2 \times 4 \times 4 \times 4 \times 4 \times 2$	1,4,4,4,4,4,1	98.15
	Tucker network	6.89	32×32	20,20,100	98.36
		23.76	32×32	10,10,30	97.96
		50.32	32×32	5,5,10	96.12
Fashion MNIST	Baseline1	1	1024	/	89.26
	TT network	56.68	$4 \times 8 \times 8 \times 4$	1,8,8,8,1	88.39
		78.06	$4 \times 4 \times 4 \times 4 \times 4$	1,8,8,8,8,1	88.64
		95.09	$4 \times 4 \times 4 \times 4 \times 4$	1,4,4,4,4,1	88.44
		93.94	$2 \times 4 \times 4 \times 4 \times 4 \times 2$	1,4,4,4,4,4,1	88.34
	Tucker network	6.89	32×32	20,20,100	89.77
		23.76	32×32	10,10,30	88.97
		50.32	32×32	5,5,10	86.00

design parameters are shown in Table 10.2. CF is the compression factor which is defined as the ratio of the number of parameters in the baseline network and that of compressed one. Ranks mean the corresponding rank of the TT layer or the Tucker layer. The CF and classification accuracy are the two main performance metrics of interest. For each dataset, the best classification accuracy and its corresponding CF are marked in bold.

From Table 10.2, we could observe TT network can get higher compression factor, while Tucker network can improve the accuracy slightly while compressing the neural network model. This is because TT network resizes the input data into a higher-order tensor which will sacrifice a small amount of precision. However Tucker network maintains the original structure and uses low-rank constraint for weight parameters to reduce the redundancy.

10.4.2 CIFAR-10 Dataset Classification

CNN is effective for many image classification tasks. In CNN, a few convolutional layers (including pooling and Relu) are usually followed by a fully connected layer which can produce an output category vector, i.e., ten categories corresponding to a ten-dimensional vector. However, the output of the last convolutional layer can be multiway tensors. In order to map the multiway tensor into vectors, two strategies are commonly adopted, flattening or Global Average Pooling (GAP). The flattening operation is also known as vectorization, such as a tensor sized of $8 \times 8 \times 32$ can be converted to a vector sized of 2048×1. The GAP is an averaging operation over all feature maps along the output channel axis. Specifically, for a third-order feature tensor $H \times W \times N_o$, the output of a GAP layer is a single vector sized of N_o by simply taking the mean value of each $H \times W$ feature map along specific output channel.

By using tensor factorization, we can map the multiway tensors into output vector without flattening operation, which maintains the original feature structures and reduces the parameters required. Here, we set a simple CNN which includes two convolutional layers (a convolutional layer with a $5 \times 5 \times 1 \times 16$ kernel, Relu, 2×2 max pooling, and a convolutional layer with $5 \times 5 \times 16 \times 32$ kernel, Relu, 2×2 max pooling) and a fully connected layer. It can be extended into GAP-CNN, TT-CNN, and Tucker-CNN by replacing its final fully connected layer by GAP layer, TT layer in Fig. 10.7, and Tucker layer in Fig. 10.6, respectively. Figure 10.18 provides a comparison of the structure of the last layer of CNN, GAP-CNN, TT-CNN, and Tucker-CNN.

The developed networks are used to tackle the classification tasks over the CIFAR-10 dataset, which is more complex than MNIST dataset. The corresponding results are shown in Table 10.3. As shown in Table 10.3, Tucker-CNN can reduce model parameters appropriately, and the corresponding precision loss can be ignored. With the same precision loss, GAP can achieve a higher compression factor. The reason may be that after convolutional operation, the redundancy has been reduced to a certain degree, and Tucker layer and TT layer cannot reduce lots of parameters anymore without the loss of precision.

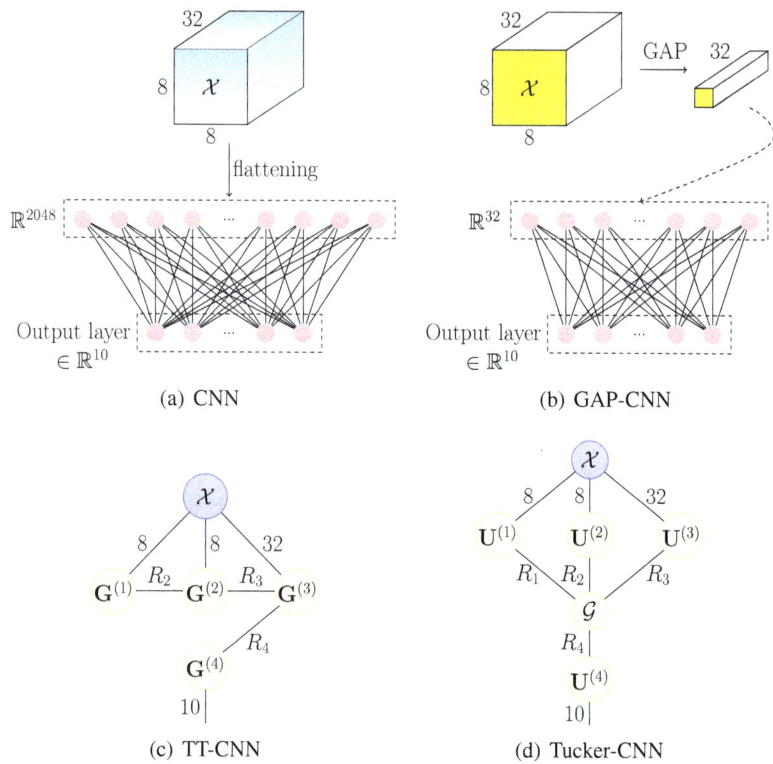

Fig. 10.18 A graphical illustration for different operations in the last layer of the CNN. (**a**) Flattening operation, (**b**) GAP operation, (**c**) TT layer where $[R_2; R_3; R_4]$ is the rank , (**d**) Tucker layer where $[R_1; R_2; R_3; R_4]$ is the rank

Table 10.3 TT-CNN and Tucker-CNN performance for CIFAR-10

Model	CF	Size	Ranks	Classification accuracy(%)
CNN	1	2048	/	70.26
GAP-CNN	64	32	/	66.11
TT-CNN	1.50	$8 \times 8 \times 32$	1,4,20,20,1	69.62
	2.86	$8 \times 8 \times 32$	1,4,20,10,1	69.87
	13.30	$8 \times 8 \times 32$	1,4,4,10,1	66.06
	24.09	$8 \times 8 \times 32$	1,4,4,5,1	64.2
Tucker-CNN	2.80	$8 \times 8 \times 32$	4,4,20,20	70.08
	5.11	$8 \times 8 \times 32$	4,4,20,10	70.07
	16.60	$8 \times 8 \times 32$	4,4,10,5	66.20
	24.03	$8 \times 8 \times 32$	2,2,10,10	66.19
	34.02	$8 \times 8 \times 32$	2,2,10,5	62.81

10.5 Summary

In this chapter, we mainly focus on applying tensor analysis in deep neural networks for more efficient computation. One way is to take tensor operations as representation of multilinear combinations in each layer. Another way is to use low-rank tensor approximation for compressing the network parameters. Especially, the selection of low-rank tensor approximations for weight tensors can balance the compression rate and classification accuracy. For example, with different network design hyper-parameters, we can obtain a higher compression rate with little accuracy degeneration or a lower compression rate with almost no accuracy degeneration. This improvement makes it possible to train and deploy neural network models on resource-constrained devices such as smartphones, wearable sensors, and IOT devices. These compact deep neural networks have been applied to human pose recognition [17], speech enhancement [21], medical image classification [22], video classification [29], etc.

However, the additional hyper-parameters, such as the size of rank and the number of tensor contraction layers, bring more uncertainty to DNNs. End-to-end tensorized neural networks need extensively combinatorial searching and much training until a good rank parameter is found. This disadvantage can be solved by pre-training a uncompressed neural networks and then calculating the rank of coefficient tensor [7, 10, 33]. However, this situation could incur huge accuracy loss and network retraining is required to recover the performance of the original network. In terms of these problems, Xu et al. [28] proposed Trained Rank Pruning for training low-rank networks; Hawkins and Zhang [8] designed an end-to-end framework which can automatically determine the tensor ranks.

Besides compression and acceleration, the concept of tensor decomposition can also help us to better understand neural networks, such as why deep networks perform better than shallow ones [4]. In deep convolutional arithmetic circuit, network weights are mapped to HT decomposition parameters (the size of output channel in each hidden convolutional layer is the size of corresponding rank in HT decomposition). In deep tensorized neural networks based on mode-n product, the physically meaningful information inferred from the gradients of the trained weight matrices (which mode plays a relatively more important role) also resolves the black box nature of neural networks [1]. This also shed new light for the theoretical research and other applications of deep learning.

References

1. Calvi, G.G., Moniri, A., Mahfouz, M., Zhao, Q., Mandic, D.P.: Compression and interpretability of deep neural networks via Tucker tensor layer: from first principles to tensor valued back-propagation (2019, eprints). arXiv–1903
2. Cao, X., Rabusseau, G.: Tensor regression networks with various low-rank tensor approximations (2017, preprint). arXiv:1712.09520

3. Chien, J.T., Bao, Y.T.: Tensor-factorized neural networks. IEEE Trans. Neural Netw. Learn. Syst. **29**, 1998–2011 (2017). https://doi.org/10.1109/TNNLS.2017.2690379
4. Cohen, N., Sharir, O., Levine, Y., Tamari, R., Yakira, D., Shashua, A.: Analysis and design of convolutional networks via hierarchical tensor decompositions (2017, eprints). arXiv–1705
5. Garipov, T., Podoprikhin, D., Novikov, A., Vetrov, D.: Ultimate tensorization: compressing convolutional and FC layers alike (2016, eprints). arXiv–1611
6. Goldberg, Y.: A primer on neural network models for natural language processing. J. Artif. Intell. Res. **57**, 345–420 (2016)
7. Guo, J., Li, Y., Lin, W., Chen, Y., Li, J.: Network decoupling: from regular to depthwise separable convolutions (2018, eprints). arXiv–1808
8. Hawkins, C., Zhang, Z.: End-to-end variational bayesian training of tensorized neural networks with automatic rank determination (2020, eprints). arXiv–2010
9. Hochreiter, S., Schmidhuber, J.: Long short-term memory. Neural Comput. **9**(8), 1735–1780 (1997)
10. Jaderberg, M., Vedaldi, A., Zisserman, A.: Speeding up convolutional neural networks with low rank expansions. In: Proceedings of the British Machine Vision Conference. BMVA Press, Saint-Ouen-l'Aumône (2014)
11. Kim, Y.: Convolutional neural networks for sentence classification. In: Proceedings of the 2014 Conference on Empirical Methods in Natural Language Processing (EMNLP), pp. 1746–1751. Association for Computational Linguistics, Doha, Qatar (2014). https://doi.org/10.3115/v1/D14-1181. https://www.aclweb.org/anthology/D14-1181
12. Kossaifi, J., Khanna, A., Lipton, Z., Furlanello, T., Anandkumar, A.: Tensor contraction layers for parsimonious deep nets. In: Proceedings of the IEEE Conference on Computer Vision and Pattern Recognition Workshops, pp. 26–32 (2017)
13. Kossaifi, J., Bulat, A., Tzimiropoulos, G., Pantic, M.: T-net: Parametrizing fully convolutional nets with a single high-order tensor. In: Proceedings of the IEEE Conference on Computer Vision and Pattern Recognition, pp. 7822–7831 (2019)
14. Lebedev, V., Ganin, Y., Rakhuba, M., Oseledets, I.V., Lempitsky, V.S.: Speeding-up convolutional neural networks using fine-tuned CP-Decomposition. In: International Conference on Learning Representations ICLR (Poster) (2015)
15. Lecun, Y., Bottou, L., Bengio, Y., Haffner, P.: Gradient-based learning applied to document recognition. Proc. IEEE **86**, 2278–2324 (1998). https://doi.org/10.1109/5.726791
16. Long, J., Shelhamer, E., Darrell, T.: Fully convolutional networks for semantic segmentation. In: Proceedings of the IEEE Conference on Computer Vision and Pattern Recognition, pp. 3431–3440 (2015)
17. Makantasis, K., Voulodimos, A., Doulamis, A., Bakalos, N., Doulamis, N.: Space-time domain tensor neural networks: an application on human pose recognition (2020, e-prints). arXiv–2004
18. Mhaskar, H., Liao, Q., Poggio, T.: Learning functions: when is deep better than shallow. preprint arXiv:1603.00988
19. Newman, E., Horesh, L., Avron, H., Kilmer, M.: Stable tensor neural networks for rapid deep learning (2018, e-prints). arXiv–1811
20. Novikov, A., Podoprikhin, D., Osokin, A., Vetrov, D.P.: Tensorizing neural networks. In: Cortes, C., Lawrence, N., Lee, D., Sugiyama, M., Garnett, R. (eds.) Advances in Neural Information Processing Systems, vol. 28. Curran Associates, Red Hook (2015). https://proceedings.neurips.cc/paper/2015/file/6855456e2fe46a9d49d3d3af4f57443d-Paper.pdf
21. Qi, J., Hu, H., Wang, Y., Yang, C.H.H., Siniscalchi, S.M., Lee, C.H.: Tensor-to-vector regression for multi-channel speech enhancement based on tensor-train network. In: IEEE International Conference on Acoustics, Speech and Signal Processing (ICASSP), pp. 7504–7508. IEEE, Piscataway (2020)
22. Selvan, R., Dam, E.B.: Tensor networks for medical image classification. In: Medical Imaging with Deep Learning, pp. 721–732. PMLR, Westminster (2020)
23. Simonyan, K., Zisserman, A.: Very deep convolutional networks for large-scale image recognition. In: International Conference on Learning Representations (2015)

24. Tachioka, Y., Ishii, J.: Long short-term memory recurrent-neural-network-based bandwidth extension for automatic speech recognition. Acoust. Sci. Technol. **37**(6), 319–321 (2016)
25. Tai, K.S., Socher, R., Manning, C.D.: Improved semantic representations from tree-structured long short-term memory networks. In: Proceedings of the 53rd Annual Meeting of the Association for Computational Linguistics and the 7th International Joint Conference on Natural Language Processing (Volume 1: Long Papers), pp. 1556–1566 (2015)
26. Tran, D., Bourdev, L., Fergus, R., Torresani, L., Paluri, M.: Learning spatiotemporal features with 3d convolutional networks. In: Proceedings of the IEEE International Conference on Computer Vision, pp. 4489–4497 (2015)
27. Wang, W., Sun, Y., Eriksson, B., Wang, W., Aggarwal, V.: Wide compression: tensor ring nets. In: Proceedings of the IEEE Conference on Computer Vision and Pattern Recognition, pp. 9329–9338 (2018)
28. Xu, Y., Li, Y., Zhang, S., Wen, W., Wang, B., Dai, W., Qi, Y., Chen, Y., Lin, W., Xiong, H.: Trained rank pruning for efficient deep neural networks (2019, e-prints). arXiv–1910
29. Yang, Y., Krompass, D., Tresp, V.: Tensor-train recurrent neural networks for video classification. In: International Conference on Machine Learning, pp. 3891–3900. PMLR, Westminster (2017)
30. Ye, J., Li, G., Chen, D., Yang, H., Zhe, S., Xu, Z.: Block-term tensor neural networks. Neural Netw. **130**, 11–21 (2020)
31. Yin, M., Liao, S., Liu, X.Y., Wang, X., Yuan, B.: Compressing recurrent neural networks using hierarchical tucker tensor decomposition (2020, e-prints). arXiv–2005
32. Yu, D., Deng, L., Seide, F.: The deep tensor neural network with applications to large vocabulary speech recognition. IEEE Trans. Audio Speech Lang. Process. **21**(2), 388–396 (2012)
33. Zhang, X., Zou, J., He, K., Sun, J.: Accelerating ery deep convolutional networks for classification and detection. IEEE Trans. Pattern Anal. Mach. Intell. **38**(10), 1943–1955 (2015)

Chapter 11
Deep Networks for Tensor Approximation

11.1 Introduction

Deep learning is a sub-field of machine learning, which can learn the inherent useful features for specific tasks based on artificial neural networks. Due to the breakthrough of resources on computing and data, deep learning makes great success in many fields, such as computer version (CV) [17] and natural language processing (NLP) [38].

With the emergence of a large amount of multiway data, tensor approximation has been applied in many fields, such as compressive sensing (CS) [22], tensor completion [25], tensor principal component analysis [27], tensor regression and classification[33, 46], and tensor clustering [37]. Besides the machine learning techniques in the other chapters, there exist some studies which are dedicated to solve tensor approximation problems through deep networks [3, 9, 23, 45, 47]. The main idea of these approaches is to employ the powerful deep networks as a tool for tackling some difficult problems in tensor approximation, such as tensor rank determination [47] and simulation of specific nonlinear proximal operators [8].

Classical deep neural networks directly map the tensor as input to the approximation result as output. These networks usually consist of stacked nonlinear operational layers, such as autoencoders [31], convolutional neural networks (CNNs) [17], and generative adversarial networks (GANs) [36]. These models are usually trained end-to-end using the well-known backpropagation algorithm. They enjoy powerful fitting and learning capability but suffer from bad interpretation due to their black box characteristic.

Model-based methods usually apply interpretable iterative algorithms to approximate the multidimensional data. In order to combine the advantages of both deep networks and model-based methods, two strategies of deep learning methods have been employed, namely, deep unrolling and deep plug-and-play (PnP), which usually have theoretical guarantees. The former deep unrolling maps the popular iterative algorithms for model-based methods onto deep networks [8, 9, 28, 32, 41,

Y. Liu et al., *Tensor Computation for Data Analysis*,
https://doi.org/10.1007/978-3-030-74386-4_11

44]. It can be regarded as a neural network realization of the iterative algorithms with fixed iteration numbers, such as ISTA-Net [41] and ADMM-Net [32]. With trainable parameters, these deep unrolling methods commonly give a better performance than the original algorithms with fewer iteration times. Deep unrolling models are trained end-to-end and usually enjoy the speed of classical deep networks and the interpretation of model-based methods.

Deep PnP is usually applied in data reconstruction-related tasks like tensor completion [45] and tensor compressed sensing [40]. It regards a specific subproblem that usually appears in the model-based methods as the denoising problem and solves it using pre-trained deep networks. Unlike deep unrolling, the number of iterations of deep PnP model is not fixed. Deep PnP can use deep neural networks to capture semantic information that expresses tensor details while having the interpretability of model-based methods.

In the following content of this section, we will introduce these three different frameworks for tensor approximation in detail, namely, classical deep neural networks, deep unrolling, and deep PnP. And some applications of these frameworks are given, including tensor rank approximation [47], video snapshot compressive imaging (SCI) [9, 44], and low-rank tensor completion [45].

11.2 Classical Deep Neural Networks

Classical deep learning models are composed of a series of nonlinear computation layers. These models can be regarded as black boxes. As long as the input is available, the output can be directly obtained without caring about the internal process. Figure 11.1 shows the basic structure of classical deep learning models, and the approximation process can expressed as

$$\hat{\mathcal{X}} = \mathfrak{N}_{\text{cla}}(\mathcal{X}^0, \boldsymbol{\Theta}), \tag{11.1}$$

where $\mathfrak{N}_{\text{cla}}(\cdot, \boldsymbol{\Theta})$ is the generalized classical deep learning model, $\boldsymbol{\Theta}$ contains trainable parameters of the model, $\mathcal{X}^0$ is the tensor for approximation, and $\hat{\mathcal{X}}$ is the approximated data.

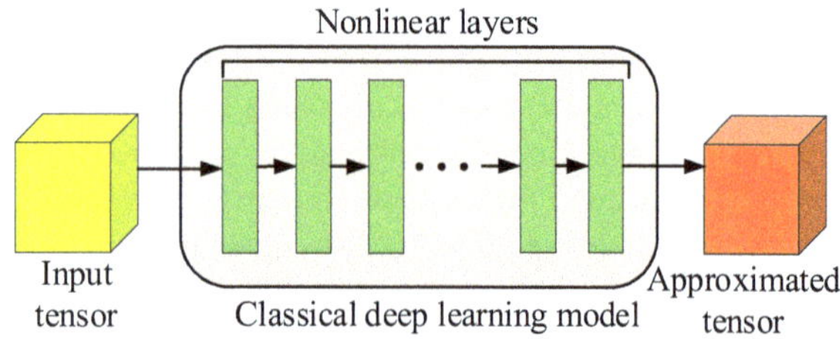

Fig. 11.1 The generalized framework of classical deep learning methods

In Fig. 11.1, the green boxes are layers of the deep neural networks which can be linear layers or linear layers combined with nonlinear activation operators. The linear layer can be fully connected layer, convolutional layer, and recurrent layer. And the activation operator can be the sigmoid function, rectified linear unit (ReLU). Furthermore, some strategies can be combined with the model in Fig. 11.1 to improve the model performance, like the residual connection in ResNet [10].

There are some classical deep learning models which are developed for tensor approximation problems, such as rank approximation [47] and video compressed sensing [13]. However, due to the black box characteristic of these classical networks, there is no good interpretation and theoretical guarantee. And these completely data-driven end-to-end manners may have risks for some undesired effects [12]. We suggest interested readers can refer to [7] to get more details about the classical deep neural networks.

11.3 Deep Unrolling

In order to describe deep unrolling more clearly, we briefly describe the generalized framework of model-based methods first. The model-based methods try to solve an optimization problem as follows:

$$\hat{\mathcal{X}} = \arg\min_{\mathcal{X}} \mathfrak{R}(\mathcal{X}), \text{ s. t. some conditions}, \tag{11.2}$$

where $\mathfrak{R}(\cdot)$ is the regularization term and $\hat{\mathcal{X}}$ denotes the approximation results. Model-based methods usually obtain $\hat{\mathcal{X}}$ using some nonlinear iteration algorithms, like iterative shrinkage thresholding algorithm (ISTA) [8, 9, 41], approximate message passing (AMP) [1, 30, 44], and alternating direction method of multipliers (ADMM) [28, 32]. If the initialized data is $\mathcal{X}^0$, for easy understanding, we formulate the k-th iteration of the generalized approximation algorithm as follows:

$$\mathcal{X}^k = \mathfrak{N}_{\text{mod}}(\mathcal{X}^{k-1}), \tag{11.3}$$

where $\mathfrak{N}_{\text{mod}}(\cdot)$ contains all calculations of one iteration of the algorithms. For example, in ADMM, $\mathfrak{N}_{\text{mod}}(\cdot)$ may contain more than two steps to obtain $\mathcal{X}^k$. Model-based methods repeat (11.3) until convergence. They have interpretation guarantees but usually cost a lot of time.

Inspired by the iterative algorithm which processes data step by step, deep unrolling maps iterative approximation algorithms onto step-fixed deep neural networks. By expressing the unrolled model as $\mathfrak{N}_{\text{unr}}(\cdot, \boldsymbol{\Theta})$, the approximation process of deep unrolling models can be expressed as follows:

$$\hat{\mathcal{X}} = \mathfrak{N}_{\text{unr}}(\mathcal{X}^0, \boldsymbol{\Theta}) = \mathfrak{N}_{\text{unr}}^K(\mathcal{X}^{K-1}, \boldsymbol{\Theta}^K), \tag{11.4}$$

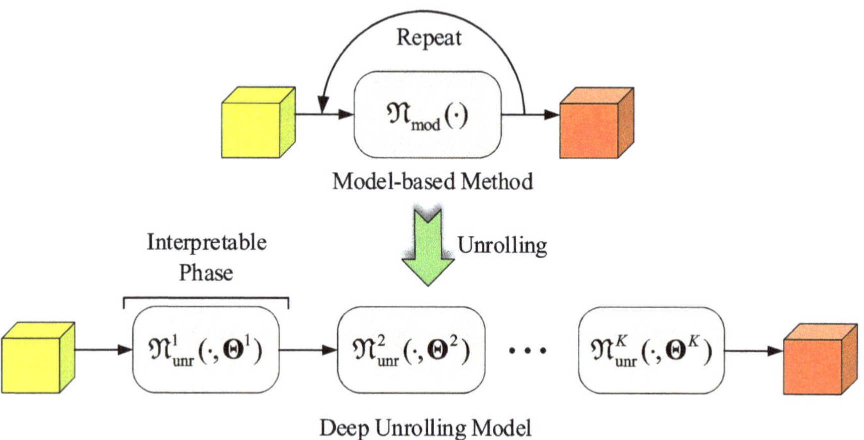

Fig. 11.2 Generalized framework of deep unrolling models

$$\mathcal{X}^k = \mathfrak{N}^k_{\text{unr}}(\mathcal{X}^{k-1}, \mathbf{\Theta}^k), \tag{11.5}$$

where $\mathcal{X}^k$ is the updated tensor after the first k modules, $\mathfrak{N}^k_{\text{unr}}(\cdot, \mathbf{\Theta}^k)$ is the k-th module/phase, $\mathbf{\Theta}^k$ is the trainable parameters of the k-th module, and $\mathbf{\Theta}$ is the collection of all the $\mathbf{\Theta}^k$ with $k = 1, \cdots, K$.

Figure 11.2 illustrates the difference between generalized model-based methods and generalized deep unrolling models. As shown in Fig. 11.2, the unrolled model usually consists of several phases with the same structure, where each phase corresponds to one iteration in the original iterative restoration algorithm.

For many inverse problems, such as tensor completion [14, 21, 42] and sparse coding [48], model-based methods usually introduce some handcrafted regularization terms, such as $\|\cdot\|_1$, $\|\cdot\|_*$ and $\|\cdot\|_0$. To solve these problems, many nonlinear operators are developed. For example, the soft thresholding function is designed for $\|\cdot\|_1$ and $\|\cdot\|_*$, and the hard thresholding function is designed for $\|\cdot\|_0$. Some studies build the deep unrolling model by training the related parameters of these nonlinear operators. Taking the soft threshold function as an example, it is designed for sparse data without value limitations [6] and can be expressed as

$$\mathfrak{T}_\theta(x) = \text{sign}(x)\max(|x| - \theta, 0), \tag{11.6}$$

where θ is the soft thresholding value. The generalized deep unrolling model with trainable soft thresholding functions can be illustrated in Fig. 11.3, where

$$\mathfrak{N}^k_{\text{unr}}(\cdot, \mathbf{\Theta}^k) = \mathfrak{T}_{\theta^k}(\mathfrak{N}^k_{\text{unr1}}(\cdot, \mathbf{\Theta}^k_1)), \quad \mathbf{\Theta}^k = \{\mathbf{\Theta}^k_1, \theta^k\}, \quad k = 1, \cdots, K \tag{11.7}$$

Fig. 11.3 Generalized framework of deep unrolling models with handcrafted thresholding functions for data priors

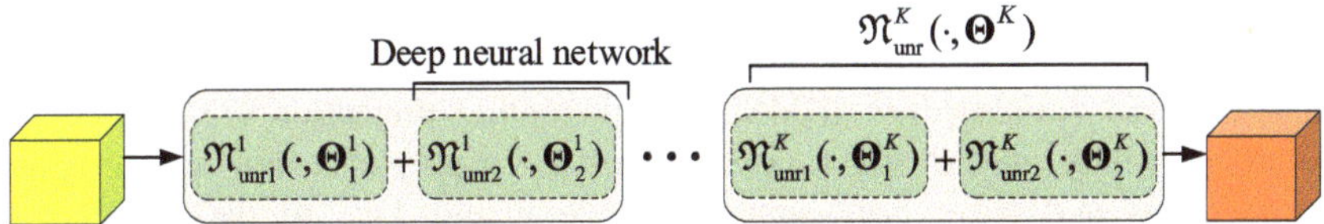

Fig. 11.4 Generalized framework of deep unrolling models with learnable operators for data priors

Obviously, $\mathfrak{N}^k_{\text{unr1}}(\cdot, \mathbf{\Theta}^k_1)$ is a part of $\mathfrak{N}^k_{\text{unr}}(\cdot, \mathbf{\Theta}^k)$, and it contains desired operators with respect to the related iterative algorithm for specific tasks, and we show some examples in Sect. 11.5.

In order to obtain better performance with a smaller phase number, the parameter θ^k is usually trained with other parameters in deep unrolling models. For example, in [8], θ^k is trained with weights related to sparse dictionaries to obtain more efficient sparse coding with only 10 to 20 phases. And in [9], θ^k is jointly trained with CNNs, which are designed for sparse transformation to achieve video snapshot compressive imaging. Besides the soft thresholding function, there are also deep unrolling models considering other thresholding functions. For example, bounded linear unit [35] is designed for ℓ_∞ norm and hard thresholding linear unit [34] is designed for ℓ_0 norm.

However, handcrafted regularizers may ignore some semantic information for data representation. To this end, some studies [4, 5, 44] try to employ deep neural networks to learn data priors directly. Figure 11.4 illustrates the generalized framework of deep unrolling model with trainable data priors. In this framework, each phase of the model can be expressed as

$$\mathcal{X}^k = \mathfrak{N}^k_{\text{unr}}(\mathcal{X}^{k-1}, \mathbf{\Theta}^k) = \mathfrak{N}^k_{\text{unr2}}(\mathfrak{N}^k_{\text{unr1}}(\mathcal{X}^{k-1}, \mathbf{\Theta}^k_1), \mathbf{\Theta}^k_2), \tag{11.8}$$

where $\mathbf{\Theta}^k = \{\mathbf{\Theta}^k_1, \mathbf{\Theta}^k_2\}$ and $\mathfrak{N}^k_{\text{unr2}}(\cdot, \mathbf{\Theta}^k_2)$ is a deep neural network which is used to represent detailed data priors and $\mathbf{\Theta}^k_2$ contains its parameters.

This unrolling strategy can be used for a lot of model-based algorithms. For example, in [4], $\mathfrak{N}^k_{\text{unr2}}(\cdot, \mathbf{\Theta}^k_2)$ is used to fit the gradient for unrolled gradient descent algorithm. In [44], $\mathfrak{N}^k_{\text{unr2}}(\cdot, \mathbf{\Theta}^k_2)$ is developed to estimate the difference between original data and the corrupted data in each iteration. Technically, framework in Fig. 11.4 is more flexible than the framework in Fig. 11.3. $\mathfrak{N}^k_{\text{unr2}}(\cdot, \mathbf{\Theta}^k_2)$ can be regarded as a trainable thresholding function which can inherit different priors by learning from different training data. In this way, the learned thresholding function can better fit various tasks and datasets.

11.4 Deep Plug-and-Play

In tensor analysis, the optimization problem usually consists of multiple components, such as data fitting items and structure-enforcing items like sparsity, low rankness, and smoothness. ADMM and alternating projection are two popular solutions, which separate the original optimization problems into several related subproblems and solve each subproblem iteratively. Due to the existence of regularizer, e.g., structure-enforcing ones, there commonly exists one subproblem as follows:

$$\min_{\mathcal{X}} \ \Re(\mathcal{X}) + \frac{1}{2\sigma^2}||\mathcal{X} - \mathcal{V}||_{\mathrm{F}}^2, \tag{11.9}$$

where $\Re(\mathcal{X})$ is the regularizer for a specific task. It is usually known as the proximal operator, such as the soft thresholding operator for ℓ_1 norm.

It is obvious that problem (11.9) can be regarded as a regularized image denoising problem, where σ determines the noise level [45]. Driven by this equivalence, some studies try to use plug-and-play (PnP) which is a strategy that employs advanced denoisers as a surrogate of the proximal operator in the ADMM or other proximal algorithms. It is a highly flexible framework that employs the power of advanced image priors [2, 40, 45].

In PnP framework, at each iteration, problem (11.9) can be solved by a denoiser $\mathfrak{N}_{\mathrm{den}}(\cdot)$ expressed as

$$\hat{\mathcal{X}} = \mathfrak{N}_{\mathrm{den}}(\mathcal{V}). \tag{11.10}$$

The denoisers could be different for different regularizers, and some state-of-the-art denoisers can be used, such as BM3D[11] and NLM[11]. Moreover, to explore the deep information of data, deep neural networks are employed to be the denoiser [29], known as deep PnP. Figure 11.5 shows the basic framework of deep PnP models, and the k-th iteration can be expressed as

$$\mathcal{X}^k = \mathfrak{N}_{\mathrm{PnP}}\left(\mathcal{X}^{k-1}\right) = \mathfrak{N}_{\mathrm{den}}^k\left(\mathfrak{N}_{\mathrm{mod}}'\left(\mathcal{X}^{k-1}\right), \boldsymbol{\Theta}^k\right), \tag{11.11}$$

where $\mathfrak{N}_{\mathrm{mod}}'(\cdot)$ is a part of $\mathfrak{N}_{\mathrm{mod}}(\cdot)$ expected for the subproblem (11.9), and $\mathfrak{N}_{\mathrm{den}}^k(\cdot, \boldsymbol{\Theta}^k)$ is the denoiser in the k-th iteration and it is determined by σ. The difference between $\mathfrak{N}_{\mathrm{mod1}}(\cdot)$ and $\mathfrak{N}_{\mathrm{unr1}}^k(\cdot, \boldsymbol{\Theta}_1^k)$ is that $\mathfrak{N}_{\mathrm{unr1}}^k(\cdot, \boldsymbol{\Theta}_1^k)$ usually contains

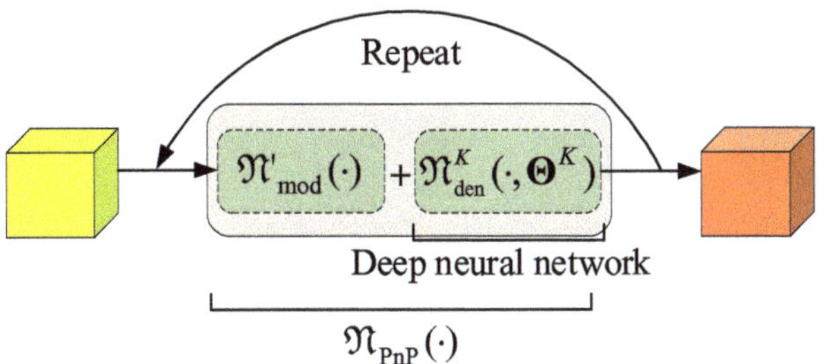

Fig. 11.5 Basic framework of plug-and-play (PnP) models

trainable parameters which are trained with other parameters. But there is usually no trainable parameters in $\mathfrak{N}'_{\text{mod}}(\cdot)$.

Unlike deep unrolling models, deep PnP models do not need to be trained end-to-end. The employed deep network-based denoisers are usually pre-trained. And similar to the model-based methods, deep plug-and-play methods have no stable iteration number like deep unrolling models.

11.5 Applications

In this section, we provide some applications to show the effectiveness of above-mentioned three frameworks. Firstly, classical deep neural networks are employed to predict the rank of the tensor. Secondly, the deep unrolling models are applied for video SCI. Finally, we introduce a deep PnP framework for tensor completion.

11.5.1 Classical Neural Networks for Tensor Rank Estimation

Tensor factorization is an important technique for high-order data analysis. One issue of CP decomposition is that a predefined rank should be known in advance. However, the tensor rank determination is an NP-hard problem.

As one of the most popular neural networks, CNN can extract features which are hidden deeply, even when the measurements are noisy [17–19]. The tensor rank is one kind of data features, which represents the degree of interdependence between entries. Motivated by this, the CNN is applied to solve the CP-rank estimation problem for multiway images [47].

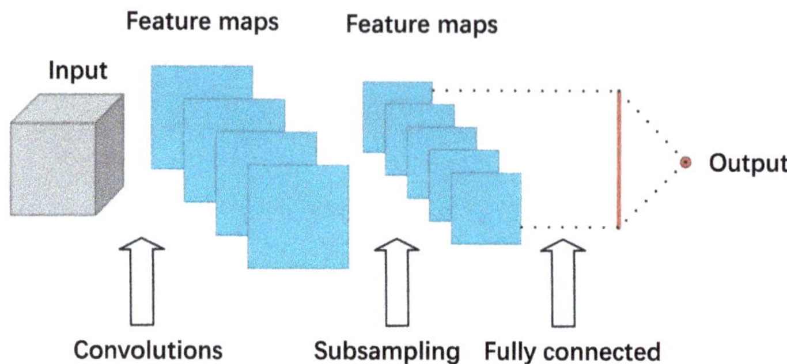

Fig. 11.6 An illustration of TRN model

11.5.1.1 Tensor Rank Learning Architecture

Given noisy measurement $\mathcal{Y}$ of tensor $\mathcal{X}$, the tensor rank network (TRN) aims to estimate the rank of tensor $\mathcal{X}$ using CNNs by minimizing the loss function as

$$\mathfrak{L} = \frac{1}{N_{\mathrm{t}}} \sum_{n=1}^{N_{\mathrm{t}}} \left(R_n - \hat{R}_n \right)^2, \tag{11.12}$$

where N_{t} is the number of samples for training, R_n is the optimal rank value of n-th sample $\mathcal{X}_n$, and $\hat{R}_n$ represents the estimated rank.

The mathematical expression of the TRN is as follows:

$$\hat{R} = \mathfrak{N}_{\mathrm{R}}(\mathcal{Y}, \boldsymbol{\Theta}), \tag{11.13}$$

where $\mathcal{Y} \in \mathbb{R}^{I \times J \times K}$ is the input, $\mathfrak{N}_{\mathrm{R}}$ is a nonlinear function as proposed by LeCun et al. [20], $\boldsymbol{\Theta}$ contains trainable parameters in the network, and $\hat{R}$ is estimated rank. Figure 11.6 provides an illustration of the TRN model, which contains a convolutional layer and a fully connected layer.

11.5.1.2 Tensor Rank Network with Pre-decomposition Architecture

Considering CP decomposition, the optimization problem for rank minimization can be formulated as

$$\min_{\mathbf{w}} \quad \lambda \|\mathbf{w}\|_0 + \frac{1}{2} \|\mathcal{Y} - \sum_{r=1}^{R} w_r \mathbf{a}_r^{(1)} \circ \mathbf{a}_r^{(2)} \circ \mathbf{a}_r^{(3)} \|_{\mathrm{F}}^2, \tag{11.14}$$

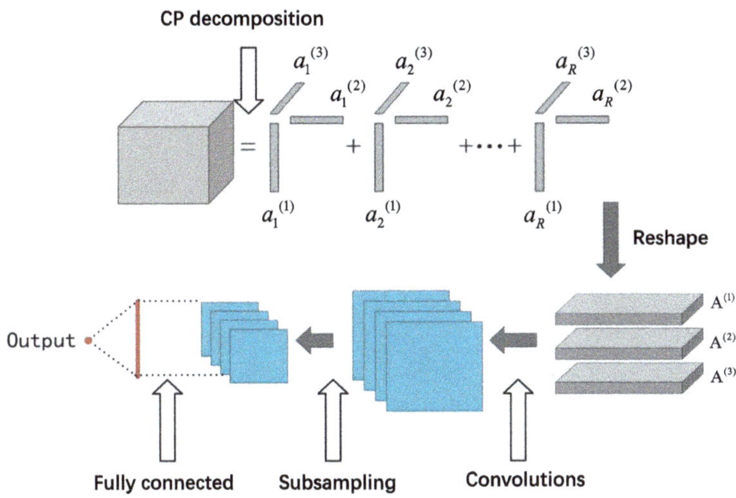

Fig. 11.7 An illustration of TRN-PD model

where $\mathbf{w} = [w_1, \cdots, w_r, \cdots, w_R]^{\mathrm{T}}$ is the weight vector and $R = \|\mathbf{w}\|_0$ denotes the number of nonzero entries in $\mathbf{w}$.

To solve the rank estimation problem (11.14), we assume a predefined rank bound $\bar{R}$ and perform a pre-decomposition to obtain $\bar{\mathbf{w}}$, $\bar{\mathbf{a}}^{(n)}$, $n = 1, 2, 3$, as follows:

$$\min_{\bar{\mathbf{w}}, \{\bar{\mathbf{a}}_r^{(1)}\}, \{\bar{\mathbf{a}}_r^{(2)}\}, \{\bar{\mathbf{a}}_r^{(3)}\},} \|\mathcal{Y} - \sum_{r=1}^{\bar{R}} \bar{w}_r \bar{\mathbf{a}}_r^{(1)} \circ \bar{\mathbf{a}}_r^{(2)} \circ \bar{\mathbf{a}}_r^{(3)}\|_{\mathrm{F}}^2. \tag{11.15}$$

This optimization problem can be solved by alternative least squares (ALS) [16]. The obtained matrices $\bar{\mathbf{A}}^{(1)}$, $\bar{\mathbf{A}}^{(2)}$, $\bar{\mathbf{A}}^{(3)}$ can be stacked into a tensor $\bar{\mathcal{Y}}$ along the third mode.

As Fig. 11.7 illustrates, we can input $\bar{\mathcal{Y}}$ to the network (11.13) to get the predicted rank $\hat{R}$. In fact, the network (11.13) is used to solve the following optimization model:

$$\begin{aligned} \min_{\hat{R}, \hat{\mathbf{w}}, \mathbf{a}_r^{(n)}} \quad & \hat{R} = \|\hat{\mathbf{w}}\|_0, \\ \text{s. t.} \quad & \sum_{r=1}^{\|\hat{\mathbf{w}}\|_0} \hat{w}_r \mathbf{a}_r^{(1)} \circ \mathbf{a}_r^{(2)} \circ \mathbf{a}_r^{(3)} = \sum_{r=1}^{\bar{R}} \bar{w}_r \bar{\mathbf{a}}_r^{(1)} \circ \bar{\mathbf{a}}_r^{(2)} \circ \bar{\mathbf{a}}_r^{(3)}. \end{aligned} \tag{11.16}$$

This CP-rank estimation method is called tensor rank network with pre-decomposition (TRN-PD), which is summarized in Algorithm 49.

Algorithm 49: TRN-PD

Input: $\mathcal{Y} \in \mathbb{R}^{I_1 \times I_2 \cdots I_N}$ and model $\mathfrak{G}(\boldsymbol{\Theta}, \mathcal{Y})$
Output: $\hat{R}$
use CP-ALS with rank $\bar{R}$ to get $\bar{\mathbf{A}}^{(1)}, \bar{\mathbf{A}}^{(2)}, \cdots, \bar{\mathbf{A}}^{(N)}$
reshape $\bar{\mathbf{A}}^{(1)}, \bar{\mathbf{A}}^{(2)}, \cdots, \bar{\mathbf{A}}^{(N)}$ to get a new tensor $\bar{\mathcal{Y}}$
use model $\mathfrak{N}_{\mathrm{R}}(\boldsymbol{\Theta}, \bar{\mathcal{Y}})$ to predict $\hat{R}$

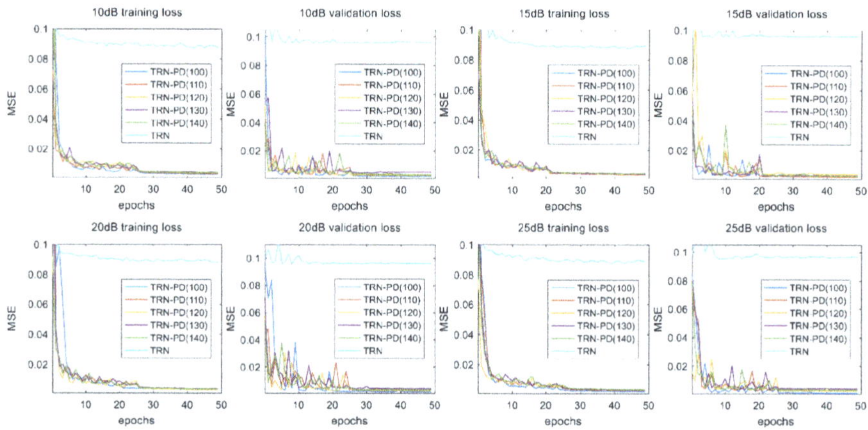

Fig. 11.8 The performance comparison between TRN and TRN-PD on the $50 \times 50 \times 50$ synthetic dataset at different noise levels

11.5.1.3 Experimental Analysis

In this experiment, all the methods are applied on synthetic datasets. We generate tensors with different ranks. The rank R of the synthetic $50 \times 50 \times 50$ tensor ranges from 50 to 100, and the rank R of synthetic $20 \times 20 \times 20$ tensor ranges from 20 to 70. Each tensor is generated with added random Gaussian noise. In this way, we get four groups of datasets at different noise levels. More specifically, the values of signal-to-noise ratio (SNR) are set to 10, 15, 20, and 25 dB, respectively. The numbers of training sets are 2000 and 3000 on the $50 \times 50 \times 50$ synthetic dataset and $20 \times 20 \times 20$ synthetic dataset, respectively. We use mean square error (MSE) to evaluate the estimation accuracy, which is defined as $\text{MSE} = \frac{1}{M}\sum_{m=1}^{M}(R_m - \hat{R}_m)^2$, where R_m is the real rank of testing data and $\hat{R}_m$ is the predicted rank.

For the dataset with the size $50 \times 50 \times 50$, we set the predefined rank bound of pre-decomposition R from 100 to 140. Figure 11.8 is the illustration of learning process on $50 \times 50 \times 50$ tensor in the first 50 iterations. TRN-PD(R) denotes the TRN-PD method with the predefined rank bound of pre-decomposition R. In training process for the network, we perform training and validation alternatively. The figures of training loss and validation loss present the accuracy performance on training dataset and validation dataset, respectively. From the results, TRN-PD works well on this dataset, but TRN can hardly converge.

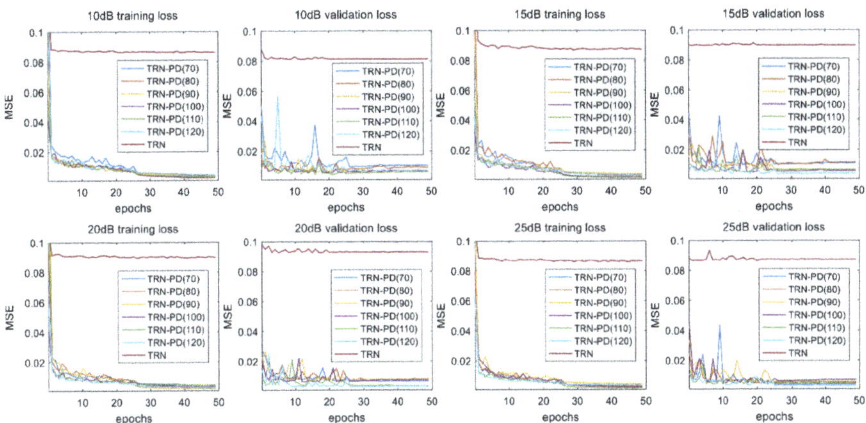

Fig. 11.9 Illustration of MSE performance comparison between TRN and TRN-PD on the $20 \times 20 \times 20$ synthetic dataset at different noise levels

We set the predefined rank bound of pre-decomposition R from 70 to 120 on the dataset with the size $20 \times 20 \times 20$. Figure 11.9 is the illustration of training process on the $20 \times 20 \times 20$ tensor. The results show TRN method is hard to find the high-order interactions of data. But TRN-PD performs well on this dataset. It proves that the pre-decomposition is able to exact features and helps the deep network in rank prediction.

11.5.2 Deep Unrolling Models for Video Snapshot Compressive Imaging

Snapshot compressive imaging is a compressive sensing-based computational imaging system [24], which can use two-dimensional camera to capture three-dimensional data such as videos and hyperspectral images. The sampling process of three-dimensional videos can be formulated as

$$\mathbf{Y} = \sum_{b=1}^{B} \mathcal{C}(:, :, b) \circledast \mathcal{X}(:, :, b), \tag{11.17}$$

where $\mathcal{X} \in \mathbb{R}^{M \times N \times B}$ is the original signal, $\mathcal{C} \in \mathbb{R}^{M \times N \times B}$ is the sampling tensor, and $\mathbf{Y} \in \mathbb{R}^{M \times N}$ is the measurement.

Recovery of $\mathcal{X}$ from $\mathbf{Y}$ is usually achieved by solving an optimization problem as follows:

$$\min_{\mathcal{X}} \ \Re(\mathcal{X}) + \frac{1}{2} \|\mathbf{y} - \mathbf{A}\mathbf{x}\|_{\mathrm{F}}^2, \tag{11.18}$$

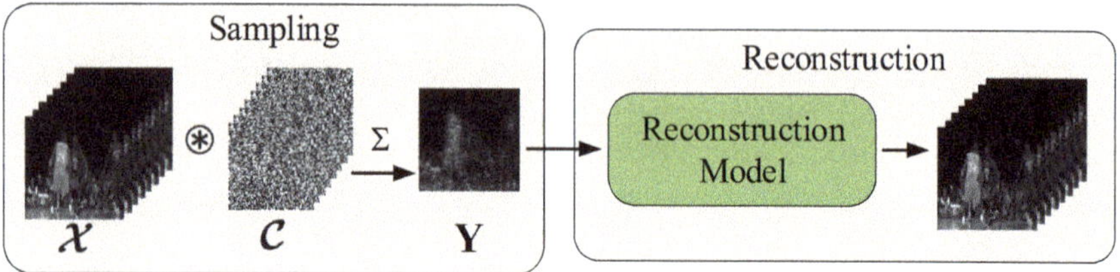

Fig. 11.10 Sampling and reconstruction of video SCI

where $\Re(\mathcal{X})$ is the regularizer for reconstruction, $\mathbf{y} = \text{vec}(\mathbf{Y}) \in \mathbb{R}^{MN}$, and $\mathbf{A} \in \mathbb{R}^{MN \times MNB}$ and $\mathbf{x} \in \mathbb{R}^{MNB}$ are built by

$$\mathbf{A} = \left[\text{diag}(\text{vec}(\mathcal{C}(:,:,1))), \cdots, \text{diag}(\text{vec}(\mathcal{C}(:,:,B)))\right], \tag{11.19}$$

$$\mathbf{x} = \text{vec}(\mathcal{X}) = \left[\text{vec}(\mathcal{X}(:,:,1))^{\text{T}}, \cdots, \text{vec}(\mathcal{X}(:,:,B))^{\text{T}}\right]^{\text{T}}. \tag{11.20}$$

Figure 11.10 shows the sampling and reconstruction of video SCI. The reconstruction model in Fig. 11.10 can be either model-based method or deep unrolling method.

In the following content of this subsection, we introduce a deep unrolling model named Tensor FISTA-Net [9] which applies an existing sparse regularizer.

11.5.2.1 Tensor FISTA-Net

Tensor FISTA-Net applies the sparse regularizer as $\|\mathfrak{D}(\mathcal{X})\|_1$ to be $\Re(\mathcal{X})$, where $\mathfrak{D}(\mathcal{X})$ is an unknown transformation function which transforms $\mathcal{X}$ into a sparse domain. Tensor FISTA-Net is built by unrolling the classical fast iterative shrinkage thresholding algorithm (FISTA) to K phases, and each phase consists of three steps. The k-th phase can be expressed as

$$\mathcal{Z}^k = \mathcal{X}^{k-1} + t^k(\mathcal{X}^{k-1} - \mathcal{X}^{k-2}), \tag{11.21}$$

$$\mathcal{R}^k = \mathcal{Z}^k - \rho^k \mathcal{C} \circledast \mathcal{A}^k, \tag{11.22}$$

$$\mathcal{X}^k = \arg\min_{\mathcal{X}} \frac{1}{2} \left\|\mathfrak{D}(\mathcal{X}) - \mathfrak{D}(\mathcal{R}^k)\right\|_{\text{F}}^2 + \|\mathfrak{D}(\mathcal{X})\|_1, \tag{11.23}$$

where $\mathcal{X}^k$ is the output of the k-th phase, t^k and ρ^k are trainable parameters, and $\mathcal{A}^k$ is formulated as

$$\mathcal{A}^k(:,:,i) = \sum_{b=1}^{B} \mathcal{C}(:,:,b) \circledast \mathcal{Z}^k(:,:,b) - \mathbf{Y} \text{ for } i = 1, \cdots, B. \tag{11.24}$$

We emphasize that (11.22) is equivalent to $\mathbf{r}^k = \mathbf{z}^k - \rho^k \mathbf{A}^{\mathrm{T}}(\mathbf{A}\mathbf{z}^k - \mathbf{y})$, where $\mathbf{r}^k = \mathrm{vec}(\mathcal{R}^k)$ and $\mathbf{z}^k = \mathrm{vec}(\mathcal{Z}^k)$.

Tensor FISTA-Net regards $\mathfrak{D}(\cdot)$ as a nonlinear transform and solves (11.23) by replacing $\mathfrak{D}(\cdot)$ with a CNN $\mathfrak{N}_{\mathrm{tra}}^k(\cdot)$ and applying the residual connection of ResNet [10]. In detail, $\mathcal{X}^k$ is obtained as

$$\bar{\mathcal{X}}^k = \tilde{\mathfrak{N}}_{\mathrm{tra}}^k(\mathfrak{T}_{\theta^k}(\mathfrak{N}_{\mathrm{tra}}^k(\mathcal{R}^k))), \tag{11.25}$$

$$\mathcal{X}^k = \mathcal{R}^k + \bar{\mathcal{X}}^k, \tag{11.26}$$

where $\mathfrak{T}_{\theta^k}$ is the soft thresholding operator like (11.6) and θ^k is a trainable parameter and $\tilde{\mathfrak{N}}_{\mathrm{tra}}^k(\cdot)$ is the inverse-transformation function of $\mathfrak{N}_{\mathrm{tra}}^k(\cdot)$. $\mathfrak{N}_{\mathrm{tra}}^k(\cdot)$ and $\tilde{\mathfrak{N}}_{\mathrm{tra}}^k(\cdot)$ are designed to be two-layer CNNs. (11.26) is the residual connection to improve the performance. In this case, (11.21), (11.22), $\mathfrak{N}_{\mathrm{tra}}^k(\cdot)$, and $\tilde{\mathfrak{N}}_{\mathrm{tra}}^k(\cdot)$ constitute $\mathfrak{N}_{\mathrm{unr1}}^k(\cdot, \mathbf{\Theta}_1^k)$ in Fig. 11.3, where $\mathbf{\Theta}_1^k$ consists of t^k, ρ^k, θ^k and parameters in $\mathfrak{N}_{\mathrm{tra}}^k(\cdot)$ and $\tilde{\mathfrak{N}}_{\mathrm{tra}}^k(\cdot)$.

Finally, the tensor FISTA-Net is trained using the loss function as follows:

$$\mathfrak{L} = \mathfrak{L}_{\mathrm{fid}} + 0.01\mathfrak{L}_{\mathrm{inv}} + 0.001\mathfrak{L}_{\mathrm{spa}}, \tag{11.27}$$

where $\mathfrak{L}_{\mathrm{fid}}$ is the fidelity of reconstructed frames, $\mathfrak{L}_{\mathrm{inv}}$ is the accuracy of inverse-transformation functions, and $\mathfrak{L}_{\mathrm{spa}}$ denotes the sparsity in transformed domains. They can be computed by

$$\mathfrak{L}_{\mathrm{fid}} = \frac{1}{N_{\mathrm{t}}} \sum_{n=1}^{N_{\mathrm{t}}} \left\| \mathcal{X}_n^K - \mathcal{X}_n \right\|_{\mathrm{F}}^2, \tag{11.28}$$

$$\mathfrak{L}_{\mathrm{inv}} = \frac{1}{N_{\mathrm{t}} K} \sum_{n=1}^{N_{\mathrm{t}}} \sum_{k=1}^{K} \left\| \tilde{\mathfrak{N}}_{\mathrm{tra}}(\mathfrak{N}_{\mathrm{tra}}(\mathcal{X}_n^k)) - \mathcal{X}_n^k \right\|_{\mathrm{F}}^2, \tag{11.29}$$

$$\mathfrak{L}_{\mathrm{spa}} = \frac{1}{N_{\mathrm{t}} K} \sum_{n=1}^{N_{\mathrm{t}}} \sum_{k=1}^{K} \left\| \mathfrak{N}_{\mathrm{tra}}(\mathcal{X}_n^k) \right\|_1, \tag{11.30}$$

where N_{t} is the size of training set.

11.5.2.2 Experimental Results

In this subsection, we compare Tensor FISTA-Net with other two state-of-the-art model-based methods:

- GAP-TV [39]: using generalized alternating projection (GAP) algorithm to solve total variation (TV) minimization problem for video SCI.

Table 11.1 Performance comparison of GAP-TV, DeSCI, and tensor FISTA-net on video SCI

Method	Kobe	Aerial	Vehicle
	PSNR (dB)/SSIM		
GAP-TV[39]	27.67/0.8843	25.75/0.8582	24.92/0.8185
DeSCI[24]	**35.32/0.9674**	24.80/0.8432	**28.47**/0.9247
Tensor FISTA-Net [9]	29.46/0.8983	**28.61/0.9031**	27.77/0.9211

Table 11.2 Time analysis of GAP-TV, DeSCI, and tensor FISTA-net on video SCI

Method	Kobe	Aerial	Vehicle
	Time (s)		
GAP-TV[39]	12.6	12.6	12.3
DeSCI[24]	15217.9	15783.2	15114.0
Tensor FISTA-Net [9]	17.5	17.6	17.5

- DeSCI [24]: reconstructing video based on the idea of weighted nuclear norm.

In this subsection, we set $M = N = 256$ and $B = 8$. And we use three training video datasets, namely, NBA, Central Park Aerial, and Vehicle Crashing Tests in [9]. All video frames are resized into 256×256 and their luminance components are extracted. Each training set contains 8000 frames, i.e., 1000 measurements.

Tensor FISTA-Net is trained for 500 epochs with batch size 2. The Adam algorithm is adopted for updating parameters and the learning rate is 0.0001 [15]. Three datasets, namely, Kobe dataset, Aerial dataset, and Vehicle Crash dataset, are used for testing. Each test set contains four video groups with the size of $256 \times 256 \times 8$.

Table 11.1 shows the average PSNR and SSIM of GAP-TV, DeSCI, and Tensor FISTA-Net on three different test sets where the best are marked in bold. And it can be noticed that Tensor FISTA-Net has better reconstruction results on test set Aerial. Table 11.2 shows the reconstruction time of three methods on three test sets tested on an Intel Core i5-8300H CPU. We conclude that although DeSCI has better performance on two test sets, namely, Kobe and Vehicle Crashing, it costs a lot of time. And Tensor FISTA-Net can have comparable performance on these two datasets while reconstructing videos fast. Furthermore, we emphasize that Tensor FISTA-Net can be accelerated using GPU devices.

11.5.3 Deep PnP for Tensor Completion

Tensor completion problem can be formulated as

$$\min_{\mathcal{X}} \Re(\mathcal{X}) \quad \text{s. t.} \quad \mathscr{P}_{\mathbb{O}}(\mathcal{X}) = \mathscr{P}_{\mathbb{O}}(\mathcal{T}), \tag{11.31}$$

where $\Re(\cdot)$ is the regularizer for completion, $\mathcal{X}$ is the underlying tensor, and $\mathcal{T}$ is the observed incomplete tensor, and $\mathscr{P}_{\mathbb{O}}(\mathcal{X})$ means projecting $\mathcal{X}$ onto the indices set $\mathbb{O}$ of the observations. Low-rank regularizers [25, 26] are usually applied for tensor completion. However, such regularizers always capture the global structures of tensors but ignore the details [45]. In this subsection, we introduce a deep PnP model dubbed deep PnP prior for low-rank tensor completion (DP3LRTC) [45] which can explore the local features using deep networks.

11.5.3.1 Deep Plug-and-Play Prior for Low-Rank Tensor Completion

DP3LRTC tries to recover a tensor by solving an optimization problem as follows:

$$\min_{\mathcal{X}} \|\mathcal{X}\|_{\text{TNN}} + \lambda \Re_{\text{unk}}(\mathcal{X}) \quad \text{s. t.} \quad \mathscr{P}_{\mathbb{O}}(\mathcal{X}) = \mathscr{P}_{\mathbb{O}}(\mathcal{T}) \tag{11.32}$$

where $\|\cdot\|_{\text{TNN}}$ is the tensor tubal nuclear norm and $\Re_{\text{unk}}(\cdot)$ denotes an unknown regularizer. DP3LRTC uses deep networks to represent the information of $\Re_{\text{unk}}(\cdot)$ and explores the local features of tensors.

The optimization problem (11.32) can be solved using ADMM. Introducing two auxiliary variables $\mathcal{Y}$ and $\mathcal{Z}$, the k-th iteration of the ADMM algorithm can be expressed as follows:

$$\mathcal{Y}^k = \underset{\mathcal{Y}}{\text{argmin}}\, \|\mathcal{Y}\|_{\text{TNN}} + \frac{\beta}{2} \left\|\mathcal{X}^{k-1} - \mathcal{Y} + \mathcal{T}_1^{k-1}/\beta\right\|_{\text{F}}^2, \tag{11.33}$$

$$\mathcal{Z}^k = \underset{\mathcal{Z}}{\text{argmin}}\, \lambda \Re(\mathcal{Z}) + \frac{\beta}{2} \left\|\mathcal{X}^{k-1} - \mathcal{Z} + \mathcal{T}_2^{k-1}/\beta\right\|_{\text{F}}^2, \tag{11.34}$$

$$\mathcal{X}^k = \underset{\mathcal{X}}{\text{argmin}}\, 1_{\mathbb{S}}(\mathcal{X}) + \frac{\beta}{2} \left\|\mathcal{X} - \mathcal{Y}^k + \mathcal{T}_1^{k-1}/\beta\right\|_{\text{F}}^2 + \frac{\beta}{2} \left\|\mathcal{X} - \mathcal{Z}^k + \mathcal{T}_2^{k-1}/\beta\right\|_{\text{F}}^2 \tag{11.35}$$

$$\mathcal{T}_1^k = \mathcal{T}_1^{k-1} - \beta(\mathcal{X}^k - \mathcal{Y}^k), \tag{11.36}$$

$$\mathcal{T}_2^k = \mathcal{T}_2^{k-1} - \beta(\mathcal{X}^k - \mathcal{Z}^k) \tag{11.37}$$

where λ and β are hyperparameters and $\mathcal{T}_1^k$ and $\mathcal{T}_2^k$ are intermediate variables. $\mathcal{X}^k$ is the output of the k-th iteration. $1_{\mathbb{S}}(\mathcal{X})$ is the indicator function which is defined as follows:

$$1_{\mathbb{S}}(\mathcal{X}) = \begin{cases} 0, & \text{if } \mathcal{X} \in \mathbb{S}, \\ \infty & \text{otherwise}, \end{cases} \tag{11.38}$$

where $\mathbb{S} := \{\mathcal{X} | \mathscr{P}_{\mathbb{O}}(\mathcal{X}) = \mathscr{P}_{\mathbb{O}}(\mathcal{T})\}$. The subproblems with respect to $\mathcal{Y}^k$ and $\mathcal{X}^k$, namely, (11.33) and (11.35), are easy to solve with closed-form solutions.

Algorithm 50: The ADMM algorithm for solving (11.32)

Input: $\mathcal{X}^0, \mathcal{Y}^0, \mathcal{Z}^0, \mathcal{T}_1^0, \mathcal{T}_2^0, \beta, \sigma, l_{\max}, l = 1$
Output: Reconstructed tensor: $\mathcal{X}$
1 **while** *not converged and* $l \leqslant l_{\max}$ **do**
2 Updata $\mathcal{Y}$ via (11.33);
3 Updata $\mathcal{Z}$ via (11.39);
4 Updata $\mathcal{X}$ via (11.35);
5 Updata $\mathcal{T}_1$ and $\mathcal{T}_2$ via (11.36) and (11.37);
6 **end**
7 Return $\mathcal{X}$.

DP3LRTC regards (11.34) as a denoising problem like (11.9) and solves it using a classical deep network-based denoiser dubbed FFDNet[43] as follows:

$$\mathcal{Z}^k = \mathfrak{N}_{\text{FFD}}(\mathcal{X}^{k-1} + \mathcal{T}_2^{k-1} \Big/ \beta, \sigma), \tag{11.39}$$

where $\mathfrak{N}_{\text{FFD}}(\cdot)$ denotes FFDNet and $\sigma = \sqrt{\lambda/\beta}$ is related to the error level between the estimation and ground truth. We emphasize that $\mathfrak{N}_{\text{FFD}}(\cdot)$ is pre-trained. Algorithm 50 concludes the process to solve (11.32). And in this case, (11.33), (11.35), (11.36), and (11.36) constitute $\mathfrak{N}_{\text{mod1}}(\cdot)$ in Fig. 11.5.

11.5.3.2 Experimental Results

To show the performance of DP3LRTC, we compare it with three state-of-the-art model-based methods which are solved by ADMM:

- SNN [21]: tensor completion based on sum of nuclear norm with Tucker decomposition.
- TNN [42]: tensor completion based on tubal nuclear norm with t-SVD.
- TNN-3DTV [14]: it combines tubal nuclear norm and three-dimensional total variation as the regularization terms.

All hyperparameters in these four methods are manually selected from a set $\{10^{-4}, 10^{-3}, 10^{-2}, 10^{-1}, 1, 10\}$ to obtain the highest PSNR. For DP3LRTC, manually selected parameters are β and σ. We test these four methods on a $256 \times 256 \times 3$ colorful image named *Lena*.

Figure 11.11 shows the histogram of PSNRs of completed *Lena* image by four methods at different sampling ratios. Obviously, DP3LRTC has higher PSNR than other methods at most sampling ratios. Figure 11.12 shows the completed *Lena* image by four methods at the sampling ratio of 10%. Furthermore, Fig. 11.12 also contains the completion times of four methods using an AMD Ryzen 5 3500X CPU. It can be noticed that with the deep PnP strategy, DP3LRTC can not only get the

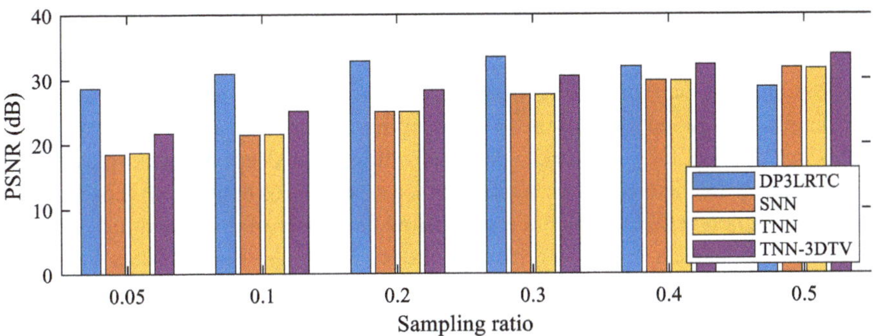

Fig. 11.11 PSNR of completed *Lena* image by DP3LRTC, SNN, TNN, and TNN-3DTV at different sampling ratios of 0.05, 0.1, 0.2, 0.3, 0.4, and 0.5

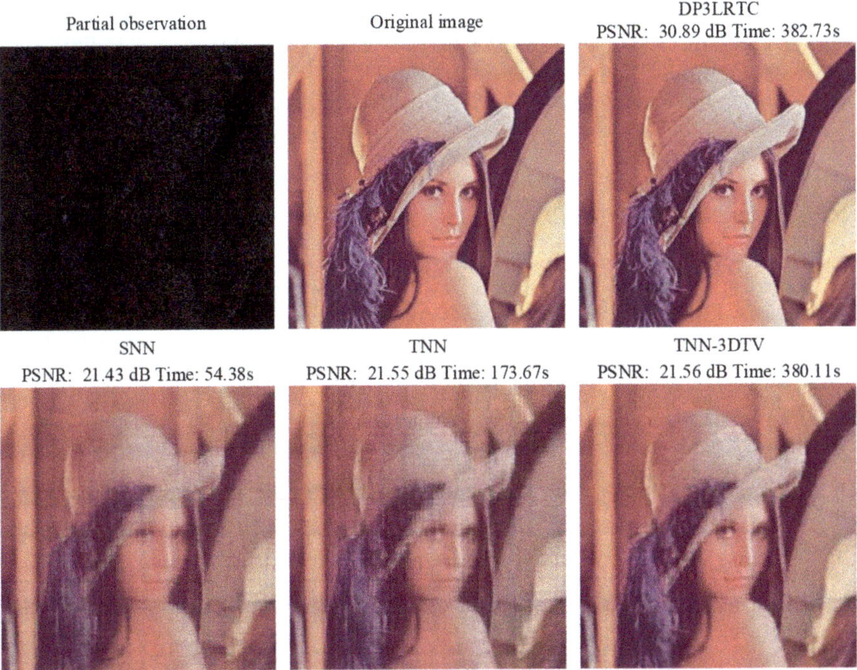

Fig. 11.12 Recovered *Lena* image by DP3LRTC, SNN, TNN, and TNN-3DTV at the sampling ratio of 0.1

global information but also capture the details of images, which brings it a better completion performance. Furthermore, DP3LRTC does not consume a lot of time to complete a tensor.

11.6 Summary

In this chapter, we summarize deep learning methods for tensor approximation. In details, we group existing deep learning-based methods into three categories: classical deep neural networks, deep unrolling, and deep plug-and-play. Some applications are discussed, including tensor rank approximation, video SCI, and tensor completion. Experimental results show the power of deep learning on tensor approximation tasks. It can be noticed that deep learning-based methods show their advantages in most cases.

Besides the discussed applications, there are some other deep learning models for applications like tensor decomposition [23] and one-rank tensor approximation [3]. Moreover, in the future, some multidimensional data-related complex tasks, such as dynamic magnetic resonance imaging (dMRI) and tensor robust principal component analysis (RPCA), can be explored. In addition, since the completely data-driven end-to-end manners of classical deep neural networks may have risks for some undesired effects [12], deep unrolling or deep PnP can be used to build deep leaning models for exact recovery due to their theoretical guarantees.

References

1. Borgerding, M., Schniter, P., Rangan, S.: AMP-Inspired deep networks for sparse linear inverse problems. IEEE Trans. Signal Process. **65**(16), 4293–4308 (2017)
2. Chan, S.H., Wang, X., Elgendy, O.A.: Plug-and-play ADMM for image restoration: fixed-point convergence and applications. IEEE Trans. Comput. Imag. **3**(1), 84–98 (2016)
3. Che, M., Cichocki, A., Wei, Y.: Neural networks for computing best rank-one approximations of tensors and its applications. Neurocomputing **267**, 114–133 (2017)
4. Diamond, S., Sitzmann, V., Heide, F., Wetzstein, G.: Unrolled optimization with deep priors (2017, e-prints). arXiv–1705
5. Dong, W., Wang, P., Yin, W., Shi, G., Wu, F., Lu, X.: Denoising prior driven deep neural network for image restoration. IEEE Trans. Pattern Anal. Mach. Intell. **41**(10), 2305–2318 (2018)
6. Donoho, D.L., Maleki, A., Montanari, A.: Message-passing algorithms for compressed sensing. Proc. Natl. Acad. Sci. **106**(45), 18914–18919 (2009)
7. Goodfellow, I., Bengio, Y., Courville, A., Bengio, Y.: Deep Learning, vol. 1. MIT Press, Cambridge (2016)
8. Gregor, K., LeCun, Y.: Learning fast approximations of sparse coding. In: Proceedings of the 27th International Conference on International Conference on Machine Learning, pp. 399–406 (2010)
9. Han, X., Wu, B., Shou, Z., Liu, X.Y., Zhang, Y., Kong, L.: Tensor FISTA-Net for real-time snapshot compressive imaging. In: Proceedings of the AAAI Conference on Artificial Intelligence, vol. 34, pp. 10933–10940 (2020)
10. He, K., Zhang, X., Ren, S., Sun, J.: Deep residual learning for image recognition. In: Proceedings of the IEEE Conference on Computer Vision and Pattern Recognition, pp. 770–778 (2016)
11. Heide, F., Steinberger, M., Tsai, Y.T., Rouf, M., Pajak, D., Reddy, D., Gallo, O., Liu, J., Heidrich, W., Egiazarian, K., et al.: Flexisp: a flexible camera image processing framework. ACM Trans. Graph. **33**(6), 1–13 (2014)

12. Huang, Y., Würfl, T., Breininger, K., Liu, L., Lauritsch, G., Maier, A.: Some investigations on robustness of deep learning in limited angle tomography. In: International Conference on Medical Image Computing and Computer-Assisted Intervention, pp. 145–153. Springer, Berlin (2018)
13. Iliadis, M., Spinoulas, L., Katsaggelos, A.K.: Deep fully-connected networks for video compressive sensing. Digital Signal Process. **72**, 9–18 (2018)
14. Jiang, F., Liu, X.Y., Lu, H., Shen, R.: Anisotropic total variation regularized low-rank tensor completion based on tensor nuclear norm for color image inpainting. In: 2018 IEEE International Conference on Acoustics, Speech and Signal Processing (ICASSP), pp. 1363–1367. IEEE, Piscataway (2018)
15. Kingma, D.P., Ba, J.: Adam: a method for stochastic optimization. In: 3rd International Conference on Learning Representations, ICLR 2015, San Diego, May 7–9, 2015, Conference Track Proceedings (2015)
16. Kolda, T.G., Bader, B.W.: Tensor decompositions and applications. SIAM Rev. **51**(3), 455–500 (2009)
17. Krizhevsky, A., Sutskever, I., Hinton, G.: Imagenet classification with deep convolutional neural networks. In: Advances in Neural Information Processing Systems, pp. 1097–1105 (2012)
18. Lawrence, S., Giles, C.L., Tsoi, A.C., Back, A.D.: Face recognition: a convolutional neural-network approach. IEEE Trans. Neural Netw. **8**(1), 98–113 (1997)
19. LeCun, Y.: Convolutional networks for images, speech, and time series. In: The Handbook of Brain Theory and Neural Networks, pp. 255–258 (1995)
20. LeCun, Y., Boser, B.E., Denker, J.S., Henderson, D., Howard, R.E., Hubbard, W.E., Jackel, L.D.: Handwritten digit recognition with a back-propagation network. In: Advances in Neural Information Processing Systems, pp. 396–404 (1990)
21. Liu, J., Musialski, P., Wonka, P., Ye, J.: Tensor completion for estimating missing values in visual data. IEEE Trans. Pattern Anal. Mach. Intell. **35**(1), 208–220 (2012)
22. Liu, Y., De Vos, M., Van Huffel, S.: Compressed sensing of multichannel EEG signals: the simultaneous cosparsity and low-rank optimization. IEEE Trans. Biomed. Eng. **62**(8), 2055–2061 (2015)
23. Liu, B., Xu, Z., Li, Y.: Tensor decomposition via variational auto-encoder (2016, e-prints). arXiv–1611
24. Liu, Y., Yuan, X., Suo, J., Brady, D.J., Dai, Q.: Rank minimization for snapshot compressive imaging. IEEE Trans. Pattern Anal. Mach. Intell. **41**(12), 2990–3006 (2018)
25. Liu, Y., Long, Z., Huang, H., Zhu, C.: Low CP rank and tucker rank tensor completion for estimating missing components in image data. IEEE Trans. Circuits Syst. Video Technol. **30**(4), 944–954 (2019)
26. Long, Z., Liu, Y., Chen, L., Zhu, C.: Low rank tensor completion for multiway visual data. Signal Process. **155**, 301–316 (2019)
27. Lu, C., Feng, J., Chen, Y., Liu, W., Lin, Z., Yan, S.: Tensor robust principal component analysis: exact recovery of corrupted low-rank tensors via convex optimization. In: Proceedings of the IEEE Conference on Computer Vision and Pattern Recognition, pp. 5249–5257 (2016)
28. Ma, J., Liu, X.Y., Shou, Z., Yuan, X.: Deep tensor admm-net for snapshot compressive imaging. In: Proceedings of the IEEE International Conference on Computer Vision, pp. 10223–10232 (2019)
29. Meinhardt, T., Moller, M., Hazirbas, C., Cremers, D.: Learning proximal operators: using denoising networks for regularizing inverse imaging problems. In: Proceedings of the IEEE International Conference on Computer Vision, pp. 1781–1790 (2017)
30. Metzler, C., Mousavi, A., Baraniuk, R.: Learned D-AMP: principled neural network based compressive image recovery. In: Advances in Neural Information Processing Systems, pp. 1772–1783 (2017)
31. Mousavi, A., Patel, A.B., Baraniuk, R.G.: A deep learning approach to structured signal recovery. In: The 53rd Annual Allerton Conference on Communication, Control, and Computing, pp. 1336–1343. IEEE, Piscataway (2015)

32. Sun, J., Li, H., Xu, Z., et al.: Deep ADMM-Net for compressive sensing MRI. In: Advances in Neural Information Processing Systems, pp. 10–18 (2016)
33. Tan, X., Zhang, Y., Tang, S., Shao, J., Wu, F., Zhuang, Y.: Logistic tensor regression for classification. In: International Conference on Intelligent Science and Intelligent Data Engineering, pp. 573–581. Springer, Berlin (2012)
34. Wang, Z., Ling, Q., Huang, T.: Learning deep l0 encoders. In: AAAI Conference on Artificial Intelligence, pp. 2194–2200 (2016)
35. Wang, Z., Yang, Y., Chang, S., Ling, Q., Huang, T.S.: Learning a deep ℓ_∞ encoder for hashing. In: Proceedings of the Twenty-Fifth International Joint Conference on Artificial Intelligence, pp. 2174–2180. AAAI Press, Palo Alto (2016)
36. Yang, G., Yu, S., Dong, H., Slabaugh, G., Dragotti, P.L., Ye, X., Liu, F., Arridge, S., Keegan, J., Guo, Y., et al.: DAGAN: deep de-aliasing generative adversarial networks for fast compressed sensing MRI reconstruction. IEEE Trans. Med. Imag. **37**(6), 1310–1321 (2017)
37. Yin, M., Gao, J., Xie, S., Guo, Y.: Multiview subspace clustering via tensorial t-product representation. IEEE Trans. Neural Netw. Learn. Syst. **30**(3), 851–864 (2018)
38. Young, T., Hazarika, D., Poria, S., Cambria, E.: Recent trends in deep learning based natural language processing. IEEE Comput. Intell. Mag. **13**(3), 55–75 (2018)
39. Yuan, X.: Generalized alternating projection based total variation minimization for compressive sensing. In: 2016 IEEE International Conference on Image Processing (ICIP), pp. 2539–2543. IEEE, Piscataway (2016)
40. Yuan, X., Liu, Y., Suo, J., Dai, Q.: Plug-and-play algorithms for large-scale snapshot compressive imaging. In: Proceedings of the IEEE/CVF Conference on Computer Vision and Pattern Recognition, pp. 1447–1457 (2020)
41. Zhang, J., Ghanem, B.: ISTA-Net: interpretable optimization-inspired deep network for image compressive sensing. In: Proceedings of the IEEE Conference on Computer Vision and Pattern Recognition, pp. 1828–1837 (2018)
42. Zhang, Z., Ely, G., Aeron, S., Hao, N., Kilmer, M.: Novel methods for multilinear data completion and de-noising based on tensor-SVD. In: Proceedings of the IEEE Conference on Computer Vision and Pattern Recognition, pp. 3842–3849 (2014)
43. Zhang, K., Zuo, W., Zhang, L.: FFDNet: toward a fast and flexible solution for CNN-based image denoising. IEEE Trans. Image Process. **27**(9), 4608–4622 (2018)
44. Zhang, Z., Liu, Y., Liu, J., Wen, F., Zhu, C.: AMP-Net: denoising based deep unfolding for compressive image sensing. IEEE Trans. Image Process. **30**, 1487–1500 (2020)
45. Zhao, X.L., Xu, W.H., Jiang, T.X., Wang, Y., Ng, M.K.: Deep plug-and-play prior for low-rank tensor completion. Neurocomputing **400**, 137–149 (2020)
46. Zhou, H., Li, L., Zhu, H.: Tensor regression with applications in neuroimaging data analysis. J. Am. Stat. Assoc. **108**(502), 540–552 (2013)
47. Zhou, M., Liu, Y., Long, Z., Chen, L., Zhu, C.: Tensor rank learning in CP decomposition via convolutional neural network. Signal Process. Image Commun. **73**, 12–21 (2019)
48. Zubair, S., Wang, W.: Tensor dictionary learning with sparse tucker decomposition. In: 2013 18th International Conference on Digital Signal Processing (DSP), pp. 1–6. IEEE, Piscataway (2013)

Chapter 12
Tensor-Based Gaussian Graphical Model

12.1 Background

In the era of big data, an unprecedented amount of high-dimensional data are available in domains like electronic engineering and computer science. Exploring the relationships between entries of these data can be very helpful in a series of fields including machine learning [3], signal processing [2], and neuroscience [10]. For example, by exploiting the user-item-time tensor in recommendation system, we are able to obtain the relationships among different users, items, and time. However, the processing of such large-scale data is not an easy task, and the high-dimensionality feature makes it even more intractable. To address this issue, we can resort to a promising tool named Gaussian graphical model (GGM), which combines probability theory and graph theory together to represent dependencies of variables intuitively.

As the name indicates, the most important assumption is the Gaussian distribution over the variables. It enables us to make reasonable suppositions on missing variables to help finding potential relationships. For example, given I variables $x_i,\ i = 1, \cdots, I$, they are supposed to follow a Gaussian distribution $\mathrm{N}(\boldsymbol{\mu}, \boldsymbol{\Sigma})$ under the GGM framework, where $\boldsymbol{\mu}$ is the mean of $\mathbf{x}$ and $\boldsymbol{\Sigma}$ is the covariance matrix. In order to characterize the correlations among variables $x_i,\ i = 1, \cdots, I$, we can either estimate the covariance matrix $\boldsymbol{\Sigma}$ or estimate the precision matrix, which is defined as $\boldsymbol{\Omega} = \boldsymbol{\Sigma}^{-1}$. The zeros in covariance matrix represent marginally independencies, and the zeros in precision matrix denote conditional independencies. Here, we focus on the estimation of precision matrices in GGM.

Specifically, each zero entry $\boldsymbol{\Omega}(i, j) = 0,\ i \neq j$, for $i, j = 1, \cdots, I$, indicates a conditional independent relationship between the i-th and j-th variables when the rest are given. For example, if the (i, j)-th element in the precision matrix is zero, then x_i and x_j are conditionally independent, given other variables. To demonstrate

Y. Liu et al., *Tensor Computation for Data Analysis*,
https://doi.org/10.1007/978-3-030-74386-4_12

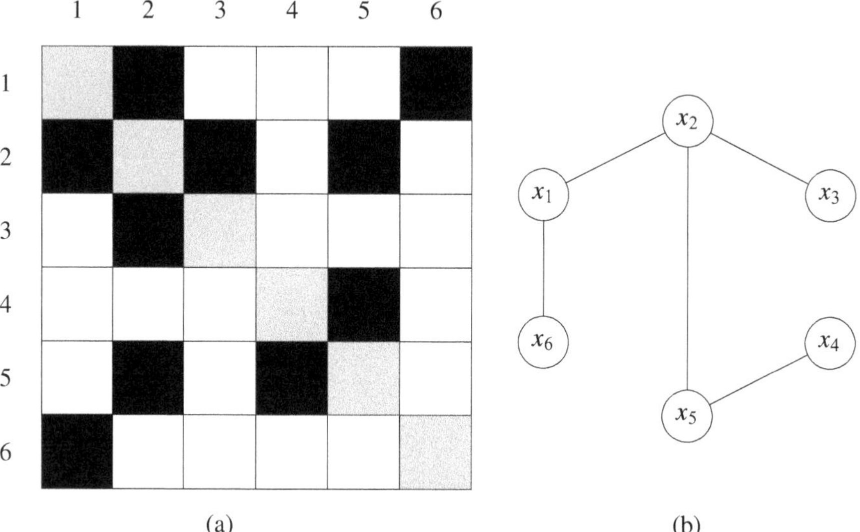

Fig. 12.1 One-to-one correspondence between the precision matrix and graph: the left picture is a precision matrix in which black block (off-diagonal) indicates nonzero entries and each nonzero entry corresponds to an edge in graph; the right figure is the graph drawn according to the relationship shown on the left. (**a**) Precision matrix. (**b**) Graph

these relationships more explicitly, graph representation are often implemented, which is expressed by $\mathbb{G} = (\mathbb{V}, \mathbb{E})$, where $\mathbb{V}$ and $\mathbb{E}$ are the sets of nodes and edges, respectively. To be specific, the nodes in a graph are associated with the random variables, and the edges indicate the correlations between the nodes/variables. Since zeros in precision matrix $\mathbf{\Omega}$ not only reflect conditional independencies between variables given the rest but also correspond to missing edges in the graph, graphical model can intuitively represent complex relationships between variables through the presence or absence of edges. For ease of expression, the edge between i-th node and j-th node is denoted as (i, j). Therefore, we can have $\mathbf{\Omega}(i, j) = 0,\ i \neq j$, for $i, j = 1, \cdots, I$ if and only if $(i, j) \notin \mathbb{E}$, as shown in Fig. 12.1. In this sense, the task of building the sparse graphical model and investigating correlations among variables can be transformed into detecting nonzero entries in its corresponding precision matrix.

According to the data structure, the existing Gaussian graphical models can be categorized into three groups, including vector-variate data [1, 4, 6, 11, 15, 18, 19], the matrix-variate data [8, 17, 21], and the tensor-variate data [7, 9, 12, 16].

12.2 Vector-Variate-Based Gaussian Graphical Model

Given N samples $\mathbf{x}_n \in \mathbb{R}^P$, we consider the independently and identically P-variate Gaussian distribution for $\mathbf{x}_n$ with the probability density function (PDF)

$$\mathrm{Prob}(\mathbf{x}_n) = \frac{1}{(\sqrt{2\pi})^P (\det \boldsymbol{\Sigma})^{\frac{1}{2}}} \exp\left(-\frac{(\mathbf{x}_n - \boldsymbol{\mu})^{\mathrm{T}} \boldsymbol{\Sigma}^{-1} (\mathbf{x}_n - \boldsymbol{\mu})}{2}\right). \tag{12.1}$$

where $\boldsymbol{\mu}$ is the mean vector, $\boldsymbol{\Sigma}$ is the covariance matrix, and $\det \boldsymbol{\Sigma}$ is the determinant of $\boldsymbol{\Sigma}$. The vector-based GGM exploits the relationships between variables by estimating the precision matrix $\boldsymbol{\Omega} = \boldsymbol{\Sigma}^{-1}$ using the maximum likelihood estimation and characterizing conditional independence in the underlying variables. By taking the logarithm of probability density function (12.1) and maximizing it, we can formulate the corresponding optimization model as

$$\max_{\boldsymbol{\Omega}} \quad \log\det \boldsymbol{\Omega} - \mathrm{tr}(\mathbf{S}\boldsymbol{\Omega}), \tag{12.2}$$

where $\mathbf{S} = \frac{1}{N} \sum_{n=1}^{N} (\mathbf{x}_n - \boldsymbol{\mu})(\mathbf{x}_n - \boldsymbol{\mu})^{\mathrm{T}}$ denotes the sample covariance matrix.

However, for high-dimensional data with $N \leq P$, the sample covariance matrix $\mathbf{S}$ is not invertible. To solve this problem, sparsity is often imposed on the precision matrix [6], and problem (12.2) is transformed into

$$\max_{\boldsymbol{\Omega}} \quad \log\det \boldsymbol{\Omega} - \mathrm{tr}(\mathbf{S}\boldsymbol{\Omega}) - \lambda \|\boldsymbol{\Omega}\|_1, \tag{12.3}$$

where λ is the tradeoff parameter between the objective function and sparsity constraint.

In order to solve the optimization problem (12.3), different methods are employed, such as the interior point method [19], block coordinate descent (BCD) algorithm [1], Glasso method [6], neighborhood selection [11], and linear programming procedure [18]. Among all these methods, Glasso is the most popular one. We will concentrate on it and give its detailed solution in what follows.

Using the fact that the derivative of $\log\det \boldsymbol{\Omega}$ equals $\boldsymbol{\Omega}^{-1}$, the gradient of Lagrangian function of (12.3) with respect to $\boldsymbol{\Omega}$ vanishes when the following equation holds:

$$\boldsymbol{\Omega}^{-1} - \mathbf{S} - \lambda\, \mathrm{Sign}(\boldsymbol{\Omega}) = 0, \tag{12.4}$$

where $\mathrm{Sign}(\boldsymbol{\Omega})$ is the sub-gradient of $\|\boldsymbol{\Omega}\|_1$ and can be denoted as

$$\mathrm{Sign}(\boldsymbol{\Omega})_{i,j} = \begin{cases} \mathrm{sign}(\boldsymbol{\Omega}(i,j)), & \boldsymbol{\Omega}(i,j) \neq 0, \\ -1 \text{ or } 1, & \boldsymbol{\Omega}(i,j) = 0, \end{cases}$$

where $\mathrm{sign}(\boldsymbol{\Omega}(i,j))$ is the sign of $\boldsymbol{\Omega}(i,j)$.

The solution about $\boldsymbol{\Omega}$ and its inverse $\boldsymbol{\Sigma} = \boldsymbol{\Omega}^{-1}$ can be acquired by sequentially solving several regression problems. To be specific, we divide $\boldsymbol{\Omega} \in \mathbb{R}^{P\times P}$ into two parts: the first part is composed of its first $P-1$ rows and columns, denoted as $\boldsymbol{\Omega}_{11}$, and the second part is composed of the rest elements. The same procedure can be done to $\boldsymbol{\Sigma} \in \mathbb{R}^{P\times P}$. By denoting $\mathbf{U}$ as the estimation of $\boldsymbol{\Sigma}$, we can have

$$\begin{pmatrix} \mathbf{U}_{11} & \mathbf{u}_{12} \\ \mathbf{u}_{12}^{\mathrm{T}} & u_{22} \end{pmatrix} \begin{pmatrix} \boldsymbol{\Omega}_{11} & \boldsymbol{\omega}_{12} \\ \boldsymbol{\omega}_{12}^{\mathrm{T}} & \omega_{22} \end{pmatrix} = \begin{pmatrix} \mathbf{I} & 0 \\ 0 & 1 \end{pmatrix}.$$

This implies

$$\mathbf{u}_{12} = \mathbf{U}_{11}\boldsymbol{\beta}, \tag{12.5}$$

where $\boldsymbol{\beta} = -\boldsymbol{\omega}_{12}/\omega_{22}$.

Taking advantage of this block formulation, the upper right block of problem (12.4) can be extracted as

$$\mathbf{u}_{12} - \mathbf{s}_{12} + \lambda\,\mathrm{Sign}(\boldsymbol{\beta}) = 0. \tag{12.6}$$

By substituting (12.5) into (12.6), we have

$$\mathbf{U}_{11}\boldsymbol{\beta} - \mathbf{s}_{12} + \lambda\,\mathrm{Sign}(\boldsymbol{\beta}) = 0. \tag{12.7}$$

To solve this problem, we can equivalently estimate a lasso regression in the form of

$$\min_{\boldsymbol{\beta}} \frac{1}{2}\|\mathbf{y} - \mathbf{Z}\boldsymbol{\beta}\|_{\mathrm{F}}^{2} + \lambda\|\boldsymbol{\beta}\|_{1}, \tag{12.8}$$

where $\mathbf{y}$ is the output variable and $\mathbf{Z}$ is the predictor matrix. Its gradient can be derived as (12.9)

$$\mathbf{Z}^{\mathrm{T}}\mathbf{Z}\boldsymbol{\beta} - \mathbf{Z}^{\mathrm{T}}\mathbf{y} + \lambda\,\mathrm{Sign}(\boldsymbol{\beta}) = 0. \tag{12.9}$$

Notice that when $\mathbf{Z}^{\mathrm{T}}\mathbf{Z} = \mathbf{U}_{11}$ and $\mathbf{Z}^{\mathrm{T}}\mathbf{y} = \mathbf{s}_{12}$, the solution of (12.7) is equal to that of (12.8). We summarize the above steps in Algorithm 51.

Besides the large number of variables collected, structural information encoded in them plays an important role in some research areas. For example, in neuroscience, electroencephalography (EEG) is often used to analyze the relationships between brain signal and the trigger of certain events, e.g., alcohol consumption. In real experiments, we can obtain the EEG recordings of each subject from multichannel electrodes during a period of time. Therefore, the observation of each subject is a channel by time matrix. If we directly use vector-variate-based GGM to

---

Algorithm 51: Graphical lasso (Glasso)

Input: $\mathbf{S}$, λ
Start with $\mathbf{U} = \mathbf{S} + \rho\mathbf{I}$. The diagonal entries of $\mathbf{U}$ remains unchanged in what follows.
Repeat
For $p = 1, \cdots, P$
1. Partition the matrix $\mathbf{U}$ into two part:
   Part 1: all but the p-th row and column
   Part 2: the p-th row and column
2. Solve the Eq. (12.7) by solving the lasso regression problem (12.8)
3. Update $\mathbf{u}_{12} = \mathbf{U}_{11}\boldsymbol{\beta}$

End for
Until convergence
Compute $\boldsymbol{\Omega} = \mathbf{U}^{-1}$
Output: $\boldsymbol{\Omega}$

---

analyze these EEG data, it would suffer from performance degradation due to the neglect of structural information.

12.3 Matrix-Variate-Based Gaussian Graphical Model

To address the problems caused by the loss of structural information in vector-variate-based GGM, matrix-variate-based one is developed [8, 17, 21]. Suppose we observe N independently and identically distributed (i.i.d.) matrix data $\mathbf{X}_n \in \mathbb{R}^{P\times Q}$, $n = 1, \cdots, N$ following matrix normal distribution $\mathrm{MN}_{P\times Q}(\mathbf{M}, \boldsymbol{\Sigma}_1, \boldsymbol{\Sigma}_2)$, where $\mathbf{M} \in \mathbb{R}^{P\times Q}$ and $\boldsymbol{\Sigma}_1 \otimes \boldsymbol{\Sigma}_2$ are the mean matrix and separable covariance matrix of $\mathbf{X}_n$, respectively. The PDF of $\mathbf{X}_n$ is

$$\mathrm{Prob}(\mathbf{X}_n|\mathbf{M}, \boldsymbol{\Sigma}_1, \boldsymbol{\Sigma}_2) = \frac{\exp\{-\frac{1}{2}\mathrm{tr}(\boldsymbol{\Sigma}_2{}^{-1}(\mathbf{X}_n - \mathbf{M})^{\mathrm{T}}\boldsymbol{\Sigma}_1^{-1}(\mathbf{X}_n - \mathbf{M}))\}}{(2\pi)^{\frac{PQ}{2}}\det \boldsymbol{\Sigma}_2{}^{\frac{P}{2}}\det \boldsymbol{\Sigma}_1{}^{\frac{Q}{2}}}. \quad (12.10)$$

The precision matrices of the row and column variables in the sample matrix are defined as $\boldsymbol{\Omega}_1 = \boldsymbol{\Sigma}_1^{-1}$ and $\boldsymbol{\Omega}_2 = \boldsymbol{\Sigma}_2^{-1}$, respectively. In matrix-variate-based GGM, maximum likelihood estimation is also involved to find the optimal precision matrix of N samples $\mathbf{X}_n$, and the objection function can be formulated as

$$\max_{\boldsymbol{\Omega}_1,\boldsymbol{\Omega}_2} \frac{NQ}{2}\log\det\boldsymbol{\Omega}_1 + \frac{NP}{2}\log\det\boldsymbol{\Omega}_2 - \frac{1}{2}\sum_{n=1}^{N}\mathrm{tr}(\mathbf{X}_n\boldsymbol{\Omega}_2\mathbf{X}_n^{\mathrm{T}}\boldsymbol{\Omega}_1).$$

Similar to vector-variate GGM, sparsity is also considered on precision matrices $\boldsymbol{\Omega}_1$ and $\boldsymbol{\Omega}_2$, which leads to the following sparse matrix-variate GGM:

$$\begin{aligned}\min_{\boldsymbol{\Omega}_1,\boldsymbol{\Omega}_2} \quad & -\frac{NQ}{2}\log\det\boldsymbol{\Omega}_1 - \frac{NP}{2}\log\det\boldsymbol{\Omega}_2 \\ & +\frac{1}{2}\sum_{n=1}^{N}\operatorname{tr}\left(\mathbf{X}_n\boldsymbol{\Omega}_2\mathbf{X}_n^{\mathrm{T}}\boldsymbol{\Omega}_1\right) + \lambda_1\,\mathrm{R}_1(\boldsymbol{\Omega}_1) + \lambda_2\,\mathrm{R}_2(\boldsymbol{\Omega}_2)\end{aligned} \tag{12.11}$$

For the penalties R_1 and R_2, there are multiple choices, such as the most commonly used ℓ_1 norm penalty [14], nonconvex SCAD (smoothly clipped absolute deviation) penalty [5], adaptive ℓ_1 penalty [22], MCP (minimax concave penalty) [20], and truncated ℓ_1 penalty [13].

Here we follow the idea in [8], which considers the estimation of precision matrix under ℓ_1 penalized likelihood framework as follows:

$$\begin{aligned}\min_{\boldsymbol{\Omega}_1,\boldsymbol{\Omega}_2} \quad & -\frac{NQ}{2}\log\det\boldsymbol{\Omega}_1 - \frac{NP}{2}\log\det\boldsymbol{\Omega}_2 + \frac{1}{2}\sum_{n=1}^{N}\operatorname{tr}(\mathbf{X}_n\boldsymbol{\Omega}_2\mathbf{X}_n^{\mathrm{T}}\boldsymbol{\Omega}_1) \\ & +\lambda_1\|\boldsymbol{\Omega}_1\|_{1,\mathrm{off}} + \lambda_2\|\boldsymbol{\Omega}_2\|_{1,\mathrm{off}},\end{aligned} \tag{12.12}$$

where $\|\cdot\|_{1,\mathrm{off}}$ is the off-diagonal ℓ_1 norm, i.e., $\|\boldsymbol{\Omega}_1\|_{1,\mathrm{off}} = \sum_{p\neq i}|\boldsymbol{\Omega}_1(p,i)|$, $\|\boldsymbol{\Omega}_2\|_{1,\mathrm{off}} = \sum_{q\neq j}|\boldsymbol{\Omega}_2(q,j)|$.

By this optimization model (12.12), we can acquire precision matrix satisfying sparsity condition. However, it is obvious that the optimization model (12.12) is nonconvex. Because of the biconvex property, we can minimize $\boldsymbol{\Omega}_1$ and $\boldsymbol{\Omega}_2$ separately in iterations of the solution [8, 17]. In practice, we fix one precision matrix and solve the other. For example, if we fix $\boldsymbol{\Omega}_1$, it will be a convex function with respect to $\boldsymbol{\Omega}_2$ and vice versa. Therefore, the optimization model (12.12) can be transformed into two subproblems as follows:

$$\min_{\boldsymbol{\Omega}_2}\frac{1}{Q}\operatorname{tr}(\boldsymbol{\Omega}_2\cdot\frac{1}{Q}\sum_{n=1}^{N}\mathbf{X}_n^{\mathrm{T}}\boldsymbol{\Omega}_1\mathbf{X}_n) - \frac{1}{Q}\log\det\boldsymbol{\Omega}_2 + \sum_{q\neq j}\lambda_2|\boldsymbol{\Omega}_2(q,j)|, \tag{12.13}$$

$$\min_{\boldsymbol{\Omega}_1}\frac{1}{P}\operatorname{tr}(\boldsymbol{\Omega}_1\cdot\frac{1}{P}\sum_{n=1}^{N}\mathbf{X}_n^{\mathrm{T}}\boldsymbol{\Omega}_2\mathbf{X}_n) - \frac{1}{P}\log\det\boldsymbol{\Omega}_1 + \sum_{p\neq i}\lambda_1|(\boldsymbol{\Omega}_1(p,i)|. \tag{12.14}$$

Note that each of those two optimization models can be treated as a vector-variate GGM with ℓ_1 penalty imposed on the precision matrix [1, 6, 11]. According to [8], we can iteratively solve the above minimization problems using Glasso [6].

12.4 Tensor-Based Graphical Model

In more generalized cases, the data to be processed can be in tensor form. Simply rearranging the tensor into a matrix or a vector will lose vital structural information, rendering unsatisfactory results. Besides, if we directly reshape the tensor data to fit in the vector- or matrix-variate-based GGM, it will require a huge precision matrix, which is computationally expensive.

To address the above issue, similar to the generalization from vector-based processing to matrix-based processing, tensor-variate-based GGM, a natural generalization of matrix-variate-based GGM, has been proposed [7, 9, 12, 16]. In particular, given N i.i.d. K-th order observations $\mathcal{T}_1, \cdots, \mathcal{T}_N \in \mathbb{R}^{I_1 \times \ldots \times I_K}$ from tensor normal distribution $\mathrm{TN}(\mathbf{0}; \mathbf{\Sigma}_1, \cdots, \mathbf{\Sigma}_K)$, its PDF is

$$\mathrm{Prob}(\mathcal{T}_n | \mathbf{\Sigma}_1, \cdots, \mathbf{\Sigma}_K) = (2\pi)^{-\frac{I}{2}} \left\{ \prod_{k=1}^{K} |\mathbf{\Sigma}_k|^{-\frac{I}{2I_k}} \right\} \exp(-\|\mathcal{T}_n \times_{1:K} \mathbf{\Sigma}^{-\frac{1}{2}}\|_{\mathrm{F}}^2/2), \tag{12.15}$$

where $I = \prod_{k=1}^{K} I_k$. Let $\mathbf{\Sigma}^{-\frac{1}{2}} := \{\mathbf{\Sigma}_1^{-\frac{1}{2}}, \cdots, \mathbf{\Sigma}_K^{-\frac{1}{2}}\}$, where $\mathbf{\Sigma}_1, \ldots, \mathbf{\Sigma}_K$ are the covariance matrices corresponding to all modes, respectively. $\times_{1:K}$ denotes the tensor mode product from mode-1 to mode-K, i.e., $\mathcal{T} \times_{1:K} \{\mathbf{A}_1, \cdots, \mathbf{A}_K\} = \mathcal{T} \times_1 \mathbf{A}_1 \times_2 \cdots \times_K \mathbf{A}_K$. In this case, the covariance matrix of the tensor normal distribution is separable in the sense that it is the Kronecker product of small covariance matrices, each of which corresponds to one way of the tensor. It means that the PDF for $\mathcal{T}_n$ satisfies $\mathrm{Prob}(\mathcal{T}_n | \mathbf{\Sigma}_1, \cdots, \mathbf{\Sigma}_K)$, if and only if the PDF for $\mathrm{vec}(\mathcal{T}_n)$ satisfies $\mathrm{Prob}(\mathrm{vec}(\mathcal{T}_n) | \mathbf{\Sigma}_K \otimes \cdots \otimes \mathbf{\Sigma}_1)$, where $\mathrm{vec}(\cdot)$ is the vectorization operator and $\otimes$ is the matrix Kronecker product.

The negative log-likelihood function for N observations $\mathcal{T}_1, \cdots, \mathcal{T}_N$ is

$$\mathrm{H}(\mathbf{\Omega}_1, \cdots, \mathbf{\Omega}_K) = -\sum_{k=1}^{K} \frac{I}{2I_k} \log\det \mathbf{\Omega}_k + \frac{1}{2}\mathrm{tr}(\mathbf{S}(\mathbf{\Omega}_K \otimes \ldots \otimes \mathbf{\Omega}_1)) \tag{12.16}$$

where $\mathbf{\Omega}_k = \mathbf{\Sigma}_k^{-1}$, $k = 1, \cdots, K$, $\mathbf{S} = \frac{1}{N} \sum \mathrm{vec}(\mathcal{T}_n)\mathrm{vec}(\mathcal{T}_n)^{\mathrm{T}}$.

To encourage sparsity on each precision matrix, a penalized log-likelihood estimator is considered, forming the following sparse tensor graphical optimization model [9]:

$$\min_{\mathbf{\Omega}_k, k=1,\cdots,K} \sum_{k=1}^{K} \frac{1}{I_k} \log\det \mathbf{\Omega}_k + \frac{1}{I}\mathrm{tr}(\mathbf{S}(\mathbf{\Omega}_K \otimes \ldots \otimes \mathbf{\Omega}_1)) + \sum_{k=1}^{K} \lambda_k \, \mathrm{R}_k(\mathbf{\Omega}_k) \tag{12.17}$$

where R_k is a penalty function weighted by a parameter λ_k.

The penalty function also has a variety of choices, such as Lasso penalty in [7, 9], ℓ_0 penalty in [16], and adaptive Lasso penalty in [7]. Note that when $K = 1$ and

$K = 2$, it degenerates into the sparse Gaussian graphical model [6] and sparse matrix graphical model [8], respectively.

If we employ Lasso penalty on precision matrices, the problem of interest will be transformed into

$$\min_{\mathbf{\Omega}_k, k=1,\cdots,K} -\sum_{k=1}^{K} \frac{1}{I_k} \log\det \mathbf{\Omega}_k + \frac{1}{I}\mathrm{tr}(\mathbf{S}(\mathbf{\Omega}_K \otimes \ldots \otimes \mathbf{\Omega}_1)) + \sum_{k=1}^{K} \lambda_k \left\|\mathbf{\Omega}_k\right\|_{1,\mathrm{off}} \tag{12.18}$$

where $\lambda_k \left\|\mathbf{\Omega}_k\right\|_{1,\mathrm{off}}$ represent the ℓ_1 penalty imposed on each mode of $\mathcal{T}$.

Similar to the situation in sparse matrix graphical model, the optimization model in (12.18) is also biconvex. By updating one precision matrix at a time while fixing all other precision matrices, we can solve it alternatively still by Glasso. Therefore, for $k = 1, \cdots, K$, the solution of the optimization model in (12.18) can be converted to alternatively optimize

$$\min_{\mathbf{\Omega}_k} -\frac{1}{I_k} \log\det \mathbf{\Omega}_k + \frac{1}{I_k}\mathrm{tr}(\mathbf{S}_k\mathbf{\Omega}_k) + \lambda_k \left\|\mathbf{\Omega}_k\right\|_{1,\mathrm{off}}, \tag{12.19}$$

where $\mathbf{S}_k = \frac{I_k}{NI}\sum_{n=1}^{N} \mathbf{V}_{(n)}^k {\mathbf{V}_{(n)}^k}^{\mathrm{T}}$, ${\mathbf{V}_{(n)}}^k$ is the mode-n unfolding of $\mathcal{V}^k$, $\mathcal{V}^k = \mathcal{T}_n \times_{1:\mathrm{K}} \{\mathbf{\Omega}_1^{1/2}, \cdots, \mathbf{\Omega}_{k-1}^{1/2}, \mathbf{I}_{I_k}, \mathbf{\Omega}_{k+1}^{1/2}, \cdots, \mathbf{\Omega}_K^{1/2}\}$, $\times_{1:\mathrm{K}}$ represents the tensor product operation, and $\left\|\mathbf{\Omega}_k\right\|_{1,\mathrm{off}} = \sum_{p\neq q} |\mathbf{\Omega}(p,q)|$ is the off-diagonal ℓ_1 norm.

By applying Glasso for the optimization model in (12.19), we can obtain the solution of precision matrices $\mathbf{\Omega}_k$, $k = 1, \cdots, K$. This procedure is called Tlasso [9], and the details are illustrated in Algorithm 52.

Algorithm 52: Sparse tensor graphical model via tensor lasso (Tlasso)

Input: Tensor samples $\mathcal{T}_1, \cdots, \mathcal{T}_N$, tuning parameters $\lambda_1, \cdots, \lambda_K$, max number of iterations T. $\mathbf{\Omega}_1^{(0)}, \cdots, \mathbf{\Omega}_K^{(0)}$ are randomly set as symmetric and positive difinite matrices. $t = 0$.
While $t \neq T$ **do:**
  $t = t + 1$
  For $k = 1, \cdots, K$**:**
    Update $\mathbf{\Omega}_k^{(t)}$ in (12.19) via Glasso, given $\mathbf{\Omega}_1^{(t)}, \cdots, \mathbf{\Omega}_{k-1}^{(t)}, \mathbf{\Omega}_{k+1}^{(t-1)}, \mathbf{\Omega}_K^{(t-1)}$.
    Set $\|\mathbf{\Omega}_k^{(t)}\|_{\mathrm{F}} = 1$.
  End for
End while
Output: $\hat{\mathbf{\Omega}}_k = \mathbf{\Omega}_k^{(T)}$, $k = 1, \cdots, K$.

However, Tlasso requires each precision matrix satisfies $\|\mathbf{\Omega}_k\|_{\mathrm{F}}^2 = 1$, $k = 1, \cdots, K$, which cannot be simultaneously achieved by normalizing the data when $K > 3$. To address this problem, Xu et al. proposed a sparse tensor-variate Gaussian graphical model (STGGM) [16], which imposes ℓ_0 penalty instead of Lasso penalty on precision matrices and achieves the same minimax-optimal convergence rate more efficiently compared with Tlasso.

The optimization model of STGGM can be written as follows:

$$\begin{aligned}
\min_{\boldsymbol{\Omega}_k \succ 0, k=1,\cdots,K} \quad & \mathrm{H}_k(\boldsymbol{\Omega}_1, \cdots, \boldsymbol{\Omega}_K) \\
\text{s. t.} \quad & \|\boldsymbol{\Omega}_k\|_0 \leq S_k,\ k = 1, \cdots, K,
\end{aligned} \tag{12.20}$$

where $\mathrm{H}_k(\boldsymbol{\Omega}_1, \cdots, \boldsymbol{\Omega}_K) = -\sum_{k=1}^{K} \frac{1}{I_k} \log\det \boldsymbol{\Omega}_k + \frac{1}{I}\mathrm{tr}(\mathbf{S}(\boldsymbol{\Omega}_K \otimes \ldots \otimes \boldsymbol{\Omega}_1))$, $\boldsymbol{\Omega}_k \succ 0$ means $\boldsymbol{\Omega}_k$ is positive definite and S_k is the parameter which controls the sparsity of $\boldsymbol{\Omega}_k$.

Since the sparsity term in (12.20) is not convex with respect to $\boldsymbol{\Omega}_k$, $k = 1, \cdots, K$, we resort to alternating minimization to update each variable with the others fixed. In this case, we can obtain K subproblems in the following format:

$$\begin{aligned}
\min_{\boldsymbol{\Omega}_k \succ 0} \ & \mathrm{H}_k(\boldsymbol{\Omega}_k) \\
\text{s. t.} \quad & \|\boldsymbol{\Omega}_k\|_0 \leq S_k
\end{aligned} \tag{12.21}$$

The details can be found in Algorithm 53. Let $\boldsymbol{\Omega}_{\neq k} = \boldsymbol{\Omega}_K \otimes \cdots \otimes \boldsymbol{\Omega}_{k+1} \otimes \boldsymbol{\Omega}_{k-1} \otimes \cdots \otimes \boldsymbol{\Omega}_1$ and $\mathrm{trunc}(\boldsymbol{\Omega}, \mathbb{S})$ operation in Algorithm 53 is defined as

$$\mathrm{trunc}(\boldsymbol{\Omega}, \mathbb{S})_{i,j} = \begin{cases} \boldsymbol{\Omega}(i, j) & \text{if } (i, j) \in \mathbb{S} \\ 0 & \text{if } (i, j) \notin \mathbb{S} \end{cases}$$

where $\boldsymbol{\Omega} \in \mathbb{R}^{I \times J}$, $\mathbb{S} \subseteq \{(i, j) : i = 1, \cdots, I,\ j = 1, \cdots, J\}$.

Algorithm 53: Alternating gradient descent (AltGD) for STGGM

Input: initial points $\boldsymbol{\Omega}_{\neq k}^{(1)}$, $k = 1, \cdots, K$, max number of iterations M, $m = 0$, sparsity $S_k > 0$, learning rate ϕ_k, tensor samples $\mathcal{T}_1, \cdots, \mathcal{T}_N$.
For $m = 1, \cdots, M$ **do**
  For $k = 1, \cdots, K$ **do**
    $\mathrm{temp}_{\boldsymbol{\Omega}_k} = \boldsymbol{\Omega}_k^{(m)} - \phi_k \nabla_k \, \mathrm{q}_k(\boldsymbol{\Omega}_k)$
    $\boldsymbol{\Omega}_k^{(m+1)} = \mathrm{trunc}(\mathrm{temp}_{\boldsymbol{\Omega}_k}, \mathrm{temp}_{S_k})$, where temp_{S_k} is the support set of the largest S_k magnitudes of $\mathrm{temp}_{\boldsymbol{\Omega}_k}$.
  end for
end for
Output: $\hat{\boldsymbol{\Omega}}_k = \boldsymbol{\Omega}_k^{(M)}$, $(k = 1, \cdots, K)$.

12.5 Applications

The aim of GGM is to study the correlations among variables, which is important for revealing the hidden structure within data. In this section, we mainly discuss two applications employing tensor-variate-based GGM, i.e., environmental prediction and mice aging study.

12.5.1 *Environmental Prediction*

Environmental prediction provides us with abundant information about future weather, which enables us to prepare for the flood, sand storm, debris flow, or hail. The analysis of data obtained from environmental instruments can help us better understand the causes of environment changes.

In this part, we focus on meteorological dataset from US National Center for Environmental Prediction (NCEP).[1] It contains climate data collected globally since 1948, including temperature, precipitation, relative humidity, and so on. For simplicity, we mainly analyze the influence of geographic location on wind velocity by extracting the wind speed data 'uwnd.sig995' from 1948 to 1957. The wind velocity data in the dataset are indexed by longitude, latitude, and time, which is represented naturally as a third-order tensor. The wind speed is recorded in the unit of 0.1 m/s. As shown in Fig. 12.2, we select the data from the area where the longitude and latitude are within (90 °W, 137.5 °W) and (17.5 °N, 40 °N), respectively, and form a 20×10 grid by partitioning the region with 2.5° spacings.

To reveal the relationship between wind velocity and geographic location, we choose the velocity of the first 10 days in 1948 to form a $20 \times 10 \times 10$ sample, and the same procedure is done on the rest 9 years. By doing so, ten third-order wind speed samples during 10 years are obtained with dimensions corresponding to longitude, latitude, and time. In order to better manipulate the data, we scale the value of data by ten times. The tuning parameters for estimating the precision matrices $\mathbf{\Omega}$ are set in the same way as the work in [9]. The results are shown in Fig. 12.3.

Note that in Fig. 12.3b, the wind velocity across node 8, node 9, and node 10 are different from the others. This indicates a different velocity pattern in corresponding areas, which is in line with the facts: the area between (0, 23.5 °N) is tropical zone, and area between (23.5 °N, 66.5 °N) belongs to the northern temperate zone. The climates of these two regions are typically different due to the latitude gaps. In our experiment, the region of node 8 divides tropical zone and temperate zone, which confirms the pattern differences in Fig. 12.3b.

---

[1] ftp://ftp2.psl.noaa.gov/Datasets/ncep.reanalysis.dailyavgs/surface/.

Fig. 12.2 Region of interest in our experiment within (90 °W, 137.5 °W) and (17.5 °N, 40°N), which corresponds to parts of Pacific Ocean, the USA, and Mexico. The red line denotes 23.5 °N, which is the dividing line between tropical and north temperate zones. The black grid represents the 20×10 partition of this area, and the Arabic numbers on the left side are the index of different latitude regions

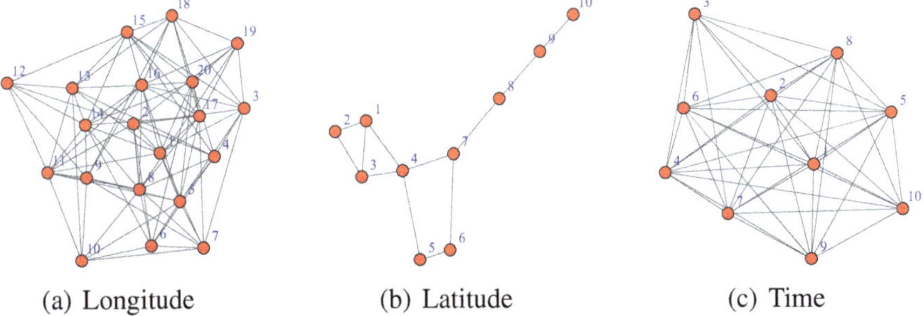

Fig. 12.3 The graphical structure of precision matrices predicted by Tlasso based on the ten samples. Graphs (a), (b), and (c) illustrate the velocity relationships in longitude, latitude, and time, respectively. Red dot denotes the corresponding variable in each mode

12.5.2 *Mice Aging Study*

The age-related diseases such as Alzheimer's disease and Parkinson's disease have always been an agonizing problem in medical research for humans. The study of aging process may shed light on the trigger factors of those diseases. To this end, a series of pioneering experiments on mice have been carried out due to the genetic and physiological resemblance to humans.

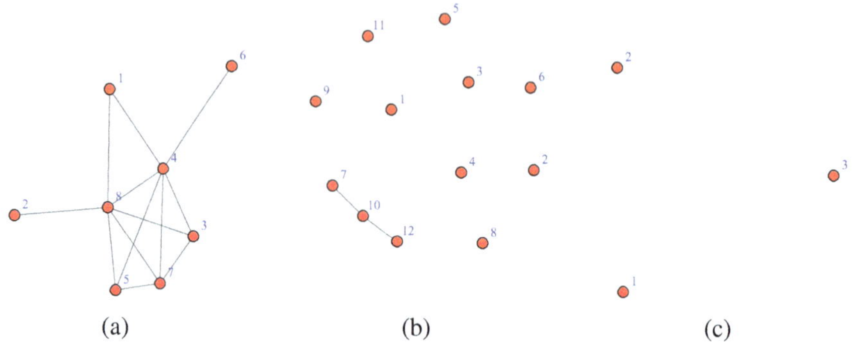

Fig. 12.4 The graphical structure of precision matrices in three modes. The numbers in three figures denote different genes, tissues, and mice, respectively. (**a**) Genes. (**b**) Tissues. (**c**) Mice

In this part, we mainly focus on gene expression database[2] for mice aging study. The intention of analysis of these data is to investigate the relationships between genes and aging process. In this dataset, gene expression level is collected from mice across ten age groups, such as 1 month, 6 months, 16 months, and 24 months. Ten mice are considered in each group. The expression levels of 8932 genes in 16 tissues for every mouse are recorded.

For ease of illustration, we only select 8 gene expression levels in 12 tissues from 3 1-month mice to form a sample of the size $8 \times 12 \times 3$. To be specific, the 8 chosen genes are "Igh-6," "Slc16a1," "Icsbp1," "Cxcr4," "Prdx6," "Rpl37," "Rpl10," and "Fcgr1," and the 12 selected tissues are "adrenal glands," "cerebellum," "cerebrum," "eye," "hippocampus," "kidney," "lung," "muscle," "spinal cord," "spleen," "striatum," and "thymus." Under the same conditions as the work in [9], the analytical results are given in Fig. 12.4.

Figure 12.4a shows the relationship between the selected eight genes. We can notice that "Slc16a1" (node 2) and "Rpl37" (node 6) are relatively independent compared with other genes. The reason may be the biological functions of those two genes are not closely related compared with others. For example, the biological functions of "Igh-6" (node 1), "Icsbp1" (node 3), "Cxcr4" (node 4), and "Fcgr1" (node 8) are all related to the immune system of the mice and thus closely connected, while the "Slc16a1" (node 2) is a gene that involved with the regulation of central metabolic pathways and insulin secretion. Figure 12.4b demonstrates the relationship among 12 different tissues. It indicates the potential internal connections between "lung" (node 7), "spleen" (node 10), and "thymus" (node 12), which can provide some clues for the studies in related fields.

[2] http://cmgm.stanford.edu/~kimlab/aging_mouse.

12.6 Summary

As a powerful tool to discover the underlying correlations within variables from large amounts of data, Gaussian graphical model has attracted lots of attentions in neuroscience, life sciences, and social networks. In this chapter, we start with vector-variate data-based model and mainly focused on its generalization for tensor data. The superior performance of tensor-variate-based model is demonstrated on wind pattern analysis and mice aging study.

However, existing tensor-variate-based Gaussian graphical models assume the considered data are conditional independent within modes, which may not be suitable for some practical situations. In the future, more flexible and efficient methods are expected to be developed for diverse data.

References

1. Banerjee, O., El Ghaoui, L., d'Aspremont, A.: Model selection through sparse maximum likelihood estimation for multivariate gaussian or binary data. J. Mach. Learn. Res. **9**, 485–516 (2008)
2. Beal, M.J., Jojic, N., Attias, H.: A graphical model for audiovisual object tracking. IEEE Trans. Pattern Anal. Mach. Intell. **25**(7), 828–836 (2003)
3. Brouard, C., de Givry, S., Schiex, T.: Pushing data into CP models using graphical model learning and solving. In: International Conference on Principles and Practice of Constraint Programming, pp. 811–827. Springer, Berlin (2020)
4. d'Aspremont, A., Banerjee, O., El Ghaoui, L.: First-order methods for sparse covariance selection. SIAM J. Matrix Anal. Appl. **30**(1), 56–66 (2008)
5. Fan, J., Li, R.: Variable selection via nonconcave penalized likelihood and its oracle properties. J. Am. Stat. Assoc. **96**(456), 1348–1360 (2001)
6. Friedman, J., Hastie, T., Tibshirani, R.: Sparse inverse covariance estimation with the graphical lasso. Biostatistics **9**(3), 432–441 (2008)
7. He, S., Yin, J., Li, H., Wang, X.: Graphical model selection and estimation for high dimensional tensor data. J. Multivar. Anal. **128**, 165–185 (2014)
8. Leng, C., Tang, C.Y.: Sparse matrix graphical models. J. Am. Stat. Assoc. **107**(499), 1187–1200 (2012)
9. Lyu, X., Sun, W.W., Wang, Z., Liu, H., Yang, J., Cheng, G.: Tensor graphical model: non-convex optimization and statistical inference. IEEE Trans. Pattern Anal. Mach. Intell. **42**(8), 2024–2037 (2019)
10. Ma, C., Lu, J., Liu, H.: Inter-subject analysis: a partial Gaussian graphical model approach. J. Am. Stat. Assoc. **116**(534), 746–755 (2021)
11. Meinshausen, N., Bühlmann, P., et al.: High-dimensional graphs and variable selection with the lasso. Ann. Stat. **34**(3), 1436–1462 (2006)
12. Shahid, N., Grassi, F., Vandergheynst, P.: Multilinear low-rank tensors on graphs & applications (2016, preprint). arXiv:1611.04835
13. Shen, X., Pan, W., Zhu, Y.: Likelihood-based selection and sharp parameter estimation. J. Am. Stat. Assoc. **107**(497), 223–232 (2012)
14. Tibshirani, R.: Regression shrinkage and selection via the lasso. J. R. Stat. Soc. B **58**(1), 267–288 (1996)
15. Wu, W.B., Pourahmadi, M.: Nonparametric estimation of large covariance matrices of longitudinal data. Biometrika **90**(4), 831–844 (2003)

16. Xu, P., Zhang, T., Gu, Q.: Efficient algorithm for sparse tensor-variate gaussian graphical models via gradient descent. In: Artificial Intelligence and Statistics, pp. 923–932. PMLR, Westminster (2017)
17. Yin, J., Li, H.: Model selection and estimation in the matrix normal graphical model. J. Multivariate Anal. **107**, 119–140 (2012)
18. Yuan, M.: High dimensional inverse covariance matrix estimation via linear programming. J. Mach. Learn. Res. **11**, 2261–2286 (2010)
19. Yuan, M., Lin, Y.: Model selection and estimation in the gaussian graphical model. Biometrika **94**(1), 19–35 (2007)
20. Zhang, C.H., et al.: Nearly unbiased variable selection under minimax concave penalty. Ann. Stat. **38**(2), 894–942 (2010)
21. Zhou, S., et al.: Gemini: graph estimation with matrix variate normal instances. Ann. Stat. **42**(2), 532–562 (2014)
22. Zou, H.: The adaptive lasso and its oracle properties. J. Am. Stat. Assoc. **101**(476), 1418–1429 (2006)

Chapter 13
Tensor Sketch

13.1 Introduction

The emergency of large-scale datasets in recent years poses challenge to subsequent data acquisition, storage, processing, and analysis. To address the issue, a number of dimensionality reduction methods are developed to efficiently map the original data into low-dimensional space using random samplings or projections while preserving the intrinsic information [7, 8, 37]. For example, the well-known compressed sensing theory states that it is of high probability to recover the K principle elements of an input vector $\mathbf{x} \in \mathbb{R}^I$ from its random samplings $\mathbf{y} = \mathbf{S}\mathbf{x}$, if $\mathbf{S} \in \mathbb{R}^{J \times I}$ is a random matrix with $J = \mathrm{O}(K \log I)$ rows [15, 16].

Besides compressive sensing, there is another subclass of randomization methods, dubbed *sketching*. Here we consider the overdetermined least squares problem

$$\mathbf{x}^* = \underset{\mathbf{x} \in \mathbb{R}^I}{\operatorname{argmin}} \|\mathbf{A}\mathbf{x} - \mathbf{y}\|_{\mathrm{F}},$$

where $\mathbf{A} \in \mathbb{R}^{J \times I}$ $(I \leq J)$, and $\mathbf{y} \in \mathbb{R}^J$ is the given measurement. By applying sketching, one can obtain the desired coefficient $\mathbf{x}$ by computing a dimensionality-reduced problem

$$\tilde{\mathbf{x}} = \underset{\mathbf{x} \in \mathbb{R}^I}{\operatorname{argmin}} \|\mathbf{S}(\mathbf{A}\mathbf{x} - \mathbf{y})\|_{\mathrm{F}},$$

which satisfies

$$\|\mathbf{A}\tilde{\mathbf{x}} - \mathbf{y}\|_{\mathrm{F}} \leq (1 \pm \epsilon)\|\mathbf{A}\mathbf{x}^* - \mathbf{y}\|_{\mathrm{F}}$$

Y. Liu et al., *Tensor Computation for Data Analysis*,
https://doi.org/10.1007/978-3-030-74386-4_13

with high probability for a small $\epsilon > 0$, where $\mathbf{S} \in \mathbb{R}^{L \times J}$ is the sketching matrix. We can see that sketching tries to model the relationship between $\mathbf{x}$ and $\mathbf{y}$ with less samples based on the sketching matrix $\mathbf{S}$. It sacrifices estimation accuracy to a certain extent for achieving computational and storage complexity reduction.

An intuitive way to choose the sketching matrix $\mathbf{S}$ is to use i. i. d. Gaussian random matrix. However, in this case, $\mathbf{S}$ is dense, and it is inefficient for matrix multiplication. As a result, the sketching matrices are commonly drawn from certain families of random maps (e.g., oblivious sparse norm approximating projections transform [28], fast Johnson–Lindenstrauss transform [2]). In this way, it yields input-sparsity computation [11, 32], improved approximation guarantees [33], and much less storage requirement. Another distinguishing feature of sketching is that unlike compressed sensing which relies on strict assumption such as sparsity, the quality of sketching is almost independent with the input data structure [9, 40], although certain structure can scale up the computation speed.

One simple and effective sketching method is count sketch (CS), which is first designed to estimate the most frequent items in a data stream [10]. Its key idea is to build sketching matrices from random hash functions and project the input vector into lower-dimensional sketched space. In [29], Pagh introduces tensor sketch (TS) for compressing matrix multiplication. TS can be regarded as a specialized case for CS where the hash function used by TS is built by cleverly integrating multiple hash functions from CS. In this manner, TS can be represented by the polynomial multiplication and hence can be fast computed by the fast Fourier transform (FFT). Compared with CS, TS achieves great acceleration and storage reduction for rank-1 matrices/tensors. TS is successfully used in many statistical learning tasks such as kernel approximation [1, 5, 30], tensor decomposition [26, 36, 38, 44], Kronecker product regression [13, 14], and network compression [21, 31, 34].

Despite the effectiveness of CS and TS, they have drawbacks under certain cases. For example, when the input data are higher-order tensors, CS and TS both map the multidimensional data into the vector space in preprocessing, which partly ignores the multidimensional structure information. Besides, CS is memory-consuming since the vectorization of the multidimensional data would lead to a long vector and the hash function is required to have the same size of the input vector. To solve these problems, higher-order count sketch (HCS) is proposed [34] to directly sketch the input tensor into a lower-dimensional one with the same order; hence the multidimensional structure information is preserved. Compared with CS, HCS achieves better approximation accuracy and provides efficient computation directly in the sketched space for some tensor contractions and tensor products.

In the following, we will provide a detailed illustration of CS, TS, and HCS. The remaining part is organized as follows. Section 13.2, 13.3, and 13.4 introduce basic conceptions of CS, TS, and HCS, respectively. In Sect. 13.5, we present some applications of tensor sketches in tensor decomposition, Kronecker product regression, and network approximation. Finally, the chapter is concluded in Sect. 13.6.

13.2 Count Sketch

Count sketch is first proposed as a randomized projection strategy to estimate the most frequent items in a data stream [10]. It has successful applications such as estimation of internet packet streams [12] and feature hashing for large-scale multitask learning [30].

Before we introduce the definition of count sketch, we first give a definition of K-wise independent.

Definition 9.1 **(K-Wise Independent [26])** A set of hash functions $\mathbb{H} = \{\mathrm{h} \mid \mathrm{h} : [I] \mapsto [J]\}$[1] is K-wise independent if for any K distinct $i_1, \cdots, i_K \in [I]$ and $\mathrm{h} \in \mathbb{H}$, $\mathrm{h}(i_1), \cdots, \mathrm{h}(i_K)$ are independent random variables with $\mathrm{Prob}(\mathrm{h}(i_k) = j) = 1/J$ for $i_k \in [I]$, $j \in [J]$.

The count sketch is defined as follows:

Definition 9.2 (Count Sketch [10]) Let $\mathrm{h} : [I] \mapsto [J]$ and $\mathrm{s} : [I] \mapsto \{\pm 1\}$ be two 2-wise independent hash functions. The count sketch of $\mathbf{x} \in \mathbb{R}^I$ produces a projection $\mathbf{y} = \mathrm{CS}(\mathbf{x}) \in \mathbb{R}^J$ with its elements computed by

$$\mathbf{y}(j) = \sum_{\mathrm{h}(i)=j} \mathrm{s}(i)\mathbf{x}(i), \tag{13.1}$$

which is equal to the following equation in matrix form:

$$\mathbf{y} = \mathbf{S}\mathbf{x}, \tag{13.2}$$

where $\mathbf{S} \in \mathbb{R}^{J \times I}$ is a sparse sketching matrix such that $\mathbf{S}(\mathrm{h}(i), i) = \mathrm{s}(i)$. The calculation complexity of $\mathbf{y}$ is about $\mathrm{O}(\mathrm{nnz}(\mathbf{x}))$ with $\mathrm{nnz}(\mathbf{x})$ representing the number of nonzero elements of $\mathbf{x}$ [11].

The recovery rule is

$$\hat{\mathbf{x}}(i) = \mathrm{s}(i)\mathbf{y}(\mathrm{h}(i)). \tag{13.3}$$

To make the estimation more robust, one can take D independent sketches and compute the median of them [10]. The following theorem gives the error analysis of the CS recovery:

Theorem 13.1 ([10]) *For any $\mathbf{x} \in \mathbb{R}^I$, given two 2-wise independent hash functions $\mathrm{h} : [I] \mapsto [J]$ and $\mathrm{s} : [I] \mapsto \{\pm 1\}$, $\hat{\mathbf{x}}(i)$ provides an unbiased estimator of $\mathbf{x}(i)$ with $\mathrm{Var}(\hat{\mathbf{x}}(i)) \leq \|\mathbf{x}\|_2^2/J$.*

Below we present a toy example showing how count sketch works.

---

[1] In this chapter, $[N]$ denotes the set $\{0, 1, \cdots, N-1\}$ for any $N \in \mathbb{R}^+$.

A toy example: approximation of vector inner product Given two vectors $\mathbf{x}_1 = [1, 0, 1, 0, 1]^T$, $\mathbf{x}_2 = [2, 0, 0, 1, 2]^T$, we adopt three Hash functions h_1, h_2, h_3 and three vectors $\mathbf{s}_1, \mathbf{s}_2, \mathbf{s}_3$ with its elements satisfying the Rademacher distribution as follows:

$$
\begin{aligned}
h_1(j) &= \mathrm{mod}(j + 2, 3) & \mathbf{s}_1 &= [1, 1, -1, 1, -1]^T \\
h_2(j) &= \mathrm{mod}(2j, 3) & \mathbf{s}_2 &= [-1, 1, -1, 1, 1]^T \\
h_3(j) &= \mathrm{mod}(\mathrm{mod}(j, 5), 3) & \mathbf{s}_3 &= [-1, 1, -1, -1, 1]^T,
\end{aligned}
$$

where mod denotes the modular operation. The count sketch of $\mathbf{x}_1$ and $\mathbf{x}_2$ is obtained as follows:

$$
\begin{aligned}
CS_1(\mathbf{x}_1) &= [-1, -1, 1]^T & CS_1(\mathbf{x}_2) &= [-2, 0, 3]^T \\
CS_2(\mathbf{x}_1) &= [-1, -1, 1]^T & CS_2(\mathbf{x}_2) &= [1, 0, 2]^T \\
CS_3(\mathbf{x}_1) &= [-1, 1, -1]^T & CS_3(\mathbf{x}_2) &= [-3, 2, 0]^T.
\end{aligned}
$$

Using the fact that inner product is preserved under the same sketches [30], we can approximate the inner product $\langle \mathbf{x}_1, \ \mathbf{x}_2 \rangle$ by

$$
\begin{aligned}
\langle \mathbf{x}_1, \mathbf{x}_2 \rangle &\approx \mathrm{median}(\langle CS_1(\mathbf{x}_1), CS_1(\mathbf{x}_2) \rangle, \cdots, \langle CS_3(\mathbf{x}_1), CS_3(\mathbf{x}_2) \rangle) \\
&= 5,
\end{aligned}
$$

which is slightly overestimated since the ground truth is 4. As the number of sketches increases, the estimation would get more accurate.

13.3 Tensor Sketch

As illustrated above, CS is originally designed for vector operations. But in real applications, the collected data usually has multiple dimensions. The standard CS may not be efficient for high-order data processing.

Given any $\mathbf{X} \in \mathbb{R}^{I \times I}$, $\mathbf{y} = CS(\mathbf{X}) \in \mathbb{R}^J$ can be computed by

$$
\mathbf{y}(j) = \sum_{H(i_1, i_2) = j} S(i_1, i_2) \mathbf{X}(i_1, i_2), \tag{13.4}
$$

where $H : [I] \times [I] \mapsto [J]$ and $S : [I] \times [I] \mapsto \{\pm 1\}$ are two 2-dimensional, 2-wise independent Hash functions. For a dense matrix $\mathbf{X}$, clearly both the computation of CS and storage for Hash functions are polynomial to the data dimension. As the number of modes increases, it becomes infeasible for practical use. Therefore, considering efficient sketching methods for matrix or tensor operations is necessary.

In [30], Pham et al. propose TS by generalizing CS into higher-order data. First, we consider the case that $\mathbf{X} = \mathbf{x} \circ \mathbf{x} \in \mathbb{R}^{I \times I}$ is a rank-1 matrix. If we choose H and S by

$$\mathrm{H}(i_1, i_2) = \mathrm{mod}(\mathrm{h}_1(i_1) + \mathrm{h}_2(i_2), J), \quad \mathrm{S}(i_1, i_2) = \mathbf{s}_1(i_1)\mathbf{s}_2(i_2), \tag{13.5}$$

where $\mathrm{h}_1, \mathrm{h}_2 : [I] \mapsto [J]$ and $\mathrm{s}_1, \mathrm{s}_2 : [I] \mapsto \{\pm 1\}$ are two pairs of 2-wise independent Hash functions; then $\mathbf{y} = \mathrm{TS}(\mathbf{X}) \in \mathbb{R}^J$ is computed by

$$\mathbf{y}(j) = \sum_{\mathrm{H}(i_1, i_2) = j} \mathrm{S}(i_1, i_2)\mathbf{X}(i_1, i_2), \tag{13.6}$$

which can be efficiently computed by Pham and Pagh [30]

$$\begin{aligned} \mathbf{y} &= \mathrm{CS}_1(\mathbf{x}) \boxtimes_J \mathrm{CS}_2(\mathbf{x}) \\ &= \mathrm{F}^{-1}(\mathrm{F}(\mathrm{CS}_1(\mathbf{x})) \circledast \mathrm{F}(\mathrm{CS}_2(\mathbf{x}))), \end{aligned} \tag{13.7}$$

where $\boxtimes_J$ and $\circledast$ denote the mode-J circular convolution and Hadamard product, respectively. Operator F and F^{-1} denote FFT and its inverse operation. The calculation of $\mathrm{TS}(\mathbf{X})$ requires $\mathrm{O}(\mathrm{nnz}(\mathbf{x}) + J\log J)$ time and $\mathrm{O}(I)$ storage, which is a great improvement compared with $\mathrm{CS}(\mathbf{X})$.

The extension to higher-order case is straightforward. Given $\mathcal{X} = \underbrace{\mathbf{x} \circ \cdots \circ \mathbf{x}}_{N \text{ times}} \in \mathbb{R}^{I \times \cdots \times I}$, N pairs of 2-wise independent Hash functions $\mathrm{h}_1, \cdots, \mathrm{h}_N : [I] \mapsto [J]$, $\mathrm{s}_1, \cdots, \mathrm{s}_N : [I] \mapsto \{\pm 1\}$, then the tensor sketch $\mathrm{TS}(\mathcal{X}) \in \mathbb{R}^J$ can be calculated in $\mathrm{O}(N(\mathrm{nnz}(\mathbf{x}) + J\log J))$ time by

$$\mathrm{TS}(\mathcal{X}) = \mathrm{F}^{-1}(\mathrm{F}(\mathrm{CS}_1(\mathbf{x})) \circledast \cdots \circledast \mathrm{F}(\mathrm{CS}_N(\mathbf{x}))). \tag{13.8}$$

It is noted that the above discussion is based on the assumption of rank-1 matrix or tensor. If it is not the case, the computation of TS takes $\mathrm{O}(\mathrm{nnz}(\mathbf{X}))$ time by (13.6) for a matrix. Although the computational complexity is the same as the CS, the storage requirement is reduced from $\mathrm{O}(I^2)$ to $\mathrm{O}(I)$. Therefore, TS is more efficient when dealing with high-order data.

It is shown that tensor sketching provides an oblivious subspace embedding [6] and can be applied in the context of low-rank approximation, principal component regression, and canonical correlation analysis of the polynomial kernels [5]. Its other variants include recursive TS [1], polynomial TS [18], sparse TS [42], etc.

13.4 Higher-Order Count Sketch

Higher-order count sketching (HCS) can be regarded as a naive generalization of CS, which tries to perform random projection while preserving the multidimensional information in original tensors [35]. Specifically, given an input tensor, HCS produces a lower dimensional one of the same order. Figure 13.1 presents an illustration for CS, TS, and HCS. Its definition is as follows:

Definition 9.3 (Higher-Order Count Sketch) Given $\mathcal{X} \in \mathbb{R}^{I_1 \times \cdots \times I_N}$, N independent Hash functions $\mathrm{h}_n : [I_n] \mapsto [J_n]$, $\mathrm{s}_n : [I_n] \mapsto \{\pm 1\}$, and $\mathcal{Y} = \mathrm{HCS}(\mathcal{X}) \in \mathbb{R}^{J_1 \times \cdots \times J_N}$ is computed as

$$\mathcal{Y}(j_1, \cdots, j_N) = \sum_{\substack{\mathrm{h}_1(i_1)=j_1 \\ \cdots \\ \mathrm{h}_N(i_N)=j_N}} \mathrm{s}_1(i_1) \cdots \mathrm{s}_N(i_N) \mathcal{X}(i_1, \cdots, i_N), \tag{13.9}$$

which can be rewritten as the Tucker form

$$\begin{aligned} \mathcal{Y} &= \mathcal{X} \times_1 \mathbf{H}_1 \times_2 \mathbf{H}_2 \cdots \times_N \mathbf{H}_N \\ &= [\![\mathcal{X}; \mathbf{H}_1, \mathbf{H}_2, \cdots, \mathbf{H}_N]\!], \end{aligned} \tag{13.10}$$

where $\mathbf{H}_n \in \mathbb{R}^{J_n \times I_n}$ is a sparse matrix satisfying $\mathbf{H}_n(\mathrm{h}_n(i_n), i_n) = \mathrm{s}_n(i_n)$ $(n = 1, \cdots, N)$. Specially, when $\mathcal{X} = \mathbf{u}_1 \circ \cdots \circ \mathbf{u}_N \in \mathbb{R}^{I_1 \times \cdots \times I_N}$ is a rank-1 tensor,

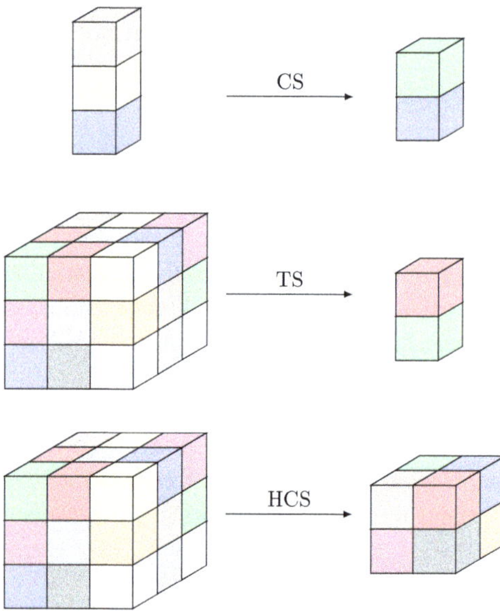

Fig. 13.1 A schematic diagram of CS, TS, and HCS. CS takes an input vector and produces a lower dimensional one. Both TS and HCS take tensors as input. For TS, the output is a vector. For HCS, the output is a lower dimensional tensor of the same order

Table 13.1 Computational and storage complexity of CS, TS, and HCS for a third-order tensor $\mathcal{X} \in \mathbb{R}^{I\times I\times I}$. J is the Hash length

Sketching methods	CS	TS	HCS
$\mathcal{X} = \mathbf{x} \circ \mathbf{x} \circ \mathbf{x}$	$\mathrm{O}(\mathrm{nnz}(\mathcal{X}))$	$\mathrm{O}(\mathrm{nnz}(\mathbf{x}) + J\log J)$	$\mathrm{O}(\mathrm{nnz}(\mathbf{x}) + J^3)$
General tensor $\mathcal{X}$	$\mathrm{O}(\mathrm{nnz}(\mathcal{X}))$	$\mathrm{O}(\mathrm{nnz}(\mathcal{X}))$	$\mathrm{O}(\mathrm{nnz}(\mathcal{X}))$
Storage for Hash functions	$\mathrm{O}(I^3)$	$\mathrm{O}(I)$	$\mathrm{O}(I)$

$\mathcal{Y} = \mathrm{HCS}(\mathcal{X})$ can be computed by the outer product of CS of each factor vector as

$$\mathcal{Y} = \mathrm{CS}_1(\mathbf{u}_1) \circ \cdots \circ \mathrm{CS}_N(\mathbf{u}_N). \tag{13.11}$$

The original tensor can be recovered as

$$\begin{aligned} \hat{\mathcal{X}} &= \mathcal{Y} \times_1 \mathbf{H}_1^{\mathrm{T}} \times_2 \mathbf{H}_2^{\mathrm{T}} \cdots \times_N \mathbf{H}_N^{\mathrm{T}} \\ &= [\![\mathcal{Y}; \mathbf{H}_1^{\mathrm{T}}, \mathbf{H}_2^{\mathrm{T}}, \cdots, \mathbf{H}_N^{\mathrm{T}}]\!]. \end{aligned} \tag{13.12}$$

And the approximation error is bounded by the theorem below.

Theorem 13.2 ([34]) *For any tensor* $\mathcal{X} \in \mathbb{R}^{I_1\times\cdots\times I_N}$, *denote* $\mathcal{X}_n$ *as an n-th order tensor obtained by fixing* $N - n$ *modes of* $\mathcal{X}$. $\hat{\mathcal{X}}$ *is an unbiased estimator of* $\mathcal{X}$ *with the variance of each entry bounded by*

$$\mathrm{Var}(\hat{\mathcal{X}}(i_1, \cdots, i_N)) = \mathrm{O}\left(\sum_{n=1}^{N} \frac{C_n^2}{J^n}\right), \tag{13.13}$$

where C_n denotes the maximum Frobenius norm of all $\mathcal{X}_n$.

HCS has several superiorities over CS and TS aside from the exploitation of multidimensional structure within tensors. Compared to CS, it reduces the storage complexity for Hash functions from $\mathrm{O}(\prod_{n=1}^{N} I_n)$ to $\mathrm{O}(\sum_{n=1}^{N} I_n)$. Compared to TS, operations such as tensor contractions and tensor products can be computed directly in the sketched space. We summarize the computational and storage complexity for different sketching methods in Table 13.1, where we refer a non-rank-1 tensor as the *general tensor*.

13.5 Applications

In this section, we will demonstrate some applications of tensor-based sketching methods in tensor decompositions, Kronecker product regression, and network approximation. Meanwhile, some experiments are conducted for comparison and validation.

It should be noted that the discussions here cannot cover everything. Readers are encouraged to refer to related papers (most with available codes) for more details.

13.5.1 Tensor Sketch for Tensor Decompositions

Sketching serves as an efficient tool to efficiently compute tensor decompositions. The main idea is to approximate the original tensor with dimensionality-reduced sketched components. In such cases, the computation and storage requirements can be greatly reduced without compromising too much accuracy.

13.5.1.1 Tensor Sketch for CP Decomposition

For example, a tensor sketch-based robust tensor power method (RTPM) is proposed for fast CP decomposition [39]. RTPM is an iterative method for CP decomposition proposed in [3]. It obtains one rank-1 component sequentially until the CP rank reaches the upper bound or the approximation error decreases below a threshold. The update of RTPM requires two tensor contractions $\mathcal{T}(\mathbf{u}, \mathbf{u}, \mathbf{u})$ and $\mathcal{T}(\mathbf{I}, \mathbf{u}, \mathbf{u})$, defined as

$$\begin{aligned} \mathcal{T}(\mathbf{u}, \mathbf{u}, \mathbf{u}) &= \langle \mathcal{T}, \mathbf{u} \circ \mathbf{u} \circ \mathbf{u} \rangle \\ \mathcal{T}(\mathbf{I}, \mathbf{u}, \mathbf{u}) &= [\langle \mathcal{T}, \mathbf{e}_1 \circ \mathbf{u} \circ \mathbf{u} \rangle, \cdots, \langle \mathcal{T}, \mathbf{e}_N \circ \mathbf{u} \circ \mathbf{u} \rangle]^{\mathrm{T}}, \end{aligned} \tag{13.14}$$

where $\mathcal{T} \in \mathbb{R}^{I \times I \times I}$ is a third-order symmetric[2]tensor and $\mathbf{u} \in \mathbb{R}^I$ is a vector. $\langle \mathcal{M}, \mathcal{N} \rangle$ denotes the tensor inner product; $\mathbf{e}_n$ is the standard basis vector along the n-th mode.

The naive implementation of RTPM is not efficient enough, since it requires $\mathrm{O}(NLTI^3)$ time and $\mathrm{O}(I^3)$ storage [39], where L and T are parameters with regard to iterations and N is the target CP rank. The major computation burden lies in the calculation of the two tensor contractions in (13.14). Therefore, a fast RTPM (FRTPM) is developed based on TS. In[39] it is proved that the inner product of any two tensors can be approximated by

$$\langle \mathcal{M}, \mathcal{N} \rangle \approx \langle \mathrm{TS}(\mathcal{M}), \mathrm{TS}(\mathcal{N}) \rangle\,. \tag{13.15}$$

---

[2]For simplicity we focus on symmetric tensor here, i.e., $\mathcal{T}_{i,j,k}$ remains the same for all permutations of i, j, k. An extension to asymmetric case can be seen in [4].

Based on (13.15), the two tensor contractions can be approximated by

$$\begin{aligned}\mathcal{T}(\mathbf{u},\mathbf{u},\mathbf{u}) &\approx \langle \mathrm{TS}(\mathcal{T}), \mathrm{TS}(\mathbf{u}\circ\mathbf{u}\circ\mathbf{u})\rangle \\ &= \left\langle \mathrm{TS}(\mathcal{T}), \mathrm{F}^{-1}(\mathrm{F}(\mathrm{CS}_1(\mathbf{u})) \circledast \mathrm{F}(\mathrm{CS}_2(\mathbf{u})) \circledast \mathrm{F}(\mathrm{CS}_3(\mathbf{u})))\right\rangle\end{aligned} \tag{13.16}$$

$$\begin{aligned}\mathcal{T}(\mathbf{I},\mathbf{u},\mathbf{u})_i &\approx \langle \mathrm{TS}(\mathcal{T}), \mathrm{TS}(\mathbf{e}_i\circ\mathbf{u}\circ\mathbf{u})\rangle \\ &= \left\langle \mathrm{F}^{-1}(\mathrm{F}(\mathrm{TS}(\mathcal{T})) \circledast \mathrm{F}(\mathrm{CS}_2(\mathbf{u}))^{\mathrm{H}} \circledast \mathrm{F}(\mathrm{CS}_3(\mathbf{u})^{\mathrm{H}})), \mathrm{CS}_1(\mathbf{e}_i)\right\rangle \\ &:= \langle \mathbf{s}, \mathrm{CS}_1(\mathbf{e}_i)\rangle \\ &= \mathrm{s}_1(i)\mathbf{s}(\mathrm{h}_1(i)),\end{aligned} \tag{13.17}$$

where (13.17) holds due to the unitary property of FFT. Both $\mathrm{TS}(\mathcal{T})$ and $\mathbf{s}$ are irrelevant to i and can be calculated beforehand. Both (13.16) and (13.17) can be computed in $\mathrm{O}(I + J\log J)$ time. FRTPM requires $\mathrm{O}(NLT(I + DJ\log J))$ time and $\mathrm{O}(DJ)$ storage, where D denotes the number of independent sketches and J denotes the sketching dimension. In this way, the computation and space complexity can be greatly reduced in comparison with the naive RTPM.

To validate the effectiveness of FRTPM, we conduct experiments on a synthetic symmetric tensor $\mathcal{T} \in \mathbb{R}^{100\times100\times100}$ and a real-world asymmetric dataset. For the synthetic data-based experiment, the approximation performance is evaluated by the residual norm, which is defined as

$$\mathrm{RN}(\mathcal{T}, \hat{\mathcal{T}}) = \|\mathcal{T} - \hat{\mathcal{T}}\|_{\mathrm{F}},$$

where $\hat{\mathcal{T}}$ is the recovered tensor. For the real-world data-based experiment, we extract the background of the video dataset Shop[3]and use it as the input tensor and use peak signal-to-noise ratio (PSNR) for evaluation. For both synthetic and the real-world datasets, the CP rank for decomposition is set to be 10. All input, output, and sketch building times are excluded since they are one-pass.

The comparison results of RTPM and FRTPM on synthetic and real-world tensors are in Tables 13.2 and 13.3, respectively. We also show the low-rank approximation of the dataset Shop in Fig. 13.2. We can clearly see that FRTPM yields faster computation with an acceptable approximation error. Also as the sketching dimension number increases, the approximation becomes more accurate.

[3]https://github.com/andrewssobral/lrslibrary.

Table 13.2 Residual norm and running time on rank-10 approximation of a synthetic symmetric tensor $\mathcal{T} \in \mathbb{R}^{100\times100\times100}$ by RTPM and FRTPM. σ denotes the level of Gaussian noise

			Residual norm				Running time (s)			
		Sketch dim	1500	2000	2500	3000	1500	2000	2500	3000
$\sigma = 0.01$	FRTPM	D=2	1.0070	0.9991	0.9683	0.7765	1.1099	1.2477	1.8701	2.2574
		D=4	0.9265	0.6641	0.5702	0.5171	2.2053	2.6157	3.8553	4.1976
		D=6	0.6114	0.5590	0.4783	0.4294	3.4393	3.6519	5.2013	5.5934
	RTPM	/	0.0998				49.4443			
$\sigma = 0.1$	FRTPM	D=2	1.0503	1.0506	1.0388	0.8760	1.1551	1.2745	1.9331	2.1459
		D=4	1.0372	0.7568	0.6922	0.5990	2.1082	2.4586	3.5557	4.0272
		D=6	0.7579	0.6371	0.5860	0.5382	3.1546	3.7927	5.3187	5.4595
	RTPM	/	0.3152				50.5022			

Table 13.3 PSNR and running time on rank-10 approximation of dataset Shop by RTPM and FRTPM

		PSNR					Running time (s)				
	Sketch dim	5000	5500	6000	6500	7000	5000	5500	6000	6500	7000
FRTPM	D=5	15.5824	20.2086	20.4758	19.0921	21.2039	4.9433	5.1242	5.8073	5.7232	6.0740
	D=10	23.7788	24.2276	24.5138	24.9739	24.8809	9.1855	9.9487	10.7859	11.4765	14.4035
	D=15	24.8358	25.1802	25.0410	25.5410	25.7796	16.2121	18.1471	20.1453	21.3052	22.7188
RTPM	/	28.9158					268.9146				

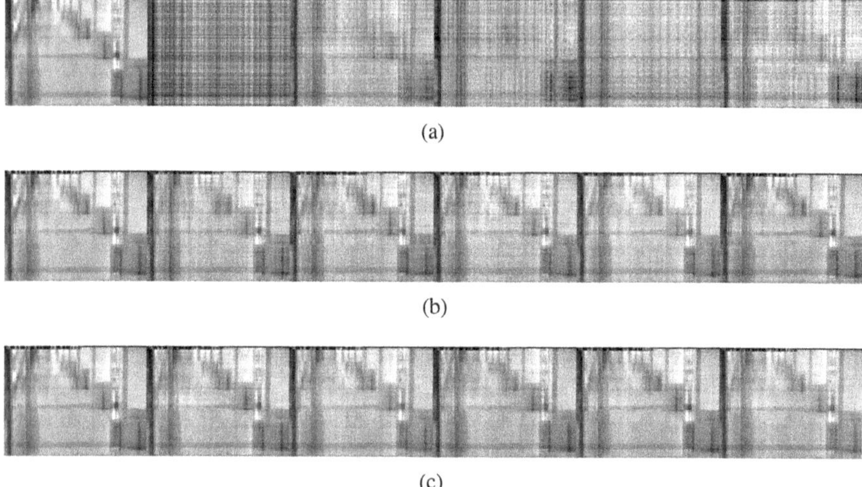

Fig. 13.2 Rank-10 approximation on dataset Shop. The first column: RTPM (ground truth). The second to eighth columns: sketching dimension of 5000, 5500, $\cdots$, 7000, respectively. (**a**) $D = 5$. (**b**) $D = 10$. (**c**) $D = 15$

13.5.1.2 Tensor Sketch for Tucker Decomposition

Consider the Tucker decomposition of a tensor $\mathcal{X} \in \mathbb{R}^{I_1 \times \cdots \times I_N}$:

$$\mathcal{X} = \mathcal{G} \times_1 \mathbf{U}^{(1)} \times_2 \cdots \times_N \mathbf{U}^{(N)} = [\![\mathcal{G}; \mathbf{U}^{(1)}, \cdots, \mathbf{U}^{(N)}]\!], \tag{13.18}$$

where $\mathcal{G} \in \mathbb{R}^{R_1 \times \cdots \times R_N}$ is the core tensor and $\mathbf{U}^{(n)} \in \mathbb{R}^{I_n \times R_n}$ $(n = 1, \cdots, N)$ are factor matrices. Higher-order orthogonal iteration (HOOI) is a commonly used algorithm for Tucker decomposition (see Algorithm 4 in Chap. 2). The standard HOOI involves solving two alternating least squares (ALS) problems iteratively:

$$\begin{aligned} \mathbf{U}^{(n)} &= \underset{\mathbf{U}}{\arg\min} \| [\![\mathcal{G}; \mathbf{U}^{(1)}, \cdots, \mathbf{U}^{(n-1)}, \mathbf{U}, \mathbf{U}^{(n+1)}, \cdots, \mathbf{U}^{(N)}]\!] - \mathcal{X} \|_{\mathrm{F}}^2 \\ &= \underset{\mathbf{U}}{\arg\min} \|(\otimes_{\substack{i=N \\ i \neq n}}^{1} \mathbf{U}^{(i)}) \mathbf{G}_{(n)}^{\mathrm{T}} \mathbf{U}^{\mathrm{T}} - \mathbf{X}_{(n)}^{\mathrm{T}} \|_{\mathrm{F}}^2, \text{ for } n = 1, \cdots, N \end{aligned} \tag{13.19}$$

$$\begin{aligned} \mathcal{G} &= \underset{\mathcal{Z}}{\arg\min} \| [\![\mathcal{Z}; \mathbf{U}^{(1)}, \cdots, \mathbf{U}^{(N)}]\!] - \mathcal{X} \|_{\mathrm{F}}^2 \\ &= \underset{\mathcal{Z}}{\arg\min} \|(\otimes_{n=N}^{1} \mathbf{U}^{(n)}) \mathrm{vec}(\mathcal{Z}) - \mathrm{vec}(\mathcal{X}) \|_{\mathrm{F}}^2, \end{aligned} \tag{13.20}$$

where $\otimes_{n=N}^{1} \mathbf{U}^{(n)}$ is abbreviated for $\mathbf{U}^{(N)} \otimes \cdots \otimes \mathbf{U}^{(1)}$.

For fast computation, Malik et al. develop two randomized algorithms by approximating (13.19) and (13.20) using tensor sketch in [26]. Their results are based on the notice that least squares regression (LSR) problem

$$\mathbf{x}^* = \arg\min_{\mathbf{x}\in\mathbb{R}^I} \|\mathbf{A}\mathbf{x} - \mathbf{y}\|_{\mathrm{F}} \tag{13.21}$$

can be approximated by

$$\tilde{\mathbf{x}} = \arg\min_{\mathbf{x}\in\mathbb{R}^I} \|\mathbf{TS}(\mathbf{A}\mathbf{x} - \mathbf{y})\|_{\mathrm{F}} \tag{13.22}$$

with $\|\mathbf{A}\tilde{\mathbf{x}} - \mathbf{y}\|_{\mathrm{F}} \leq (1+\epsilon)\|\mathbf{A}\mathbf{x}^* - \mathbf{y}\|_{\mathrm{F}}$ [14]. Therefore, (13.19) and (13.20) can be approximated by

$$\begin{aligned}\mathbf{U}^{(n)} &= \arg\min_{\mathbf{U}} \|\mathrm{TS}^{(n)}(\otimes_{i=N, i\neq n}^{1}\mathbf{U}^{(i)})\mathbf{G}_{(n)}^{\mathrm{T}}\mathbf{U}^{\mathrm{T}} - \mathrm{TS}^{(n)}(\mathbf{X}_{(n)}^{\mathrm{T}})\|_{\mathrm{F}}^2 \\ &= \arg\min_{\mathbf{U}} \|\mathrm{F}^{-1}((\odot_{i=N, i\neq n}^{1}(\mathrm{F}(\mathrm{CS}_1^{(i)}(\mathbf{U}^{(i)})))^{\mathrm{T}})^{\mathrm{T}})\mathbf{G}_{(n)}^{\mathrm{T}}\mathbf{U}^{\mathrm{T}} \\ &\quad - \mathrm{TS}^{(n)}(\mathbf{X}_{(n)}^{\mathrm{T}})\|_{\mathrm{F}}^2\end{aligned} \tag{13.23}$$

$$\begin{aligned}\mathcal{G} &= \arg\min_{\mathcal{Z}} \|\mathrm{TS}^{(N+1)}(\otimes_{n=N}^{1}\mathbf{U}^{(n)})\mathrm{vec}(\mathcal{Z}) - \mathrm{TS}^{(N+1)}(\mathrm{vec}(\mathcal{X}))\|_{\mathrm{F}}^2 \\ &= \arg\min_{\mathcal{Z}} \|\mathrm{F}^{-1}((\odot_{n=N}^{1}(\mathrm{F}(\mathrm{CS}_2^{(n)}(\mathbf{U}^{(n)})))^{\mathrm{T}})^{\mathrm{T}})\mathrm{vec}(\mathcal{Z}) \\ &\quad - \mathrm{TS}^{(N+1)}(\mathrm{vec}(\mathcal{X}))\|_{\mathrm{F}}^2,\end{aligned} \tag{13.24}$$

where $\odot_{n=N}^{1}\mathbf{U}^{(n)}$ is short for $\mathbf{U}^{(N)} \odot \cdots \odot \mathbf{U}^{(1)}$, $\mathrm{CS}_1^{(n)} : \mathbb{R}^{I_n} \to \mathbb{R}^{J_1}$ and $\mathrm{CS}_2^{(n)} : \mathbb{R}^{I_n} \to \mathbb{R}^{J_2}$ ($J_1 < J_2$, $n = 1, \cdots, N$) are two different count sketch operators. $\mathrm{TS}^{(n)} : \mathbb{R}^{\prod_{i\neq n}^{I_i}} \to \mathbb{R}^{J_1}$ and $\mathrm{TS}^{(N+1)} : \mathbb{R}^{\prod_n^{I_n}} \to \mathbb{R}^{J_2}$ are two different tensor sketch operators built on $\mathrm{CS}_1^{(n)}$ and $\mathrm{CS}_2^{(n)}$, respectively.

Substituting (13.23) and (13.24) into Algorithm 4, we obtain the first algorithm named Tucker-TS, summarized in Algorithm 54.

The second algorithm Tucker-TTMTS applies tensor sketch to a given tensor which can be represented in chains of n-mode tensor-times-matrix form (TTM). Recall that the standard HOOI requires computing

$$\mathcal{Y} = \mathcal{X} \times_1 \mathbf{U}^{(1)\mathrm{T}} \cdots \times_{n-1} \mathbf{U}^{(n-1)\mathrm{T}} \times_{n+1} \mathbf{U}^{(n+1)\mathrm{T}} \cdots \times_N \mathbf{U}^{(N)\mathrm{T}}, \tag{13.25}$$

which can be reformed as

$$\mathbf{Y}_{(n)} = \mathbf{X}_{(n)} \otimes_{i=N, i\neq n}^{1} \mathbf{U}^{(i)}. \tag{13.26}$$

Algorithm 54: Tucker-TS

Input: $\mathcal{X} \in \mathbb{R}^{I_1 \times I_2 \times \cdots \times I_N}$, target rank $(R_1, R_2, \cdots, R_N)$, sketching dimensions (J_1, J_2)
Output: $\mathcal{G}, \mathbf{U}^{(1)}, \mathbf{U}^{(2)}, \cdots, \mathbf{U}^{(N)}$
initialize $\mathbf{U}^{(n)} \in R^{I_n \times R_n}$ for $n = 2, \cdots, N$ using HOSVD
Construct $\text{TS}^{(n)} : \mathbb{R}^{\prod_{i \neq n}^{I_i}} \rightarrow \mathbb{R}^{J_1}$ for $n \in [N]$ and and $\text{TS}^{(N+1)} : \mathbb{R}^{\prod_{n}^{I_n}} \rightarrow \mathbb{R}^{J_2}$
repeat
  For $n = 1, \cdots, N$ **do**
    Compute $\mathbf{U}^{(n)}$ by (13.23),
  End for
  Compute $\mathcal{G}$ by (13.24);
  Orthogonalize each $\mathbf{U}^{(n)}$ and update $\mathcal{G}$
until fit ceases to improve or maximum iterations exhausted

Applying tensor sketch, it becomes

$$\tilde{\mathbf{Y}}_{(n)} = (\text{TS}^{(n)}(\mathbf{X}_{(n)}^{\text{T}}))^{\text{T}}\text{TS}^{(n)}(\otimes_{\substack{i=N \\ i \neq n}}^{1} \mathbf{U}^{(i)}). \tag{13.27}$$

Similarly

$$\text{vec}(\mathcal{G}) = (\text{TS}^{(N+1)}(\otimes_{i=N}^{1} \mathbf{U}^{(i)}))^{\text{T}}\text{TS}^{(N+1)}(\text{vec}(\mathcal{X})). \tag{13.28}$$

We summarize it in Algorithm 55.

Algorithm 55: Tucker-TTMTS

Input: $\mathcal{X} \in \mathbb{R}^{I_1 \times I_2 \times \cdots \times I_N}$, target rank $(R_1, R_2, \cdots, R_N)$, sketching dimensions (J_1, J_2)
Output: $\mathcal{G}, \mathbf{U}^{(1)}, \mathbf{U}^{(2)}, \cdots, \mathbf{U}^{(N)}$
initialize $\mathbf{U}^{(n)} \in R^{I_n \times R_n}$ for $n = 2, \cdots, N$ using HOSVD
Construct $\text{TS}^{(n)} : \mathbb{R}^{\prod_{i \neq n}^{I_i}} \rightarrow \mathbb{R}^{J_1}$ for $n \in [N]$ and and $\text{TS}^{(N+1)} : \mathbb{R}^{\prod_{n}^{I_n}} \rightarrow \mathbb{R}^{J_2}$
repeat
  For $n = 1, \cdots, N$ **do**
    Compute $\mathbf{Y}_{(n)}$ by (13.27),
  End for
  Compute $\mathcal{G}$ by (13.28);
until fit ceases to improve or maximum iterations exhausted

To validate the effectiveness of tensor sketch in Tucker decomposition, we use the codes provided by [26] for low-rank approximation of both sparse and dense tensors. For the sparse data-based experiment, we use a synthetic tensor, while for the dense data-based experiment, we use the extracted background of real-world dataset Escalator.[4] All other configuration is set by default.

[4] https://github.com/andrewssobral/lrslibrary.

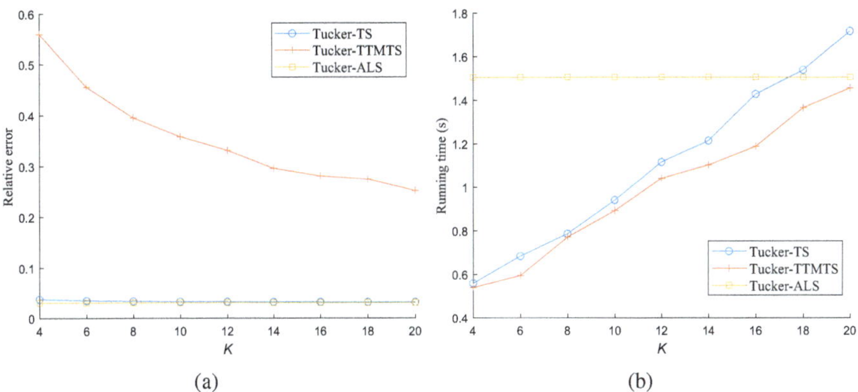

(a)

(b)

Fig. 13.3 Comparison of Tucker-TS, Tucker-TTMTS, and Tucker-ALS on a rank (10, 10, 10) synthetic sparse tensor $\mathcal{T} \in \mathbb{R}^{500\times 500\times 500}$. K is a parameter with regard to the sketching dimension. (**a**) Relative norm against different values of K. (**b**) Running time against different values of K

Fig. 13.4 Rank-10 Tucker-TS approximation on dataset escalator. The first column: Tucker-ALS (ground truth). The second to ninth columns: Tucker-TS with $K = 4, 6, \cdots, 20$

In Fig. 13.3, we compare the relative error and running time of Tucker-TS, Tucker-TTMTS, and Tucker-ALS on the sparse synthetic tensor $\mathcal{T} \in \mathbb{R}^{500\times 500\times 500}$ with Tucker rank (10, 10, 10). It shows that Tucker-TS achieves similar relative norm compared to Tucker-ALS. On the other hand, Tucker-TTMTS is the fastest at the cost of higher relative error.

Figure 13.4 displays the Rank-10 approximation of Tucker-ALS and Tucker-TS on dataset Escalator. It shows that the approximation is acceptable even when the sketching parameter K is small and grows better as K increases.

13.5.2 Tensor Sketch for Kronecker Product Regression

Recall the over-determined least squares problem $\min_{\mathbf{x}\in\mathbb{R}^I} \|\mathbf{A}\mathbf{x} - \mathbf{y}\|_F$ at the beginning of this chapter. In [14], Diao et al. consider the context when $\mathbf{A}$ is a Kronecker product of some matrices, e.g., $\mathbf{A} = \mathbf{B} \otimes \mathbf{C}$. By applying tensor sketch, one obtains $\tilde{\mathbf{x}} = \arg\min_{\mathbf{x}\in\mathbb{R}^I} \|\mathbf{TS}((\mathbf{B} \otimes \mathbf{C})\mathbf{x} - \mathbf{y})\|_F$ with $\|\mathbf{A}\tilde{\mathbf{x}} - \mathbf{y}\|_F \leq (1+\epsilon)\|\mathbf{A}\mathbf{x}^* - \mathbf{y}\|_F$ in $\mathrm{O}(\mathrm{nnz}(\mathbf{B}) + \mathrm{nnz}(\mathbf{C}) + \mathrm{nnz}(\mathbf{y}) + \mathrm{poly}(N_1 N_2/\epsilon))$ time, where N_1, N_2 are column size of $\mathbf{B}$, $\mathbf{C}$, respectively, $\mathrm{poly}(\cdot)$ denotes some low-order polynomial.

Considering more general cases, given $\mathbf{A}^{(n)} \in \mathbb{R}^{P_n \times Q_n}$ for $n = 1, \cdots, N$, we denote $\mathbf{A} = \mathbf{A}^{(1)} \otimes \cdots \otimes \mathbf{A}^{(N)} \in \mathbb{R}^{P\times Q}$ with $P = \prod_{n=1}^N P_n$, $Q = \prod_{n=1}^N Q_n$.

Based on the fact that applying Tensor sketch to the Kronecker product of some matrices equals convolving the count sketches of them [5], we can accelerate the computation using FFT. Therefore, for $q_n \in [Q_n], n = 1, \cdots, N$, the tensor sketch of each column of $\mathbf{A}$ is

$$\begin{aligned} &\mathrm{TS}(\mathbf{A}^{(1)}(:, q_1) \otimes \cdots \otimes \mathbf{A}^{(N)}(:, q_N)) \\ &= \mathrm{F}^{-1}(\mathrm{F}(\mathrm{CS}_1(\mathbf{A}^{(1)}(:, q_1)) \circledast \cdots \circledast \mathrm{F}(\mathrm{CS}_N(\mathbf{A}^{(N)}(:, q_N)))). \end{aligned} \tag{13.29}$$

Since $\mathbf{A}$ has Q columns, the computation for $\mathrm{TS}(\mathbf{A})$ takes the time $\mathrm{O}(Q \sum_{n=1}^{N} \mathrm{nnz}(\mathbf{A}^{(n)}) + NQJ\log(J))$, while the explicit calculation of the Kronecker product requires at least $\mathrm{O}(P)$ time. It can be proved that when $J = 8(Q+1)^2(2+3^N)/(\epsilon^2\delta)$ [14]

$$\mathrm{Prob}(\|\mathbf{A}\tilde{\mathbf{x}} - \mathbf{y}\|_{\mathrm{F}} \leq (1+\epsilon) \min_{\mathbf{x}\in\mathbb{R}^Q} \|\mathbf{A}\mathbf{x} - \mathbf{y}\|_{\mathrm{F}}) \geq 1 - \delta. \tag{13.30}$$

The TS-Kronecker product regression is summarized in Algorithm 56.

Algorithm 56: TS-Kronecker product regression

Input: $\mathbf{A} = \mathbf{A}^{(1)} \otimes \cdots \otimes \mathbf{A}^{(N)} \in \mathbb{R}^{P\times Q}$, $\mathbf{y} \in \mathbb{R}^P$, ϵ, δ
Set $J = 8(Q+1)^2(2+3^N)/(\epsilon^2\delta)$
Construct Tensor Sketch matrix $\mathrm{TS} \in \mathbb{R}^{J\times P}$
Compute $\tilde{\mathbf{x}} \leftarrow \min_{\mathbf{x}\in\mathbb{R}^Q} \|\mathrm{TS}(\mathbf{A}\mathbf{x} - \mathbf{y})\|_{\mathrm{F}}$
Output: $\tilde{\mathbf{x}}$

To test the algorithm performance, we use the codes in [27] on MNIST dataset [23]. The experiment aims to see how much the distance between handwritten digits can be preserved by TS. After preprocessing (see [27] for details), we obtain a rank-10 CPD-based approximation of a tensor $\mathcal{A} = [\![\mathbf{A}^{(1)}, \mathbf{A}^{(2)}, \mathbf{A}^{(3)}]\!] \in \mathbb{R}^{32\times 32\times 100}$ containing handwritten "4"s and another tensor $\mathcal{B} = [\![\mathbf{B}^{(1)}, \mathbf{B}^{(2)}, \mathbf{B}^{(3)}]\!] \in \mathbb{R}^{32\times 32\times 100}$ for handwritten "9"s. The reason for choosing this particular pair of digits is that handwritten "4"s and "9"s can be hard to distinguish, especially after the low-rank approximation.

Denote $\mathbf{A} := \mathbf{A}^{(1)} \odot \mathbf{A}^{(2)} \odot \mathbf{A}^{(3)} \in \mathbb{R}^{102400\times 10}$ and $\mathbf{B} := \mathbf{B}^{(1)} \odot \mathbf{B}^{(2)} \odot \mathbf{B}^{(3)} \in \mathbb{R}^{102400\times 10}$. We take

$$\left| \frac{\|\mathrm{TS}(\mathbf{A} - \mathbf{B})\|_{\mathrm{F}}}{\|\mathcal{A} - \mathcal{B}\|_{\mathrm{F}}} - 1 \right| \tag{13.31}$$

as the performance metric. The result is shown in Fig. 13.5. Clearly we can see along with the increasing sketching dimension, the estimation becomes more accurate.

In [14], similar extensive work on regularized spline regression is presented. However, this is out of the scope for our discussion.

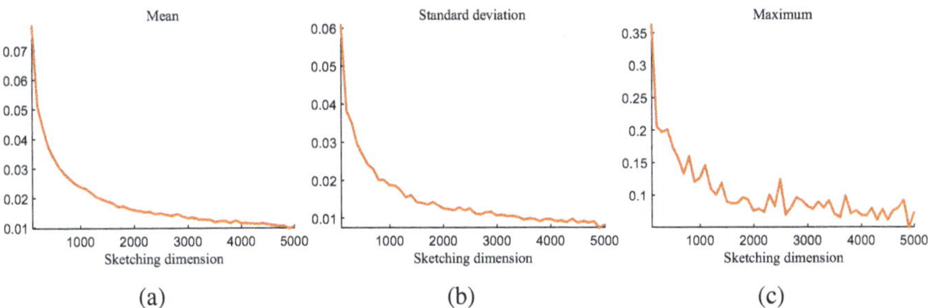

Fig. 13.5 The performance in terms of the metric of (13.31) against the increasing sketching dimension characterized by the (**a**) mean value, (**b**) standard deviation and (**c**) maximum value over 1000 trials

13.5.3 Tensor Sketch for Network Approximations

Deep neural network receives a lot of attentions in many fields due to its powerful representation capability (see Chaps. 10 and 11). However, with the networks growing deeper, it brings a large number of network parameters, which would lead to a huge computational and storage burden. Therefore, it is helpful if we can approximate the original networks with smaller ones while preserving the intrinsic structures. In this subsection, we introduce two works using tensor sketch for network approximations.

13.5.3.1 Tensor Sketch in Convolutional and Fully Connected Layers

The traditional convolutional (CONV) and fully connected (FC) layers of convolutional neural network (CNN) can be represented by

$$\begin{aligned} \mathcal{I}_{\text{out}} &= \text{Conv}(\mathcal{I}_{\text{in}}, \mathcal{K}) \\ \mathbf{h}_{\text{out}} &= \mathbf{W}\mathbf{h}_{\text{in}} + \mathbf{b}, \end{aligned} \tag{13.32}$$

where $\mathcal{I}_{\text{in}} \in \mathbb{R}^{H_1 \times W_1 \times I_1}$ and $\mathcal{I}_{\text{out}} \in \mathbb{R}^{H_2 \times W_2 \times I_2}$ are the tensors before and after the CONV layer, respectively, and $\mathbf{h}_{\text{in}} \in \mathbb{R}^{I_1}$ and $\mathbf{h}_{\text{out}} \in \mathbb{R}^{I_2}$ are the vectors before and after the FC layer, respectively. $\mathcal{K} \in \mathbb{R}^{I_1 \times H \times W \times I_2}$ and $\mathbf{W} \in \mathbb{R}^{I_2 \times I_1}$ denote the convolutional kernel and the weight matrix, respectively. When the input dimensions of $\mathcal{I}_{\text{in}}$ are high, the high computation and storage requirement of CNNs can bring burden for hardware implementation.

In [21], it is shown the operations in (13.32) can be approximated by a randomized tensor sketch approach. Specifically, the mode-n sketch of tensor $\mathcal{X} \in \mathbb{R}^{I_1 \times \cdots \times I_N}$ is defined by the mode-n tensor-matrix product as

$$\mathcal{S}^{(n)} = \mathcal{X} \times_n \mathbf{U}, \tag{13.33}$$

where $\mathbf{U} \in \mathbb{R}^{K \times I_n}$ $(K \ll I_n)$ is a random scaled sign matrix [21].

Then the sketched convolutional layer (SK-CONV) is constructed by

$$\hat{\mathcal{I}}_{\text{out}} = \frac{1}{2D}\sum_{d=1}^{D}\text{Conv}(\mathcal{I}_{\text{in}},\ \mathcal{S}_1^{(d)} \times_4 \mathbf{U}_1^{(d)^{\text{T}}} + \mathcal{S}_2^{(d)} \circledcirc \mathbf{U}_2^{(d)^{\text{T}}}), \tag{13.34}$$

where $\mathcal{S}_1^{(d)} \in \mathbb{R}^{I_1 \times H \times W \times K}$ and $\mathcal{S}_2^{(d)} \in \mathbb{R}^{K \times H \times W \times I_2}$ are two convolutional kernels and $\mathbf{U}_1^{(d)} \in \mathbb{R}^{K \times I_2}$ and $\mathbf{U}_2^{(d)} \in \mathbb{R}^{KHW \times I_1 HW}$ are two independent random scaled sign matrices for $d \in [D]$. The estimation is made more robust by taking D sketches and computing the average. $\mathcal{S}_2^{(d)} \circledcirc \mathbf{U}_2^{(d)^{\text{T}}} \in \mathbb{R}^{I_1 \times H \times W \times I_2}$ denotes a tensor contraction computed element-wisely by

$$(\mathcal{S}_2^{(d)} \circledcirc \mathbf{U}_2^{(d)^{\text{T}}})(p, q, r, s) = \sum_{k=1}^{K}\sum_{h=1}^{H}\sum_{w=1}^{W} \mathcal{S}_2^{(d)}(k, h, w, s)\mathbf{U}_2^{(d)}(khw, pqr). \tag{13.35}$$

And $\hat{\mathcal{I}}_{\text{out}}$ is proved to be an unbiased estimator of $\mathcal{I}_{\text{out}}$ with bounded variance [21]. The sketched FC layer (SK-FC) can be built in a similar way as follows:

$$\begin{aligned} \hat{\mathbf{h}}_{\text{out}} &= \frac{1}{2D}\sum_{d=1}^{D}(\mathbf{S}_1^{(d)} \times_1 \mathbf{U}_1^{(d)^{\text{T}}} + \mathbf{S}_2^{(d)} \times_2 \mathbf{U}_2^{(d)^{\text{T}}})\mathbf{h}_{\text{in}} + \mathbf{b} \\ &= \frac{1}{2D}\sum_{d=1}^{D}(\mathbf{U}_1^{(d)^{\text{T}}}\mathbf{S}_1^{(d)} + \mathbf{S}_2^{(d)}\mathbf{U}_2^{(d)})\mathbf{h}_{\text{in}} + \mathbf{b}, \end{aligned} \tag{13.36}$$

where $\mathbf{S}_1^{(d)} \in \mathbb{R}^{K \times I_1}$, $\mathbf{S}_2^{(d)} \in \mathbb{R}^{I_2 \times K}$ are two weight matrices. $\mathbf{U}_1^{(d)} \in \mathbb{R}^{K \times I_2}$ and $\mathbf{U}_2^{(d)} \in \mathbb{R}^{K \times I_1}$ are two independent random scaled sign matrices for $d \in [D]$. $\hat{\mathbf{h}}_{\text{out}}$ is also unbiased with bounded variance [21].

We summarize the number of parameters and computation time for SK-CONV, SK-FC, and the corresponding baselines in Table 13.4.

Table 13.4 Number of parameters and computation time for SK-CONV and SK-FC and the corresponding baselines (assuming $KN \leq I_1 I_2/(I_1 + I_2)$ holds)

Layer name	Number of parameters	Computation time
CONV	HWI_1I_2	$\text{O}(H_2W_2I_1I_2)$
SK-CONV	$DHWK(I_1 + I_2)$	$\text{O}(DH_2W_2K(I_1 + I_2))$
FC	I_1I_2	$\text{O}(I_1I_2)$
SK-FC	$DK(I_1 + I_2)$	$\text{O}(DK(I_1 + I_2))$

13.5.3.2 Higher-Order Count Sketch for Tensor Regression Network

In [22], tensor regression layer (TRL) is proposed to replace the flattening operation and fully connected layers of traditional CNNs with learnable low-rank regression weights. Given the input activation $\mathcal{X} \in \mathbb{R}^{I_0 \times I_1 \times \cdots \times I_N}$, regression weight tensor $\mathcal{W} \in \mathbb{R}^{I_1 \times I_2 \times \cdots \times I_N \times I_{N+1}}$ with I_0 and I_{N+1} denote the batch size and the number of classification classes, respectively; the prediction output $\mathbf{Y} \in \mathbb{R}^{I_0 \times I_{N+1}}$ is computed element-wisely by

$$\mathbf{Y}(i, j) = \left\langle \mathbf{X}_{[1]}(i, :)^{\mathrm{T}}, \mathbf{W}_{[N+1]}(j, :)^{\mathrm{T}} \right\rangle + b. \tag{13.37}$$

In [34], Yang et al. show that HCS can be applied to further sketch the weights and input activation before regression is performed. More specifically, suppose the tensor weight admits Tucker decomposition $\mathcal{W} = [\![\mathcal{G}; \mathbf{U}^{(1)}, \mathbf{U}^{(2)}, \cdots, \mathbf{U}^{(N+1)}]\!]$, then HCS can be applied by

$$\begin{aligned} \mathrm{HCS}(\mathcal{W}) &= [\![\mathcal{G}; \mathrm{HCS}(\mathbf{U}^{(1)}), \mathrm{HCS}(\mathbf{U}^{(2)}), \cdots, \mathbf{U}^{(N+1)}]\!] \\ &= [\![\mathcal{G}; \mathbf{H}_1\mathbf{U}^{(1)}, \mathbf{H}_2\mathbf{U}^{(2)}, \cdots, \mathbf{U}^{(N+1)}]\!]. \end{aligned} \tag{13.38}$$

where the construction of $\mathbf{H}_n (n = 1, 2, \ldots, N)$ is introduced in Sect. 13.4. Notice that the final mode of $\mathcal{W}$ is not sketched so as to match the number of classification classes. Therefore, (13.37) can be approximated by

$$\tilde{\mathbf{Y}} = \mathrm{HCS}(\mathcal{X})_{[1]}^{\mathrm{T}} \mathrm{HCS}(\mathcal{W})_{[N+1]} + \mathbf{b}. \tag{13.39}$$

We compare the performance of TRL and HCS-TRL models on dataset FMNIST [43] ($I_{N+1} = 10$) based on the online codes.[5]The TRL model is composed of two convolutional and maxpooling layers. Again, we assume the regression weight tensor admits Tucker decomposition with the target Tucker rank set as (4, 4, 10, 10). By default, the input activation $\mathcal{X}$ fed to the TRL is of size $I_0 \times 7 \times 7 \times 32$. The prediction output $\mathbf{Y} \in \mathbb{R}^{I_0 \times 10}$ is computed by (13.37). The number of network parameters of the regression layer is 15,690.

For the HCS-TRL model, we set the sketching dimensions as (3, 3, 10) and compute the prediction by (13.39). The number of network parameters reduces to $3 \times 3 \times 10 \times 10 + 10 = 910$, and the corresponding compression ratio is 5.8%.

We summarize the two network structures in Fig. 13.6 and comparison results in Table 13.5. It shows that HCS-TRL compresses the network parameters almost without affecting the classification accuracy.

[5]https://github.com/xwcao/LowRankTRN.

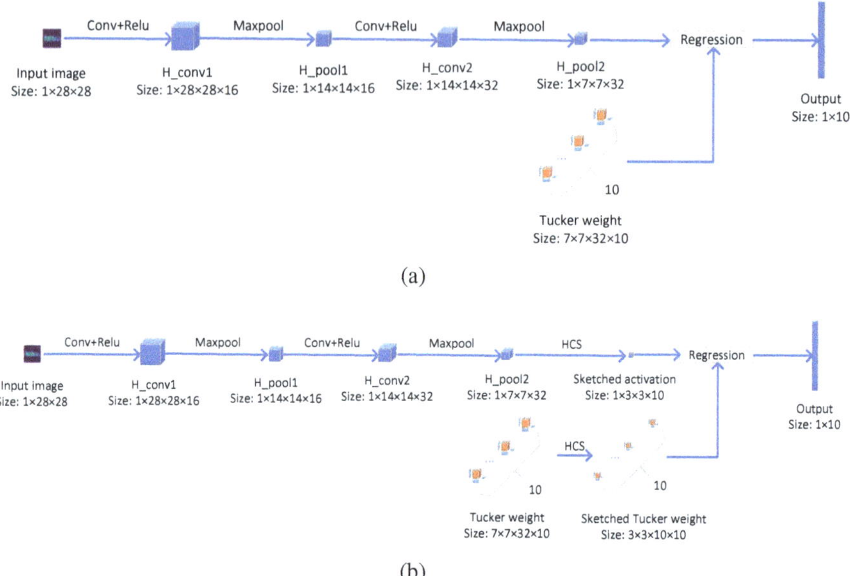

Fig. 13.6 Network structures of the TRL (**a**) and HCS-TRL (**b**). The batch size is set to 1 for ease of illustration

Table 13.5 The comparison of TRL and HCS-TRL on dataset FMNIST

Model	Compression ratio (%)	Test accuracy (%)
TRL	/	87.08
HCS-TRL	5.8	85.39

13.6 Conclusion

In this chapter, we have introduced some fundamentals of an efficient dimensionality reduction technique dubbed *sketching*. It can significantly accelerate the computation progress and reduce the storage requirements in many data processing techniques. Starting with count sketch, we mainly focus on generalized sketching methods for tensor data and their applications in tensor decomposition, tensor regression, and tensor neural network.

Sketching also serves as an important tool in the context of kernel PCA [6], sparse approximation [42], and so forth. Latest studies have also successfully combined sketching with adaptive sampling [25], parallel factorization [45], graph signal processing [17], etc.

Although the sketching method is widely used, there are still limitations in complex scenes. For example, sketching methods are inapplicable for multiple, adaptively chosen queries[41]. Besides, as stated before, the quality of sketching is almost independent with the input structure, which is a double-edged sword: it offers worst-case guarantees but cannot further enhance performance for structured inputs. Therefore, learning-based sketches also open a fresh research domain for enhancing the performance of sketching methods for data of interest [19, 20, 24].

References

1. Ahle, T.D., Kapralov, M., Knudsen, J.B., Pagh, R., Velingker, A., Woodruff, D.P., Zandieh, A.: Oblivious sketching of high-degree polynomial kernels. In: Proceedings of the Fourteenth Annual ACM-SIAM Symposium on Discrete Algorithms, pp. 141–160. SIAM, Philadelphia (2020)
2. Ailon, N., Chazelle, B.: Approximate nearest neighbors and the fast johnson-lindenstrauss transform. In: Proceedings of the Thirty-Eighth Annual ACM Symposium on Theory of Computing, pp. 557–563 (2006)
3. Anandkumar, A., Ge, R., Hsu, D., Kakade, S., Telgarsky, M.: Tensor decompositions for learning latent variable models. J. Mach. Learn. Res. **15**, 2773–2832 (2014)
4. Anandkumar, A., Ge, R., Janzamin, M.: Guaranteed non-orthogonal tensor decomposition via alternating rank-1 updates (2014, preprint). arXiv:1402.5180
5. Avron, H., Nguyen, H., Woodruff, D.: Subspace embeddings for the polynomial kernel. In: Ghahramani, Z., Welling, M., Cortes, C., Lawrence, N., Weinberger, K.Q. (eds.) Advances in Neural Information Processing Systems, vol. 27. Curran Associates, Red Hook (2014). https://proceedings.neurips.cc/paper/2014/file/b571ecea16a9824023ee1af16897a582-Paper.pdf
6. Avron, H., Nguyundefinedn, H.L., Woodruff, D.P.: Subspace embeddings for the polynomial kernel. In: Proceedings of the 27th International Conference on Neural Information Processing Systems - Volume 2, NIPS'14, pp. 2258–2266. MIT Press, Cambridge (2014)
7. Bhojanapalli, S., Sanghavi, S.: A new sampling technique for tensors (2015, preprint). arXiv:1502.05023
8. Bingham, E., Mannila, H.: Random projection in dimensionality reduction: applications to image and text data. In: Proceedings of the Seventh ACM SIGKDD International Conference on Knowledge Discovery and Data Mining, pp. 245–250 (2001)
9. Bringmann, K., Panagiotou, K.: Efficient sampling methods for discrete distributions. In: International Colloquium on Automata, Languages, and Programming, pp. 133–144. Springer, Berlin (2012)
10. Charikar, M., Chen, K., Farach-Colton, M.: Finding frequent items in data streams. Autom. Lang. Program. **2380**, 693–703 (2002)
11. Clarkson, K.L., Woodruff, D.P.: Low-rank approximation and regression in input sparsity time. J. ACM **63**(6), 1–45 (2017)
12. Demaine, E.D., López-Ortiz, A., Munro, J.I.: Frequency estimation of internet packet streams with limited space. In: European Symposium on Algorithms, pp. 348–360. Springer, Berlin (2002)
13. Diao, H., Jayaram, R., Song, Z., Sun, W., Woodruff, D.: Optimal sketching for kronecker product regression and low rank approximation. In: Advances in Neural Information Processing Systems, pp. 4737–4748 (2019)
14. Diao, H., Song, Z., Sun, W., Woodruff, D.P.: Sketching for kronecker product regression and p-splines. In: Proceedings of International Conference on Artificial Intelligence and Statistics (AISTATS 2018) (2018)
15. Donoho, D.L.: Compressed sensing. IEEE Trans. Inf. Theory **52**(4), 1289–1306 (2006)
16. Eldar, Y.C., Kutyniok, G.: Compressed Sensing: Theory and Applications. Cambridge University Press, Cambridge (2012)
17. Gama, F., Marques, A.G., Mateos, G., Ribeiro, A.: Rethinking sketching as sampling: a graph signal processing approach. Signal Process. **169**, 107404 (2020)

18. Han, I., Avron, H., Shin, J.: Polynomial tensor sketch for element-wise function of low-rank matrix. In: International Conference on Machine Learning, pp. 3984–3993. PMLR, Westminster (2020)
19. Hsu, C.Y., Indyk, P., Katabi, D., Vakilian, A.: Learning-based frequency estimation algorithms. In: International Conference on Learning Representations (2019)
20. Indyk, P., Vakilian, A., Yuan, Y.: Learning-based low-rank approximations. In: Advances in Neural Information Processing Systems, vol. 32. Curran Associates, Red Hook (2019). https://proceedings.neurips.cc/paper/2019/file/1625abb8e458a79765c62009235e9d5b-Paper.pdf
21. Kasiviswanathan, S.P., Narodytska, N., Jin, H.: Network approximation using tensor sketching. In: Proceedings of the Twenty-Seventh International Joint Conference on Artificial Intelligence, IJCAI-18, pp. 2319–2325. International Joint Conferences on Artificial Intelligence Organization (2018). https://doi.org/10.24963/ijcai.2018/321
22. Kossaifi, J., Lipton, Z.C., Kolbeinsson, A., Khanna, A., Furlanello, T., Anandkumar, A.: Tensor regression networks. J. Mach. Learn. Res. **21**, 1–21 (2020)
23. LeCun, Y., Bottou, L., Bengio, Y., Haffner, P.: Gradient-based learning applied to document recognition. Proc. IEEE **86**(11), 2278–2324 (1998)
24. Liu, S., Liu, T., Vakilian, A., Wan, Y., Woodruff, D.P.: Extending and improving learned CountSketch (2020, preprint). arXiv:2007.09890
25. Ma, J., Zhang, Q., Ho, J.C., Xiong, L.: Spatio-temporal tensor sketching via adaptive sampling (2020, preprint). arXiv:2006.11943
26. Malik, O.A., Becker, S.: Low-rank tucker decomposition of large tensors using tensorsketch. Adv. Neural Inf. Process. Syst. **31**, 10096–10106 (2018)
27. Malik, O.A., Becker, S.: Guarantees for the kronecker fast johnson–lindenstrauss transform using a coherence and sampling argument. Linear Algebra Appl. **602**, 120–137 (2020)
28. Nelson, J., Nguyên, H.L.: OSNAP: faster numerical linear algebra algorithms via sparser subspace embeddings. In: 2013 IEEE 54th Annual Symposium on Foundations of Computer Science, pp. 117–126. IEEE, Piscataway (2013)
29. Pagh, R.: Compressed matrix multiplication. ACM Trans. Comput. Theory **5**(3), 1–17 (2013)
30. Pham, N., Pagh, R.: Fast and scalable polynomial kernels via explicit feature maps. In: Proceedings of the 19th ACM SIGKDD International Conference on Knowledge Discovery and Data Mining, KDD'13, vol. 128815, pp. 239–247. ACM, New York (2013)
31. Prasad Kasiviswanathan, S., Narodytska, N., Jin, H.: Deep neural network approximation using tensor sketching (2017, e-prints). arXiv:1710.07850
32. Sarlos, T.: Improved approximation algorithms for large matrices via random projections. In: 2006 47th Annual IEEE Symposium on Foundations of Computer Science (FOCS'06), pp. 143–152. IEEE, Piscataway (2006)
33. Sarlós, T., Benczúr, A.A., Csalogány, K., Fogaras, D., Rácz, B.: To randomize or not to randomize: space optimal summaries for hyperlink analysis. In: Proceedings of the 15th International Conference on World Wide Web, pp. 297–306 (2006)
34. Shi, Y.: Efficient tensor operations via compression and parallel computation. Ph.D. Thesis, UC Irvine (2019)
35. Shi, Y., Anandkumar, A.: Higher-order count sketch: dimensionality reduction that retains efficient tensor operations. In: Data Compression Conference, DCC 2020, Snowbird, March 24–27, 2020, p. 394. IEEE, Piscataway (2020). https://doi.org/10.1109/DCC47342.2020.00045
36. Sun, Y., Guo, Y., Luo, C., Tropp, J., Udell, M.: Low-rank tucker approximation of a tensor from streaming data. SIAM J. Math. Data Sci. **2**(4), 1123–1150 (2020)
37. Vempala, S.S.: The Random Projection Method, vol. 65. American Mathematical Society, Providence (2005)
38. Wang, Y., Tung, H.Y., Smola, A., Anandkumar, A.: Fast and guaranteed tensor decomposition via sketching. UC Irvine **1**, 991–999 (2015). https://escholarship.org/uc/item/6zt3b0g3
39. Wang, Y., Tung, H.Y., Smola, A.J., Anandkumar, A.: Fast and guaranteed tensor decomposition via sketching. In: Advances in Neural Information Processing Systems, pp. 991–999 (2015)
40. Weinberger, K.Q., Saul, L.K.: Distance metric learning for large margin nearest neighbor classification. J. Mach. Learn. Res. **10**, 207–244 (2009)

41. Woodruff, D.P., et al.: Sketching as a tool for numerical linear algebra. Found. Trends® Theor. Comput. Sci. **10**(1–2), 1–157 (2014)
42. Xia, D., Yuan, M.: Effective tensor sketching via sparsification. IEEE Trans. Inf. Theory **67**(2), 1356–1369 (2021)
43. Xiao, H., Rasul, K., Vollgraf, R.: Fashion-mnist: a novel image dataset for benchmarking machine learning algorithms (2017, preprint). arXiv:1708.07747 (2017)
44. Yang, B., Zamzam, A., Sidiropoulos, N.D.: Parasketch: parallel tensor factorization via sketching. In: Proceedings of the 2018 SIAM International Conference on Data Mining, pp. 396–404. SIAM, Philadelphia (2018)
45. Yang, B., Zamzam, A.S., Sidiropoulos, N.D.: Large scale tensor factorization via parallel sketches. IEEE Trans. Knowl. Data Eng. (2020). https://doi.org/10.1109/TKDE.2020.2982144

Appendix A
Tensor Software

As the multiway data widely exists in a lot of data processing applications, software that can directly perform tensor-based machine learning is in need within many fields. After years of development, there are various packages on different platforms. Here we conduct a relatively complete survey of the existing packages and the details are summarized in Table A.1.

Python and MATLAB are two main platforms for these packages. TensorLy [8] and tntorch[4, 6, 7] are for Python, and TensorToolbox[1–3] and Tensorlab [5, 7, 10, 11] are for MATLAB. There are also packages in C, C++, and OpenMP like mptensor and SPLATT [9].

Those packages are not only developed for different platforms but also focus on different specific fields. Most of the existing packages can finish basic tensor operations and decompositions. Apart from that, tntorch and hottbox can solve tensor regression, classification, and statistics analysis problems. TensorToolbox and Tensorlab can deal with structured tensor. Data fusion problems can be solved by Tensorlab, and sensitivity analysis can be achieved on tntorch. According to Table A.1, it is worth noting that there is no package which integrates all operations together on one platform. One needs to turn to different platforms and refer to corresponding user's manuals in order to achieve different goals.

We also put the links of corresponding software and user's manual in Table A.2 at the end of Appendix.

Y. Liu et al., *Tensor Computation for Data Analysis*,
https://doi.org/10.1007/978-3-030-74386-4

Table A.1 Comparison of different tensor software

Name	Platform	User's manual	Visualization	Applications
TensorLy	Python	✓		Tensor decomposition (CP, Tucker); tensor regression; tensor regression layer network
tntorch	Python	✓	✓	Tensor decomposition (CP, Tucker, mixing mode); tensor completion; tensor classification; tensor subspace learning; global sensitivity analysis; optimization
TensorD	Python	✓		Tensor decomposition (CP, Tucker, NCP, NTucker)
Scikit-tensor	Python	✓		Tensor operation and decomposition
hottbox	Python	✓(user friendly)	✓	Tensor decomposition (CP, Tucker, TT); ensemble learning; tensor classification (LS-STM)
tensor toolbox	MATLAB	✓	✓	Basic operations (matricization, multiplication, etc.) and decompositions (mainly focus on CP, Tucker) on tense/sparse/symmetrical/structural tensor
Tensorlab	MATLAB	✓	✓	Detailed decomposition algorithm (CP, Tucker, BTD) of tensor; tensor fusion algorithm and examples
TT toolbox	MATLAB and Python	✓		Basic operations on tensor train decomposition (matricization, rounding, low-rank approximation)
TTeMPS	MATLAB			Basic operation of tensor train and low-rank algorithm, especially Riemann optimization
TDALAB	MATLAB	✓	✓	CP, Tucker, BCD, MBSS, MPF Application: Tucker discriminant analysis and cluster analysis
HT toolbox	MATLAB	✓		Code construction and experimental operations for hierarchical Tucker decomposition

(continued)

Table A.1 (continued)

Name	Platform	User's manual	Visualization	Applications
The Universal Tensor Network Library (Uni10)	C++		✓	Basic operations for tensor; find the optimal tensor network shrinkage based on computation and memory constraints, process and store the details of the graphical representation of the network
mptensor	C++	✓		Basic tensor operations (tensor transpose, tensor slice, tensor singular value decomposition, QR decomposition, tensor matrix multiplication); high-order tensor renormalization group algorithm (HO-TRG)
n-way toolbox	MATLAB			PARAFAC, Tucker, N-PLS, GRAM, and TLD
SPLATT	C & OpenMP			Parallel sparse tensor decomposition; adopt data structure, reduce memory usage; CP decomposition algorithm for large-scale data
iTensor	C++	✓	✓	Matrix product state, tensor network state calculation, tensor chain basic operation; block sparse tensor representation
cuTensor	C++	✓		Fast calculation of t-SVD-related operations

Table A.2 Links of software and user's manual

Name	Link of software	Link of user's manual
TensorLy	https://github.com/tensorly/tensorly	https://github.com/JeanKossaifi/tensorly-notebooks
tntorch	https://tntorch.readthedocs.io/en/latest/	https://github.com/VMML/tntorch/tree/master/docs/tutorials
TensorD	https://github.com/Large-Scale-Tensor-Decomposition/tensorD	https://github.com/Large-Scale-Tensor-Decomposition/tensorD/tree/master/docs
Scikit-tensor	https://github.com/mnick/scikit-tensor	N/A

(continued)

Table A.3 (continued)

Name	Link of software	Link of user's manual
hottbox	https://github.com/hottbox/hottbox	https://github.com/hottbox/hottbox-tutorials
tensor toolbox	http://www.tensortoolbox.com/	Embedded in MATLAB
Tensorlab	http://www.tensorlab.net/	https://www.tensorlab.net/userguide3.pdf
TT toolbox	https://github.com/oseledets/TT-Toolbox https://github.com/oseledets/ttpy	https://github.com/oseledets/TT-Toolbox/blob/master/quick_start.pdf
TTeMPS	http://anchp.epfl.ch/TTeMPS	N/A
TDALAB	https://github.com/andrewssobral/TDALAB	https://pdfs.semanticscholar.org/ff4d/d9c218080fbedce17ea517c287b28db704d2.pdf?_ga=2.261533262.2146657559.1586745756-2003922837.1583151361
HT toolbox	https://www.epfl.ch/labs/anchp/index-html/software/htucker/	N/A
The Universal Tensor Network Library (Uni10)	https://gitlab.com/uni10/uni10	N/A
mptensor	https://github.com/smorita/mptensor	https://smorita.github.io/mptensor/modules.html
n-way toolbox	https://github.com/andrewssobral/nway	N/A
SPLATT	http://glaros.dtc.umn.edu/gkhome/splatt/Overview	N/A
iTensor	http://itensor.org/	https://itensor.org/docs.cgi?page=tutorials
cuTensor	https://github.com/NVIDIA/CUDALibrarySamples	https://docs.nvidia.com/cuda/cutensor/index.html

References

1. Bader, B.W., Kolda, T.G.: Algorithm 862: MATLAB tensor classes for fast algorithm prototyping. ACM Trans. Math. Softw. **32**(4), 635–653 (2006). https://doi.org/10.1145/1186785.1186794
2. Bader, B.W., Kolda, T.G.: Efficient MATLAB computations with sparse and factored tensors. SIAM J. Sci. Comput. **30**(1), 205–231 (2007). https://doi.org/10.1137/060676489
3. Bader, B.W., Kolda, T.G., et al.: Matlab tensor toolbox version 3.1 (2019). https://www.tensortoolbox.org
4. Ballester-Ripoll, R., Paredes, E.G., Pajarola, R.: Sobol tensor trains for global sensitivity analysis. Reliab. Eng. Syst. Saf. **183**, 311–322 (2019)
5. Cichocki, A., Mandic, D., De Lathauwer, L., Zhou, G., Zhao, Q., Caiafa, C., Phan, H.A.: Tensor decompositions for signal processing applications: from two-way to multiway component analysis. IEEE Signal Process. Mag. **32**(2), 145–163 (2015)

6. Constantine, P.G., Zaharatos, B., Campanelli, M.: Discovering an active subspace in a single-diode solar cell model. Stat. Anal. Data Min. ASA Data Sci. J. **8**(5–6), 264–273 (2015)
7. Kolda, T.G., Bader, B.W.: Tensor decompositions and applications. SIAM Rev. **51**(3), 455–500 (2009)
8. Kossaifi, J., Panagakis, Y., Anandkumar, A., Pantic, M.: Tensorly: tensor learning in python. J. Mach. Learn. Res. **20**(26), 1–6 (2019)
9. Smith, S., Karypis, G.: SPLATT: the surprisingly parallel spArse tensor toolkit (2016). http://cs.umn.edu/~splatt/
10. Sorber, L., Barel, M.V., Lathauwer, L.D.: Numerical solution of bivariate and polyanalytic polynomial systems. SIAM J. Numer. Anal. **52**(4), 1551–1572 (2014)
11. Sorber, L., Domanov, I., Van Barel, M., De Lathauwer, L.: Exact line and plane search for tensor optimization. Comput. Optim. Appl. **63**(1), 121–142 (2016)

Index

Y. Liu et al., *Tensor Computation for Data Analysis*,
https://doi.org/10.1007/978-3-030-74386-4

U

V

W